Lecture Notes in Mechanical Engineering

Series Editors

Francisco Cavas-Martínez, Departamento de Estructuras, Universidad Politécnica de Cartagena, Cartagena, Murcia, Spain

Fakher Chaari, National School of Engineers, University of Sfax, Sfax, Tunisia

Francesco Gherardini, Dipartimento di Ingegneria, Università di Modena e Reggio Emilia, Modena, Italy

Mohamed Haddar, National School of Engineers of Sfax (ENIS), Sfax, Tunisia

Vitalii Ivanov, Department of Manufacturing Engineering Machine and Tools, Sumy State University, Sumy, Ukraine

Young W. Kwon, Department of Manufacturing Engineering and Aerospace Engineering, Graduate School of Engineering and Applied Science, Monterey, CA, USA

Justyna Trojanowska, Poznan University of Technology, Poznan, Poland

Francesca di Mare, Institute of Energy Technology, Ruhr-Universität Bochum, Bochum, Nordrhein-Westfalen, Germany

D1807367

Lecture Notes in Mechanical Engineering (LNME) publishes the latest developments in Mechanical Engineering—quickly, informally and with high quality. Original research reported in proceedings and post-proceedings represents the core of LNME. Volumes published in LNME embrace all aspects, subfields and new challenges of mechanical engineering. Topics in the series include:

- Engineering Design
- Machinery and Machine Elements
- Mechanical Structures and Stress Analysis
- Automotive Engineering
- Engine Technology
- Aerospace Technology and Astronautics
- Nanotechnology and Microengineering
- Control, Robotics, Mechatronics
- MEMS
- Theoretical and Applied Mechanics
- Dynamical Systems, Control
- Fluid Mechanics
- Engineering Thermodynamics, Heat and Mass Transfer
- Manufacturing
- Precision Engineering, Instrumentation, Measurement
- Materials Engineering
- Tribology and Surface Technology

To submit a proposal or request further information, please contact the Springer Editor of your location:

China: Ms. Ella Zhang at ella.zhang@springer.com
India: Priya Vyas at priya.vyas@springer.com
Rest of Asia, Australia, New Zealand: Swati Meherishi
at swati.meherishi@springer.com
All other countries: Dr. Leontina Di Cecco at Leontina.dicecco@springer.com

To submit a proposal for a monograph, please check our Springer Tracts in Mechanical Engineering at http://www.springer.com/series/11693 or contact Leontina.dicecco@springer.com

Indexed by SCOPUS. All books published in the series are submitted for consideration in Web of Science.

More information about this series at http://www.springer.com/series/11236

Mokhtar Awang · Seyed Sattar Emamian
Editors

Advances in Material Science and Engineering

Selected articles from ICMMPE 2020

 Springer

Editors
Mokhtar Awang
Department of Mechanical Engineering
Universiti Teknologi Petronas
Seri Iskander, Perak, Malaysia

Seyed Sattar Emamian
Department of Mechanical Engineering
Universiti Malaya
Kuala Lumpur, Malaysia

ISSN 2195-4356   ISSN 2195-4364 (electronic)
Lecture Notes in Mechanical Engineering
ISBN 978-981-16-3643-1   ISBN 978-981-16-3641-7 (eBook)
https://doi.org/10.1007/978-981-16-3641-7

This Springer imprint is published by the registered company Springer Nature Singapore Pte Ltd.
The registered company address is: 152 Beach Road, #21-01/04 Gateway East, Singapore 189721,
Singapore

Contents

Tig Welding of Dissimilar Steel: A Review

N. Echezona[1], S. A. Akinlabi[2(✉)], T. C. Jen[1], O. S. Fatoba[3], S. Hassan[2],
and E. T. Akinlabi[4]

[1] Department of Mechanical Engineering Science, University of Johannesburg, Johannesburg,
South Africa
[2] Department of Mechanical Engineering, Walter Sisulu University, Butterworth Campus,
Butterworth, South Africa
[3] College of Aeronautics and Engineering, Kent State University, Ohio, USA
[4] Pan African University for Life and Earth Science, Ibadan, Nigeria

Abstract. Dissimilar welding of metals has gained lots of interest from researchers worldwide due to its numerous benefits. The fact that dissimilar weld joints are projected to give more versatility in the production and design of industrial and commercial components cannot be disputed. This paper presents a review of the recent works done on TIG welding of dissimilar steel with the focus on properties, structure, and performance relationships. It also includes the influence of welding parameters on the quality of joints. The papers for investigation in this work were considered from 2012 to date. A total of 29 research papers were chosen for our consideration after an in-depth examination of several such papers. There is a growing need for dissimilar welding of steel due to its various benefits such as the production of lightweight machine parts, production of less expensive engineering components with acceptable corrosion resistance, high strength, and recyclability in the power generation, chemical, petrochemical, and automotive industries. Welding of dissimilar steels is very challenging due to the dissimilarities in their chemical, physical, and metallurgical properties. Although other welding processes such as solid-state welding processes and some fusion welding techniques such as laser welding and electron beam welding have produced high-quality dissimilar steel joints, their application has been limited due to their high cost. Consequently, the TIG welding technique has been widely employed for welding of dissimilar materials because of its low cost, stability, and high quality of weld produced. In TIG welding, process parameters such as the current, gas flow rate, voltage, welding speed play a vital role in the standard of the weld produced. Several types of research have proved successful TIG welding of dissimilar steel with reliable joint performance as illustrated in this study. The review concluded that TIG welding of dissimilar steel is capable of producing joints with acceptable properties in the production and manufacturing industry.

1 Introduction

The importance of welding in the production and manufacturing industries cannot be overemphasized. It is a permanent joining process that involves the application of heat and/or pressure in joining two or more metals or alloys together [1, 2]. Welding provides

M. Awang and S. S. Emamian (Eds.): *Advances in Material Science and Engineering*, LNME, pp. 1–9, 2021.
https://doi.org/10.1007/978-981-16-3641-7_1

several benefits which the most significant is the production of joints with high performance and undeniably offers outstanding structural integrity when compared to other joining processes. Welding of dissimilar metals especially dissimilar steel materials has become very important in the manufacturing industries due to its advantages such as the production of lightweight machine parts in the automotive industries, production of engineering components with satisfactory corrosion resistance in the chemical, petrochemical, and power generation industries. Several researches have been carried out using solid-state and fusion welding methods to weld dissimilar metals [6, 7]. Solid-state processes and some fusion welding processes such as the laser beam and electron beam welding are capable of producing joints with high-quality. However, the main limitations in solid-state welding of dissimilar steel are its high cost and inability to weld intricate parts. In laser beam and electron beam welding, the setback is the low-gap bridging capability, low density, and high cost. Due to these challenges, TIG welding is widely employed in the welding of dissimilar steels in the production and manufacturing industries. Of all fusion welding methods, TIG is the most widely used for welding of dissimilar metals because of its stability and high quality of weld produced [1]. Tungsten inert gas welding (TIG) also called Gas tungsten arc welding (GTAW) is an arc welding process that uses a non-consumable tungsten electrode to produce a weld [2–4] as shown in Fig. 1. It makes use of shielding gas to shield the weld zone from atmospheric gases or air. Helium and argon gases are the most widely used in TIG welding. TIG welding can be done with or without filler material. When joining similar metals of large thickness, filler materials having the same chemical composition as the base metal are used. The major challenge faced by manufacturing industries today is the control of welding input parameters. The welding input parameters such as current, travel speed, voltage, gas flow rate, current polarity, and root gap affect the outcome of the weld. Poor combinations of these input parameters will lead to the production of welds with low quality. To obtain welds with the needed bead geometry and quality with reduced residual stress that is damaging, these welding input parameters must not only be studied but optimized to obtain welds of the highest quality. Welding of dissimilar steel is very challenging due to the differences in the chemical, physical, and metallurgical properties of component steel which affects the weld joint properties [5].

Some of the variables that influence the quality of weld in TIG welding include:

- Welding Current: This is one of the most vital variables in TIG welding. Welding current affects the bead geometry, depth of fusion, and burn off rate. Some considerations are made before selecting a current for a work-piece such as the thickness of the material, joint design, electrode type, and size [7].
- Current Polarity: The type of current used affects the weldment properties. The current utilized in Tungsten Inert Gas welding are classified into three: The Direct current reverse polarity (DCRP), Direct current straight polarity (DCSP), and the Alternating current (AC). The most commonly used is the DCSP (electrode is fixed to the negative terminal) due to the positive effect it has on weld joints. It produces weld with good penetration and a narrow profile.
- Weld Geometry: This is utilized in the selection of the welding technique. There are different types of joints such as the lap joint, butt, and T-joint. Weld geometry has a direct effect on the quality of the weld. There are different welding positions such as

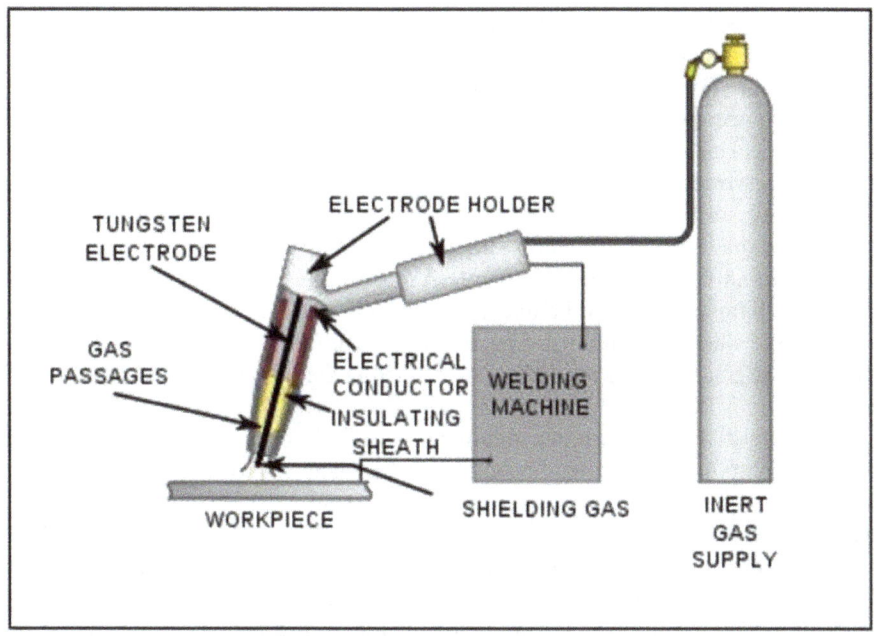

Fig. 1. Tungsten arc welding system setup [6]

vertical, horizontal, flat, etc. The most widely used are horizontal and vertical. When the welding position becomes challenging, the chances of producing weld of low-quality increases.

• Welding Speed: The highest weld penetration is attained at a particular welding speed and reduces with the variation in speed. In arc welding, an increment in the welding speed leads to reduced heat input per unit length, reduction in the electrode burn-off rate, and weld reinforcement reduction.

• Gas Flow Rate: This controls the weld features such as the weld bead, weld height. An increment in the gas flow rate leads to a change in the bead geometry of the weld joint.

• Welding Voltage: This determines the form of the fusion and weld reinforcement. It is the difference of electrical potential between the surface of the molten weld pool and the tip of the welding wire. The depth of penetration reaches its peak at the optimal voltage. Too much increase in the voltage leads to porosity and formation of spatter in fillet weld while a very low voltage will lead to overlapping at the edges of the weld bead.

2 TIG Welding of Dissimilar Steel

Most chemical, petrochemical, and power generation industries use dissimilar steel joints due to its several benefits. In power generation for instance where an operation is done at high temperature, the steels used are susceptible to corrosion and oxidation. Due to this, austenitic stainless steel is more desirable. The only set back to the use of austenitic stainless steel is due to its high cost compared to other steel materials. As a result of this, the joining of austenitic stainless steel to other less expensive steel materials becomes very important. Ferritic steels are less expensive but have poor corrosion resistance. It is mostly joined with austenitic stainless steel in the power generation, chemical and petrol chemical industries. Due to the application of the joints of these two sheets of steel in industries where operations are done at high temperatures, Jahanzeb et al. [1] used SS400 ferritic steel and 316L austenitic stainless steel as components of base material to study the dissimilar weld joints. They observed that the differences in the micro-hardness within the Heat affected zone (HAZ) of the ferritic steel were pronounced while the differences in micro-hardness within the base metal, weld zone, and HAZ of the austenitic steel were minimal. The widmanstatten ferrite and refinements of grain in the heat-affected zone of the ferritic steel aided in the improvement of the hardness value.

Literature studies on dissimilar welding of metals show that apart from the sudden change in composition along the fusion line, the mechanical properties also deteriorate along the fusion zone because of the local high stresses that are formed. These are connected with the discrepancies in the thermal expansion between the stainless steel and low carbon ferritic steel. Researchers have reported that this challenge is the main reason for failure in low carbon ferritic steel and austenitic stainless steel joints. In a view to reduce or remove the problem of non- uniform hardness distribution encountered across the fusion zone, Osoba et al. [8] analyzed dissimilar welding of AISI 304L Austenitic stainless steel to AISI 1005 low carbon ferritic steel using the TIG welding technique. The result obtained showed that the micro-hardness profile in the FZ of the as-welded specimen exhibited a sharp change from the AISI 304L area to the AISI 1005 area. They concluded that quality welds can be achieved between AISI 1005 steel and AISI 304L using ER 309L electrodes.

One of the major challenges faced by manufacturing industries is corrosion. It has a significant effect on the essential mechanical properties of materials. It has become a very big challenge especially for the automotive industries where some parts of the vehicles such as the mufflers of a vehicle exhaust system are normally exposed to corrosion failure. In a bid to provide a solution to this problem by improving corrosion resistance and also making it feasible economically, Joseph et al. [9] investigated and analyzed TIG welding of dissimilar metals 439 stainless steel and 409 stainless steel using electrode 309L.

The result obtained showed that the joints of the dissimilar metal exhibited high corrosion resistant property which makes it possible to be used for different applications liable to corrosion areas. Guilherme et al. [10] studied the corrosion behavior of dissimilar joint between ferritic AISI 444 stainless steel and AISI 316L using the TIG welding process. AISI 444 has become an alternative to replace AISI 316L because it is affordable and has good corrosion resistance. Result obtained revealed that the joint of the dissimilar metal experiences galvanic corrosion with high degeneration of the heat-affected zone of the AISI 444 tube. The dissimilar joint exhibited improved corrosion resistance when compared to the AISI 316L welded joint at temperature up to 70 °C. Wang et al. [11] studied the corrosion behavior and microstructural evolution of dissimilar 304 and 430 stainless steel using TIG welding. The experimental result showed that the increase in the fusion ratio of AISI 304, reduced the corrosion resistance of the welded joint. More increase in the fusion ratio resulted in the improved corrosion resistance of the welded joint. The weld microstructure majorly consisted of ferrite and martensite phases.

Dissimilar steel joints have achieved significant success in a weight reduction of machine parts in various manufacturing processes. Though most structural steels are characterized by high ultimate tensile strength, they are hard to weld, their fatigue properties and ability to go through plastic deformation without failing is low. In applications where toughness and safety are extremely vital such as the body of a car, TRIP (transformation induced plasticity), dual-phase, and TWIP (twinning induced plasticity) steel grades are applied. Due to the large plasticity of TWIP and TRIP advanced high strength steel, they are more suitable for the production of car bodies and have great tensile properties that aid in the protection of passengers and reduce the weight of a vehicle. Previous researches focused on the dissimilar and similar welding of TWIP and TRIP steel mostly with the laser welding process. Rossini et al. [12], in a bid to assist in the application of advanced high strength steel in the automotive industry, examined dissimilar laser welding of DP, TWIP, TRIP, and 22M n B5 steels. The result revealed that dissimilar laser welding of the combination of 22M n B5, TRIP, and DP maybe a quick fix for parts made of these steel grades. They however advised against laser welding of TWIP steel with other AHSS s except with the use of appropriate filler metal. Due to the advantages of the TIG welding process in the welding of dissimilar metals, Kornel et al. [13] investigated the weldability of high alloy TWIP and TRIP steels using automatized TIG without filler metal. The study showed that welding TRIP-TRIP joints can be achieved with satisfactory quality, but it is easily affected by shielding gas. TWIP-TWIP joint exhibited the best mechanical properties and was endorsed for automation. However, welding TRIP-TWIP joints without filler material yielded a poor result and therefore not recommended. This is different from the result of Eszter et al. [14], they successfully welded TRIP800 to TWIP1000 steel sheet. The tensile sample exhibited a ductile behavior and the TRIP steel side of the joint exhibited improved micro-hardness.

With the knowledge of the benefits of Tungsten inert gas welding such as low HAZ, non-existence of slag, and the role of process parameters such as gas flow rate, voltage, welding speed, current in achieving these benefits, there is need for optimal selection of these input parameters to achieve the needed weld quality. Several research efforts have been directed towards parametric optimization and the influence of input parameters on TIG-welded dissimilar steel joints. Bahar [15] produced a butt joint between SS304 and mild steel 1018 plate at different combinations of process parameters in Tungsten inert gas welding. The experimental design focused on the optimization of the bending strength using the Taguchi method. The result revealed that welding current had the most influence on the bending strength of the specimen. An increment in the welding current improved the bending strength of the specimen. Similarly, Subhani et al. [16] investigated the influence of TIG welding input parameters on the quality of the weld with a focus on the tensile properties of dissimilar joints of SS 310 and Mild Steel to find the optimal process parameter. The experimental result showed that welding current had the greatest influence on the tensile strength of the joint. Increasing the welding current increases tensile strength. The optimum parameter setting for tensile strength was also determined. Chaudhari et al. [17] analyzed welding input parameters and optimized

TIG welding of dissimilar metals between SS 304 and Mild steel to maximize hardness and tensile strength using the Taguchi approach. They concluded that welding current had the most effect on the tensile strength and hardness. Optimum parameter settings were obtained for both hardness and tensile strength. Anand and Mittal [18] conducted parameter optimization of TIG-welded dissimilar metals between Austenitic stainless steel 316 and Mild steel. Analysis of variance and Signal-to-noise ratio was utilized in studying the influence of the welding parameter on the hardness and tensile strength of the weld joint. The result showed that the current had the greatest influence on the tensile strength and hardness of the weld joint. Optimum parameter settings were obtained for both properties. Singh et al. [19] successfully joined dissimilar metals of En 31 steel and Mild steel using the Tungsten inert gas welding process to investigate the influence of input parameters on the hardness of the joint. The signal-to-noise ratio and analysis of variance showed current to be the most influential on the hardness of the joint. Most of the work reviewed on the optimization of welding parameters in TIG welding of dissimilar steel showed current to be the most influential parameter but that was not the case in the study done by Ramana et al. [20], they optimized weld parameters for dissimilar SS 304L-SS 430 using robotic TIG welding to produce joints with improved impact strength. Response surface methodology was adopted for the experiment and the result showed that the wire feed rate had the most significant influence on the impact strength of the joints. Optimum parameter settings for impact strength of the joint was obtained. This result is in agreement with the study done by Reddy and Ramana [21], they optimized process parameters to increase hardness in welding of dissimilar stainless-steel SS 304L and SS 430 using Robotic TIG welding. Response surface methodology was employed for the study and the result showed that wire feed had the most influence on the hardness of the joint. Optimum parameter settings for hardness was obtained.

In power plant industries where the boiler pressure element operates on a high range of mechanical and thermal loads, high temperature creep resistance and large creep strength materials are desired. As a result of this, the steel industry invented new types of steel characterized by better thermal conductivity, creep properties, and thermal expansion. Tigga et al. [22] investigated the mechanical properties and microstructure of two of the invented steel T91 martensitic and S304H austenitic using semi-automatic TIG welding. The influence of three different filler materials (SS, Inconel 617, and grade 91 filler wires) on the joint was investigated. The experimental result showed that the SS filler wire is suitable for the joining of dissimilar T91 and S304H while the Inconel wire is not suitable even though the joints produced by Inconel filler wire showed adequate strength. They concluded that the joint produced by using the three different filler wires meet the requirement for Ultra-super critical boiler (USC). In boilers' header assembly, P22 steel is used as a tube while P91 is used as a header. During steam piping, the pipes undergoing temperatures above 550 °C are made from tempered martensitic P91 steels because of their superb combination of oxidation-resistant and mechanical properties. The other parts of the piping are normally made from low alloy Cr-M_o steels such as P22 steels [23]. This makes necessary welding between P22 and P91. Kulkarni et al. [24] studied the influence of five different oxide fluxes TiO_2, MoO_3, Cr_2O_3, SiO_2, and CuO on the bead geometry of P91 steel-P22 steel joint using Activated Flux Tungsten inert gas welding. The result showed that better depth penetration was attained successfully in a single pass in the weld joints that contained MoO_3, TiO_2, and Cr_2O_3. This result is in good agreement with the study done by Sharma et al. [25], they studied the influence of four oxide fluxes (MoO_3, TiO_2, Cr_2O_3, and SiO_2) on A-TIG-welded joints of P92 steel and 304H austenitic stainless-steel plates with the intent of getting the flux that will provide the best depth penetration. The result revealed that TiO_2 provided better depth penetration than all the other oxide fluxes and no defect was detected on the welded joint. Microstructural analysis showed the weld joint to consist of austenite grain boundaries, lath martensite, polygonal austenite, and ferrite. Pandey et al. [26] studied the evolution of δ-ferrite and mechanical properties of dissimilar weld joints of P91 and P92 steel using the multi-pass Gas tungsten arc welding (GTAW) and Autogenous Tungsten inert gas welding (A-TIG). The study revealed that ferrite patches formed in the A-TIG weld FZ due to its faster cooling rate, higher heat input and larger percentage weight of ferrite stabilizer when compared to that of Gas Tungsten arc weld FZ. Peak hardness was observed in the Autogenous-TIG weld fusion due to the larger percentage weight of carbon.

The selection of appropriate filler metals for dissimilar welding of steel is very important and is one of the parameters that can highly influence the joint performance of TIG-welded dissimilar steel. Several researches have applied different filler metals to join dissimilar steels using the TIG welding process. Devaraju [27] attempted to join Duplex stainless steel to 316L Austenitic stainless steel using Tungsten inert gas process. 316L SS was used as filler metal and different microstructural and mechanical tests were conducted on the weld joint to figure out the mechanical properties. The results showed that the X-ray test observed no defect. There was an improvement in the yield and tensile strength. He concluded that TIG welding is suitable for welding the selected dissimilar metal and is endorsed for real-time application. Khalifeh et al. used the same 316L filler metal but compared it with ER309L, ER310, and ER308L to investigate its effect on TIG-welded AISI 304L/St37 steel. ER316L and ER309L gave a better combination between the metallurgical and mechanical properties of the welded joint. They concluded that good welds could be achieved using any of the four filler metals. Jeraldnavinsavio et al. [28] studied mechanical properties and microstructure of dissimilar TIG-welded AISI 430 and AISI 316L plates using ER2594 and ER310 filler materials. The experimental result showed the macrostructure of the weld joint to be sound with no porosity. Weld produced using ER2594 showed the highest ultimate strength. They concluded that TIG welding is suitable for welding dissimilar metals and the filler materials can produce weld of good quality when used on dissimilar metals.

3 Conclusion

From the literature survey of TIG welding of dissimilar steels, the following conclusions can be drawn:

- Dissimilar steel joints have significant capability of producing lightweight machine parts, high corrosion resistant components and it is progressively integrated into the automotive and power generation structures.
- The TIG welding process can successfully join dissimilar steel materials and produce joints of acceptable properties.

- Welding current is the most important input parameter in TIG welding of dissimilar steel.
- Taguchi method and response surface methodology is efficient for parametric optimization.

Acknowledgements. The authors wish to thank the welder Mr Tendai and acknowledge the financial support of the Pan African University for Life and Earth Sciences, Ibadan, Nigeria.

References

1. Jahanzeb, N., Shin, J., Singh, J., Heo, Y., Choi, S.: Materials science & engineering a effect of microstructure on the hardness heterogeneity of dissimilar metal joints between 316L stainless steel and SS400 steel. Mater. Sci. Eng. A **700**, 338–350 (2017). https://doi.org/10.1016/j.msea.2017.06.002
2. Britto Joseph, G., Mageshwaran, G., Rajesh, K., Jeevahan, J.: Study and analysis of welding of dissimilar metals 409 stainless steel and 439 stainless steel by TIG welding. J. Chem. Pharm. Sci. **9**(3), 1046–1050 (2016)
3. Roy, P.: A study on TIG welding process and its basic features as well as parametric optimization. Int. J. Innov. Res. Sci. Technol. **2**(06), 71–74 (2015)
4. Choudhury, N.: Design optimization of process parameters for TIG welding based on Taguchi method. Int. J. Curr. Eng. Technol., 12–16 (2013). https://doi.org/10.14741/ijcet/spl.2.2014.03
5. Laha, K., Chandravathi, K.S., Parameswaran, P., Goyal, S., Mathew, M.D.: A comparison of creep rupture strength of ferritic/austenitic dissimilar weld joints of different grades of Cr-Mo ferritic steels. Metall. Mater. Trans. A Phys. Metall. Mater. Sci. **43**(4), 1174–1186 (2012). https://doi.org/10.1007/s11661-011-0957-8
6. Mehta, H.R.: Analyzing effects of weld parameters for increasing the strength of welded joint on mild steel material by using the TIG welding process. Int. J. Sci. Technol. Eng. **2**(12), 480–483 (2016)
7. Tewari, S.P., Gupta, A., Prakash, J.: Effect of welding parameters on the weldability of material. Int. J. Eng. Sci. Technol. **2**(4), 512–516 (2010)
8. Osoba, L., Ekpe, I., Elemuren, R.: Analysis of dissimilar welding of austenitic stainless steel to low carbon steel. Int. J. Metall. Mater. Sci. Eng. **5**, 1–12 (2015)
9. Britto Joseph, G., Das, K., Murugesan, G., Prabhaharan, R.: Dissimilar welding of SS 409 and mild steel by gas tungsten arc welding (TIG) method by ER 309 L electrodes. Appl. Mech. Mater. **813–814**, 420–424 (2015). https://doi.org/10.4028/www.scientific.net/amm.813-814.420
10. Guilherme, L.H., Alberto, C., Rovere, D., Kuri, S.E., De Oliveira, M.F.: Corrosion behaviour of a dissimilar joint TIG weld between austenitic AISI 316L and ferritic AISI 444 stainless steels. Weld. Int. **7116** (2016). https://doi.org/10.1080/09507116.2015.1096476
11. Wang, C., Yu, Y., Yu, J., Zhang, Y., Zhao, Y., Yuan, Q.: Microstructure evolution and corrosion behavior of dissimilar 304/430 stainless steel welded joints. J. Manuf. Process **50**, 183–191 (2020). https://doi.org/10.1016/j.jmapro.2019.12.015
12. Rossini, M., Spena, P.R., Cortese, L., Matteis, P., Firrao, D.: Investigation on dissimilar laser welding of advanced high strength steel sheets for the automotive industry. Mater. Sci. Eng. A **628**, 288–296 (2015). https://doi.org/10.1016/j.msea.2015.01.037
13. Korn, J., et al.: TIG welding of advanced high strength steel sheets. In: International Scientific Conference on Advances in Mechanical Engineering, pp. 313–318 (2016)

14. Kalácska, E., Májlinger, K., Fábián, E.R., Pasquale, R.S.: MIG-welding of dissimilar advanced high strength steel sheets. Mater. Sci. Forum **885**, 80–85 (2017). https://doi.org/10.4028/www.scientific.net/MSF.885.80

15. E. Technologies, Bahar, D.: Optimization of process parameters for tungsten inert gas (TIG) welding to join a butt weld between stainless steel (SS 304) and mild steel (MS 1018) Int. J. Mater. Sci. Technol. **10**(1), 1–8 (2017)

16. Subhani, S.M., Suresh Kumar, D., Haq, A.U., Satyanarayana, K.: Evaluation of mechanical properties for TIG welding aspects of SS 310 and MS materials. Mater. Today Proc. **19**, 737–741 (2019). https://doi.org/10.1016/j.matpr.2019.08.121

17. Chaudhari, V., Bodkhe, V., Deokate, S., Mali, B., Mahale, R.: Parametric optimization of TIG welding on SS 304 and MS using Taguchi approach. Int. Res. J. Eng. Technol., 880–885 (2019)

18. Anand, K.R., Mittal, V., Scholar, P.G.: Parameteric optimization of Tig welding on joint of stainless steel (316) & mild steel using Taguchi technique. Int. Res. J. Eng. Technol. **4**(5), 366–370 (2017)

19. Singh, A., Singh, J., Ku, R.: A study of microstructure and hardness in En 31steel and mild steel welded joints using TIG welding. Int. J. Eng. Comput. **6**(10), 2849–2854 (2016)

20. Venkata Ramana, M., Ravi Kumar, B.V.R., Krishna, M., Venkateshwar Rao, M., Kumar, V.S.: Optimization and influence of process parameters of dissimilar SS304L – SS430 joints produced by Robotic TIG welding. Mater. Today Proc. **23**, 479–482 (2020). https://doi.org/10.1016/j.matpr.2019.05.388

21. Reddy, V.M.N.: Optimization of process parameters in welding of dissimilar steels using robot TIG welding. IOP Conf. Ser.: Mater. Sci. Eng., 330 (2018). https://doi.org/10.1088/1757-899X/330/1/012096

22. Shray, S., Kant, D., Panneerselvam, K.: Materials today: proceedings microstructure & mechanical properties of dissimilar material joints between T91 martensitic & S304H austenitic steels using different filler wires. Mater. Today Proc. (2020). https://doi.org/10.1016/j.matpr.2020.03.055

23. Vidyarthy, R.S., Dwivedi, D.K.: Optimization of A-TIG process parameters using response surface methodology. Mater. Manuf. Process. **6914** (2017). https://doi.org/10.1080/10426914.2017.1303154

24. Kulkarni, A., Dwivedi, D.K., Vasudevan, M.: Materials science & engineering a study of mechanism, microstructure and mechanical properties of activated flux TIG welded P91 steel-P22 steel dissimilar metal joint. Mater. Sci. Eng. A **731**, 309–323 (2018). https://doi.org/10.1016/j.msea.2018.06.054

25. Sharma, P., Dwivedi, D.K.: A-TIG welding of dissimilar P92 steel and 304H austenitic stainless steel: mechanisms, microstructure and mechanical properties. J. Manuf. Process. **44**, 166–178 (2019). https://doi.org/10.1016/j.jmapro.2019.06.003

26. Pandey, C., Mohan, M., Kumar, P., Saini, N.: Dissimilar joining of CSEF steels using autogenous tungsten-inert gas welding and gas tungsten arc welding and their effect on δ-ferrite evolution and mechanical properties. J. Manuf. Process. **31**, 247–259 (2018). https://doi.org/10.1016/j.jmapro.2017.11.020

27. Devaraju, A.: An experimental study on TIG welded joint between duplex stainless steel and 316L austenitic stainless steel. Int. J. Mech. Eng. **2**(10), 1–4 (2015)

28. Jeraldnavinsavio, D., Farid, A.M., Ramanamurthy, E.V.V., Porchilamban, S., Ravikumar, S.: ScienceDirect evaluation of mechanical properties and micro structural characterization of dissimilar TIG welded AISI 316L and AISI 430 plates using ER310 and ER2594 filler. Mater. Today Proc. **16**, 1212–1218 (2019). https://doi.org/10.1016/j.matpr.2019.05.216

Development of a Parabolic Trough Solar Water Heater for Domestic Houses in Ongwediva

Sam Shaanika$^{(\boxtimes)}$, Mutiu Erinosho, Fillemon Nangolo, and Ester Angula

Department of Mechanical and Industrial Engineering,
University of Namibia, P.O. Box 3624, Ongwediva, Namibia
sshaanika@unam.na

Abstract. In this paper, the existing solar water heating systems were studied with their applications in order to find better design approaches. Solar energy is free, environmentally clean, and therefore it is accepted as one of the most promising alternative energy sources. Nowadays, plenty of hot water is used for domestic, commercial and industrial purposes. Various resources such as coal, diesel and gas are used to heat water and sometimes for steam production. Solar energy is the main alternative to replace the conventional energy sources. Namibia have enough solar energy that can be captured and used in domestic water heating and rural areas. A design of a "Parabolic Trough Solar Water Heater" (PTSWH) that incorporates a solar tracker was designed since it is a better system that can concentrate solar energy on water pipes compared to the flat solar water heater. The size of the system depends on the availability of solar radiation, the temperature requirement of customers, the geographical condition and the arrangement of the solar system. The PTSWH system was designed using the above parameters.

Nomenclature

ρ Density of water [kg/m^3]
V Required volume of hot water per day [m^3]
T_f Outgoing water temperature [K]
T_i Incoming water temperature [K]
T_1 First Surface glass cover temperature [K]
T_2 Second Surface glass cover temperature [K]
T_a Ambient temperature [K]
A_C Collector area [m^2]
A_r Absorber area
I_T Total solar radiation [kWh/m^2]
τ Transmittance of the surface
α Absorptance of glass cover
T_f Mean film temperature [K]

M. Awang and S. S. Emamian (Eds.): *Advances in Material Science and Engineering*, LNME, pp. 10–16, 2021.
https://doi.org/10.1007/978-981-16-3641-7_2

1 Introduction

Ever since the beginning of the world, the planet has been sustained and nurtured by the sun acting as a source of heat and providing light to the living organisms. Many people have discovered a lot of ways to harness energy from different kinds of energy sources like oil, coal, geothermal and nuclear energy. Solar energy is one of the energies in abundance that will forever be used, due to the nuclear fusion occurring in the core of the sun. Due to increasing demand for energy and rising cost of fossil type fuels, solar energy is considered an attractive source of renewable energy that can be used for water heating in both homes and industries. Heating water consumed approximately 20% of total energy consumption for an average family [1]. Concentrated solar collectors have proved that energy can be trapped and concentrated on any important material that conduct heat and generate enormous amount of heat [2]. Namibia is one of the blessed countries in terms of solar radiation, and there is a potential of utilizing electromagnetic energy from the sun and turns it into different forms of energies. A design and fabrication of a parabolic trough solar water heater was done by Singh et al. [3], and the design was mainly focused on the heat collector since it is the prime component that plays a significant role in bringing the need for the design. However, the whole system did not incorporate a solar tracker that made operations to be manual. It was only designed to measure the solar radiation flux density from the hemisphere above within a wavelength range from 0.3 μm to 3 μm [4]. In the case of evacuated-tube collectors that consists of rows of parallel and transparent glass tubes, each tube consists of a glass outer tube and an inner tube covered with a selective coating that absorbs solar energy effectively but inhibits radiative heat loss [5].

Another collector that was considered is the flat-plate collectors which are used extensively for domestic water heating applications. It is a simple process and has no moving parts so it involves less maintenance. It is an insulated with weatherproofed box containing a dark absorber plate under one or more transparent covers [6]. ISO 9459-4:2013 specified a method of evaluating the annual energy performance of solar water heaters using a combination of test results for component performance and a mathematical model to determine the annual load cycle task performance under specified weather and load conditions. The procedure is applicable to solar water heaters with integral backup or preheating into a conventional storage or instantaneous water heater and to integral collector storage water heaters [7]. Solar trackers are rising in popularity, but not everyone understands the complete benefits and potential drawbacks of the system. Hydraulic trackers are good at accuracy and performance compared to pneumatic trackers, but however, both trackers are very expensive [8]. It is for this reason that they are not widely used in for domestic application.

Most of solar water heaters available in domestic houses in Ongwediva are indirect or flat solar collectors and panels that cannot trap most of the reflections from the sun. A Parabolic Trough Solar Water Heaters (PTSWH) have been designed before for domestic use without a solar tracker. This means that maximum heat concentration is not attained. This project is aimed to solve these problems by designing an integrated system that can give almost consistent heat concentration and can also provide large quantity of hot water for use in domestic houses. In addition, this system will perform much better than the regular solar water systems installed already.

2 Conceptual Design

This section gives a full conceptual design of solar water heating system for residential area of Ongwediva. A research done by Savita has proved that it was vital to have glass as an insulation to prevent too much heat loss [9]. And a glass with emissivity of 0.82 to 0.85 is usually used as it is of good thermal performance. With this design, a value of 0.83 was assumed as the emissivity of glass used as an insulator, and it is single evacuated; having a copper tube passing through it. Designing a system involves a lot of important criteria that have to be taken into consideration and these are whether a system is affordable, reliable, environmental friendly, performance and how easy the design can be for maintenance.

2.1 Design Approach/Possible Solutions

In this design, two alternative solutions were considered. The first alternative was to consider the PTSWH and the second approach was focusing on the selection of a suitable tracking system to track the sun for high performance. Both alternatives were done based on literature review as to enhance performance and cost reduction. The first approach was decided on the shape of the absorber tube for the provision of more volume of hot water. Copper was selected as an absorber material since it is the most frequent material used in heat exchangers between fluids and it has high absorptivity and less emissivity. Depending on the availability of copper size, a diameter of 22 mm was chosen.

2.2 Material Selection

Material selection is one of the important aspects of design and it plays a vital role in ensuring that the design will last long and will operate for a long time in harsh conditions. During material selection, factors like flexibility in construction, economic analysis, how easy the design can be maintained and material properties were considered. Materials that were selected for this design are stated as follows:

(i)   *Parabolic Concentrator collector*
      Aluminium coated plates were selected as a reflective material because of its high reflectivity of $(\rho) = 0.80$ according to the *"Progress in Energy and Combustion Science"* [10]. This allows 80% of all the radiation to be reflected onto the absorber tube that increases thermal efficiency compared to flat solar water heaters [5].
(ii) *Glass Tube*
      A single-glazed glass is recommended for the design on "Solar Heating Systems". It is cheaper and more heat is allowed, and it can be used as a glass cover to prevent heat loss from the copper tube [11]. The glass has a transmittance of $\tau = 0.84$.
(iii) *Storage Tank*
      Polyethene with silver coating is mostly used in heat storage tanks, and it can go up to more than four (4) hours without losing a lot heat [12]. These kind of storage tanks are similar to the electric geysers and a tank of 80 l was chosen for this design work.

(iv) *Copper Absorber Tube*
Copper is well known for having a good thermal conductivity and it is not only used for thermal conduction but also in electrical devices, thus copper is easy to get and cheaper than other thermal conductive materials due to these reasons. Copper was chosen due to the availability, cost and its thermal properties.
(v) *Support for the PTSWH*
For the support stand and the concentrated support, mild steel was used because of availability and easy weldability.

2.3 Hot Water Demand

The hot water demand is very much important in knowing how often people need it. According to the 2010 Housing and Planning National Census, it was found out that there was an average of 4 people living within a house [13]. And according to the research that was done on Energy Consumption for Ongwediva, on geysers consumption and the amount of water heated for use, it was found that 20 l of hot water is required or used by each person per day [14].

3 Result and Discussion

Design analysis and calculations are done based on data from literature review and following some standards of design. First calculations are done on the number of litres to be delivered and followed by the amount of heat required. Calculations of the collector area required to deliver hot water at the required outlet temperature of 75 °C, heat loss, and efficiency of the whole system were done.

3.1 Tank and Heat Required Calculations

Volume of water required in every household of Ongwediva:

$$\text{Volume } V = (20 \, l \times 4 \text{ people}) \tag{1}$$

The result of the amount of water used is 0.08 m³/day, and the heat required to raise the 80 l of water to a temperature of 75 °C.

$$Q = \dot{m} \, C_p (T_o - T_i) \tag{2}$$

$$Q = m \, C_p (T_o - T_i) \tag{3}$$

Equations (2) and (3) are both giving the same value, but Eq. (2) is in terms of mass flow rate while Eq. (3) is in terms of mass only.

$$\rho = \frac{m}{V} \tag{4}$$

The calculation revealed the heat Q of 4.688 kWh/day and a mass flow rate $\dot{m}$ of 0.0202 kg/s.

3.2 Collector Area Calculations

The parabolic trough solar water heaters are known to perform very well compared to flat solar water heaters [5] and according to the U.S. Department of energy, parabolic efficiencies are usually from 50% to 70% [11]. The aperture area was calculated using the Eq. (5) since the efficiency of this design was assumed to be 0.60, and the amount of energy required to heat the water of 0.08 m^3 volume are known.

$$\eta = \frac{Q}{IA_c} \tag{5}$$

Table 1 shows the average irradiation, ambient temperature, calculated collector area needed to heat up the 80 l of water and the irradiance due to Optical Efficiency (OE) for glass cover for every month for the year 2016 from the University of Namibia (UNAM) Weather Station.

Table 1. Results from UNAM weather station and area calculations

Month	Irradiation (kW/m^2)	Ambient Temp. (K)	Aperture Area (m^2)	OE (kW/m^2)
January	7.923	303.42	1.0140	5.655
February	5.892	307.42	0.7541	4.206
March	6.345	308.92	0.8121	4.529
April	5.786	306.72	0.7405	4.130
May	6.795	304.02	0.8697	4.850
June	6.659	301.92	0.8523	4.753
July	7.898	298.32	1.0108	5.638
August	8.105	311.52	1.0373	5.785
September	6.236	310.22	0.7981	4.451
October	6.783	311.32	0.8681	4.842
November	8.567	312.72	1.0965	6.115
December	7.876	307.62	1.0080	5.622

From Table 1, it shows that in order to have a consistent efficiency of 0.60 throughout the year, a maximum collector area of 1.0965 m^2 is required to get an out let of 75 °C. From the data, a collector area of 1.17 m^2 was selected to compensate for the whole year. Using the selected area, heat losses and new efficiencies were calculated to see how such a collector will perform.

To calculate the collector efficiency, a new energy balance equation was used, which involves multiplying the heat removal factor.

$$Q_u = A_c F_r [I_T(\tau \cdot \alpha) - U_L(T_i - T_a)] = m \cdot C_p (T_{f,o} - T_{f,i}) \tag{6}$$

By rearranging:

$$F_R = \frac{m \cdot C_p\left(T_{f,o} - T_{f,i}\right)}{A_c\left[I_T(\tau \cdot \alpha) - U_L\left(T_{f,i} - T_a\right)\right]}.$$ (7)

The efficiency of the collector is then obtained by dividing the term $(I_T A_c)$ throughout the energy equation, which gives:

$$\eta = F_R\left[(\tau \cdot \alpha) - \frac{U_L(T_i - T_a)}{I_T}\right] \cdot 100\%$$ (8)

The results calculated for heat removal factor and collector efficiency are shown in Table 2.

Table 2. PTSC efficiency and heat removal

Month	I_T	m^2	F_R	η
January	5.655	1.17	0.707	50.57
February	4.206	1.17	0.949	68.00
March	4.529	1.17	0.881	63.15
April	4.130	1.17	0.967	69.25
May	4.850	1.17	0.825	58.97
June	4.753	1.17	0.842	60.17
July	5.638	1.17	0.711	50.73
August	5.785	1.17	0.690	49.44
September	4.451	1.17	0.897	64.25
October	4.842	1.17	0.824	59.07
November	6.115	1.17	0.653	46.77
December	5.622	1.17	0.711	50.87

Namibia is still facing a major challenge of supplying sufficient amount of energy to its people around the country. Most of people around Ongwediva still use electric water heating systems to supply hot water, and this has a huge impact on people electric bills. Alternative solutions to such problems is using renewable energy source that can replenish themselves. One of these renewable sources is solar energy from the sun. Solar energy is one of the most growing energy industry and an analysis on energy saved by using a parabolic solar collector was done. If residents in every house hold of Ongwediva decide to use a PTSWH, they can save an amount of N$ 4639.01 per year. This could reduce money spent on electric geysers and improve living standards as money saved could be used to buy other necessities.

4   Conclusion

Namibia is one of the countries that still rely on imports of electricity to sustain it daily electricity consumption for its people. Despite this import reliance, Namibia is one of the blessed countries when it comes to solar energy. The maximum average of solar radiation received in Ongwediva according to the data from the UNAM Weather Station is 903 W/m^2 on a daily basis. This gives an opportunity to people to be able to save electrical energy by using solar related devices. Even though Namibia has all this opportunity, people still face a lot of challenges of implementing solar thermal installations due to the fact they cannot afford a one-time investment. One-time investment is beneficial because as years goes the system start paying back for itself. A Parabolic Trough Solar Heater was designed to heat water to a temperature of 75 °C, and also to output enough amount hot water into an 80 l tank for a family of 4 people. The energy demand required to heat the water was calculated to be 4.688 kWh/day and a collector area of 1.17 m^2 was chosen in according to the highest value to carter for every month. From the thermal performance calculated, it was found out that the PTSWH had the highest efficiency of 69.25% which is a good sign that concentrated collectors integrated with a solar tracker are good in thermal performance compared to flat and evacuated tube solar water heaters.

However, an economic analysis was done on the system to see how much it can take a client to return back the money, and it was calculated to be an approximation of 2 years.

References

1. Energy Consumption and Supply. http://www.energyworldconsumption/solar.com. Accessed 11 Apr 2017
2. Gaitan, D.J., Poly, C., Obispo, S.L.: Design, Construction, and Test of a Miniature Parabolic Trough Solar Concentrator, pp. 3–34 (2012)
3. Singh, S.K., Singh, A.K., Yadav, S.K.: Design and fabrication of parabolic trough solar water heater for hot water generation. IJERT 1, 1–9 (2012)
4. What is a Pyranometer. http://www.azosensors.com/article.aspx?ArticleID=248. Accessed 10 Apr 2017
5. Shelke, V.G., Patil, C.V., Sontakk, K.R.: Solar water heating systems: a review. Int. J. Sci. Eng. Res. (IJSER) 3, 2–16 (2014)
6. Pachkawade, M.A., Nimkar, P.R., Chavhan, B.K.: Design and fabrication of low cost solar water heater. Int. J. Emerg. Trends Eng. Develop. 6(3), 10–16 (2013)
7. ISO 9459-4:2013 Solar heating—domestic houses. https://www.iso.org/standard/41695.html. Accessed 09 May 2017]
8. Solar tracking, solar systems. http://www.solar-tracking.com/. Accessed 28 October 2017
9. Sivita, J.H.: Performance of parabolic trough solar water heater, Nigeria (2013)
10. Kalogirou, S: Solar thermal collectors (2004)
11. U.S.D. Energy: Parabolic-trough solar water heating for power generation (2008)
12. Gabriel, J., Toren, I.: Tank selection and application. In: Materials and Properties, Mexico, pp. 56–70 (2005)
13. Agency, N.N.: Namibia 2010 Housing and Planning National Census Report, Windhoek (2010)
14. Jaden, S.: Assessment of feasibility for the replacement of electrical water heaters with solar water heaters, pp. 30–42 (2009)

Design of Specimen Table for Laser Deposition Operation

Mutiu Erinosho$^{(\boxtimes)}$, Ester Angula, Fillemon Nangolo, and Sam Shaanika

Department of Mechanical and Industrial Engineering, University of Namibia, P.O. Box 3624, Ongwediva, Namibia
`merinosho@unam.na`

Abstract. The KUKA robotic arm is used for laser cladding, repair and maintenance of parts or engine components. The KUKA robot sprays metal powders on the testing substrate or component to be repaired or cladded and solidified after melting. However, during this process, not all metallic powders is utilized and as such, some were wasted. About 40% approximately are wasted for a given operation. Due to the wastage and cost of these powders, four tables were developed to recover the wasted powders, but only one was selected due to the convenience and ease of powder retrieval. The selected concept which is Concave Hollow Table was put through a stress analysis with MSST and MDET failure criteria and the maximum displacement analysis obtained was 1.91902 mm on a high load of 200 kg that was used.

1 Introduction

It is widely understood that George Devol designed the first programmable robotic arm in 1954 in the United States, and in 1962, the General Motors however employed a Unimate robotic arm for car production in the assembly line. In 1969, industrial robot was designed with six axes arm. Their arms can perform extensive variety of jobs through various attachment at the end. In 1973, KUKA with six-axis robotic arm was invented for commercial purpose [1]. The programmable commercial robotic arm for manufacturing and research means is a device that operates just like the human arm which can rotate in several directions [1]. Today, the industrial sector is relying on industrial robotic arms such as the KUKA robot to assemble, manufacture and repair parts and they are getting wide spread. Cladding is a firm process used in the industries for the repairs of worn parts in order to improve their surface properties. High quality clad having good dilution ratio and surface evenness are produced [2] if the process parameters are well utilized. The cladding process is used controllably to add one material to another surface in order to enhance properties. And, this is done by feeding the surface with metallic powder and get melted. It is basically a fusion and solidification method, which involves the interfaces between the laser beam, the powders to be used, gases and the receiving substrate [3, 4]. It was reported that higher flow rates of deposited powders require higher gas pressures; this can be regulated to the measured the gas pressure [3]. The wasted powders during this process are not yet studied. The reuse of metallic powder metal came to play for

M. Awang and S. S. Emamian (Eds.): *Advances in Material Science and Engineering*, LNME, pp. 17–22, 2021.
https://doi.org/10.1007/978-981-16-3641-7_3

container-to-container consistency, the powder is consistent throughout a manufacturing lot. During the characterization of powder for additive manufacturing process, samples of 17-4 stainless steel powder was recovered after each of eight different builds. This was done in order to determine the changes in the powder properties as a result of the recycled powder [5]. Ti6Al4V ELI powder re-use was also investigated by another authors using a Renishaw AM250 system. A single batch of powder was tested for 38 builds until no powder left for the test. After the characterizations, it was concluded that the change in powder is not obvious to affect the settings [6]. Nickel powder was reusable after sieving in order to prevent congestion of the nozzle. It was reported that the observed microstructure was similar with the number of reused powder and different particle sizes while keeping the same process parameters. Prior thermal treatment, the hardness values were reported to be steady [7]. A flow simulation on how the collisions of particles from the powder feeder to the tip of the nozzles was revealed. It was reported that particles collide with the inner wall of the coaxial nozzle unevenly and on getting to the nozzle tip, the location and direction of the particles were dispersed accordingly [8]. It was revealed by another authors that the powder particle velocity was increased with an increase in the carrier-gas flow rate and decreased with a rise in the powder catchment efficiency [9]. A laser processing head for rapid manufacturing on vertical surfaces was investigated. It was successfully used on several tubular geometries and the maximum powder catchment efficiency was found to be approximately 40% [10]. The major problem is that some powders used for laser deposition process are expensive and the rate at which they are wasted is not pleasing. It was reported that metallic powders cost around $ 120-$ 220 (N$ 1797.60–3295.60)/kg depending on the metal and with a reported powder wastage of around 20–40% (N$ 359.52–N$ 719.04) of the total powder used. There is a need to find a way to reuse the wasted powder [11, 12]. The powder dropped on the ground cannot be reused, and it may be contaminated by dust particles and cross contamination of metals from previous use of the process. However, a movable table with a locking device wheel will be developed and designed for the recovery of wasted powder and without contamination.

2   Design Concepts and Selection

In the design, a total number of four concepts were designed based on the design requirements established. The design concepts served as a means to address the problem stated. All the designs are based on a structure that meets the dimensional requirements 700 × 800 × 800 mm, which also has a center plate to mount a vice. Concept 1 is a hollow table that allows the falling of powders and trapped for the retrieval. Here, the captured powder is a bit tricky to retrieve and labour intensive. Concept 2 is a rework of concept 1 and is a hollow table with a drawer in order to provide solution to the retrieval problem faced in concept 1. The drawer is removable and can be tipped over to remove the powders. Concept 3 is a slanted channel table; aimed to make the process of powder retrieval more stream lined by having a slanted inside where the powder slides to a specific corner. This corner has an opening where powder can flow into an external container as shown in Fig. 1 (a). Concept 4 is a concave hollow table with a sphere shaped inside which act as a funnel of sorts through which wasted powders are collected. The powders are collected through the center opening as shown in Fig. 1 (b).

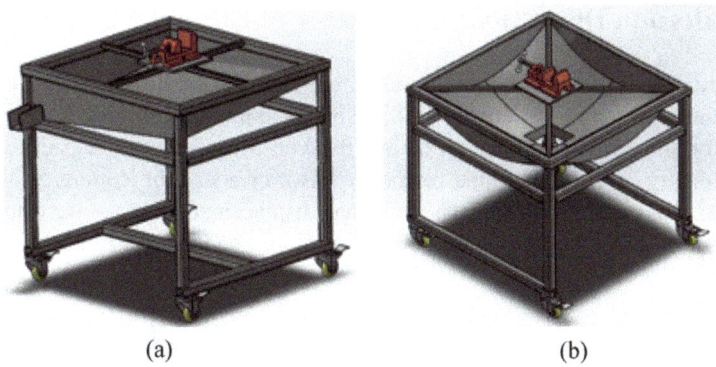

(a)   (b)

Fig. 1. (a) The slanted channeled table, (b) concaved hollow table

All the tables have locking tires for easy movement from one section to another. This opening can also be closed to make a fixed volume if a container is not available at the moment of operation.

2.1 Selection

The selection criteria was done for the four concepts and a narrowed decision was made as shown in Table 1.

Table 1. Table selection matrix

Selection criteria	Concept 1	Concept 2	Concept 3	Concept 4
Dimensions 700 × 800 × 800 mm	✓	✓	✓	✓
Wheels	✓	✓	✓	✓
Steps for powder retrieval	3	3	1	1
Volume	0.085 m^3	0.085 m^3	Variable	Variable
Clamp	✓	✓	✓	✓

The selection matrix was narrowed down to two concepts which are concepts 3 and 4 due to the fact that they both channel/funnel powder to an external storage container with ease. Concept 4 was selected finally when checked for aesthetics and adaptability. Also, due to be able to funnel powder to an external container and the being able to have a fixed volume when the opening is closed. The stress analysis was conducted on the table to assess the stresses and displacement and to determine the deformation under loading and the failure of the specimen table.

3   Results and Discussion

The failure was analysed using the two common methods applicable to ductile materials; Maximum Shear Stress Theory (MSST) and Maximum Distortion Energy Theory (MDET) on the maximum shear stress and the Von-Mises stresses respectively. They are additionally used to determine the safety factor of a state of loading for the given material which is ASTM 36 Steel. The stress and displacement analyses were done using SolidWorks as shown in Figs. 2 and 3 with the color-coded visualization chart.

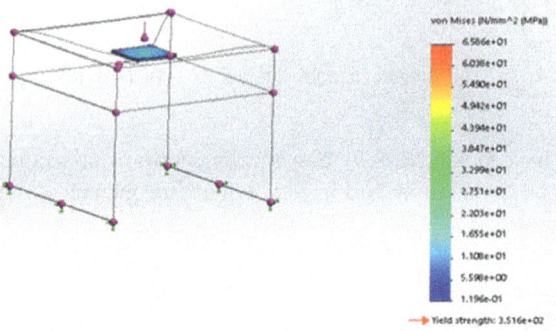

Fig. 2.   Stress analysis on the table.

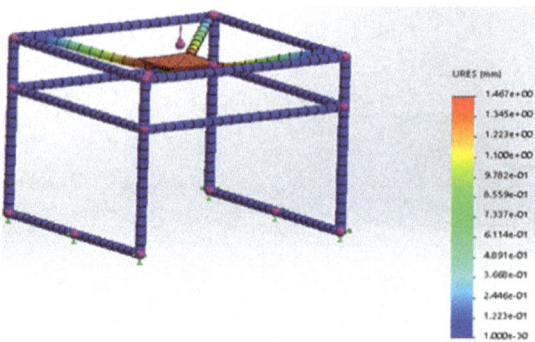

Fig. 3.   Displacement analysis on the table

   The stresses (Von-Mises and Maximum Principal) were obtained for the different loading conditions; self-weighed and from 20 kg to 200 kg in 20 kg intervals, as well as the displacement of the center plate under the same loads as shown in Table 2.
   The results showed that the frame is capable of supporting a properly distributed load even up to 200 kg. Failure occurs when Eq. (1) is satisfied for Maximum Shear Stress Theory.

$$\frac{(\sigma_1 - \sigma_2)}{2} \geq \frac{sy}{2} \tag{1}$$

Table 2. Load analysis results

Load		Equivalent stress (MPa)		Max principal stress (MPa)		Displacement (mm)
Mass (kg)	Weight (N)	Max	Min	Max	Min	
0	0	0.0439	0.000102	0.0377	−0.00991	0.000978
20	196.2	5.68	0.0153	2.64	−0.263	0.191902
40	392.4	11.40	0.0306	5.28	−0.525	0.383805
60	588.6	17.00	0.0459	7.92	−0.788	0.575707
80	784.8	22.70	0.0612	10.6	−1.050	0.767609
100	981.0	28.40	0.0765	13.2	−1.310	0.959512
120	1177.2	34.10	0.0918	15.8	−1.580	1.15141
140	1373.4	39.80	0.1070	18.5	−1.840	1.34332
160	1569.6	45.50	0.1220	21.1	−2.100	1.53522
180	1765.8	51.10	0.1380	23.8	−2.360	1.72712
200	1962.0	56.80	0.1530	26.4	−2.630	1.91902

where σ_1 and σ_2 are the maximum and minimum principal stresses respectively and Sy is the yield strength of the material. The safety factor using Eq. (2) was calculated for load of 200 kg. The values for stress and yield strength were obtained from the SolidWorks simulation results.

$$N = \frac{sy}{(\sigma_1 - \sigma_2)} \qquad (2)$$

Where "N" is the safety factor, and for the 200 kg, the material still does not fail and is at a safety factor of 8.6. Failure occurs when Eq. (3) is satisfied for Maximum Distortion Energy Theory.

$$\sigma \geq s_y \qquad (3)$$

Safety factor using Eq. (4) was also calculated for load of 200 kg. The values for stress and yield strength were obtained from the SolidWorks simulation results.

$$N = \frac{sy}{\sigma\,eq} \qquad (4)$$

Where σ_{eq} is the Von-Mises/equivalent stress and with this, the safety factor is 4.4 for the 200 kg load. At the load size of 200 kg, the maximum displacement of the center plate was found to be 1.9192 mm, which is an acceptable displacement.

In summary, it can be deduced that the frame does not fail for 200 kg load, and, it does not meet the failure criterion of both MSST and MDET. An average heavy-duty vice of 30 kg and any part to be used for laser cladding, ranges from 1–100 kg will give a maximum loading size of 130 kg which is less than the 200 kg load used to run the failure criterion.

4 Conclusion

The design of a movable table with a locking device wheel that can collect the wasted metallic powders during the laser deposition operation was achieved. The design requirements were drawn which help to guide and shape the design of the specimen table. Four functional concepts were created which was later narrowed down to one final concept with a concave hollow table due to the ease of powder retrieval. A stress analysis was done on the specimen table, and the result showed that the table is more capable of withstanding load of 200 kg. In both Maximum Shear Stress Theory and Maximum Distortion Energy Theory failure criteria, it was proved that the table will not yield or fail and safety factor of 8.6 and 4.4 respectively were calculated for this design.

References

1. Design Robotics: Robotic Arms in Manufacturing | Design Robotics (2019). https://www.des ignrobotics.net/robotic-arms-in-manufacturing/. Accessed 27 Nov 2019
2. Coherent: Cladding with High Power Diode Lasers (2019). https://www.fst.nl/systems/laser-cladding/laser-cladding-coherent/. Accessed 21 Dec 2019
3. Tang, L., Ruan, J., Landers, R., Liou, F., et al.: Variable powder flow rate control in laser metal deposition processes. J. Manuf. Sci. Eng. **130**(4), 11 (2008). https://doi.org/10.1115/1.2953074
4. Lasercladding.co.uk: Laser Cladding Technology (2019). http://www.lasercladding.co.uk/Laser-Cladding-Process.aspx. Accessed 27 Nov 2019
5. Slotwinski, J., Garboczi, E., Stutzman, P.: Characterization of metal powders used for additive manufacturing. J. Res. Natl. Ins. Stand. Tech **119**, 460–493 (2014)
6. Renishaw Apply Innovation: Investigating Ti6Al4V metal powder re-use in additive manufacturing. White Paper, pp. 1–9 (2016)
7. Renderos, M., Girot, F., Lamikiz, A., et al.: Ni based powder reconditioning and reuse for LMD process. Phys. Procedia **83**, 769–777 (2016)
8. Liu, H., et al.: Numerical simulation of powder transport behavior in laser cladding with coaxial powder feeding. Sci. China Phys. Mech. Astron. **58**(10), 1 (2015). https://doi.org/10.1007/s11433-015-5705-4
9. Liu, S., Zhang, Y., Kovacevic, R.: Numerical simulation and experimental study of powder flow distribution in high power direct diode laser cladding process. Lasersa Manuf. Mater. Process. **2**(4), 199–218 (2015). https://doi.org/10.1007/6-015-0015-2
10. Paul, C., Mishra, S., Kumar, A., et al.: Laser rapid manufacturing on vertical surfaces: analytical and experimental studies. Surf. Coat. Technol. **224**, 18–28 (2013)
11. Katinas, C., Shang, W., Shin, Y.C., et al.: Modeling particle spray and capture efficiency for direct laser deposition using a four nozzle powder injection system. J. Manuf. Sci. Eng. **140**(4), 10 (2018). https://doi.org/10.1115/1.4038997
12. Bauers, M.: Raw material pricing and additive manufacturing, engineering and physical sciences research centre (2014)

Design of a Stacking System for a Small-Scale Warehouse

Mutiu Erinosho$^{(\boxtimes)}$, Fillemon Nangolo, Ester Angula, and Ignatius Shahonya

Department of Mechanical and Industrial Engineering,
University of Namibia, P.O. Box 3624, Ongwediva, Namibia
merinosho@unam.na

Abstract. Material handling system is the art and science of conveying, elevating, positioning, transporting, packaging and storing of materials. It's not today that stacking rack systems have been in use in the warehouses. However, in some warehouses, the stacking racks available needs the use of forklift for the retrieval of goods. This paper is focused on the design of a mini warehouse stacking system that allows the retrieval of goods and equipped with features that allows easy flow of material from the loading end and without the use of forklift with the aisles. A 3D CAD model was developed and analysis was carried out and the results showed that material could easily and efficiently be moved from the loading end to the receiving end and be retrieved with the help of a lifting mechanism with ease. This design will improve the safety of the workers, goods and the efficiency of the general warehouse.

1 Introduction

In a production setup, an uncontrolled production leads to stack of unused inventory due to unpredictable demand and supply within the system. The abandoned inventory gets accumulated till damage occur or get expired. The stacking system is a very big challenge when there is inadequate space for storage and retrieval of goods. Old and new stocks are piled together which can lead to time consuming if a particular stock is needed to be retrieved [1]. As a result of high volume of goods, new products are stored in any way with the older ones which is always difficult to pick out the oldest items. In this instance, first-in-first-out is important for easy flow of goods on the shelves [2]. This has created a big problem as the older items may just sit and gather dust, while the newly manufactured items are sent out well before the older ones since the products are not arranged in terms of the manufacturing date. Extra work and time are utilised to reorganise and rearrange the stock [3]. Many manufacturing and distribution organizations keep up huge warehouses to store in-process inventories or parts supplied by outside providers. The benefit of running a successful warehousing lies in having the correct items in the timely retrieval spot at the perfect time. Utilising these products with ease is very essential for a warehousing company [4]. Although, warehousing is a tedious and non-value-adding activity due to the extra work and time required to store things in storage spaces and recover them later when needed. However, just-in time producing theory is recommended to get rid of any temporary storage and maintain a

M. Awang and S. S. Emamian (Eds.): *Advances in Material Science and Engineering*, LNME, pp. 23–29, 2021.
https://doi.org/10.1007/978-981-16-3641-7_4

strategy wherein goods are delivered uniquely [3]. The optimization of automated storage and retrieval systems (AS/RS) was investigated using multiple objective ant colony. The path to dispatch, product height, storage space usage and the factor of inquiry were considered for the distribution of products in the systems with no corridors and one single elevator for multiple products. The simulation results predicted the structural and operational requirements before and after the construction [5]. Some authors studied the problem to the optimal slot sizes in pallet rack system in order to maximize the occupied space. The "k" boxes of "m" pallet racks in a warehouse was chosen. A dynamic programming heuristic that finds the optimal slot sizes for the given number of "k" of pallet racks was suggested in their study [6]. The size of a new flow type warehouse and a new U-Type warehouse by minimizing the total picker distance was revealed. Number of dock factor was considered in their research for determining the warehouse size, and it was reported to have affected the average travel distance of picker and the input-output traffic of trucks [7]. An elaborative review was done on warehouse operation planning problems and were grouped according to the basic warehouse functions. The link between the researchers and warehouse consultants was revealed through the available planning models and methods for warehouse operations [8]. The order picking process was revealed to be the most labour-intensive and costly activity in every warehouse, and the cost of order picking was valued to be about 55% of the total warehouse operational cost. The zoning, optimal layout design, order batching, storage assignment methods and routing methods were critically reviewed and the system has proposed new research way forward [9]. The pyramid of decision problems faced when establishing a warehousing system were discussed by some authors. The warehousing models, systems and management as well as the relation between inventory control decisions and product allocation and assignment problems were revealed. It was also concluded that a higher warehouse service level and shorter response times might lead to additional savings [10]. High density storage solution racking system was recommended for great amounts of good with only few different storage units. This system was developed in order to increase the usage and reduce the handling costs. Dynamic rack solution was introduced to speed–up and reduce the handling cost. The installed system has promoted the increase in productivity and improved safety measures [11]. The material stacking systems available in some warehouses are not capable of handling all the goods they store because the stacking racks available are not enough hence goods are left lying around and also warehouse layouts are not strategic enough to ensure efficient operations. On the stacking racks available, retrieval of material can be sometimes difficult and most of the times workers are forced to improvise. Due to this, workers are forced to perform work at high elevations between very narrow aisles and this often causes body injuries to workers and damage to goods. However, a stacking system is designed to allow plain sailing navigation through narrow aisles, easy retrieval of goods from elevated heights and easy maintenance and service on the racks.

2 Design Methodology

The racking system is important for the solution to optimize the efficiency of material flow. It is necessary to select a stacking rack that is most suitable in the mini warehouse. Three types of stacking racks were considered which are: gravity flow, push back and cantilever rack. The efficiency, cost maintenance and durability are considered for racking decision matrix. Gravity flow rack was deemed suitable for the system because it carried more weight compared to other two due to its high density. The push back rack was eliminated due to its low density and storage rate. The cantilever rack was left out because it requires a lot of space to accommodate picking and restocking [12, 13]. Figure 1 shows the stacking system with multiple steps of seven rows and eight columns.

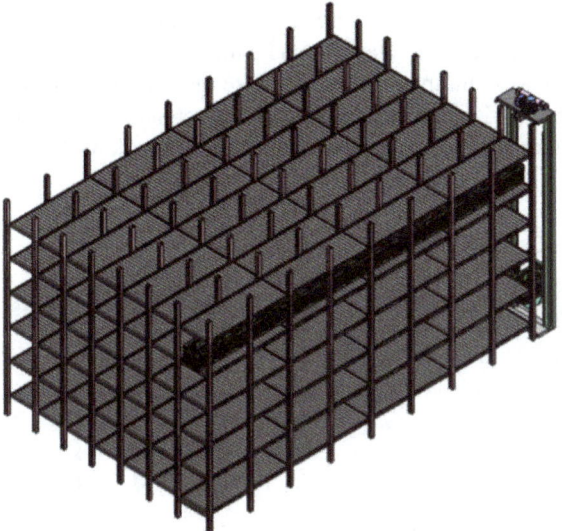

Fig. 1. Stacking system

It has a lifting mechanism with a loading tray, load beam, roller arrangement and electric motor. The system provides easy retrieving means by using the tray that moves left to right and up and down. The lifting mechanism is the device that lift the goods placed on the loading tray up and down the stacking system. This is controlled by the electric motor. Electric motor plays an important role of creating the lifting force on the lifting mechanism. Figure 2 shows the lifting mechanism.

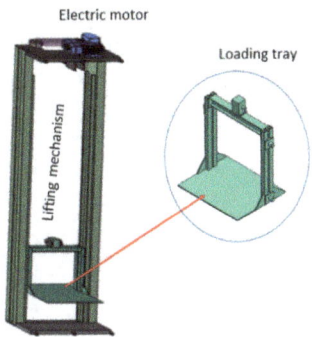

Fig. 2. Lifting mechanisms

A stepper motor is considered in this design because of its low cost, high reliability and ability to produce high torque at low speeds [13].

3   Results and Discussion

The design requirements and deliverables for the stacking system was designed with the following requirements: Efficient flow rate, delivery and retrieval of goods and the safety of workers when retrieving goods. Three types of lifting mechanisms were considered which are cascade lift, self-propelled lift and a moveable lift respectively. Figure 3 is a cascade lift and it was eliminated at the expense of its complex cable routing and the need of an external device that ensures that it does not jam, and also its complex design structure that makes final stage to move up first and come down last.

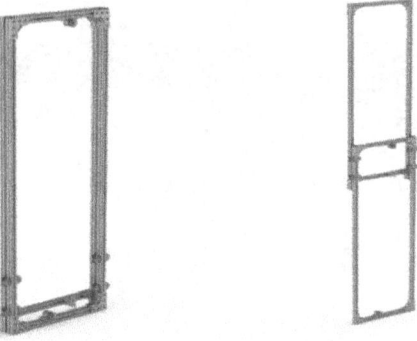

Fig. 3. Cascade lift mechanism

The self-propelled lift was also deemed unworthy because of its limitation in height, instability at high elevations and high maintenance cost. A simple decision matrix in was developed in Table 1 in order to determine the lifting mechanism that is best suited for this design project.

Table 1. Lift mechanism decision matrix

		Cascade		Self-propelled		Moveable	
Selection criteria	Weight	Score	Weighted score	Score	Weighted score	Score	Weighted score
Efficiency	5	3	15	2	10	4	20
Cost	5	1	5	4	20	4	20
Maintenance	5	2	10	4	20	4	20
Durability	5	2	10	3	15	4	20
Total			**40**		**65**		**80**

The selection criteria and their weights were selected considering the design require-ments while the scores were assigned based on observations from various literature reviewed [13].

A shelf is subjected to stress and displacement analyses using Solid works, and the failure was analyzed using Maximum Distortion Energy Theory (MDET) on the maximum shear stress and the Von-Mises stresses respectively. The MDET predicts failure when the strain energy in a material exceeds the shear strain energy of the material per unit volume at the moment that it fails under a uniaxial tensile test.

For MDET, failure occurs when Eq. (1) is satisfied

$$\sigma_{eq} \geq S_y \tag{1}$$

The safety factor N is given by Eq. (2).

$$N = \frac{S_y}{\sigma_{eq}} \tag{2}$$

Where σ_{eq} is the Von-Mises/equivalent stress.

The static displacement result is shown in Fig. 4, where the major displacement is localised in the middle of the shelf. Buckling takes place on all the horizontal beams where the load effect is obvious.

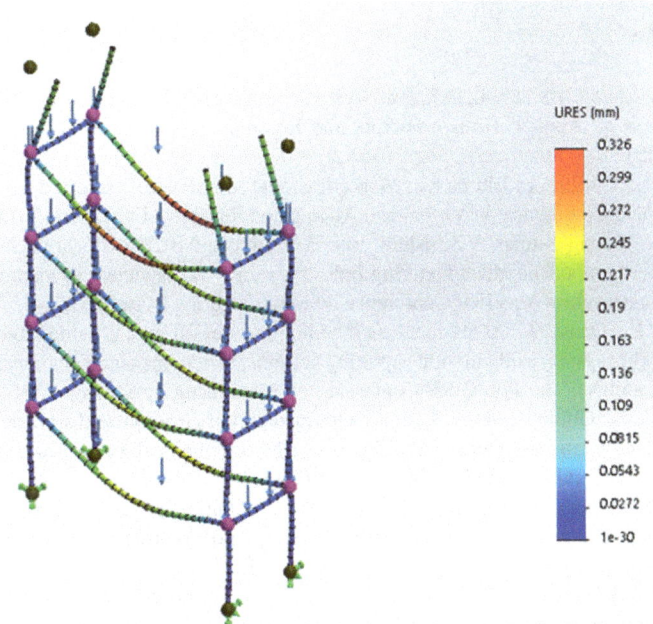

Fig. 4. Static displacement

Here, the safety factor achieved is 4.02 for the 2000 N load applied. It can be seen that the frame does not fail for 2000 N load as it does not meet the failure criterion. The frame failed when the critical load of 8360 N was applied. With the stacking system loaded at a maximum weight of 2000 N, the top shelf deflected the most with 0.326 mm. However, based on the result obtained, the maximum weight that can be loaded per shelf is 150 kg. The rest of the shelves will deflect at a maximum distance lower than that of the top shelf.

4   Conclusion

The design of a stacking system was a success. The racking system is important in the warehouse in order to optimize the efficiency of material flow. The gravity flow rack was regarded to be suitable for the system because of its high density. The use of forklift will be reduced for loading and retrieving of goods due to the lifting mechanism. In this design, the warehouse workers will be free from the tedious work of loading and retrieving process and without the need to drive into the rack. The easy assembling of the stacking system ensures easy maintenance culture. For future work, there are some aspects of the stacking system that could be improved. The introduction of vertical beams could improve the strength of the structure and secondly, the pulley system on the lift mechanism could be replaced with a chain mechanism in order to add to the efficiency of the lifting device.

References

1. Lee, J.A., Chang, Y.S., Shim, H.J., et al.: A study on picking process time. In: 6th International Conference on Applied Human Factors and Ergonomics and the Affiliated Conferences, AHFE (2015)
2. Ginters, E., Cirulis, A., Blums, G.: Markerless outdoor AR-RFID solution for logistics. In: International Conference on Virtual and Augmented Reality in Education (2013)
3. Balaji, K., Senthi Kumar, V.S.: Multi criteria inventory ABC classification in an automobile rubber components manufacturing industry, variety management in manufacturing. In: Proceedings of the 47th CIRP Conference on Manufacturing Systems (2014)
4. Weisner, K., Deuse, J.: Assessment methodology to design an ergonomic and sustainable order picking system using motion capturing systems, variety management in manufacturing. In: Proceedings of the 47th CIRP Conference on Manufacturing Systems (2014)
5. Brezovnik, S., Gotlih, J., Balic, J., et al.: Optimization of an automated storage and retrieval systems by swarm intelligence. In: 25th DAAAM International Symposium on Intelligent Manufacturing and Automation, DAAAM (2014)
6. Sukhov, P., Batsyn, M., Terentev, P.: A dynamic programming heuristic for optimizing slot sizes in a warehouse. In: Information Technology and Quantitative Management, ITQM (2014)
7. Cakmak, E., Gunay, N.S., Aybakan, G., Tanyas, M.: Determining the size and design of flow type and u type warehouses. In: 8th International Strategic Management Conference (2012)
8. Gu, J., Goetschalckx, M., McGinnis, F.: Research on warehouse operation: a comprehensive review. Eur. J. Oper. Res. 177(1), 1–21 (2007)
9. Koster, R., Le-Duc, T., Roodbergen, K.J.: Eur. J. Oper. Res. 182(2), 481–501 (2007)
10. Van den Berg, J.P., Zijm, W.H.M.: Int. J. Prod. Econ. 59, 519–528 (1999)
11. Vujanac, R., Miloradovic, N., Vulovic, S.: Int. J. Eng. 14 (2016)

12. Vujanac, R., Miloradovic, N., Slavkovic, R.: Metalurgia Int. **18**(8), 49–55 (2013)
13. Vujanac, R.V.: Development of methodology for design and calculation of carrying elements of storage systems. Master Thesis, Faculty of Mechanical Engineering University of Kragujevac, Serbia (2007)

The Prospective Direction of Solar Energy in Namibia

Mutiu Erinosho(✉), Ester Angula, Fillemon Nangolo, and Sam Shaanika

Department of Mechanical and Industrial Engineering,
University of Namibia, P.O. Box 3624 Ongwediva, Namibia
merinosho@unam.na

Abstract. The fluctuating output of the hydropower station, which is the main generating plant of the country due to drought, the ageing of coal fired power station, and the increase in demand of electricity have caused a gap between the energy demand and supply. To mitigate the problem, inexhaustible and sustainable energy sources should be implemented. Thus, solar energy, due to its abundance in nature is one the ways chosen to meet the demand and fill this gap. This study examines the internal strengths and weaknesses, and external opportunities and threats to expand domestic solar technologies usage. Hence, the strengths in terms of solar potential, low population density of the country, the dependence on biomass, the government support were explored whereas the high upfront cost of solar technologies, expensive off-grid solar off-grid plants, and the lack of human capital were found to be the weaknesses that need to be addressed and fixed. Suggestions were further revealed on the ways to enhance solar technologies based on the strengths discussed.

1 Introduction

Electricity is an inevitable source of power that is very crucial to development from domestic daily activities to heavy industrial processes. However none of these is possible without the electricity supply [1]. The population increase of the country has caused a disproportionate increment of power consumption with regard to its generation [2]. Currently, the most reliable generating plant in the country is the Ruacana Hydro Power Station, which is situated on the Kunene River; whose performance is dependent on the river flow. Moreover, the other generating plants such as Van Eck coal-fired power station built on the outskirt of the city of Windhoek, Anixas diesel-powered power station at Walvis Bay, and other renewable energy plants have contributed to the total installed capacity of the facilities. Table 1. depicts the individual installed capacity of the plants and the total installed capacity, as well as the generating plant.

It is worth mentioning that most of the renewable energy plants are photovoltaic plants, which is also a type of solar energy plant. According to the 2018 NamPower report, the maximum hourly demand for the system with the mine inclusive, is at 639 MW. With this, the country has to rely on interconnection with the external energy producers through energy trade such Eskom, Zesco, Zesa, and so on for the remaining energy beyond the installed capacity [3]. The suggested solution was to invest in power generation in order

M. Awang and S. S. Emamian (Eds.): *Advances in Material Science and Engineering*, LNME, pp. 30–35, 2021.
https://doi.org/10.1007/978-981-16-3641-7_5

Table 1. Individual installed capacity of plants and total capacity of the country [1]

Licensee	Generation plant	Resource types	Capacity (MW)
NamPower	Ruacana	Hydro	347
NamPower	Van Eck	Coil	30
NamPower	Anixas	Heavy fuel oil	22.5
NamPower total			**399.5**
Renewable energies			67.5
Total installed capacity			**467**

to reduce or eliminate reliance on import and to secure supply of electricity for the present and the future of the country's needs and requirements [1]. This means the new or future generation plants should be sustainable, zero carbon, economic, and efficient. Solar energy is the energy force that sustain life on planet earth and has the potential to provide solution to the global energy needs of human. The sunlight is readily available and poses no threat to our environment and global climate systems from pollution emissions [4].

The aim of this brief review is to provide information on the various implementation ways of solar energy system technologies and to describe the challenges faced by solar technologies and suggest ways to overcome them based on case studies as well as describing the policies that promote the success of solar technologies in Namibia. In this review study, the Strengths-Weaknesses-Opportunities-Threats (SWOT) approach was briefly reviewed in order to investigate the internal strengths and weaknesses, and the external opportunities and threats that may influence the successful implementation of solar energy technologies in Namibia.

2 Solar Potential

The solar irradiance of Namibia ranges between 6 kW/m^2/day, which represents the highest values in the world and this however, positions the country as a potential place for harvesting solar energy which is readily available and abundant [5]. Figure 1 shows the potential regions of photovoltaic plants based on the solar irradiance of the country.

These regions represent a wide possibility of solar technologies application since solar energy can be harvested as heat, light or both. Recent studies in renewable energies, especially in the sector of solar energy has led to improved efficiencies of solar technologies used to generate electricity. Solar photovoltaic plants convert light into electricity using silicon based material and concentrated solar plants used the heat to boil water using reflecting mirrors that focused the heat and light of the sun on a located point, and then the steam generated under pressure is used to power turbines. Namibia is one of the least populated countries in the world; its population density is around three per square kilometre. With this, it is a challenge to electrify isolated communities around the country since the extension of the grid to certain locations is not economically feasible [6]. It is therefore important to restructure the way to access energy. The solar irradiance in most parts of the country is suitable for the implementation of off-grid solar

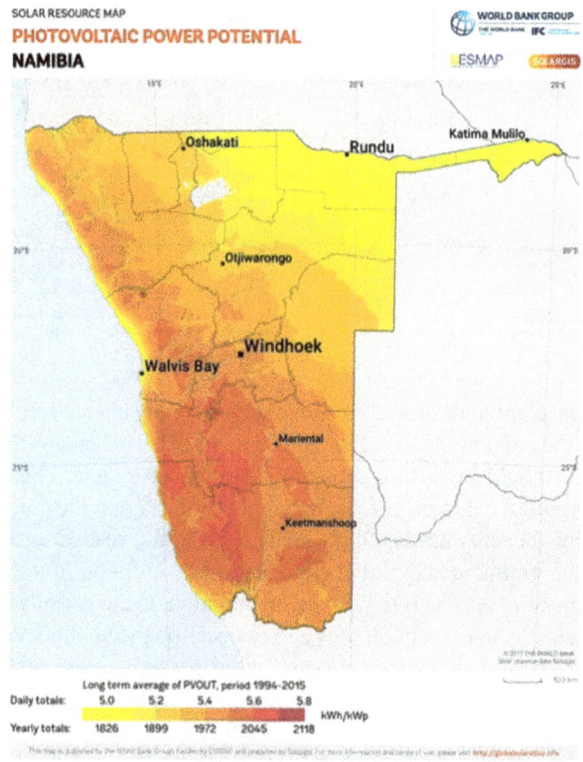

Fig. 1. Regions with their respective photovoltaic power potential [7]

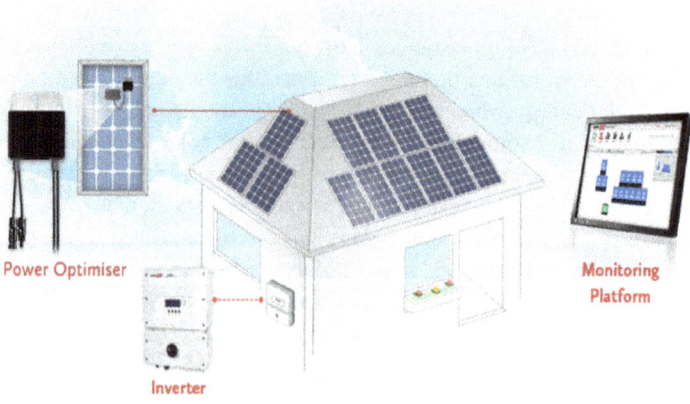

Fig. 2. Ultimate solar edge [8]

plants. A careful study and implementation of off-grid solar systems might be the way to ensure energy access to isolated communities and initiate development in all parts of the country [5]. Figure 2 shows a house with solar panel systems with inverter and power optimizer.

The Solar PV needs smarter inverters to reduce grid integration costs, large volumes of small-scale PV will require innovation to be managed and integrated [9]. But today, the society is bringing to the market technologies and practices that can provide the energy needed with fewer or no harmful impacts from pollution [10].

3   The Dependence on Biomass

Most isolated communities that do not have access to electricity still rely on wood as their primary source of energy for their basic activities [11]. The recent years of drought due to climate change has reduced the availability of this resource. The cutting down of trees contributes to the climate change; which has negatively affected the earth's ecosystem. However, solar technologies for cooking have the potential to substitute conventional cookers and heating systems and thereby reducing the emission of CO_2. The government on their own are unable to ensure a secure supply of electricity to the entire country due to its high cost and other priority's departments [12]. Therefore, a change in policies that allow private investors, through independent power producers (IPPs), to invest in the energy sector and produce electricity was implemented. The reformation allowed IPPs to sell the bulk output electricity to NamPower Corporation, the state owned company in charge of energy generation and transmission. This initiative is perhaps the greatest milestone towards securing electricity supply for country's needs. Approximately 19 IPPs will generate 175 MW of renewable energy by 2020 using solar PV plants [13]. However, the creation of the solar system will be a major subsidy to off-grid solar technologies [14]. This new model will allow IPPs to directly sell their output electricity to their consumers and will attracts foreign entities to invest into the Namibian energy sector [15].

4   High Upfront Cost of Solar Technologies and Expensive Off-Grid Solar Plants

Despite the fact that the cost of solar technologies considerably decreased, the upfront cost was still high [16]. The government acknowledged the inability to secure full electricity based on the national budget. The high cost is distinguished because people that were in need electricity supplies are usually off-grid and cannot afford to pay. Thus, solar technologies have to compete with the existing technologies in order to be economically viable. Since the extension of the grid is not achievable, there is a need for storage devices added with the high upfront cost. There is also lack of human capital due to the fact that technical skills of the labour force is yet to be developed and furnished for proper use and maintenance of this technology empowered by the government [16].

5 Opportunities and Threats

The United Nations Development Programme and the International Renewable Energy Agency were some of the international organizations that work in close collaboration with the government of Namibia to promote renewable energies. It was reported that solar energy alone cannot sustain the electricity demand of the country. However, this international supports have alleviated the main issues faced by the Renewable Energies in the country, and also provide expertise and render solutions to the hurdles hindering the development [17]. Fossil fuels are concerned in most energy industries and their utilization has a negative impact on the environment. Carbon dioxide and other Greenhouse gas emissions have gradually affected the planet and caused climate revolution. The UN agency sets a scheme aimed at reducing Carbon dioxide and other Greenhouse gas emissions. This international benefits favour the development of solar energy in Namibia [18]. The availability and affordability of electricity from the Southern African Power Pool is the main threat to the future of solar technologies. 44,000 MW of electricity was proposed to be generated from an hydropower station [19]. However, if this is implemented, the tariff of electricity will drop and eventually affect the future of solar energy. The future of solar technologies in Namibia does not only depends on the implementation of photovoltaic and concentrated solar panel, but also on the various ways of acquiring the energy. The acceptation and widespread of solar technologies is crucially influenced by the combined effect of research funding and government incentives. China, Japan, and some other countries have shown remarkable growth in solar energy technologies due to the government involvement through the support of these new technologies [20]. Raising public awareness and encouraging the usage of solar technologies is a major step to guarantee the adaptation of the devices from industrial applications to individual daily activities of the society [21].

6 Conclusion

This review paper focused more on the SWOT model that were used to analyze the future direction of solar system technologies in Namibia. Today, the availability of solar resources represented one of the main enhancers of solar technologies and the low population density has promoted their diversification. Solar devices can substitute current tools that use biomass to meet the energy demand in rural areas more eco-friendly. In addition, the government support through reform of policies and incentives provision will guarantee the adaptation and widespread of solar energy usage within the country. However, the financial strain due to the high upfront cost of solar technologies, increased cost of off-grid due to storage devices, and the lack of human capital, especially in rural areas were the critical internal obstacles reviewed. On the other hand, the international support and incentives to control and reduce atmospheric pollution are external factors that will promote renewable energies in general.

References

1. Press statement on the power supply situation in Namibia. http://www.mme.gov.na/files/publications/2e0_Media Statement on Electricity Supply.pdf. Accessed 01 Aug 2019

2. Kannan, N., Vakeesan, D.: Solar energy for future world: a review. Renew. Sustain. Energy Rev. **62**, 1092–1105 (2016)
3. NamPower Annual Report of NamPower (2018). https://www.nampower.com.na/public/docs/annual-reports/Nampower%202018%20Annual%20Report%2028.01.19-print.pdf. Accessed 03 July 2019
4. Brunet, C., Savadogo, O.: Baptiste P et al shedding some light on photovoltaic solar energy in Africa – a literature review. Renew. Sustain. Energy Rev. **96**, 325–342 (2018)
5. Diaz-Maurin, F., Chiguvare, Z. Gope, G.: Scarcity in abundance: the challenges of promoting energy access in the Southern African region (2018)
6. Moller, L. Namibia: energy policy general information on Namibia.
7. Solaris: Solar resource maps and GIS data for 180+ countries | Solargis. https://solargis.com/maps-and-gis-data/download/namibia. Accessed 21 Aug 2019
8. Ultimate solar edge. https://www.accordelectrical.com.au/ultimate-solar-edge/. Assessed 19 October 2019
9. Technology Innovation to Accelerate Energy Transitions (2019). file:///C:/Users/admin/Documents/innovation/Technology%20innovation%20to%20accelerate%20energy%20transitions.pdf
10. Carbonnier, G., Grinevald, J.: Energy and Development. Int. Dev. Policy **2**, 9–28 (2011)
11. REEECAP 3.3 energy-related impacts of climate change in rural Namibia executive summary. http://nei.nust.na/sites/default/files/downloads/Energy-related%20Impacts%20of%20Climate%20Change%20on%20Rural%20Namibia%20-%20DRFN%20-%20Executive%20Summary.pdf. Accessed 12 July 2019
12. Republic of Namibia, ministry of mines and energy, electricity supply industry national policy for independent power producers (IPPS) in Namibia (2017). https://www.ecb.org.na/images/docs/Investor_Portal/NATIONAL_IPP_POLICY_OF_NAMIBIA_-_August_2017.pdf. Accessed 15 Aug 2019
13. Renewable energy licensees licensee address location technology (Type) Capacity (Size) 1 Ark Industries. https://www.ecb.org.na/images/docs/Licensing/licenses-issued/RE_Licensees_(REFIT_Programme).pdf. Accessed 01 July 2019
14. Ministry of Mines and Energy: Ministry of Mines and Energy - Solar Revolving Fund. http://www.mme.gov.na/directorates/efund/srf/. Accessed 19-Aug-2019
15. The Namibian: Cabinet approves revised electricity supply model - the Namibian (2019). https://www.namibian.com.na/187861/archive-read/Cabinet-approves-revised-electricity-supply-model. Accessed 15 Aug 2019
16. Kabir, E., Kumar, P., Kumar, S., et al.: Solar energy: potential and future prospects. Renew. Sustain. Energy Rev. **82**, 894–900 (2018)
17. Ministry of Mines and Energy Ministry of Mines and Energy - Renewable Energy. http://www.mme.gov.na/directorates/energy/renewable/. Accessed 19 Aug 2019
18. New Era. Namibia signs up for ICAO's carbon offsetting scheme - New Era Live. https://neweralive.na/posts/namibia-signs-up-for-icaos-carbon-offsetting-scheme. Accessed 19 Aug 2019
19. Grand Inga Hydroelectric Project: An Overview | International Rivers. https://www.internationalrivers.org/resources/grand-inga-hydroelectric-project-an-overview-3356. Accessed 19 Aug 2019
20. Chen, W.M., Kim, H., Yamaguchi, H.: Renewable energy in Eastern Asia: renewable energy policy review and comparative SWOT analysis for promoting renewable energy in Japan, South Korea, and Taiwan. Energy Policy**74**, 319–329 (2014)
21. Lei, Y., Lu, X., Shi, M., et al.: SWOT analysis for the development of photovoltaic solar power in Africa in comparison with China. Environ. Impact Assess. Rev. **77**, 122–127 (2019)

Development of a Refrigeration System Test Rig

Ester Angula[✉], Fillemon Nangolo, Mutiu Erinosho, and Sam Shaanika

Department of Mechanical and Industrial Engineering,
University of Namibia, Ongwediva, Namibia
eangula@unam.na

Abstract. Thermodynamics is an important aspect of mechanical engineering, and the fundamental part of it is refrigeration. It is often beneficial for students of mechanical engineering to observe practically at the working principle of a refrigeration system. Thus, this paper presents the development of a vapour compression refrigeration system test rig to be used for student's practical work in the mechanical engineering laboratory at engineering campus, University of Namibia. The refrigeration system was designed to have a refrigeration capacity of 25 kW and to provide cooling at a desired temperature of 16 °C. The calculations for the refrigeration system were based on a thermodynamic analysis of the components, which makes use of the principles of conservation of energy. The system's theoretical Coefficient of Performance (COP) was found to be 9.19. A 3D CAD model of the test rig was developed and the layout of the test rig was kept simple with all components clearly identified.

Nomenclature

Symbols

$\dot{Q}$	Refrigeration or condenser capacity [W]
$\dot{m}$	Mass flow rate [kg/s]
h	Enthalpy [kJ/kg]
W	Compressor power [kW]

Subscripts

e	Evaporator
c	Condenser
$1, 2, 4, a$	Locations

1 Introduction

Thermodynamics is a fundamental branch of mechanical engineering, and refrigeration system is one of the fundamental aspects of thermodynamics. A refrigeration system can be defined as a combination of equipment connected in sequential order to produce

M. Awang and S. S. Emamian (Eds.): *Advances in Material Science and Engineering*, LNME, pp. 36–40, 2021.
https://doi.org/10.1007/978-981-16-3641-7_6

an effect whereby heat is extracted from a heat source and deposited into a heat sink, for maintaining the temperature of the heat source below that of its surrounding [1]. These systems consist of four major components: a compressor for pumping the refrigerant around the system, a condenser for rejecting heat to the surroundings, an expansion device for reducing the pressure of the refrigerant, and an evaporator for absorbing heat from the heat source and producing the cooling effect [2]. Vapour compression systems are the most common refrigeration systems in use. In vapour compression systems, the refrigerant is compressed to a high pressure and temperature vapour after it has produced the refrigerating effect. Several researchers designed and constructed refrigeration system test rigs in order to demonstrate the fundamental refrigeration principles in thermodynamics, as well as the effects of refrigerants on the refrigeration system performance. Oyelami and Bolaji [3] designed and constructed a refrigeration system test rig to investigate the performance of Liquefied Petroleum Gas (LPG) as a refrigerant using a vapour compression. The refrigeration system test rig was used as an experimental apparatus for demonstrating fundamental refrigeration principles in thermodynamics. The first and second laws of thermodynamics were incorporated in their performance test, and the COP of the system was determined. The effects of R12 and R134a refrigerants on the COP of the vapour compression refrigeration system were investigated by Nagalakshmi and Marurhiprasad Yadav [5]. It was found that when R12 is used, the COP was slightly greater than when the system used R134a; however, Authors recommended R134a as the refrigerant of choice due to its non-toxicity and its zero ozone depletion potential. The performance of the refrigeration system test rigs depends on the performance requirements [5]. In reviewed studies, the designed refrigeration system test rigs have different requirements and therefore different operation conditions. From academic point of view, it is crucial for mechanical engineering students to acquire a good understanding of refrigeration system from both theoretical and practical point of view before graduation. Therefore, this paper aims to develop refrigeration system test rig that students of mechanical engineering department can use to study the aspects of refrigeration systems.

2 Methods

The refrigeration system test rig was designed by employing the design approached of Bolaji and Falade [4] who designed, constructed and carried out a performance test of a vapour compression refrigeration system for use as an experimental apparatus for demonstrating fundamental refrigeration principles in thermodynamics. The system requirements, standard parameters, design variables, and design constraints, which included safety constraints, financial constraints, and environmental constraints were first determined. The components of the system were defined, bearing in mind that each component has an effect on the performance of other components in the system. To ensure the components were selected in accordance with the design requirements, weighted decision and calculations using established thermodynamic principles and formulae matrices were used to determine the best options. All major components were investigated with particular emphasis placed on affordability and thermal performance optimization. The apparatus were design to provide cooling to a desired temperature of 16 °C.

2.1 Component Selection

The Based on the of weighted decision matrices as well as on the criteria that were set for selection of the rig components, the plate evaporator, air-cooled condensers, capillary tube, and reciprocating compressors were selected as main components. The required length of capillary tube depends mostly on the size of the system. Refrigerant-specific rating charts for capillary tube selection are found in the 2010 ASHRAE Refrigeration Handbook [6]. The refrigerants considered for this system was R134a. For R134a, for the equivalent of 5 °C sub-cooling, a capillary tube of length of 1.5 m and inner diameter of 1.25 mm was selected. In addition to thermodynamic properties of the R134a, its environmental properties were also considered. The ASHRAE Refrigeration Handbook 2010 [6] recommends giving preference to refrigerants that have low or zero Ozone depleting potential, have low total equivalent warming impact values and provide good system efficiency. The selected refrigerant, R134a, satisfies these three criteria.

2.2 Refrigeration Piping

The refrigerant piping system is the network of pipes that connects the components of the system. The three major pipes or lines in a refrigeration system are the suction line, the discharge line and the liquid line. The design of these lines was detailed, along with the requirements for each individual line, and the pipe specifications were recommended based on refrigerant nomographs for R134a, which was found in the DuPont Refrigerant Piping Handbook [7] and the sizing of these lines was done according to ASHRAE 2010 refrigeration standards [6]. From the R134a refrigerant nomograph, the diameter for each pipe was taken from the acceptable diameters range, and the corresponding velocity was selected. In addition to that, the fittings along each pipe and their equivalent length were selected based on Daikin Refrigerant Piping Design Guide [8]. The summary of pipes dimensions is depicted in Table 1.

Table 1. Summary of pipes dimensions

Pipe/line	Diameter, mm	Velocity, m/s	Fitting type	Fitting equivalent length, mm
Suction	28.575	17.78	Two short-radius 90° elbows	792
Discharge	22.225	15.748	Two 45° elbows	274
Liquid	9.525	2.134	One sight glass	304.8

3 Results and Discussion

The component's specifications were determined through calculations. The calculations for the refrigeration system were based on a thermodynamic analysis of the components, which makes use of the principle of conservation of energy. Firstly, each of the principal states is located on the accompanying schematic and T–s diagrams in Fig. 1.

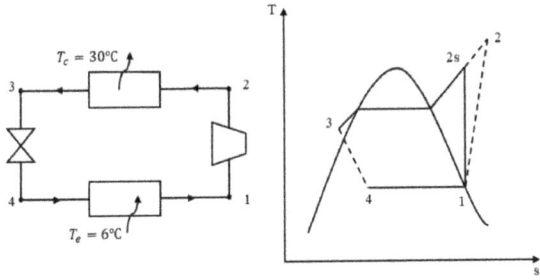

Fig. 1. Vapour compression refrigeration system schematic and T-s diagram

The calculations were performed by assuming steady state operation; kinetic and potential energy effects are negligible; saturated vapour enters the compressor and saturated vapour leaves the condenser; the compressor isentropic efficiency is 80%, and the refrigeration capacity is 25 kW, which is the rating of the available evaporator. The performance characteristics of the refrigeration system which are: the mass flow rate, condenser capacity, required compressor power and COP were determined using Eqs. (1), (2), (3), and (4) respectively.

$$\dot{m} = \frac{\dot{Q}_e}{(h_1 - h_4)} \tag{1}$$

$$\dot{Q}_c = \dot{m}(h_2 - h_3) \tag{2}$$

$$\dot{W}_c = \dot{m}(h_2 - h_1) \tag{3}$$

$$COP = \frac{\dot{Q}_e}{\dot{W}_c} \tag{4}$$

The obtained values are presented in Table 2, while the layout of the test rig along with the selected components and their specifications is shown in Fig. 2.

Table 2. Performance characteristics

Description	Unit	Value
Mass flow rate	Kg/s	0.1571
Condenser capacity	kW	27.73
Compressor power	kW	2.72
COP		9.19

The COP is found to be higher, which signifies the lower operating costs. Practically, the COP might be lower than that attained, since the only energy consumption included

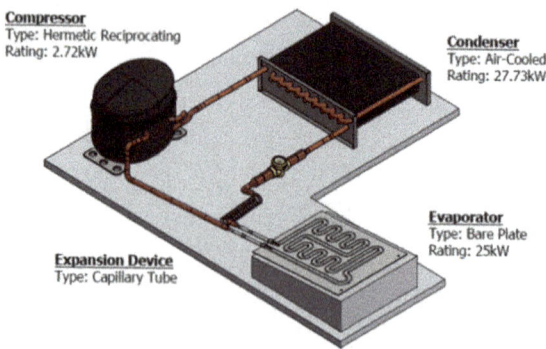

Fig. 2. 3D CAD model of refrigeration system test rig

in the calculation was for compressor, and other power-consuming auxiliaries were not considered. Furthermore, the COP is highly dependent on operating conditions, especially absolute temperature and relative temperature between sink and system, therefore, the COP of the designed test rig is depend on the location or laboratory where it is placed.

4   Conclusions

The test rig was designed and components were selected and specified to provide cooling to 16 °C. The piping system of the test rig was carried out according to ASHRAE 2010 standards for refrigeration. The layout of the test rig was kept simple and all components were classified. A 3-D CAD model and detailed drawings of the test rig were developed. The COP of the system was found to be 9.19, which indicates the lower operating costs. The apparatus is portable and can be used for laboratory experiments and classroom demonstrations. It will help students of engineering disciplines to have a thorough understanding of both the practical aspects of refrigeration and the thermodynamic processes affecting the performance of the cycle.

References

1. Wang, S.: Handbook of Air Conditioning and Refrigeration. McGraw-Hill, New York (2000)
2. Thornton, R.: Refrigerant Piping A Trane Air Conditioning Clinic Air Conditioning Clinic TRG-TRC006-EN © American Standard Inc. (2002). Slide Player, http://slideplayer.com/slide/6172042/. Accessed 16 May 2017
3. Oyelami, S., Bolaji, B.: Design and construction of a vapour compression refrigeration system as test rig to investigate the performance of Liquefied Petroleum Gas (LPG) as refrigerant. J. Sci. Eng. Re. **3**, 350–357 (2016)
4. Bolaji, B., Falade, T.: Development of an experimental apparatus for demonstrating vapour compression refrigeration system. Int. J. Therm. Env. Eng. **4**(1), 1–6 (2012)
5. Nagalakshmi, K., Marurhiprasad, Y.G.: The Design and performance analysis of refrigeration system using R12 & R134a refrigerants. Int. J. Eng. Re. Appl. **4**(2), 638–643 (2014)
6. American Society of Heating: Refrigerating and Air-Conditioning Engineers (2010)
7. Denison, G.: DuPont Refrigerant Piping Handbook. DuPont Canada Inc., Ontario (2001)
8. Daikin Applied, Refrigerant Piping Design Guide. http://hvacrknowlagecenter.homestead.com/PipeSizing.pdf. Accessed 22 Oct 2017

Scaling and Fouling of Reverse Osmosis (RO) Membrane: Technical Review

Ignatius Shahonya[✉], Fillemon Nangolo, Mutiu Erinosho, and Ester Angula

Department of Mechanical and Industrial Engineering, University of Namibia, P. O. Box 3624
Ongwediva, Namibia
ishahonya@unam.na

Abstract. An estimated 1.2 billion inhabitants are still faced with the water crises worldwide and the number keeps increasing. The population growth, rising standards of living, industrial proliferation, water source contamination and climate change remains one of the contributing factors to water scarcity. The use of reverse osmosis technology to desalinate brackish and sea water is one of the key reliable sources of alternatives water supply. However, the technology is subjected to scaling and fouling challenges. Fouling and scaling are regarded as major cost intensive trials in the RO desalination industry. Scaling is mostly stirred by the precipitation of calcium carbonates, calcium sulphates, barium sulphate and silicates on the RO membranes which clogs the membrane pores whereas, fouling is by the deposition of suspended matters, colloids and micro-organisms on the membrane. Even though there are several methodologies developed to control scaling and fouling, the feed water composition, pre-treatment, temperature, chemical composition and filtration process prior to RO membrane filtration remains different from plant to plant, making it almost impossible to predict, mitigate or control fouling and scaling.

1 Introduction

Water scarcity continues to haunt many economies as the population charts heighten. In addition to population growth, rising standards of living, industrial proliferation, water source contamination and climate change are equally the contributing factors to water shortages worldwide [1]. An estimated 1.2 billion inhabitants worldwide are experiencing this dilemma and about 500 million are expected to live in water scarce zones [2]. The sea carries about 97% of water worldwide and 2.5% is frozen; a clear indication that only 0.5% of the water is available of which, part of it is brackish underground water and other part as fresh water [3].

Consequently, many economies are obliged to find alternative ways to satisfy the demand for clean drinkable water. In the world of continuous water shortage and insecurity, desalination of brackish and sea water using reverse osmosis membrane technology is a key reliable source of clean drinking water that complies with international standards. Upon inception, RO membrane technology was highly utilised in arid areas to desalinate sea and brackish water and it was seen as a luxury; due to its high operational costs [4]. However, advances in the technology led to a dramatic reduction in

M. Awang and S. S. Emamian (Eds.): *Advances in Material Science and Engineering*, LNME, pp. 41–47, 2021.
https://doi.org/10.1007/978-981-16-3641-7_7

costs. Nowadays, the technology has become commercially viable and cost competitive among traditional methods. RO membrane technology is a preferred water purification technology employed to provide high quality water [5]. Figure 1 on the next page depicts a typical desalination plant from the intake of water and runs through the RO system to the final fresh water.

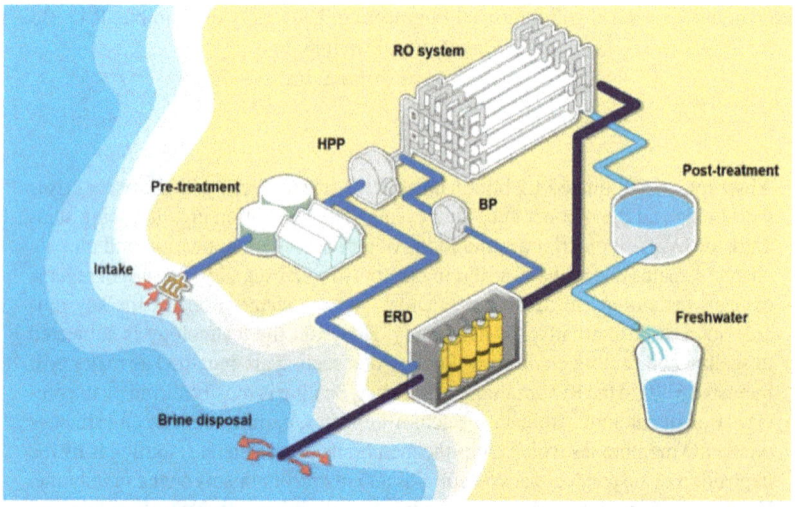

Fig. 1. Typical RO membrane desalination plant [6].

Despite the technology's success in water purification, the RO membrane technology is faced with challenges that threaten its functionality and increases its operational costs. The major challenges are scaling and fouling of the membrane. According to Characklis [7], scaling and fouling increase costs associated with the RO membrane process and its mitigation is instrumental in membrane further advancement [7].

2   Scaling and Fouling Phenomenon

2.1   Scaling

The precipitation of calcium carbonates, calcium sulphates, barium sulphate and silicates on the RO membranes which clogs the membrane pores is what is referred to as scaling (see Fig. 2 for a typical picture of it) [8]. Additionally, scaling refers to the unwanted precipitation of sparingly soluble salts onto equipment surfaces during operation due to the solubility of salt being exceeded [9]. Scaling of RO membranes is one of the tedious hurdles in the desalination industry and it is influenced by different factors. The influence of temperature and Graphene oxide on the scaling of RO was revealed and it was reported that a rise in temperature enhanced the membrane scaling as a result of intense declination of the fluxes over a period of time [10]. The scale layer tend to grow as it is exposed to the same conditions. Figure 2 below shows a typical scaling of a membrane.

Fig. 2. RO membrane scaling [8].

The growth in RO membrane technology has advanced to a stage where the high water recovery desalination system is vowing for a zero liquid discharge, as a result a high concentration of soluble salts is experienced which leads to scaling [11]. In addition, the Arabian Gulf has a calcium ions concentration in a range of 220–500 mg/L, consequently with a recovery rate of 30–70% and the concentration of the ions was 870–2800 mg/L [11]. The production of water, whether brackish or seawater is hindered by the high concentration of dissolved salts which leads to membrane scaling. The scalants namely; silica, divalent salts of carbonate, sulphate, fluoride, and phosphate are the most commonly encountered salts [4]. The impactful constituents in membrane scaling are magnesium, calcium, bicarbonate and sulphate [12]. There are two processes proposed as pathways to scaling namely; homogenous (as bulk crystallization followed by deposition) and heterogeneous (surface crystallization) [9]. Since there is always nano-particles (nucleation) from the metals that dissolves in water, the bulk crystallization is also a heterogeneous process [13]. The scaling of the membrane process is entirely based on the operating condition (cross flow and pressure) which in hand determines the mechanisms employed whether bulk or surface crystallization [14]. Prior to the growth of deposit, macrocrystals, an atmosphere for the precipitation of a crystalline substance from an impact on the site of the protruding scale is enhanced by permeation, nucleation, followed by nucleus growth, accumulation and scale deposition [15]. The scale deposition on the surface of the RO membrane is a result of different factors namely; feedwater chemical composition, super saturation degree, pH, temperature, hydrodynamics, antiscalant concentration, a sequence of antiscalant and brines mixing, etc. [16]. If any of the condition stated above changes, there will be a variation in the mechanisms of both scale formation and inhibition. Scaling is not only enhanced by nano- or micro-molecules but by macromolecules equally [17].

2.2 Fouling

Fouling is an especially persistent problem in all membrane filtration processes and referred to the deposition of suspended matters, colloids and micro-organisms on the membrane [9]. Additionally, fouling is the adhesion and growth of bacteria present in water on the surface of the membrane, which in turn blocks the membrane pores – reducing the pressure and thus decreasing membrane flux and eventually biodegrading the membrane polymer [18]. The early RO membranes were made of cellulose acetate, which suffered greatly to hydrolysis at pH range beyond 4 to 6, low rejection (especially silica) and high pressure [19]. As technology matures, most RO desalination plants utilised membranes made out of polyamide; these membranes even though prone to fouling (due to the inability to controlling deposition of foulants, in particular, microorganism on the membranes) offers better pH ranges, improved rejection and functions well under high pressure [20]. The foulants that stick on the membrane forms a gel layer called biofilm, and this layer interferes with the filtration process as it creates a secondary membrane that reduced the pressure of the flow due to friction [21]. Initially, the biofilm layer formation is caused by the glycoproteins (feed water composition differs so is the proteins and thus different microorganisms) and other organics, which boost the cell development phases taking place in the erection of the layer [22]. Figures 3 depict the biofouling, caused by the deposition and growth of microorganism.

Fig. 3. Biofouling, caused by the deposition and growth of microorganism [24].

Membrane fouling is complex, the latter is prone to be as a result of different factors involved in the water filtration process; intake water quality, pre-treatment and the entire setup of the RO desalination process and parameters [23]. Additional attributes to the latter includes the diversity of the organic and biological foulants as depicted in Fig. 3 on the next page, which at most occasions interact at a much faster rate to form an intact and rigid biofilm layer [24]. Reverse Osmosis membrane fouling is referred to as the deposition of particles or materials on the membrane surfaces which in turn blocks the membrane pores causing a decrease in the feed pressure and other parameters. There is a wide range of foulants that contribute to this dilemma however, this study is focused on organic and biological fouling. Organic fouling is the physical or chemical attachment of organic compounds, whereas biofouling is the attachment of bacteria/fungal/algal onto the membrane [25]. The degradation of the membrane structure is caused possibly by the sulphate and the anaerobic bacteria present in the feedwater prompting biofouling. On the other hand, the low concentration of metals such as Aluminium, Manganese, Iron and Lead etc. in raw water equally contribute to organic fouling [4]. The fouling development is sequenced; initially the rapid interaction between the foulants - membrane and gradually foulants - foulants [26].

3 Scaling and Fouling Control

Scaling and fouling is inevitable in RO membrane desalination plants. There are various ways to control scaling and fouling but the following controls are discussed.

3.1 Scaling

If scaling is predicted to occur in a membrane system, various approaches is taken to control it. As sealants may be difficult to remove, the prevention of scale formation is the most desirable approach. This can be achieved by two main techniques, broadly categorised into physical or chemical methods which are further described. If these methods are unsuccessful and scaling does occur, its immediate removal is desirable by methods described further in section. Lowering the recovery rate beneath the solubility of the salt is one way of preventing scaling, however it increases brine and reduce permeate [27]. This process is costly, thus a better one is the hydrodynamic process – which involves high cross flow (turbulence) velocity at the membrane surface to prevent the crystals from attaching to the membrane [28]. However, there are times when the feedwater salt saturation is high and thus the impact of this process may be minimal.

Another process applied to reduce scaling is the chemical process which involves the use of chemicals such as: citric, sulphuric, ethylenediaminetetraacetic and hydrochloric acid – these are added to destabilise the alkaline scale formation process, also the antiscalants are added to evade scaling [29]. Since scaling will continuously invade the membranes, it is irreversible and it may halt the water production process, at that instance – the desalination plant has to shut down and the membranes has to be cleaning physically, a very cost ineffective process [30].

3.2 Fouling Control

Fouling has a huge economic impact on any desalination plant thus, proper maintenance of the membrane is a prime objective. There are several methods used to ensure that fouling is countered. The brackish water is immune to filtration unlike the counterpart (sea water) as it is naturally filtered, however a minimum pre-treatment of a cartridge can filter of up to 10 μm and it is employed to remove big particles that clogged the membrane pores [31]. The surface water on the other hand requires serious filtration and pre-treatment to remove the suspended solids and the dissolved salts, typical methods are employed; coagulation, sedimentation and filtration, the particles found are agglomerated and flocculated by ferric chloride, alum and polymers [4]. Additionally, other methods are applied, the latter includes; ozonation, hydraulic (backwashing), chemical, cross flushing, pre-coating and physical [32]. Fouling control comes to effect when the flux decreased and pressure drop increases, the cleaning in most cases is unsustainable and very costly i.e. plant downtime, in the end the control is weak and the cleaning is unsuccessful, and fouling is irreversible.

4 Conclusion

An estimated 1.2 billion inhabitants worldwide have experienced this water scarcity and about 500 million are expected to live in water scarce zones. The sea carried about 97% of water worldwide and 2.5% is frozen which gives a clear indication that only 0.5% of the water is available of which part of it is brackish underground water and part as fresh water. However, the RO membrane technology is faced with challenges that threaten its functionality and increases its operational costs. Scaling and fouling are inevitable in RO membrane desalination plants and ways to stop or reduce these are very important for cost reduction. Scaling and fouling has a huge economic impact on any desalination plant thus, proper maintenance of the membrane is a prime objective, and this needs to be dealt with in the future for greater sustainability.

References

1. Abramovich, S., Hyams-Kaphzan, O., Kenigsberg, C.: The effect of long-term brine discharge from desalination plants on benthic foraminifera. PLoS ONE **15**(1), e0227589 (2020)
2. Jegathambal, P., Nisha, R.R., Parameswari, K., Subathra, M.S.P.: Desalination and removal of organic pollutants using electrobiochemical reactor. Appl. Water Sci. **9**(4), 1 (2019). https://doi.org/10.1007/s13201-019-0990-0
3. Glass, N.: The water crisis in Yemen: causes, consequences and solutions. Glob. Majority e-J. **1**(1), 17–30 (2010)
4. Boerlage, S.: Scaling and particulate fouling in membrane filtration systems, Ph.D. thesis, Wageningen University, Wageningen (2001)
5. Maddah, H., Chogle, A.: Applicability of low pressure membranes for wastewater treatment with cost study analyses. Membr. Water Treat **6**(6), 477–488 (2015)
6. Kim, J., Park, K., Yang, D.R., Hong, S.: A comprehensive review of energy consumption of seawater reverse osmosis desalination plants. Appl. Energy **254**, 113652 (2019)
7. Characklis, W.G.: Microbial Fouling: A Process Analysis: Fouling of Heat Transfer Equipment. Oxford Press, London (1981)
8. Ashfaq, M.Y., Al-Ghouti, M.A., Qiblawey, H., Rodrigues, D.F., Hu, Y., Zouari, N.: Isolation, identification and biodiversity of antiscalant degrading seawater bacteria using MALDI-TOF-MS and multivariate analysis. Sci. Total Environ. **656**, 910–920 (2019)
9. Mi, B., Elimelech, M.: Silica scaling and scaling reversibility in forward osmosis. Desalination **312**, 75–81 (2013)
10. Da'na, D.A., Al-Ghouti, M.A., Zouari, N., Qiblawey, H., Ashfaq, M.Y.: Investigating the effect of temperature on calcium sulfate scaling of reverse osmosis membranes using FTIR, SEM-EDX and multivariate analysis. Sci. Total Environ. **703**, 134726 (2020)
11. Du, X., Zhang, Z., Carlson, K.H., Lee, J., Tong, T.: Membrane fouling and reusability in membrane distillation of shale oil and gas produced water: Effects of membrane surface wettability. J. Membr. Sci. **567**, 199–208 (2018)
12. Saavedra, E.: Graphic evolution of the 24.000 hours (3 years) operating data of a RO brackish water desalination plant, in Las Palmas, Canary Islands. Spain. Desalination **76**, 15–26 (1989)
13. Popov, K., Oshchepkov, M., Afanas'eva, E., Koltinova, E., Dikareva, Y., Rönkkömäki, H.: A new insight into the mechanism of the scale inhibition: DLS study of gypsum nucleation in presence of phosphonates using nanosilver dispersion as an internal light scattering intensity reference. Collid Surf. A Physicochem. Eng. Asp. **560**, 122–129 (2019)
14. Lee, S.: Effect of operating conditions on $CaSO_4$ scale formation mechanism in nanofiltration for water softening. Water Res. **34**(15), 3854–3866 (2000)

15. Amjad, Z., Demadis, K.D.: Mineral Scales and Deposits: Scientific and Technological Approaches. Elsevier, Amsterdam (2015)
16. Liu, L.X., He, A.J.: Research progress of scale inhibition mechanism. Adv. Mater. Res. **955–959**, 2411–2414 (2014)
17. Rahman, F.: Calcium sulfate precipitation studies with scale inhibitors for reverse osmosis desalination. Desalination **319**, 79–84 (2013)
18. Filloux, E., Wang, J., Pidou, M., Gernjak, W., Yuan, Z.: Biofouling and scaling control of reverse osmosis membrane using one-step cleaning-potential of acidified nitrite solution as an agent. J. Membr. Sci. **495**, 276–283 (2015)
19. Kucera, J.: Biofouling of polyamide membranes: fouling mechanisms, current mitigation and cleaning strategies, and future prospects. Membranes **9**(9), 111 (2019)
20. Nguyen, T., Roddick, F., Fan, L.: Biofouling of water treatment membranes: a review of the underlying causes. Monit. Tech. Control Meas. Membr. **2**(4), 804–840 (2012)
21. Flemming, H., Schaule, G., Griebe, T., Schmitt, J., Tamachkiarowa, A.: Biofouling—the Achilles heel of membrane processes. Desalination **113**(2–3), 215–225 (1997)
22. Khan, M.T., De O. Manes, C., Aubry, C., Gutierrez, L., Croue, J.P.: Kinetic study of seawater reverse osmosis membrane fouling. Environ. Sci. Technol. **47**(19), 10884–10894 (2013)
23. Croué, J., Manes, C.D., Aubry, C., Khan, M.T.: Source water quality shaping different fouling scenarios in a full-scale desalination plant at the Red Sea. Water Res. **47**(2), 558–568 (2013)
24. Tang, C.Y., Kwon, Y., Leckie, J.O.: The role of foulant–foulant electrostatic interaction on limiting flux for RO and NF membranes during humic acid fouling—theoretical basis, experimental evidence, and AFM interaction force measurement. J. Membr. Sci. **326**(2), 526–532 (2009)
25. Ahmed, S., Alansari, M., Kannari, T.: Biological fouling and control at Ras Abu Jarbur RO plant - a new approach. Desalination **74**, 69–84 (1989)
26. Kang, G., Liu, Z., Yu, H., Cao, Y.: Enhancing antifouling property of commercial polyamide reverse osmosis membrane by surface coating using a brush-like polymer containing poly (ethylene glycol) chains. Desalin. Water Treat. **37**(1–3), 139–145 (2012)
27. Qiang, X., Sheng, Z., Zhang, H.: Study on scale inhibition performances and interaction mechanism of modified collagen. Desalination **309**, 237–242 (2013)
28. Sweity, A., Oren, Y., Ronen, Z., Herzberg, M.: The influence of antiscalants on biofouling of RO membranes in seawater desalination. Water Res. **47**(10), 3389–3398 (2013)
29. Kavitskaya, A., Knyazkova, T., Maynarovich, A.: Reverse osmosis of concentrated calcium sulphate solutions in the presence of iron (III) ions using composite membranes. Desalination **132**(1–3), 281–286 (2000)
30. Benyahia, F.: Membrane fouling and scaling in membrane distillation. In: Membrane-Distillation in Desalination, pp. 101–115 (2019)
31. Ang, W., Mohammad, A., Benamor, A., Hilal, N., Leo, C.: Hybrid coagulation–NF membrane process for brackish water treatment: effect of antiscalant on water characteristics and membrane fouling. Desalination **393**, 144–150 (2016)
32. Suratt, W.B., Andrews, D.R., Pujals, V.J., Richards, S.: Design considerations for major membrane treatment facility for groundwater. Desalination **131**(1–3), 37–46 (2000)

Design of a Fan Motor Lifting Mechanism for an Air-Cooled Condenser

Ester Angula[✉], Fillemon Nangolo, Mutiu Erinosho, and Jean Habiyaremye

Department of Mechanical and Industrial Engineering, University of Namibia, P.O. Box 3624, Ongwediva, Namibia
eangula@unam.na

Abstract. The Van Eck power plant is generating electricity based on the Rankine cycle in which the steam expanded in the turbine is condensed back to liquid. During this process, the plant uses the forced convection type of air-cooled condensers where the axil fans are used to maintain the air flow at a certain rated speed. The axil fan's motors and gearboxes sometimes experience mechanical and electrical failures, thus has to be replaced or repaired. However, there are challenges of delivering the motors and gearboxes as well as safely conveying them into the condenser house. Currently, there is no proper mechanism to deliver the motors and gearboxes into the condenser house. Therefore, during motors and gearboxes repair or replacement, more time is spent, and maintenance workers are vulnerable to risks of injuries, while the equipment is vulnerable to damages. To solve these problems, several concepts were generated from which the most effective and reliable concept was selected. The manual stress analysis was conducted on the selected concept in order to determine the required sectional properties. Then, the CAD model was developed in Solidworks, and simulated to analyze the stress distribution on the model, thus verifying the manual stress analysis calculations conducted in order to ensure that the design is capable to lift the intended load safely.

1 Introduction

Air-cooled condensers are commonly used in the electricity- generating power plant to condense the steam exhausted from the turbine. Condensation in air-cooled condensers occur by the aid of the forced draft air cooled condensers which uses the axial fans to force air upwards through the condenser tube bundles. The axial fans operate at different speeds, and therefore connected to an electrical motors and gearboxes. Van Eck power plant operates based on the Rankine cycle, and therefore uses the air-cooled condensers.

Unfortunately, these air-cooled condensers at Van Eck power plant were constructed at a certain height above the ground within an enclosure, thus certain machines had to be used to lift the gearboxes motors and the fans to that floor level whereas other certain mechanism had to be employed to move these loads within that enclosure.

For many years, several lifting equipment had been designed to overcome major challenges facing the industry in terms of lifting heavy products as wells as conveying them from one position to another either on the same floor level or at different floor

M. Awang and S. S. Emamian (Eds.): *Advances in Material Science and Engineering*, LNME, pp. 48–54, 2021.
https://doi.org/10.1007/978-981-16-3641-7_8

levels. Challenges of lifting these loads have been addressed by designs such as fixed gantry hoists, overhead cranes and swing jib cranes mostly used in workshops. Mabrouk and Abdelkhalek [1] indicated that, due to the gantry cranes roles in handling loads, they became indispensable for a large number of industrial facilities. Umesh et al. [2] designed a gantry hoist for a rated load of 5 kN based on theoretical calculations of strength of materials and as wells as computer-based simulations in ANSYS. While, Dhanoosha and Gowtham [3] carried out a detailed design and analysis of a jib crane. The design of the lifting equipment depends on the load to be lifted, location or industry, as well as the operation environment. Therefore, each design is always unique. Hence, the designed lifting mechanism is specific for Van Eck power plant load requirement of 150 kg and operation environment.

2 Methods

2.1 Design Requirements and Concepts Selection

In order to achieve load requirements, firstly, the structure was made of structural steel, specific ASTM A36 steel. Secondly, the factor of safety of 8 was used throughout the material selection as per strength of material analysis procedures [4]. Except from the selected beam connector, which is the universal beam section of significant length, which fails only not due to bending stresses but crushing. The selected trolley and conveyance universal beams have the maximum allowable displacement given as the length per 602 (L/602). All the parts in the concepts have the allowable yield stress of 31.25 MPa (Bending, Compressive and Tensile) as well as 15.625 MPa in shear. An exercise of 30% over-design was considered in this design work, hence using the probable design load of 1962 N throughout the stress analysis as per strength of material formulations. Lastly, all the developed concepts are based on the two criteria that, the concept shall be able to transport the motor and gearbox, and convey them into the condenser house and safety measures shall be considered on all the concepts.

The designed lifting mechanism is to deliver and lift the 150 kg gear box motor of the condenser axial fan at Van Eck power plant. The space at the entrance of the condenser house is limited and critical. Based on the requirement, four design concepts, which are fixed gantry hoist; Z-hoist; hoist trolley and cart-rail mechanisms were considered. The concept-scoring matrix, which was used to select the best concept, was developed based on the design requirements as well as on the geometry of the concept, materials required, the assembling of the product; labor and time required during operation. Overall, the hoist trolley concept was rated as the most suitable design that can effectively and safely transport and convey the gear box motor into the condenser fan house.

2.2 Detailed Design

Initially, the stress analysis of the concept frames was carried out manual. This was performed by using strength of materials formulations on the concept sketch as indicated in Fig. 1. The dimensions used to make the frame for the hoist are 600 mm for EF, E'F', BA & B'A', 1300 mm for BE, AF, B'E' & A'F', 1100 mm for DD', and 600 – 800 mm for CD'.

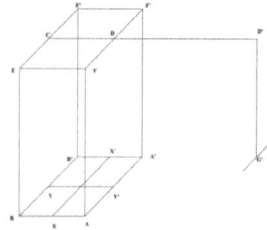

Fig. 1. Sketch of the design concept (Hoist Trolley)

The materials properties were initially chosen to be ASTM A36 (2001) which has a yield strength of 250×10^6 N/mm^2 and the tensile strength of 400×10^6 to 550×10^6 N/mm^2, and elongation of 20–25% as well as a Youngs Modulus of 200 GPa. Furthermore, the design was based on the factor of safety of 8 because it fell in the category of lifting equipment [4]. Most literatures have indicated that shear stress is approximately 50% of the material's yield strength and the allowable maximum materials deflection is indicated to be $L/602$ in beams. Based on that, the design stresses of the design project were obtained as indicated in Eqs. (1) and (2).

$$\sigma_{y(all)} = \frac{\sigma_y}{foS} = 31.25 \text{ MPa} \tag{1}$$

$$\tau_{y(all)} = 0.5\sigma_{y(all)} = 15.625 \text{ MPa} \tag{2}$$

Beam CD and DD' were analyzed based on the moving load analysis in order to fully capture all the possible loading configurations of the beam to be selected as discussed in AISC [5]. The analysis are carried out with the load P equivalent to 1962 N, and the bending moment, shear forces, as well as section modulus are computer using Eq. (3), (4), and (5) respectively.

$$R_C = R_D = P \tag{3}$$

$$M_{max} = \frac{PL}{4} \tag{4}$$

$$Z = \frac{M_{max}}{\sigma_{y(all)}} \tag{5}$$

The shear force, bending moment, section modulus and maximum deflection for beam CD were found to be 1962 N, 294300 Nmm, 9.42 cm^3 and 0.01283 mm respectively, while for DD' are 1962 N, 539550 Nmm, 17.28 cm^3 and 0.02712 mm respectively. Based on these results, the preliminary materials with the indicated cross-sectional properties are selected. The same beam is selected for both beam CD and DD', not only ensure maximum flexural strength but also smooth conveyance of the load between the two beams. The selected beams cross sectional properties are depicted in Table 1.

For both beams, the maximum deflection in the beam due to the load is less than the required maximum deflection; hence, the beam is safe for selection. Columns EB, E'B',

Table 1. Selected Beam Cross Sectional Properties for Beam *CD and DD'*

Universal Beam (150 × 100) [Area: 26.3 cm^2]							
Section		Thickness		Second moment of inertia		Elastic modulus	
Depth [mm]	Width [mm]	Flange [mm]	Web [mm]	I_{xx} [cm^4]	I_{yy} [cm^4]	Z_{xx} [cm^3]	Z_{yy} [cm^3]
150	100	9	6	1003	150	135	30.1

AF & A'F' were designed based on the displacement of the load in any direction or a swing about the load-beam connection point by at most 15° due to human errors and direct normal loading because of the hoisting process and self-weight of the beams. The columns were selected based on the flexural strength and the loading carrying capacity in compression, where the sectional area required to carry the predetermined load was determined to be 0.157 cm^2. 0.157 cm^2 Minor Beam EE' and FF' were designed based on the probable maximum load condition which is when the load is exactly below the front minor beam FF'. The shear force, bending moment and section modulus for columns EB, E'B', AF and A'F' were found to be 507.8 N, 660400 Nmm and 21.13 cm^3 respectively, while for minor beam EE' and FF' are 981 N, 392400 Nmm and 12.56 cm^3 respectively. The material cross sectional properties depicted in Table 2 were selected for both columns and minor beams.

Table 2. Selected cross-sectional areas for columns *EB, E'B', AF & A'F' and* minor beam EE' and FF'

80 × 80 × 5 Square Tube Cross-sectional Properties				
Thickness(t) [mm]	Mass [kg/m]	Sectional area [cm^2]	Second moment of inertia *(Ixx = Iyy)* [cm^4]	Elastic modulus *(Zxx = Zyy)* [cm^3]
5	11.70	14.7	177	34.2

The analysis of Members YY' and XX' is similar to that of the minor beams, and therefore same material cross sectional properties as shown in Table 2 were selected for them. Based on the calculated design yield stress and the direction of the induced load, the cross members are in tension in most cases, thus an area of 38.464 mm^2 is required. Using the mentioned area, a 30 × 30 × 5 angle iron is used for the design concept. The bolt size for the connector which connecting the beams, as well as for all the connecting and joint points were determined using both shear and tensile loading conditions which require a bolt with a diameter at least equivalent to 5.685 mm, therefore a M12 × 30 mm bolt was selected.

The detailed concept shown in Fig. 2 for the host trolley mechanism was modelled in Solidworks software. This was performed as per components dimensions obtained from the stress analysis conducted as per strength of material formulations. Furthermore, the caster wheels, hoist trolley and chain hoist were size accordingly.

Fig. 2. Hoist trolley detailed concept

3 Results and Discussion

The manual stress analysis carried out in Sect. 2.2 was verified by performing the finite element analysis in Solidworks simulation packages. The CAD model in Fig. 2 was employed in the simulation. Figure 3(a) shows the von mises stress analysis of the concept members under the predetermined load of 1962 N. Due to the load effects, a maximum stress of 7.072×10^6 N/m^2 was recorded in the concept at the point of concentrated loading, while the smallest stress occurred in the lower parts of the columns and corners of the bottom and top frames of the hoist trolley model. This is due to the deflection about the connection point of the beams which had led to significant rotation of the columns about their fixture points. The same arguments can be used for the effects of the beam self-weights due to gravitational effects. The maximum stress recorded is less than the allowable maximum yield strength of these materials; hence the affected member is safe to carry this load during its indented service time. The frame has an average stress of 4.359×10^6 N/m^2.

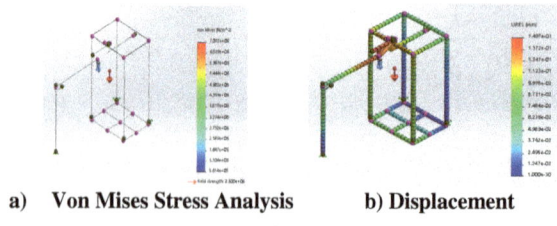

a)   **Von Mises Stress Analysis**   b) **Displacement**

Fig. 3. Hoist trolley finite element analysis

As shown in Fig. 3(b), based on the deformation induced in the members, the maximum deformation was also recorded around the beam connection region. At this area, the effect of the load is more concentrated as compared to other regions of the model' members. Based on that, the induced rotation of the members due to bending at the connection point had led to the displacement of the forward pair of stoppers by an amount of 0.004989 mm to 0.009978 mm. While, at the conveyance column connector, a displacement in the range of 0.002495 mm to 0.004989 mm with minor significant displacement of the conveyance column due to bending was recorded with an amount of 0.007484 mm.

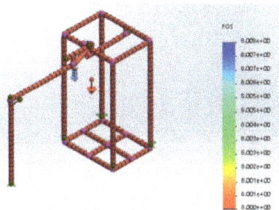

Fig. 4. Stoppers column stress analysis (factor of safety study)

Figure 4 indicates the factor of safety analysis based on the von mises stress or the yield stress of the materials used. In this study, an effort was made to ensure that all the concept structural members are within the safe limit determined by the factor of safety upper limit greater than or equals to 8. The results had shown that all the components are within the safe limit, both exactly analyzed at the factor of safety of 8 as exercised in the manual stress analysis calculations. This implies that the system has the overall carrying capacity of 1962 N on average thus it can safely carry and convey the rated condenser fan gearbox motor without failure.

4 Conclusions

Four concepts were developed to address the difficulties faced by Van Eck power plant during their maintenance of the condenser axial fan gearboxes motor. The most effective concept, the hoist trolley, was selected. This concept structures were analyzed as per strength of material formulations and on computer-based simulations. For stability and maneuverability purposes, the system was fitted with lock nut stoppers and caster wheels, respectively. Throughout the process of analyzing system, all the stresses were below the maximum allowable stress permitted on the structural members to be used on the model; therefore, the hoist trolley is capable of hoisting the design load of 200 kg for the gearbox motor. The usage of this design will minimize the risk of injuries and damages to the maintenance personnel and gearbox motor respectively.

References

1. Mabrouk, M.H., Abdelkhalek, S.M.M.: Design and implementation of light duty gantry crane. Int. J. Eng. Re. Tech. **3**(12), 586–594 (2014)
2. Umesh, G., Ravi, G., Sharath, T.S.: Design and fabrication of lightweight aluminum gantry crane. Int. J. Eng. Sci. Re. Tech. **6**, 586–594 (2017)
3. Dhanoosha, M., Gowtham, R.V.: Detailed design and analysis of a free standing I beam Jib crane. Int. Res. J. Eng. Tech. (IRJET) **3**(12), 193–203 (2016)

4. Factor of Safety - Lifting Equipments, Engineering Toolbox. https://www.engineeringtoolbox.com/Factor-of-Sfaty-for-Lifting-Equipments/. Accessed 26 Aug 2019
5. Table 3: Shear, Moments & Deflections (Simple Beam - One Concentrated Moving Load), in American Institute of Steel Construction, p. 227 (2001)
6. Universal Beams and Columns, in Products Handbook (Structural Steel), Continental Steel Pty, Ltd., p. 31 (2006)

Technical Review on Severe Plastic Deformation of Aluminium Series

Mutiu Erinosho[1]([⊠]), Ester Angula[1], Fillemon Nangolo[1], Bolanle Ikotun[2], and Oluwagbenga Johnson[3]

[1] Department of Mechanical and Industrial Engineering, University of Namibia, P O. Box 3624, Ongwediva, Namibia
merinosho@unam.na
[2] Department of Civil and Chemical Engineering, College of Science, Engineering and Technology, University of South Africa, Pretoria, South Africa
[3] Department of Mining and Metallurgical Engineering, University of Namibia, P O. Box 3624, Ongwediva, Namibia

Abstract. Severe plastic deformation (SPD) of Aluminium alloys has been of great interest to many researchers due to the improved properties that the alloy has exhibited from grain refinement up to micro and nanoscale structure. In this review, various processes of deformation have been presented such as high-pressure torsion, equal channel angular pressing, planar twist extrusion process, tube channel pressing and other deformation means. The research results from the application of various techniques of severe plastic deformation on different types of Aluminium alloys are presented in this review. This includes the final microstructural analyses after deformation, grain sizes, strength improvement, and electrical conductivity and other mechanical properties.

1 Introduction

Aluminium is an important and popular nonferrous metal which is used in many fields for instance in building, packaging, automobile and electrical materials because of its attractive properties such as light weight, good electrical conductivity, good corrosion resistance and excellent mechanical properties [1]. This nonferrous metal is at times alloyed with other metals to have a combination of properties for better performance. Its light weight affects the mechanical strength; however, in order to improve both the strength and the ductility properties of the material, special techniques such as strengthening mechanisms are required for their achievement. In case of other metals, it encompasses strengthening by grain size reduction, solid-solution strengthening and strain hardening [2].

The grain size reduction is aimed to reduce the size of the grain of a material, which would eventually alter its microstructure as well as its performance and properties. The grain size reduction is one of the key features that changes not only the mechanical but also the chemical and other physical properties of the materials. Among the methods of grain refinement, severe plastic deformation (SPD) process has been of great interest to researchers due to the extent to which grains are refined. SPD process started with the

M. Awang and S. S. Emamian (Eds.): *Advances in Material Science and Engineering*, LNME, pp. 55–63, 2021.
https://doi.org/10.1007/978-981-16-3641-7_9

work of Bridgman who developed various techniques to process materials through the combination of high hydrostatic pressure and shear deformation [3] and may impose a very high strain on a bulk solid without impeding the overall dimensions of the specimen [4]. In the recent review by Bagherpour et al. [5], the research status, classification, microstructure evolution, and applications of SPD were presented. Thus, as a result of the good properties that can be achieved by ultrafine-grained materials, various SPD techniques are still under investigation in order to find new processing ways either by modifying the current available techniques or developing a new techniques [6]. Different SPD techniques have been reported in the literature by many authors as well as grain structure refinements; for example, to mention a few [7–11]. Ehab et al. [12] investigated a commercial Aluminium 6082 alloy prepared in two different conditions. One specimen was prepared in a T651 annealed condition and the other sample was prepared after a solution treatment and this was followed by equal-channel angular pressing (ECAP). On all the tested samples, the effect of the initial condition was examined by high-pressure torsion (HPT) through 1/2, 1, 2, 5 and 10 turns respectively. Figure 1 shows the microstructure as well as the Number fraction against the Mis-orientation angle.

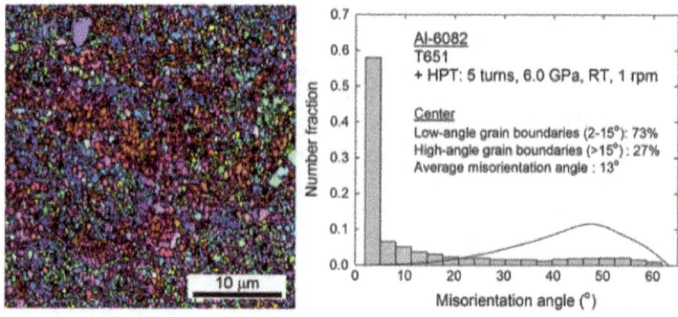

Fig. 1. Color-coded orientation map and histogram for the number fractions of the misorientation angles in the center of the tested sample from a T651 condition processed by HPT through 5 turns [12]

The microstructure was analyzed using electron backscattered diffraction. The low-angle grain boundaries of the misorientation angle of the tested sample at the center between 2° to 15° amounted to 73% while the high-angle grain boundaries measured at the center of the tested sample greater 15° amounted to 27%. The hardness values of the two initial conditions were reported and showed different result, although they remained practically constant after being processed by HPT. These hardness results have shown how HPT process has greatly increased the hardness of the tested samples [12].

Jozef et al. [13] studied the ultrafine grain structure evolution at high temperature of commercial aluminium alloy AA6082 using HPT. Thermal treatment was used to prepare two dissimilar initial structural states of the alloy. The microstructure showed that an ultrafine grain (UFG) structure was formed in the deformed disc toward the end of the first turn and it was modified locally by dynamic recrystallization and the rise in deformation temperature became more observable and the growth of new grains was also reported with more stability of the UFG microstructure of the specimens prepared by ageing and quenching before torsion deformation. In addition, the tensile strength data revealed that the strength was partly relaxed by local recrystallization and the required torque to deform the specimen was increased until the first turn was completed and subsequently maintained stability [13]. For non-heat treatable commercial aluminium alloys, there is a significant increase in tensile strength, compared to the heat treatable alloys where there is no improvement in mechanical properties by using SPD process [14].

Brodova et al. [15] studied the effect of the level of deformation under quasi-hydrostatic pressure upon torsion on the evolution of the structure of V95 aluminum alloy by making use of electron microscopy and X-ray diffraction methods. The study revealed that a crystalline structure is formed with a hardness value of 2500 MPa. Furthermore, it was found that the dynamic strain aging starts at values of strain greater than 4.8 from the X-ray diffraction analysis data. Figure 2 shows the graph of dependence of microhardness on the degree of deformation for the deformed samples.

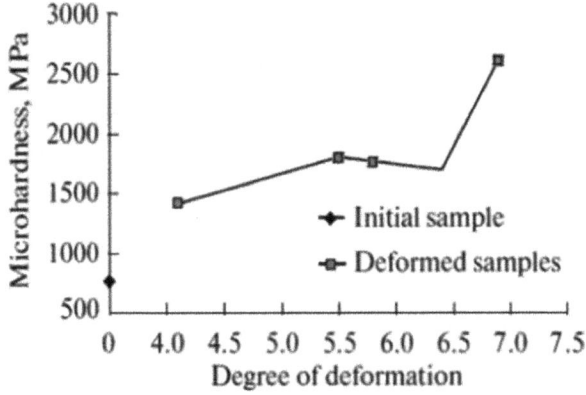

Fig. 2. Dependence of microhardness on the degree of deformation [15]

At strain values within the range from 5.5 to 6.4, the nanostructure with a minimum average size of approximately 55 to 80 nm was being formed. Besides the dynamic strain aging of the alpha solid solution, the analysis done on the alloy structure also suggested the existence of deformation-induced dissolution of tiny intermetallic compounds [15]. Severe plastic deformation technology development has shown that UFG bulk alloys with acceptable strength and ductility can be obtained using aluminium alloys 6061 and 7075. Two heat treatment effects were suggested for grain coarsening and formation of precipitates. It was recommended that artificial aging and solution heat treatment should be prepared as post heat treatment especially for low formable alloys such as 7075. For the production of Al/Al$_2$O$_3$ nanofiber composites, it was shown that equal channel angular pressing is a viable route for the process. The production of metal matrix composites with aligned nanofibers has been allowed by the process of accumulative roll bonding and it was revealed that the Al$_2$O$_3$ nanofibers were cut because of the tensile stresses applied during the rolling process. From the second cycles, Al$_2$O$_3$ nanofibers begin to be embedded in the matrix and the temperature of compaction was found to be very useful. The method of repetitive press rolls bonding method has allowed the reduction of processing time to embed Al_2O_3 nanofibers even in only one cycle to attain a strong bonding [16]. Mahmoud [17] experimentally examined the fatigue behaviours of aluminium 6061 alloy structures before and after the planar twist extrusion (PTE) process, which is another method of SPD process. Although the improvement rate was reported to decrease after some passes. However, the yield strength, the tensile strength, the hardness, and the fatigue endurance of the tested sample were increased as shown in Fig. 3 [17]. The stress-strain curve in Fig. 4 depicts the stress values for the tested sample without planar twist extrusion as well as with first and fourth passes.

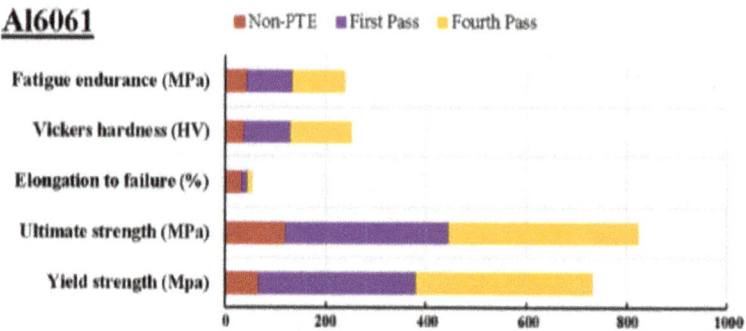

Fig. 3. Mechanical properties before and after the planar twist extrusion process [17]

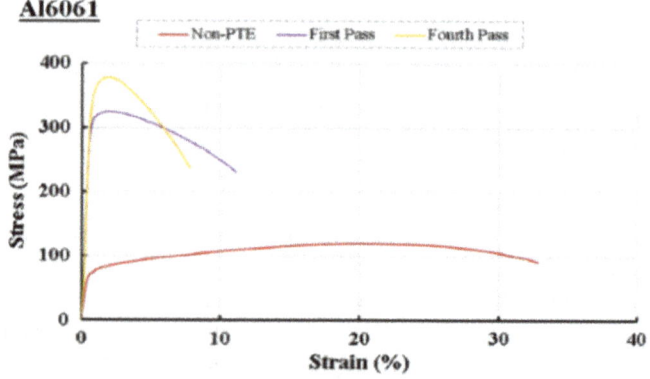

Fig. 4. Stress-strain curve of Aluminium 6061 [17]

A substantial decline in the capability for strain hardening was reported and the tensile fracture morphology changed progressively from the ductile mode in the initial state to ductile-brittle mode upon completion of the final pass. The grain size correction factor for the Aluminium 6061 was found to be approximately 3 by comparison. An establishment was made by the author on the results and stated that SPD method can be regarded as a suitable process to develop material properties and fabricating nanostructured materials [17]. Investigation of the deformation behaviour of solid solution-treated

6061 Aluminium was done by Farshidi and Kazeminezhad [18] through tube channel pressing (TCP). It was reported that there is spread of unsteady-state region in initial step of TCP process and the authors compared finite element method and experimental data and concluded that increasing the cave depth amounts to an increased equivalent strain. Figure 5 shows the equivalent strain distribution in tube when processed by mandrel with different curvature radii.

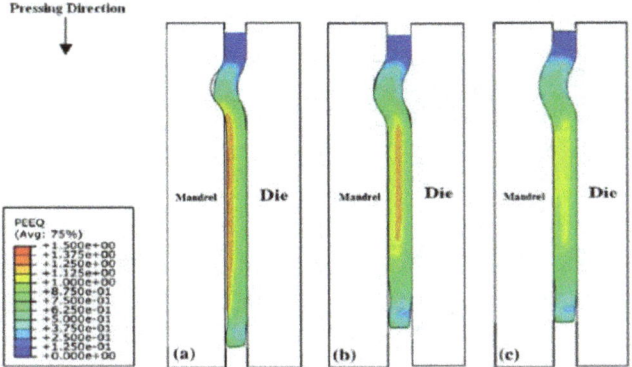

Fig. 5. Equivalent strain distribution in tube when processed by mandrel with curvature radius of (a) 5 nm, (b) 7.5 nm and (c) 10 nm [18]

It was revealed that the homogeneity of equivalent strain decreases through the tube wall thickness. The decrease in the radius of curvature of the mandrel has led to the limited unsteady-state region as well as a decrease in the strain homogeneity [18].

Joshi et al. [19] conducted a research on the mechanical properties and microstructure evolution of Al 2014 alloy through multidirectional cryo-forging. The investigation showed the formation of UFG sizes ranges from 100 to 450 nm after 4 cycles at a cumulative strain of 2.4. Furthermore, an increment of 7% in tensile strength and 3% in yield strength resulted in the investigation of the cryo-forged samples over the initial condition samples.

The fractographs of the Al 2014 alloy samples after tensile testing as shown in Fig. 6 revealed the diversified mode of fracture.

In 3-point bend test, the fracture toughness was increased by 60% for the cryo-forged Al 2014 alloy when compared to the solution-treated alloy. In addition, a change was observed from ductile tearing fracture to brittle fracture with an increase in strain steps for fracture mode of cryo-forged Al 2014 alloy after carrying out the 3-point bend test. Therefore, there was improvement in the mechanical properties of tested samples using multidirectional cryoforging [19]. The structure of Ni3Al single crystal was studied by Kuts et al. [20] after severe plastic deformation. The results of synchrotron showed that the compression of the samples and the torsion applied at different angles yielded to the disturbance of the single crystallinity of the tested sample which led to the complete disarranging of Ni3Al. Lipinska et al. [21] processed Al-Mg-Si alloy using incremental equal channel angular pressing to yield non inhomogeneous, UFG plates with low anisotropy regarding mechanical properties. This was the first attempt to

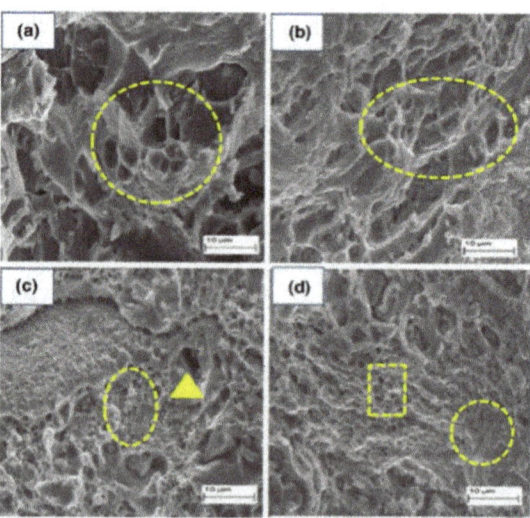

Fig. 6. Fractographs of cryo-forged Al 2014 alloy after tensile test for different processing conditions: (a) solution treated (b) 2 cycles (c) 3 cycles (d) 4 cycles [19]

process an Al-Mg-Si alloy via the means of SPD. It was revealed that the average grain size decreased from an average of 17 μm to less than 1 μm after four passes. The ultimate tensile strength (UTS) was increased more than twice its initial value, whereas the yield strength was found to be more than three times its initial value. Therefore, their research on Al-Mg-Si alloys refinement was proved to be successful with the incremental equal channel angular pressing [21].

In another research work, Al-Mg-Si alloy was nanostructured via high pressure torsion at various temperatures and in different deformation modes; which was conducted at room temperature and at high temperature of about 180°–230 °C. It was experimentally demonstrated that by introducing additional dislocation density, from 2×10^3 to 5×10^{13} m^{-2}, without changing the basic parameters of the UFG structure leads to an increment of about 15% in the strength of the alloy [22]. The structural changes in Al-Zn, Al-Mg, and Al-Mg-Zn alloys were investigated by [23] before and after SPD of these alloys by high pressure torsion with true strain of about 6. Supersaturated Al solid solution was reported to decompose and achieved the equilibrium state at room temperature. The microhardness measurements showed that the alloys were softened due to disintegration and deformation of the supersaturated solid solution [23]. Mohammadi et al. [24] studied the influence of combining ECAP and shot peening surface treatment on fatigue, tensile strength and microhardness on Aluminium 6082 alloy. The test specimens that were shot peened after one pass of the ECAP process exhibited a longer fatigue life than those which went through shot peening after more than one pass. It was also reported that the fatigue life was reduced after one pass prior the shot peening as a result of the local work softening that emerges at the subsurface layer of the test specimen [24]. The structure and mechanical properties of an Aluminium 1570 alloy were investigated using severe plastic deformation by HPT at 100 °C and 200 °C. This investigation by Murashkin et al. [25] showed that after performing the HPT at room temperature, there was a significant increase in the strength of the alloy. The UTS and the offset yield strength attained 950 MPa, and 905 MPa respectively; and the relative elongation at fracture obtained was approximately 5%. Though, at higher temperatures, it was also revealed that the HPT treatment did not noticeably affect the strength characteristics but leads to a substantial decrease in plasticity.

In a new research work by Murashkin et al. [26], the structure and mechanical properties of Aluminum 6061 alloy using ECAP in parallel channels were examined. It was reported that after four cycles of ECAP in parallel channels, the Al alloy exhibited a combination of increase in both strength and plasticity [26]. The ECAP in parallel channels process has also been used to study the effect of grain refinement on the microstructure, electrical conductivity and mechanical properties of an Al-0.6Mg-0.45Si alloy [27]. It was demonstrated that after 6 passes in parallel channel, a homogeneous ultrafine-grained microstructure of 400 to 600 nm can be reached. At 100 °C, the formation of metastable β precipitates closely beside the grain boundaries were reported [27]. Pavlova et al. [28] studied the structure and internal friction of Fe_3Al and $(FeCr)_3Al$ alloys under SPD process in which Fe-26Al-5Cr and Fe-26Al alloys were revealed to cause the formation of an UFG structure with an improved concentration of crystal defects. After the SPD process, the two alloys possessed high concentration of crystal defects which has led to the formation of peaks in the spectrum of the temperature dependence of internal friction. However, the addition of chromium increased the dislocation segments in Fe-26Al-5Cr alloys [28]. The influence of equal channel angular extrusion on the mechanical properties of AA6111 alloy was studied by Rhee et al. [29] for automotive outer body panels. Using this process combined with heat treatment, the authors discovered that both the strength and the ductility were increased [29]. Sabirov et al. [30] examined the effect of ECAP in parallel channels on Al 6061 and Al 6063 alloys. It was reported that the chemical composition and the processing temperature have a strong influence on the microstructure with an equiaxed ultra-fine grains. However, at 160 °C, a recovery rate occurred in Al 6063 alloy which subdues the formation of the ultra-fine grains in the microstructure [30].

2 Conclusion

Severe plastic deformation has been demonstrated to be a successful route to achieve considerable reduction in the grain size of various Aluminium alloys up to nanoscale structures with excellent combination of properties. The parameters that affect the final behaviour of the Aluminium alloys during the process of SPD include the temperature at which the process is conducted, the number of pass, the angle of rotation, the amount of strain induced, chemical composition of the alloy as well as the shape of the die. The improvement in both the strength and the ductility of Aluminium alloy series can be achieved via SPD techniques. The new SPD techniques has proven to be more efficient and should be an exciting area of future works.

References

1. Guan, R.-G., Tie, D.: A review on grain refinement of aluminum alloys: progresses, challenges and prospects. Acta Metallurgica Sinica (Engl. Lett.) **30**(5), 409–432 (2017)
2. Callister, W.D., Rethwisch, D.G.: Materials Science and Engineering: An Introduction, 8th edn. Wiley, United States (2018)
3. Bridgman, P.W.: The effect of hydrostatic pressure on plastic flow under shearing stress. J. Appl Phys. **17**, 692 (1946)
4. Valiev, R.Z., Estrin, Y., Horita, Z., Langdon, T.G., Zechetbauer, M.J., Zhu, Y.T.: Producing bulk ultrafine-grained materials by severe plastic deformation. JOM **58**(4), 33–39 (2006). https://doi.org/10.1007/s11837-006-0213-7
5. Bagherpour, E., Pardis, N., Reihanian, M., Ebrahimi, R.: An overview on severe plastic deformation: research status, techniques classification, microstructure evolution, and applications. Int. J. Adv. Manuf. Technol. **100**(5–8), 1647–1694 (2018). https://doi.org/10.1007/s00170-018-2652-z
6. Hu, J., Kulagin, R., Ivanisenko, Y., et al.: Finite element modeling of conform-HPTE process for a continuous severe plastic deformation path. J. Manuf. Process. **55**, 373–380 (2020)
7. Gong, S., Ravi Shankar, M.: Effect of microstructural anisotropy on severe plastic deformation during material removal at micrometer length-scales. Mater. Des. **194**, 108874 (2020)
8. Eliseev, A.A., Kalashnikova, T.A., Gurianov, D.A., Rubtsov, V.E., Ivanov, A.N., Kolubaev, E.A.: Ultrasonic assisted second phase transformations under severe plastic deformation in friction stir welding of AA2024. Mater. Today Commun. **21**, 100660 (2019)
9. Liao, Z., Polyakov, M., Diaz, O.G., et al.: Grain refinement mechanism of nickel-based superalloy by severe plastic deformation-mechanical machining case. Acta Mater. **180**, 2–14 (2019)
10. Lü, J., Yang, W., Wu, S., Zhao, X., Xiao, R.: Microstructure and mechanical properties of galvanized steel/AA6061 joints by laser fusion brazing welding. Acta Metallurgica Sinica (Engl. Lett.) **27**(4), 670–676 (2014)
11. Chen, Z.N., Kang, H.J., Fan, G.H., et al.: Grain refinement of hypoeutectic Al–Si alloy with B. Acta Mater. **120**, 168–178 (2016)
12. El-Danaf, E., Kawasaki, M., El-Rayes, M., Baig, M., Mohammed, J.A., Langdon, T.G.: Mechanical properties and microstructure evolution in an aluminum 6082 alloy processed by high-pressure torsion. J. Mater. Sci. **49**(19), 6597–6607 (2014). https://doi.org/10.1007/s10853-014-8266-4
13. Zrnik, J., Kraus, L., Scheriau, S., Pippan, R.: Structural evolution in aluminium alloy AA6082 during HPT deformation at increased temperature. In: Weiland, H., Rollett, A.D., Cassada, W.A. (eds.) ICAA13 Pittsburgh, pp. 1613–1620. Springer, Cham (2012). https://doi.org/10.1007/978-3-319-48761-8_242
14. Markushev, M.V., Murashkin, M.Y.: Evaluation of the Tensile Properties of Severely Deformed Commercial Aluminium Alloys. In: Lowe, T.C., Valiev, R.Z. (eds.) Investigations and Applications of Severe Plastic Deformation. NATO Science Series (Series 3. High Technology), vol. 80, pp. 319–325. Springer, Dordrecht (2000). https://doi.org/10.1007/978-94-011-4062-1_41
15. Brodova, I.G., Shirinkina, I.G., Petrova, A.N., Antonova, O.V., Pilyugin, V.P.: Evolution of the structure of V95 aluminum alloy upon high-pressure torsion. Physics Met. Metall. **111**(6), 630–638 (2011). https://doi.org/10.1134/S0031918X11050036
16. Pramono, A.: Investigation of Severe Plastic Deformation Processes for Aluminum Based Composites. Tallin University of Technology (2016)
17. Ebrahimi, M.: Fatigue behaviors of materials processed by planar twist extrusion. Metall. Mater. Trans. A **48**(12), 6126–6134 (2017). https://doi.org/10.1007/s11661-017-4375-4

18. Farshidi, M.H., Kazeminezhad, M.: Deformation behavior of 6061 aluminum alloy through tube channel pressing: severe plastic deformation. J. Mater. Eng. Perform. **21**(10), 2099–2105 (2012). https://doi.org/10.1007/s11665-012-0155-x

19. Joshi, A., Kumar, N., Yogesha, K.K., Jayaganthan, R., Nath, S.K.: Mechanical properties and microstructural evolution in Al 2014 alloy processed through multidirectional cryoforging. J. Mater. Eng. Perform. **25**(7), 3031–3045 (2016). https://doi.org/10.1007/s11665-016-2126-0

20. Kuts, O.A., et al.: Structure of Ni3Al single crystal after severe plastic deformation. Optoelectron. Instrum. Data Process. **55**(2), 133–137 (2019). https://doi.org/10.3103/S87566990 19020043

21. Lipinska, M., Chrominski, W., Olejnik, L., et al.: Ultrafine-grained plates of Al-Mg-Si alloy obtained by incremental equal channel angular pressing: microstructure and mechanical properties. Mater. Sci. Eng. A **48**, 4871–4882 (2017)

22. Mavlyutov, A.M., Kasatkin, I.A., Murashkin, M.Y., et al.: Influence of the microstructure on the physicomechanical properties of the aluminum alloy Al–Mg–Si nanostructured under severe plastic deformation. Phys. Solid State **57**(10), 2051–2058 (2015)

23. Mazilkin, A.A., Straumal, B.B., Protasova, S.G., et al.: Structural changes in aluminum alloys upon severe plastic deformation. Phys. Solid State **49**(5), 868–873 (2007)

24. Mohammadi, S., Irani, M., Karimi Taheri, A.: The effects of combination of severe plastic deformation and shot peening surface treatment on fatigue behavior of 6082 aluminum alloy. Russ. J. Non-Ferrous Met. **56**(2), 206–211 (2015). https://doi.org/10.3103/S10678212150 20133

25. Murashkin, M.Y., Kil'mametov, A.R., Valiev, R.Z.: Structure and mechanical properties of an aluminum alloy 1570 subjected to severe plastic deformation by high-pressure torsion. Phys. Met. Metall. **106**(1), 90–96 (2008). https://doi.org/10.1134/S0031918X08070120

26. Murashkin, M.Y., Bobruk, E.V., Kilmametov, A.R., et al.: Structure and mechanical properties of aluminum alloy 6061 subjected to equal-channel angular pressing in parallel channels. Phys. Met. Metall. **108**(4), 415–423 (2009). https://doi.org/10.1134/S0031918X09100123

27. Murashkin, M.Y., Sabirov, I., Kazykhanov, V.U., et al.: Enhanced mechanical properties and electrical conductivity in ultrafine-grained Al alloy processed via ECAP-PC. Sci. Mater. Eng. A **48**, 4501–4509 (2013)

28. Pavlova, T.S., Golovi, I.S., Gunderov, D.V., et al.: Effect of Severe Plastic Deformation on the Structure and Low-Temperature Internal Friction of Fe3Al and (Fe, Cr)3Al. Phys. Met. Metall. **105**(1), 36–44 (2008)

29. Rhee, K., Lapovok, R., Thoms, P.F.: The influence of severe plastic deformation on the mechanical properties of AA6111. JOM **57**(5), 62–66 (2005). https://doi.org/10.1007/s11 837-005-0099-9

30. Sabirov, I., Perez-Prado, M.T., Murashkin, M., et al.: Application of equal channel angular pressing with parallel channels for grain refinement in aluminium alloys and its effect on deformation behavior. Int. J. Mater. Form. **3**(Suppl. 1), 411–414 (2010). https://doi.org/10. 1007/s12289-010-0794-0

Microstructural and Mechanical Properties of Friction Stir Welding of AZ91D

O. Kayode[1]([✉]) and E. T. Akinlabi[2]

[1] Department of Mechanical Engineering Science, University of Johannesburg, Johannesburg, South Africa
[2] Pan African University for Life and Earth Sciences Institute, Ibadan, Nigeria

Abstract. In this study, a relatively new manufacturing process - friction stir welding (FSW) was used to join 3 mm thick AZ91D plates. The process was conducted at a traverse speed of 50 mm/min and rotational speed of 1000 rpm; and the weld sample's microstructural and mechanical properties were evaluated. The microstructural analysis was conducted with an optical microscope. The mechanical properties evaluation includes tensile and microhardness tests. The microstructural analysis shows a defect-free sound joint of the materials with absence of voids and wormholes. The tensile strength of the weld was enhanced significantly and a joint efficiency of 109% was achieved. However, there was no substantial effect on the ductility and hardness properties of the weld sample. In conclusion, the FSW process is an efficient and sustainable technique to join and enhance the tensile properties of similar AZ91D alloys.

1 Introduction

Friction Stir welding (FSW) is a welding technique invented at The Welding Institute (TWI), United Kingdom [1]. According to Mishra and Ma [2], the joining process is considered as a green manufacturing process due to its versatility, environment friendliness – lack of flux or fume gas, and energy efficiency. Akinlabi and Akinlabi [3] highlighted the energy, environmental and metallurgical benefits of the welding process compared to the conventional joining techniques. Based on these characteristics, FSW can be regarded as sustainable technique. Although the welding process has been arguably described as the most prominent development in metal joining over the last ten years [2], the basic idea and operational procedure is remarkably simple. FSW entails the passage of a non-consumable tool through the mating ends of two plates of metals or alloys to be joined. The details of this technique has been discussed extensively elsewhere [4–7]. Furthermore, while the process was initially invented for welding of 2xxx and 7xxx series aluminium alloys [8], it has emerged as a capable technique of joining copper, magnesium and their alloys, and other ferrous and nonferrous alloys due to its outstanding success over times. Due to the lightweight property of magnesium, coupled with their good ductility, sound damping capacities, good hot formability and cast-ability, excellent specific strength etc., various manufacturers are enticed to replacing aluminium and steel with magnesium alloys, so far the structural requirements of the previously used materials are met [9–12].

© The Author(s), under exclusive license to Springer Nature Singapore Pte Ltd. 2021
M. Awang and S. S. Emamian (Eds.): *Advances in Material Science and Engineering*, LNME, pp. 64–71, 2021.
https://doi.org/10.1007/978-981-16-3641-7_10

AZ91D alloy is the most commonly used magnesium die casting alloy, with vast applications in the transportation industry. Therefore it's not surprising that joining this alloy through FSW is beginning to receive a lot of attention, since the joining of magnesium alloys part is still much restricted [13]. However, there are challenges faced in the joining of AZ91D with FSW – the appropriate tool material, geometry and processing parameters still remains ambiguous. Kumar and Singh [14] used cylindrical threaded, grooved threaded and tapered threaded pin tools made from high speed steel (HSS) material and reported the highest tensile strength and hardness at a rotational speed of 500 rpm and traverse speed of 50 mm/min, with a cylindrical threaded pin tool. Patel et al. [15] considered tapered cylindrical, threaded straight cylindrical and straight cylindrical pin tools made from HCHCrD2 grade material and reported that there were defects in all the weld samples regardless of the pin profile. The authors experienced a tool breakage when a tapered straight cylindrical pin at 1000 rpm and 40 mm/min was used, and concluded that the tool material may not be suitable for the FSW of AZ91D plates with a thickness of 6mm. On the other hand, Ramkumar and Hadi [16] used an HSS cylindrical tool for the same thickness of AZ91D and recorded optimal mechanical properties with tensile strength up to 85% of that of the parent material at traverse and rotational speeds of 48 mm/min and 720 rpm, respectively. Kouadri-Henni and Barrallier [17] reported there were voids and cracks resulting from different combinations of rotational and traverse speeds during the FSW of AZ91D. The authors observed that low traverse speed causes dynamic recrystallization leading to fine grain structures in the stir zone (SZ) while high traverse speed causes cracks and porosity in the SZ and results to a decrease in the tensile strength of the weld. Lee et al. [18] recorded defects in the welds manufactured at a rotational speed of 1800 rpm regardless of which traverse speed was employed between the range of traverse speeds used. A mutual report from these studies is that the improvements in the mechanical properties of the welds were caused by recrystallization and refinement in grain structures. However, there is no trend and correlation between the various tools and parameters employed in these studies. The aim of this study is to explore other welding parameters and tool geometry combination towards enhancing the mechanical properties and joint integrity of AZ91D welds. The FSW of this alloy was conducted with the optimum process parameters combination obtained from an unpublished array of experiments.

2 Experimental Methods

3 mm thick AZ91D magnesium alloys with dimension 300 x 120 mm were used in this study. The nominal composition and mechanical properties of the parent material are presented in Table 1 and Table 2, respectively. The welding experiments were conducted on a 2 Ton linear numerical controlled FSW machine at the Friction Stir Welding Lab of the Department of Mechanical Engineering, Indian Institute of Technology (IIT), Kharagpur, India with a H13 tool steel tool, with square pin profile, at traverse speed of 50 mm/min and rotational speed of 1000 rpm. The pin length and diameter are 2.4 mm and 5 mm, respectively. The shoulder diameter is 15 mm.

Table 1. Chemical composition of AZ91D

Element	Si	Al	Mn	Fe	Na	K	Zn	Ca	Mg
Wt. %	0.54	9.91	0.24	0.05	0.11	0.34	0.89	0.12	87.04

Table 2. Mechanical properties of AZ91D parent material

Mechanical properties	AZ91D
Elongation (%)	7
Ultimate tensile strength (MPa)	117
Hardness (HV)	72

A cross-section of the nugget zone of size 30 x 10 x 3 mm was sectioned perpendicular to the tool traverse direction with a wire-cut electric discharge machining (EDM); and prepared according to standard metallographic procedures. The cut section was immersed and etched in a picral solution of 10 mm acetic acid, 10 ml distilled water and 4.2 g picral acid in 100 ml ethanol for 6 s. The sample was rinsed thoroughly with distilled water, cleansed with acetone to reduce the effects of water on the weld, and observed with an Olympus BX51M microscope for microstructural analysis. For the tensile testing, the samples were prepared according to ASTM E8/E8M-13a guidelines [19] and tested at a constant crosshead displacement of 1 mm/min, using a 250 kN capacity Zwick Roell static tensile testing machine. Three tensile specimens were prepared and tested for each welding condition and the average of the tensile values of the three specimens was calculated and computed as the representative UTS of the weld samples. Samples of the same size as those used for microstructural evaluation were also cut across the stir zone for microhardness measurements. The measurements were conducted according to ASTM 384–16 guidelines [20], using a TH713 Vickers Microhardness Tester manufactured by Beijing Cap High Technology Co. Ltd. The indentations were made with a load of 200 g and dwell time of 15 s at a constant interval of 1 mm in a row throughout the cross-section of the weld sample to cover all the weld microstructural zones.

3   Results and Discussion

3.1   Microstructural Evaluation

The top surface of the weld is shown in Fig. 1. It can be clearly observed that a defect-free sound weld was achieved with the process parameters employed. Figure 2 shows the macrograph where a refinement in the grain structures in the SZ, similar to reports in the reviewed literature discussed earlier in this paper can be observed in the weld map.

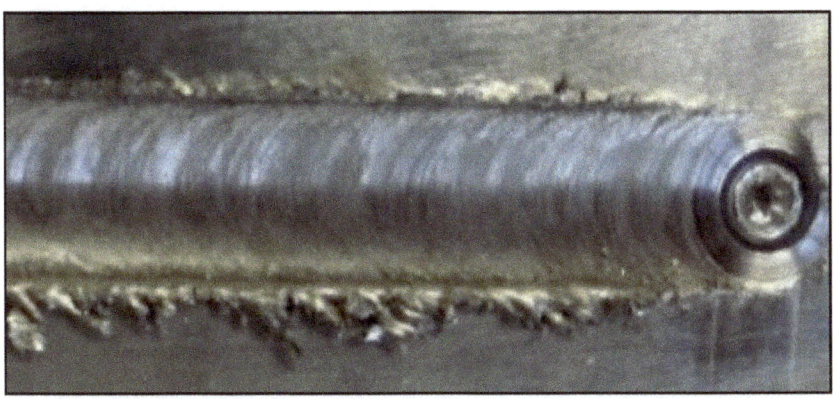

Fig. 1. Top surface of weld of AZ91D conducted at 50 mm/min and 1000 rpm

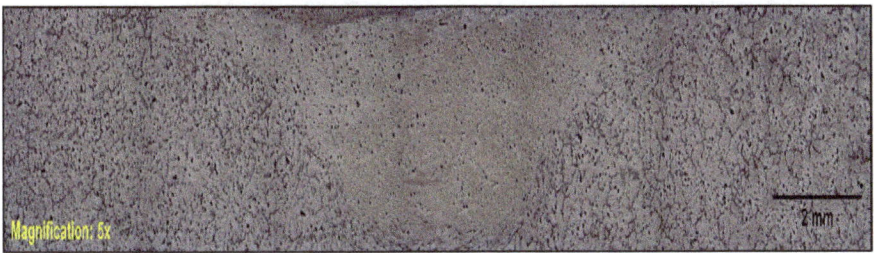

Fig. 2. Weld map of FSW of AZ91D

The cross-sectional macrograph makes it possible to visually distinguish the SZ, thermomechanically affected zone (TMAZ) and heat affected zone (HAZ), by showing a lighter contrast caused by the frictional heating and intense plastic deformation experienced in the SZ. The transition zone can be seen in a magnified view in Fig. 3. It is evident that the weld sample experienced a severe dynamic recrystallization due to the higher-angle grain boundary diffusion rate and dislocations absorbed by the grain boundaries; and the constraints imposed by the absence of easily activated slip systems in their HCP crystal structure [21].

3.2 Tensile Properties

Typical tensile behavior of the weld sample and parent material are presented in Fig. 4. As expected, both samples exhibit brittle properties by showing a slight measure of plasticity before fracture. However, it could be clearly observed that the weld sample possess superior tensile strength compared to the parent material.

The mean UTS of the weld sample is 127 MPa while that of the parent material is 117 MPa, implying an impressive joint efficiency of approximately 109%. The significant enhancement in the tensile properties of the weld sample could be attributed to grain refinement as clearly observed in the weld sample's microstructure. Yamashita et al. [22] also gave a similar report that refinement in AZ91D grain structures led to an increase

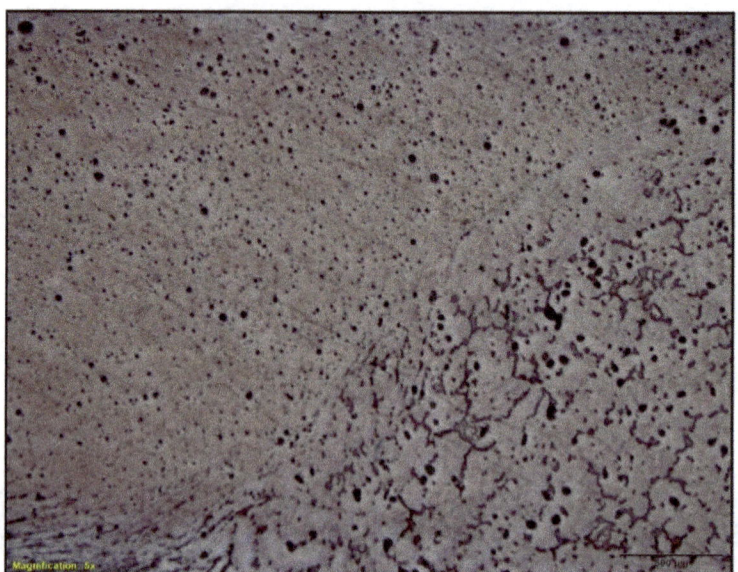

Fig. 3. TMAZ and HAZ boundaries showing a clear transition zone

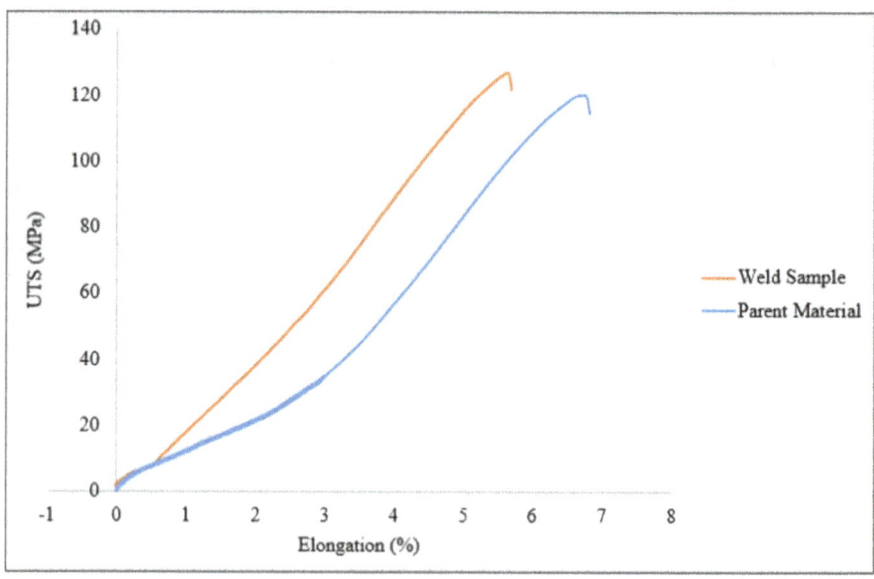

Fig. 4. Tensile behavior of parent material and weld sample

in its ductility and strength during severe plastic deformation. However, in this case, the welding process did not have a significant influence on the ductility of the weld sample. The mean elongation of the weld sample still remains 7% with the total elongation of each Mg alloy weld sample between an approximate range ± 2% of that of the parent material. The insignificant change in ductility may be attributed to magnesium HCP structure and low stacking fault energy (SFE) level. There is limited number of favorable oriented slip system in HCP structure due to the low symmetry of the HCP lattice. Therefore, this causes a difficulty in accommodating strain in the crystal structure. On the other hand, the low SFE level in Mg alloys increases the activation energies and resist the slip system. This do not favor plastic deformation occurring at elevated temperature unlike twinning deformation that occurs at or below the room temperature [23]. This result further substantiates the report by Wang et al. [24] that although there may be gain refinement in AZ91D through severe plastic deformation, it's still a major challenge to achieve a simultaneous improvement in tensile strength and ductility.

3.3 Microhardness Profiles

Figure 5 shows the microhardness profiles taken across the cross-section of both parent material and weld sample, perpendicular to the tool travel direction. The hardness values are plotted from the center of the SZ as reference to other points taken on both sides of the weld, with the advancing side (AS) and retreating side (RS) on the left-hand side and right-hand side, respectively. The weld sample exhibit a slightly higher mean hardness values than the parent material. The weld sample have a mean hardness value of 73 HV while that of the parent material is 72 HV. The range of hardness values for the samples in the order above is between 59 – 85 HV and 64 – 85 HV, respectively.

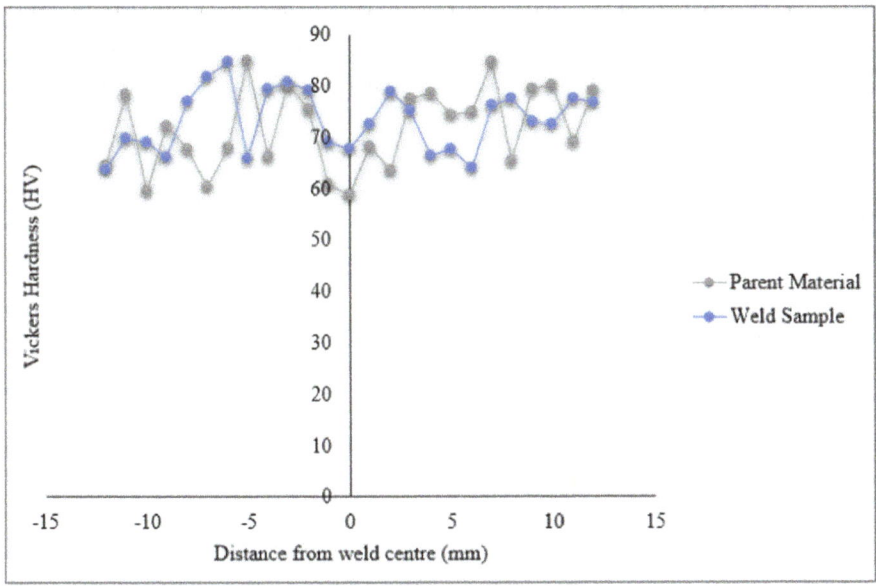

Fig. 5. Microhardness distribution of parent material and weld sample

The weld sample displayed a more relatively uniform distribution than the parent material. Similar to the tensile behavior of the weld sample, the enhanced hardness values could also be credited to the recrystallized and fine grain structure in the SZ because of its significant contributions to materials strengthening. It could be observed that the hardness values in area corresponding to the TMAZ and HAZ zones are higher than the average hardness value of the parent material due to a combined effect of grain refinement, β intermetallic and solid solution strengthening since the parent material is comprised of softer α-Mg matrix and harder eutectic β intermetallic phase which when reduced, causes more dissolution in aluminium. However, it is worthy to note that the relationship between the tensile behavior and hardness properties is still ambiguous because tensile strength is an intrinsic property of a material while hardness property depends more on the attributes of the material such as microstructure, strain hardening etc. and it's strongly influenced by the elastic recovery of the material under test.

4 Conclusions

Joining of 3 mm thick AZ91D plates through friction stir welding was conducted successfully using a H13 tool steel. A defect-free sound weld with absence of void and wormholes was obtained at a traverse speed of 50 mm/min and rotational speed of 1000 rpm. Based on the characterizations, there was a refinement in the grain structures which contributes to the mechanical properties of the weld joint. The weld joint has a superior UTS compared to the parent material with a joint efficiency of about 109%. However, there was no significant effect on the ductility of the joint. Furthermore, there was no substantial increase in the mean hardness value of the weld sample compared to the base material.

Acknowledgements. The corresponding author acknowledges the funding support of the Global Excellence Stature (GES) scholarship award by the University of Johannesburg, South Africa, and payment of article processing charge by the Pan African University for Life and Earth Sciences Institute Ibadan, Nigeria.

References

1. Thomas, W.M., Nicholas, E.D., Needham, J.C., Murch, M.G., Templesmith, P., Dawes C.B.: G.B. Patent Application No. 9125978.8 (1991)
2. Mishra, R.S., Ma, Z.Y.: Friction stir welding and processing. Mater. Sci. Eng. R. Rep. **50**(1–2), 1–78 (2005)
3. Akinlabi, E.T., Akinlabi, S.A.: Friction stir welding process: a green technology. Int. J. Mech. Aerosp. Indus. Mechatron. Eng. **6**(11), 2514–2516 (2012)
4. Mishra, R.S., De, P.S., Kumar, N.: Friction Stir Processing. In: Friction Stir Welding and Processing, pp. 259--296 Springer, Cham (2014) https://doi.org/10.1007/978-3-319-070 43-8_9
5. Lohwasser, D., Chen, Z.: Friction Stir Welding from Basics to Applications. Woodhead Publishing Limited, Cambridge (2010)
6. Mohammad, K., Besharati, G., Parviz, A. (2014) Advances in friction stir welding and processing. Woodhead Publishing Series in Welding and Other Joining Technologies.

7. Mishra, R.S., Mahoney, M.W.: Friction stir welding and processing. ASM International (2007)
8. Dawes, C., Thomas, W.: Friction stir joining of aluminium alloys. TWI Bull. **6**, 124–127 (1995)
9. Mult, E.H., Haferkamp, H., Niemeyer, M., Dilthey, U., Trager, G.: Laser and electron beam welding of magnesium materials. Weld. Cut. **52**(8), 178–180 (2000)
10. Pastor, M., Zhao, H., DebRoy, T.: Continuous wave Nd: yttrium–aluminium–garnet laser welding of AM60B magnesium alloys. J. Laser Appl. **12**(3), 91–100 (2000)
11. Munitz, A., Cotler, C., Stern, A., Kohn, G.: Mechanical properties and microstructure of gas tungsten arc welded magnesium AZ91D plates. Mater. Sci. Eng. A **302**(68–73), 6 (2001)
12. Jain, C.C., Koo, C.H.: Creep and corrosion properties of the extruded magnesium alloy containing rare earth. Mater. Trans. **2**, 265–272 (2007)
13. Asadi, P., Kazemi-Choobi, K., Elhami, A.: Welding of magnesium alloys. In: New Features on Magnesium Alloys, pp. 121–158 (2012)
14. Kumar, N., Singh, M.: Effect of different tool pin profile on the mechanical properties of magnesium based alloy AZ91 by friction stir welding. Int. J. Eng. Sci. Res. Technol. **4**(10), 25–38 (2015)
15. Patel, N., Bhatt, K.D., Mehta, V.: Influence of tool pin profile and welding parameter on tensile strength of magnesium alloy AZ91 during FSW. Procedia Technol. **23**, 558–565 (2016)
16. Ramkumar, A., Hadi, Y.: A study of microstructure and mechanical behavior of friction stir welded (FSW) joints of AZ91D magnesium alloy. Am. J. Sci. Res. **108**, 37–42 (2016)
17. Kouadri-Henni, A., Barrallier, L.: Mechanical properties, microstructure and crystallographic texture of magnesium AZ91-D alloy welded by friction stir welding (FSW). Metall. Mater. Trans. A. **45**(11), 4983–4996 (2014)
18. Lee, W.B., Kim, J.W., Yeon, Y.M., Jung, S.B.: The joint characteristics of friction stir welded AZ91D magnesium alloy. Mater. Trans. **44**(5), 917–923 (2003)
19. Standard test methods for tension testing of metallic materials, E8M-13. Copyright ©ASTM International, USA (2013)
20. Standard test method for microindentation hardness of materials, E384-16. Copyright ©ASTM International, USA (2016)
21. Tan, J.C., Tan, M.J.: Dynamic continuous recrystallization characteristics in two stage deformation of Mg–3Al–1Zn alloy sheet. Mater. Sci. Eng. A **339**(1–2), 124–132 (2003)
22. Yamashita, A., Horita, Z., Langdon, T.G.: Improving the mechanical properties of magnesium and a magnesium alloy through severe plastic deformation. Mater. Sci. Eng. A **300**(1–2), 142–147 (2001)
23. Alaneme, K.K., Okotete, E.A.: Enhancing plastic deformability of Mg and its alloys – a review of traditional and nascent developments. J. Magnes. Alloy. **5**(4), 460–475 (2017)
24. Wang, H.Y., et al.: Achieving high strength and high ductility in magnesium alloy using hard-plate rolling (HPR) process. Sci. Rep. **5**, 17100 (2015)

Microstructural and Mechanical Properties of Friction Stir Welding of AA1050

O. Kayode[1(✉)] and E. T. Akinlabi[2]

[1] Department of Mechanical Engineering Science,
University of Johannesburg, Johannesburg, South Africa
[2] Pan African University for Life and Earth Sciences Institute, Ibadan, Nigeria

Abstract. In this study, a sustainable manufacturing process – friction stir welding (FSW) was used to join 3 mm thick AA1050 plates in butt joint configuration. The process was conducted at a traverse speed of 50 mm/min and rotational speed of 800 rpm. The weld sample's microstructural and mechanical properties were evaluated. The microstructural analysis was conducted with an optical microscope. The mechanical properties investigation includes tensile and microhardness tests. The microstructural analysis show a defect-free sound joint of the materials with absence of kissing bond. The mechanical evaluations revealed that the hardness property of the weld sample was not significantly affected by the welding process, but the tensile properties were slightly reduced. However, an acceptable joint efficiency of 94% was obtained, showing that FSW is a suitable sustainable welding process for joining of similar AA1050 alloys.

1 Introduction

The industrial sector could account for about a quarter of the total energy consumption in a continent [1]. Due to this, sustainable manufacturing systems and techniques that are environmentally, socially and economically friendly are in demand globally to meet strict environmental and governments regulations imposed towards energy consumption and environmental impact of manufacturing processes. Azeez et al. [2] reported that the modern manufacturing processes strongly depends on welding processes to join ferrous and non-ferrous alloys. Kumar et al. [3] reported that the various advantages of welding compared to other joining processes include high joint efficiency, higher structural stiffness, air and water tightness, flexibility in design, time and weight savings, cheaper and faster integration etc. Friction stir welding (FSW) is an energy efficient solid-state welding technique invented in 1991 at The Welding Institute (TWI), United Kingdom (UK). The method was originally invented to join aluminium alloy, but so far it has displayed its capabilities to join similar and dissimilar alloys as evidenced in literatures [3–6]. Aside from the low energy consumption, FSW also has metallurgical and environmental benefits and has contributed to significant weight reduction and cost savings in components, making the welding process acceptable globally as a green manufacturing technology. The details of the technique has been discussed extensively elsewhere [7–10]. Furthermore, Akinlabi & Akinlabi [11] has highlighted the metallurgical, environmental and energy benefits of FSW.

© The Author(s), under exclusive license to Springer Nature Singapore Pte Ltd. 2021
M. Awang and S. S. Emamian (Eds.): *Advances in Material Science and Engineering*, LNME, pp. 72–78, 2021.
https://doi.org/10.1007/978-981-16-3641-7_11

Aluminium 1XXX series alloys possess the best corrosion resistance among all aluminium alloys series group available. As a result of this, AA1050 has excellent corrosion resistance properties. In addition, the alloy has high ductility and good weldability. Although few researchers have successfully joined similar alloys of AA1050 using FSW, there is still not yet an identifiable trend in the process parameters combination employed and there are still some major challenges reported in literatures. Uger [12] was able to join AA1050 plates though FSW with a traverse speed of 20 mm/min and rotational speed of 1500 rpm, but recorded that there were lots of flashes from the largest shoulder diameter used due to excess heat input. Mironov et al. [13] used a constant high traverse speed of 600 mm/min and a wide range of rotational speeds (200 – 2000 rpm) and reported that the welds manufactured at higher rotational speeds in range 1000 – 2000 rpm, where there is high welding temperature due to high heat input, exhibits coarse grains and low joint efficiency resulting from significant softening experienced in the stir zone. Sato et al. [14] varied the traverse speeds between 87 – 720 mm/min and the rotational speeds between 600 – 2400 rpm when joining similar AA1050 alloys using FSW. The authors reported that there was occurrence of a kissing bond with either a combination of low traverse speed and high rotational speed, or vice versa as a result of the presence of oxide layer on the butt surfaces which was not properly broken up because of low heat input. The kissing bond in turn caused the fracture of the weld when subjected to a bending test. Liu et al. [15] also gave a similar report on the observation of a kissing bond during FSW of AA1050 plates. Additionally, the authors recorded defects in the weld as a result of insufficient frictional heat generation. The disparities in the welding process parameters and the fact that a clear description of sufficient heat during the FSW of AA1050 in correlation with the process parameters is still ambiguous has prompted further studies on the FSW of AA1050. In this study, FSW of AA1050 was conducted with the aim of improving the weld joint mechanical properties and structural integrity.

2   Experimental Methods

3 mm thick AA1050 aluminium alloys with dimension 300 x 120 mm were used in this study. Table 1 and Table 2 shows the nominal composition and mechanical properties of the parent material, respectively. The welding experiments were conducted on a 2 Ton linear numerical controlled FSW machine at the Friction Stir Welding Lab of the Department of Mechanical Engineering, Indian Institute of Technology (IIT), Kharagpur, India with a H13 tool steel tool, with square pin profile, at traverse speed of 50 mm/min and rotational speed of 1000 rpm. The pin length and diameter are 2.4 mm and 5 mm, respectively. The shoulder diameter is 15 mm.

Table 1. Chemical composition of AA1050

Element	Si	Cu	Fe	Na	Zn	Ga	Al
Wt. %	0.15	0.03	0.27	0.03	0.03	0.02	99.48

Table 2. Mechanical properties of AA1050 parent material

Mechanical properties	AA1050
Elongation (%)	13
Ultimate tensile strength (MPa)	126
Hardness (HV)	50

A cross-section of the stir zone of size 30 x 10 x 3 mm was sectioned perpendicular to the tool traverse direction with a wire-cut electric discharge machining (EDM) and prepared according to standard metallographic procedures. The cut section was immersed in a solution of 20 g NaOH in 100 ml distilled water for 40 s and then dipped in a Weck's reagent solution comprising of 4 g $KMnO_4$ and 2 g NaOH in 100 ml distilled water for 10 s. The sample was rinsed thoroughly with distilled water, cleansed with acetone and observed with an Olympus BX51M microscope for microstructural analysis. For the tensile testing, the specimens were prepared according to ASTM E8/E8M-13a guidelines [16] and tested at a constant crosshead displacement of 1 mm/min, using a 250 kN capacity Zwick Roell static tensile testing machine. Three tensile specimens were prepared and tested for each welding condition and the average of the tensile values of the three specimens was calculated and computed as the representative UTS of the weld samples. Samples of the same size as those used for microstructural evaluation were also cut across the stir zone for microhardness measurements. The measurements were conducted according to ASTM 384–16 guidelines [17], using a TH713 Vickers Microhardness Tester manufactured by Beijing Cap High Technology Co. Ltd. The indentations were made with a load of 200 g and dwell time of 15 s at a constant interval of 1 mm in a row throughout the cross-section of the weld specimen to cover all the weld microstructural zones.

3   Results and Discussion

3.1   Microstructural Evaluation

A defect-free sound joint was obtained. Figure 1 shows the top surface of the weld, while the macrograph showing a cross-section of the weld sample across all weld zones (heat affected zone (HAZ), thermomechanically affected zone (TMAZ) and stir zone (SZ)) is presented in Fig. 2. It could be clearly observed that the weld map showed a homogenously distributed fine grains exhibiting the occurrence of a thorough dynamic recrystallization.

The homogeneity in the workpiece material aided the effective material flow and mixing during plastic deformation. It is evident that the heat input and plastic strain resulting from the combination of the process parameters are sufficient. Furthermore, there was absence of the kissing bond as previously reported in FSW of AA1050 [14, 15], and there was no observable contrast between the various weld zones due to dynamically recrystallized grains as shown in Fig. 3.

Fig. 1. Top surface of weld of AA1050 conducted at 50 mm/min and 800 rpm

Fig. 2. Weld map of FSW of AA1050

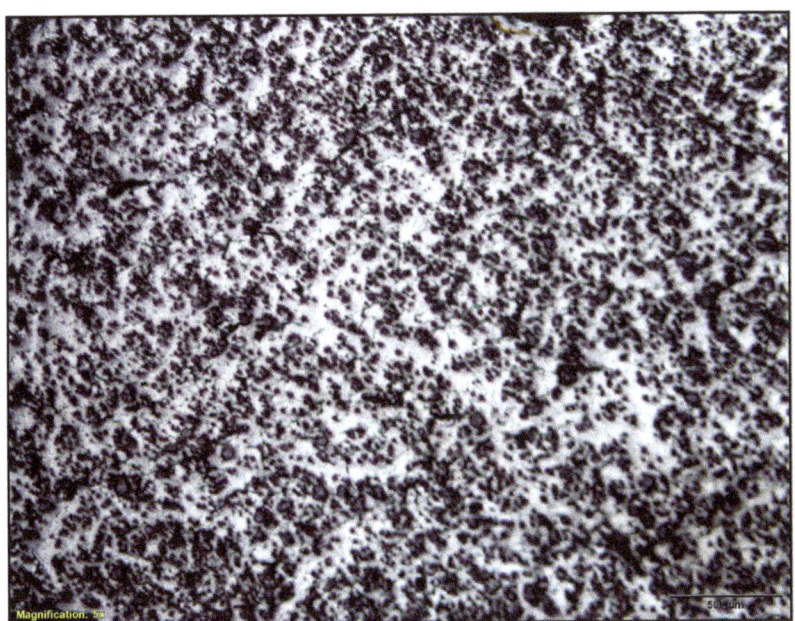

Fig. 3. SZ, TMAZ and HAZ boundaries showing dynamically recrystallized grains

3.2 Tensile Properties

Typical tensile behavior of the weld sample and parent material are presented in Fig. 4. Both samples exhibit ductile properties as expected. Although the weld sample have a reduced ultimate tensile strength (UTS) in comparison with that of the parent material, the values are very close.

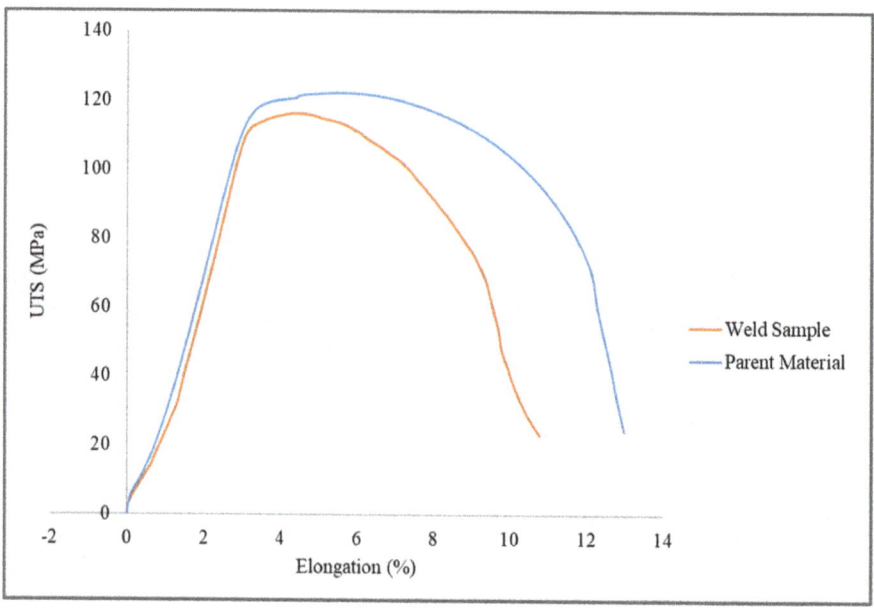

Fig. 4. Tensile behavior of parent material and weld sample

The parent material has an average UTS of approximately 126 MPa as reported in Table 1 above. The weld sample has an average UTS of 119 MPa, showing a joint efficiency of approximately 94%. Similarly, the parent material has superior average elongation at break of 13% while the weld sample has an elongation of 10%. This implies that the parent material exhibit more ductility than the weld samples. However, both samples showcased a similar ductile mode of fracture with presence of deep and fine dimple structure.

3.3 Microhardness Profiles

The Vickers microhardness profile taken across the cross-section of both parent material and weld sample perpendicular to the tool travel direction are presented in Fig. 5. The hardness values are plotted from the center of the SZ as reference to other points taken on both sides of the weld, with the advancing side (AS) and retreating side (RS) on the left-hand side and right-hand side, respectively.

The range of hardness values of the parent material is 43 – 57 HV with a mean hardness value of 50 HV. The weld sample have a slightly higher mean hardness value

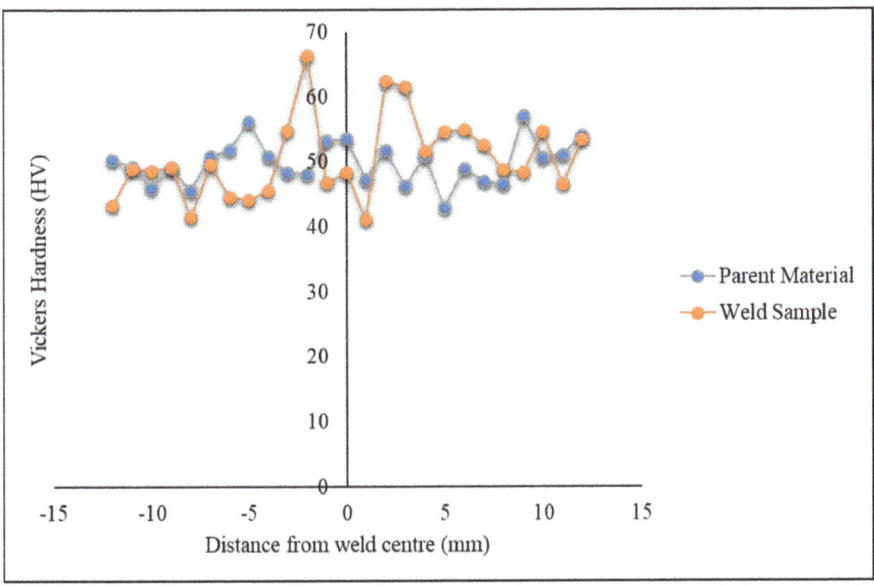

Fig. 5. Microhardness distribution of parent material and weld sample

of 51 HV but a significantly higher peak hardness value of 66 HV. However, a lower hardness value could be clearly observed in the center of the SZ of the weld sample compared to the mean hardness value of the parent material. The peak hardness value of the weld sample could be attributed to the finer grains, depicting the Hall-Petch relation [18]. It could be observed that the weld sample has a minimum hardness value in a location close to the HAZ on the AS and there are comparatively higher hardness values on the RS than those on the AS. This indicates that the RS may have experienced a higher degree of friction and plastic deformation than the AS, causing higher but adequate heat input and finer grains on the RS. As a result of these effects, there is an irregular microstructural evolution in the weld zone, which caused the obvious fluctuations in the hardness profile of the weld sample compared to the relatively stable distribution in the hardness profile of the parent material.

4 Conclusions

Joining of 3 mm thick AA1050 plates through friction stir welding was conducted successfully, using a H13 tool steel. Based on the characterizations and findings, the following conclusions can be made:

1. Friction stir welding is a sustainable manufacturing process suitable for joining AA1050 plates.
2. A defect-free sound weld with absence of void and kissing bond was obtained at rotational speed of 800 rpm and traverse speed of 50 mm/min.
3. The weld sample of similar AA1050 alloys has a good joint efficiency of approximately 94%.

4. There was no significant change in the mean hardness value of the weld sample compared to that of the parent material.
5. There is a higher degree of plastic deformation and heat input on the RS than the AS of the weld.

Acknowledgements. The corresponding author acknowledges the funding support of the Global Excellence Stature (GES) scholarship award by the University of Johannesburg, South Africa, and payment of article processing charge by the Pan African University for Life and Earth Sciences Institute Ibadan, Nigeria.

References

1. Bevilacqua, M., Ciarapica, F.E., D'Orazio, A., Forcellese, A., Simoncini, M.: Sustainability analysis of friction stir welding of AA5754 sheets. Procedia CIRP **62**, 529–534 (2017)
2. Azeez, S., Mashinini, M., Akinlabi, E.: Sustainability of friction stir welded AA6082 plates through post-weld solution heat treatment. Procedia Manuf. **33**, 27–34 (2019)
3. Kumar, N., Mishra, R.S., Yuan, W. (eds.): Friction Stir Welding of Dissimilar Alloys and Materials. Butterworth-Heinemann, Oxford (2015)
4. Verma, S., Misra, J.P.: A critical review of friction stir welding process. DAAAM Int. Sci. Book **14**, 249–266 (2015)
5. Kumar, H.M.A., Ramana, V.V.: An overview of friction stir welding (FSW): a new perspective. Res. Inventy Int. J. Eng. Sci. **4**(6), 1–4 (2014)
6. Unnikrishnan, M.A., Edwin, R.D.: Friction stir welding of magnesium alloys - a review. Adv. Mater. Sci. Eng. Int. J. (MSEJ) **2**(4), 7–18 (2015)
7. Mishra, R.S., De, P.S., Kumar, N.: Friction stir welding and processing. In: Science and Engineering, Springer, Switzerland (2014) https://doi.org/10.1007/978-3-319-07043-8
8. Lohwasser, D., Chen, Z.: Friction Stir Welding from Basics to Applications. Woodhead Publishing Limited, Cambridge (2010)
9. Mohammad, K., Besharati, G., Parviz, A.: Advances in friction stir welding and processing. Woodhead Publishing Series in Welding and Other Joining Technologies (2014)
10. Mishra, R.S., Mahoney, M.W.: Friction stir welding and processing. ASM International (2007)
11. Akinlabi, E.T., Akinlabi, S.A.: Friction stir welding process: a green technology. Int. J. Mech. Aerosp. Indus. Mechatron. Eng. **6**(11), 2514–2516 (2012)
12. Uygur, I.: Influence of shoulder diameter on mechanical response and microstructure of FSW welded 1050 al-alloy. Arch. Metall. Mater. **57**(1), 53–60 (2012)
13. Mironov, S., Inagaki, K., Sato, Y.S., Kokawa, H.: Effect of welding temperature on microstructure of friction-stir welded aluminum alloy 1050. Metall. Mater. Trans. A. **46**(2), 783–790 (2015)
14. Sato, Y.S., Takauchi, H., Park, S.H.C., Kokawa, H.: Characteristics of the kissing-bond in friction stir welded al alloy 1050. Mater. Sci. Eng. A **405**(1–2), 333–338 (2005)
15. Liu, H., Fujii, H., Maeda, M., Nogi, K.: Heterogeneity of mechanical properties of friction stir welded joints of 1050–H24 aluminum alloy. J. Mater. Sci. Lett. **22**(6), 441–444 (2003)
16. Standard test methods for tension testing of metallic materials, E8M-13. Copyright ©ASTM International, USA (2013)
17. Standard test method for microindentation hardness of materials, E384–16. Copyright ©ASTM International, USA (2016)
18. Park, S.H.C., Sato, Y.S., Kokawa, H.: Microstructural evolution and its effect on Hall-petch relationship in friction stir welding of thixomolded Mg alloy AZ91D. J. Mater. Sci. **38**(21), 4379–4383 (2003)

Total Productive Maintenance in Small and Medium-Sized Enterprises
Literature Review

T. X. Zhang and J. F. Chin[✉]

School of Mechanical Engineering, Universiti Sains Malaysia,
Engineering Campus, 14300 Nibong Tebal, Penang, Malaysia
chinjengfeng@usm.my

Abstract. The purpose of the paper is to review literatures related to total productive maintenance (TPM) in small and medium-sized enterprises (SMEs). The first part of paper presents the research background to underline the importance for SMEs to integrate holistic maintenance strategies in their productions. The second part of the paper briefly covers the history of equipment maintenance, followed by systematic discussion of TPM, including the definition of TPM, the eight pillars, the objective of TPM, and the benefits of TPM. Next, the paper summarizes the influencing factors and analyzes case studies to seek research gaps in TPM. As findings, the paper revealed that the suitability of TPM in SMEs and the need in research to pay attention to the implementation method by fully considering the changing characteristics of SMEs.

1 Introduction

Researchers increasingly hold the view that Small and medium-sized enterprises (SMEs) are important contributors to global economic growth [1], as 60%–65% of enterprises in the world belongs to this category [2]. SMEs are the main force to reduce poverty [3], especially in developing countries that rely heavily on the performance of SMEs to grow their economy [4]. These SMEs provide many employment opportunities and 70% of the gross national product [5]. Also, SMEs have relatively short incubation period and relatively minimal investment [6]. In the ecosystem of supply chain, SMEs enhance large enterprise's competitiveness by supplying large enterprises with numerous quality components [7]. Given the globalization of the economic market, large enterprises would continue to reduce their sizes and outsource many businesses to SMEs [7].

Different countries have different definitions of SME. For examples, based on the SME promotion law 2011 of China, manufacturing industry enterprises with over 300 employees and operating income of above CNY 20 million are considered medium enterprises; those with over 20 employees and operating income of above CNY 3 million are small enterprises; and those with under 20 employees or operating income of below CNY 3 million are micro enterprises [8]. In Malaysia, the Small and Medium Industries Development Corporation (SMIDEC) classifies SMEs on the bases of annual sales turnover and the number of full-time employees [9]. Medium enterprises have an annual sales turnover between RM 10 million and RM 25 million, and the number of

© The Author(s), under exclusive license to Springer Nature Singapore Pte Ltd. 2021
M. Awang and S. S. Emamian (Eds.): *Advances in Material Science and Engineering*, LNME, pp. 79–92, 2021.
https://doi.org/10.1007/978-981-16-3641-7_12

full-time employees is between 50 and 150 [9]. Small enterprises have an annual sales turnover between RM 0.25 million and RM 10 million, and the number of full-time employees is from 5 to 50 [9].

In the context of economic globalization, SMEs are facing new challenges, such as the global economic decline caused by the COVID-19 epidemic, more expensive raw material and dwelling financial resources, low production efficiency and more aggressive competition pressure from competitors. For example, the U.S. Census Small Business Pulse Survey indicated that approximately 50% of enterprises are severely and adversely affected by COVID-19, and only 15–20% of companies have sufficient cash on hand to maintain operations for 3 months [10]. Bartik et al. [11] surveyed 6000 small enterprises, 43% of which were temporarily closed, a large number of layoffs and most companies had less than a month of cash on hand. To meet these challenges, enterprises must maintain equipment reliability even when the demand is low. For SMEs, production equipment accounts for a large proportion of the total investment of the enterprise. The lack of maintenance of equipment will lead to low productivity and negatively affect the finances of enterprises [12]. In addition, lack of maintenance can accelerate the deterioration in equipment, which leads to increased production costs, lower product quality, and longer delivery cycles [13]. Therefore, it is important to introduce holistic maintenance strategies.

Some enterprises often wrongly perceived maintenance functions as an operational expense, rather than an investment to improve equipment reliability [14]. As the use of equipment becomes a necessity in mass production, the methods of equipment maintenance have also evolved [15], through the following stages: breakdown maintenance, preventive maintenance, predictive maintenance, corrective maintenance, reliability-centered maintenance, and total productive maintenance. Before 1950, breakdown maintenance was adopted by manufacturing organizations worldwide. The basic concept of breakdown maintenance is to do nothing unless the equipment stops working [16]. Nevertheless, unpredictable failure reduces the service life of the equipment and causes stoppage, spare part issues, and high repair expenses [17]. Preventive maintenance was introduced in 1951, as a forward-looking maintenance strategy [18]. It emphasizes inspection before equipment breakdowns to prevent unpredicted failures and extend the life cycle of the equipment [19]. It has two maintenance modes: periodic maintenance and maintenance based on equipment status [18]. The maintenance based on equipment status leads to the development of predictive maintenance programs [18]. Through equipment monitoring, the enterprise predicts possible breakdowns and makes predictive maintenance plans in advance, paying attention to early signs of equipment breakdown and corresponding predictive measures [20]. Well-organized predictive maintenance can eliminate all but catastrophic failures, reduce parts inventory, minimize costs, and improve equipment reliability [21]. Corrective maintenance theory was developed in 1957, which introduced the concept of preventing equipment breakdown through equipment improvement, on the reliability, maintainability, and safety of the equipment [16]. Reliability-centered maintenance was put forward in the 1960s, which was initially intended for maintaining airplanes by aircraft manufacturers, governments, and airlines [22]. It is a technique for developing cost-effective maintenance plans and standards to restore or maintain equipment conditions [23]. The aim is to improve the reliability of the equipment for it to function normally during the design life cycle [24]. Total productive maintenance (TPM) is an innovative Japanese maintenance strategy based on preventive maintenance [25], which mitigates organizational failures [26]. This maintenance strategy was first put forward by M/S Nippon Denso Co., Ltd of Japan in 1971 [27]. TPM is a remarkable method to maintain equipment; it is exceptional in improving equipment efficiency, eliminating equipment breakdown, and involving all employees in promoting the autonomous maintenance of the operator through daily activities [28]. An in-depth coverage of TPM is given in the following section.

Today, the attention of maintenance has reached an unprecedented level. For example, Europe's annual maintenance budget is approximately 150 billion euros [29]. Recently, there is a growing number of enterprises realized that the maintenance department plays an important role in improving the efficiency of the equipment and the running time [30]. Muchiri et al. [31] divided maintenance objectives into five perspectives: ensuring the equipment reaches the designed life, plant functionality (availability, reliability and product quality), plant safety and environmental safety, maintenance cost effectiveness and the efficient use of resources (energy and raw materials). Appropriate maintenance strategies can improve productivity and quality, and thus have a positive impact on enterprises' long-term profitability [32].

2   TPM

TPM extends from preventive maintenance and widely used in manufacturing [33]. TPM is a continuous improvement method that strives to integrate production and maintenance activities [34]. TPM consists of three words. "Total" signifies the participation of all staff in the enterprise, from top management to equipment operators [35]. "Productive" signifies the minimization of the issues in the production process and the maintenance of the equipment at a high level of reliability [35]. "Maintenance" signifies that the operator spends the necessary time to keep the equipment in good condition [35]. Nakajima [36] views TPM as a method to maximize equipment efficiency and defined TPM as an innovative maintenance method that optimizes equipment efficiency, eliminates failures, and promotes autonomous maintenance by operators through daily activities.

The Japan Institutes of Plant Maintenance (JIPM) [37] has listed five goals of TPM: 1. Improving equipment effectiveness; 2. Improving maintenance efficiency and effectiveness; 3. Early equipment management and maintenance prevention; 4. Training to improve the skills of all people involved; 5. Involving operators in routine maintenance. Bajaj and Kumar [38] summarized the objectives of TPM as the following five aspects: involving personnel in all levels of the organization; achieving zero breakdowns, zero defects, zero speed losses, and zero accidents; increasing production; increasing associate morale; and increasing job satisfaction. Tewari and Rawa [39] concurred on the issue that TPM is aimed at improving manufacturing quality and productivity with zero failures, zero losses, zero defects, and zero health hazards. Venkatesh [40] believed that the goal of TPM is to obtain a minimum of 90% overall equipment efficiency (OEE). Joochim and Meekaew [27] related the objective of TPM to maximizing the OEE of the equipment by eliminating equipment breakdowns, product defects, and accidents.

Captured the concept of TPM, the house of TPM is built on eight pillars [34], as shown in Fig. 1. Proposed by the JIPM, the eight pillars are the foundation for achieving the TPM objectives [41]. The eight pillars of TPM are safety, health and environment; focused maintenance; autonomous maintenance; planned maintenance; education and training; office TPM; development management; and quality maintenance [34].

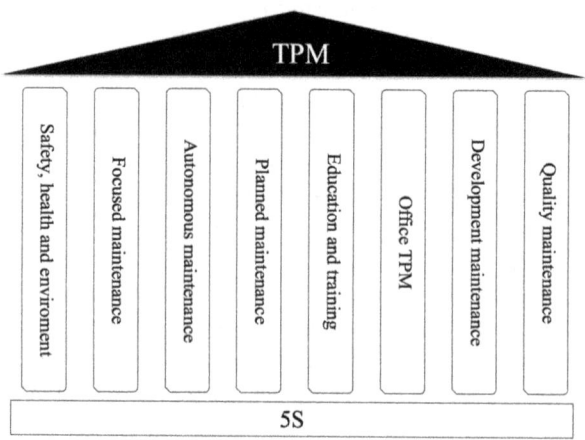

Fig. 1. TPM house [34]

5S: 5S is the foundation of TPM to ensure that employees get a working environment free from excessive interference [42]. 5S is composed of five basic elements: Sort, Set in order, Shine, Standardize and Sustain [43]. Sort means keeping the items that need to be used in the workplace [42]. Set in order mean determining the location and quantity of needed items [44]. Shine means cleaning the workplace and maintain it through preventive measures [45]. Standardize means establishing the standards to ensure the cleanliness of workplace [43]. Sustain means maintaining the established working procedures and cultivate employees' 5S habits [46].

Safety, Health and Environment: The pillar is to eliminate all factors that are harmful to employees in the workplace. When employees work in a safe working environment, their attitudes also change positively, which can help to increase productivity [42]. In addition, poetry competition, drama, safety quiz and slogans are several ways to increase employee safety awareness [25].

Focused Maintenance: Focus maintenance is to identify waste, and then reduce or eliminate waste to improve OEE of equipment [47]. Focused Maintenance emphasizes kaizen to instil a large number of small improvements [48].

Autonomous Maintenance (AM): AM is to cultivate the maintenance skills of equipment operators and empower operators to carry out routine maintenance of equipment, such as minor inspections, lubrication and cleaning [9]. It effectively prevents sudden equipment failures and enhance the sense of belonging of equipment operators to equipment [49]. AM demands a cultural change to the way maintenance is done [50], in two themes [51]. The first theme is the concept of "self", which focuses on incorporating "self-healing", "self-monitoring", "self-aware", "self-configure", and "self-protect" technologies or characteristics. The second theme focuses on automating maintenance practices within enterprises through autonomous robotics to assist or guide maintenance tasks. In additions, autonomous maintenance also helps to reduce the workload of equipment maintenance personnel, because 80% of unplanned downtime can be resolved by the trained operator [52].

Planned Maintenance (PM): PM is to focus on increasing the availability of equipment and reducing or avoiding equipment failures [13]. Corrective maintenance, maintenance prevention and breakdown maintenance are the three commonly used maintenance methods for planned maintenance [48]. Corrective maintenance aims to improve the reliability and maintainability of equipment by improving equipment and components [40]. Maintenance prevention involves giving equipment higher maintenance and reliability during the design phase of the equipment to prevent the occurrence of equipment failure radically [16]. Breakdown maintenance refers to repair the equipment after an equipment breakdown to restore it to operational condition [16].

Education and Training: Education and training is to improve expertise and morale of employees by providing the skills and technical training for them [47]. Moreover, effective education and training also helps to ensure that employees have the ability to detect equipment abnormalities [52]. Graisa [53] pointed out that the pillar creates a pleasant working environment for employees and reduce the occurrence of internal conflicts by establishing a more active working relationship.

Office TPM: Office TPM is to improve the efficiency of administrative departments to provide effective support and services to other departments [47]. Office TPM mainly eliminates nine types of losses within administrative departments: retrieved information loss, communication channel breakdown, idle loss, cost loss, set-up loss, accuracy loss, processing loss, communication loss and office equipment breakdown [54]. The elimination requires multi-departmental cooperation, reduce procedural bureaucracy and production costs [55].

Development Management: Development management is to apply the knowledge and experience gained from maintaining existing equipment to the design of new equipment, it helps reduce the time it takes to receive, install, and set up newly purchased equipment [56].

Quality Maintenance: Quality maintenance is to provide the best quality products to please customers [54]. It minimizes the possibility of repetitive defects in the equipment to achieve the goal of zero defects. Improving raw materials, basic equipment, processing methods and skills, which are an effective method to reduce repetitive defects of equipment [57].

3   Benefits of TPM to SME

TPM is an equipment-centric continuous improvement program [48]. It is the result of the integration of maintenance and manufacturing activities. Ahuja and Khamba [34] pointed out that the benefits of companies implementing TPM are reflected in six dimensions: productivity, quality, cost, delivery, safety and morale, as shown in Fig. 2. TPM keeps the equipment in perfect operational condition to avoid failures, delays and safety accidents in the production process [58]. Singh et al. [54] contend that TPM can optimize equipment efficiency, eliminate failures and promote autonomous maintenance of operators through daily activities involving all staffs. It enables enterprises to strategically use internal resources for equipment maintenance, which is critical for the effective and efficient operation of equipment [32]. TPM strategy helps to change the corporate culture, it transforms the maintenance of the equipment from passive to active, maintains the equipment at the best level of performance and reliability, and extends the life expectancy of equipment [26, 59]. The implementation of TPM by enterprises can also help to improve equipment uptime, reduce equipment adjustment time, improve product quality, enhance equipment reliability and stability, and reduce the investment demand of the enterprise to purchase new equipment [60]. Winatie, Maharani and Rimawan [61] advocated a view that implementing TPM can also reduce the enterprises' production costs and obtain a healthy and safe working environment. TPM can enhance availability, performance and quality to improve equipment OEE, whether it is a large enterprise or SMEs [62]. TPM can also strengthen the connection between the equipment maintenance department and the production department [9]. Moreover, TPM can increase the production capacity of enterprise, unlock hidden capacity of enterprises, improve team morale and job satisfaction, reduce the expenditure on equipment maintenance and obtain a higher return on investment. [48]. Moreover, TPM can also increase the accumulation of employee knowledge in term of equipment, improve internal communication, and lay the foundation for team cooperation [59]. Therefore, many prominent multinational companies, such as Proctor and Gamble, DuPont, Eastman Chemical, Ford, and AT&T, have been using TPM as a strategic approach to enhance their competitiveness [34].

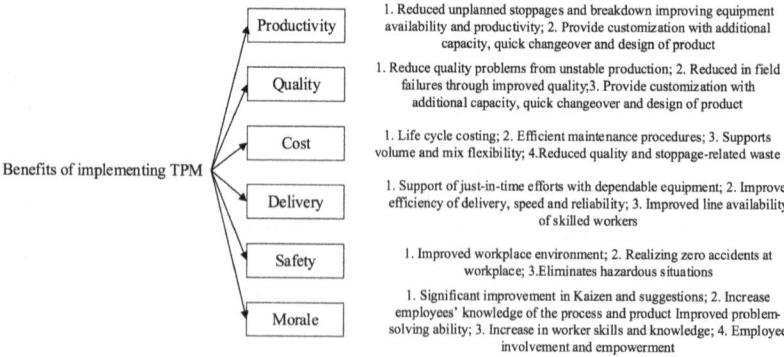

Fig. 2. Benefits of implementing TPM [34]

4 Influencing Factors of TPM Implementation in SMEs

By perusing the literature about SMEs, the factors restricting the development of SMEs are related to their idiosyncratic characteristics. Compared with large enterprises, the education level of employees in SMEs is generally lower, and the lack of communication is a common phenomenon in SMEs [63]. Other reasons include insufficient leadership and management skills of managers [64]; poor productivity and process improvement [64]; short-sighted and give priority to short-term profit [65]; lack of finance and new technologies [63]; lack of resources and long-term planning [66]; lack the motivation to learn and implement new management methods [67]; use outmoded labor-intensive technology and traditional management methods [67].

Everything has two sides, and some characteristics of SMEs have a positive impact on the development of the enterprises. Due to the low standardization and formalization of SMEs, and simple production planning and control, they have greater adaptability and flexibility to meet customer needs, which is beyond the reach of many large enterprises [68]. Moreover, SMEs have a flat management structure and top management personnel

Table 1. The factors that influence the implementation of TPM in SMEs

Researchers	Finding
Manihalla et al. [70]	1. Enterprise managers and employees fail to understand TPM correctly and form a mindset that TPM is only suitable for large enterprises; 2. Managers do not invest enough time to support the implementation of TPM. The employees of the enterprise lack skills and lack proper training
Poduval et al. [71]	The communication gap between enterprise managers and employees will hinder the implementation of TPM in SMEs
Gupta and Khanna [72]	Inappropriate or lack of regulatory policies and higher financing rate will impede the implementation of TPM in SMEs
Baglee and Knowles [32]	Introducing factors influencing the implementation of TPM in SMEs from four aspects: 1. Finance: lack of financial support; 2. Skills: lack of skilled operators; 3. Management awareness: lack of correct understanding of maintenance and advanced maintenance strategies.4. Time: lack of time to train employees
Prashanth Pai et al. [73]	The skills of employees are the most important factors affecting the implementation of TPM in SMEs. Therefore, enterprises must regularly train employees to improve their skills
AlManei, Salonitis and Xu [74]	Employee resistance to change can hinder the implementation of TPM in SMEs. The main reasons for this phenomenon are employees' complacency about their current work status and fear of failure

participate in daily management activities of the enterprise, which help enterprises to make decisions quickly [69].

Although TPM is regarded as one of the most effective methods to improve production efficiency and equipment reliability, it is a considerably challenging for SMEs to reach the same implementation depth as large enterprises [66]. In order to ensure that SMEs can successfully implement TPM, it is necessary to analyse the influencing factors, as summarized in Table 1.

To implement TPM in SMEs, management and employees should establish a clear understanding in TPM [70]. Management also need to participate in the implementation and to constantly communicate with employees during the process, and to provide essential training for employees in order to bridge the gaps in the relevant knowledge and skills [32].

5 TPM Practice in SMEs

Some literature also demonstrates that TPM is not only suitable for large enterprises [75], but also for SMEs, with success implementations as summarized in Table 2.

Table 2. Successful cases in SMEs

Researchers	Method	Result
Eugen [76]	Divided TPM implementation into four phases: preparatory stage, introduction stage, implementation stage and institutionalizing stage	The equipment became more efficient and reliable and the maintenance skills of operator were improved
Sharma and Sharma [77]	Based on the DMAIC	Increased the OEE of the equipment from 49% to 78%, reduced the defect rate from 20% to 5.4%
Jain, Bhatti and Singh [46]	A comparatively simpler model: data collection, staff training through technical seminars, trial runs, fine-tuning, and final implementation	The average OEE of the equipment increased from 44.47% to 65.13%
Nallusamy [33]	5S implementation, AM implementation, education and training implementation, and improvements in the OEE	The OEE of equipment has increased by 6%

(continued)

Table 2. (*continued*)

Researchers	Method	Result
Raut and Raut [28]	Directly aimed at critical equipment at the outset, with the implementation of 5S, AM, PM, and focused improvement, followed by education and training and OEE monitoring	The OEE of the equipment is increased from 38.21% to 82.64%
Chukwutoo and Nkemakonam [78]	Implementing TPM in three stages: the introductory stage involving top management, the preparatory stage for employees training and implementation plan, and the execution stage for carrying out the eight pillars of TPM	The OEE of the equipment was increased from 67.41% to 80.98%
Joshi and Bhatt [75]	Introducing a modified version of the TPM framework: awareness and training phases, implementation preparation phases, TPM implementation pillars phases, and sustainability development phases	The OEE of the equipment increased from 72.90% to 80.01%
Nallusamy et al. [79]	Introducing a method to implement TPM, which consists of five steps: data collection; OEE calculation before TPM implementation; root cause analysis, AM, PM, and Kobetsu Kaizen implementation; and OEE calculation after implementing TPM	The OEE of the equipment increased from 55.45% to 68.04%

One common elements promoted in these success cases are the adequate provision of training to employees before implementing TPM [75–78]. Small-scale implementation or pilot implementation is one of the effective methods of TPM implement in SMEs [9]. Moreover, SMEs can be selective in implement TPM pillars, as evidenced by some case studies [28, 33, 79]. Finally, these methods have not shown significant increase in the human and financial burden of the enterprise.

Three research gaps are identified from the review. First is the lack of innovation in the implementation of TPM process. Some researchers [76] have applied the classical 12-step method of implementing TPM in SMEs, which was introduced 30 years ago. Secondly, some researchers [28, 33, 46, 79] have introduced how to implement TPM in SMEs, but they have overlooked the issue of sustaining the effect of TPM implementation. Thirdly, in the implementation process, a number of researchers [28, 33, 46, 76–79] have ignored the goal setting, that is, what kind of goal should be achieved through the implementation of TPM.

6 Conclusions

This paper reviewed the literature of TPM specifically in SMEs, and identified several research gaps in these studies. In a highly competitive market, enterprises can implement TPM to gain a competitive advantage over competitors. As one of the advanced equipment maintenance strategies, TPM can fundamentally transform the maintenance culture of the enterprise and realize the proactive maintenance strategy transformation from the reactive maintenance strategy. As a maintenance strategy widely adopted by enterprises, TPM can effectively improve production efficiency by identifying and reducing losses in the production process. In addition, the introduction of TPM in SMEs can effectively improve the maintenance skills of equipment operators, delay equipment cracking, and extend equipment service life, thereby reducing enterprise equipment investment costs. This is also proved by actual cases, but the implementation method may be slightly different from that of large enterprises. SMEs must fully consider their own characteristics when implementing TPM. As the business status, equipment structure, working environment, working methods are subjected to various changes with the progress of technology and the development of time, continuously reviewing and improving TPM to maintain its relevancy and without diluting its principles are therefore important in the future research.

References

1. Haider, S.H., Asad, M., Fatima, M.: Entrepreneurial orientation and business performance of manufacturing sector small and medium scale enterprises of Punjab Pakistan. Eur. Bus. Manag. 3(2), 21–28 (2017)
2. Manihalla, P.P., Gopal, R.C., Rao, S.T., et al.: A survey on factors affecting total productive maintenance (TPM) in service industries. AIP Conf. Proc. 2080(060005), 1–9 (2019)
3. Singh, R.K., Garg, S.K., Deshmukh, S.G.: The competitiveness of SMEs in a globalized economy observations from China and India. Manag. Res. Rev. 33(1), 54–65 (2009)
4. Fayaz, M., Rahman, S.U., Rauf, A.: The influence of environmental factors and access to financial capital on the link between entrepreneurial orientation and SMEs performance: a case from Pakistan. Glob. Soc. Sci. Rev. 5(1), 146–153 (2020)
5. Ammenberg, J., Hjelm, O.: Tracing business and environmental effects of environmental management systems - a study of networking small and medium-sized enterprises using a joint environmental management system. Bus. Strat. Environ. 12, 163–174 (2003)
6. Dumbu, E., Chadamoyo, P.: Managerial deficiencies in the Small and Medium Enterprises (SMEs) in the craft industry: an empirical evidence of SMEs at Great Zimbabwe in chief Mugabe's area. Eur. J. Bus. Manage. 4(10), 79–85 (2012)

7. Prashanth Pai, M., Ramachandra, C.G., Srinivas, T.R., et al.: A Study on usage of total productive maintenance (TPM) in selected SMEs. IOP Conf. Ser. Mater. Sci. Eng. **376**, 1–8 (2018)
8. Tang, M.S.Q.Y.: The Study of success factors of SME'S CRM practice in China. Master thesis, Siam University (2017)
9. Jain, A., Bhatti, R.S., Singh, H.: OEE enhancement in SMEs through mobile maintenance: a TPM concept. Int. J. Qual. Reliab. Manage. **32**(5), 503–516 (2015)
10. Bohn, S., Mejia, M.C., Lafortune, J.: The economic toll of COVID-19 on small business. Public Policy Institute of California (2020)
11. Bartik, A.W., Bertrand, M., Cullen, Z.B., et al.: How are small businesses adjusting to COVID-19? Early Evidence from a Survey. NBER Working Paper (2020)
12. Alseiari, A., Farrell, P.: Technical and operational barriers that affect the successful total productive maintenance (TPM) implementation: case studies of Abu Dhabi power industry. Int. J. COMADEM **23**(2), 9–14 (2020)
13. Singh, J., Singh, H., Sharma, V.: Success of TPM concept in a manufacturing unit - a case study. Int. J. Product. Perform. Manage. **67**(3), 536–549 (2018)
14. Patterson, J.W., Fredendall, L.D.: Adapting total productive maintenance to Asten, Inc. Prod. Inventory Manage. J. **37**(4), 32–37 (1996)
15. Peng, K.: Equipment Management in the Post-maintenance era: A New Alternative to Total Productive Maintenance (TPM). CRC Press, New York (2012)
16. Gupta, A., Tyagi, V.K.: Total Productive Maintenance. Int. J. Recent Technol. Mech. Electr. Eng. **6**(6), 31–37 (2019)
17. Telang, A.D.: Preventive maintenance. In: Proceedings of the National Conference on Maintenance and Condition Monitoring, 14 February 1998, Government Engineering College, Institution of Engineers, Cochin Local Centre, Thissur (1998)
18. Eyoh, J., Kalawsky, R.S.: Evolution of maintenance strategies in oil and gas industries: the present achievements and future trend. In: The FEAST International Conference on Engineering Management, Industrial Technology, Applied Sciences, Communications and Media, 28–29 July 2019, London, UK (2019)
19. Singh, K., Ahuja, I.P.S.: Transfusion of total quality management and total productive maintenance: a literature review. Int. J. Technol. Policy Manage. **12**(4), 275–311 (2012)
20. Selcuk, S.: Predictive maintenance, its implementation and latest trends. Proc. Inst. Mech. Eng. Part B J. Eng. Manuf. **231**(9), 1670–1679 (2016)
21. Poor, P., Basl, J., Zenisek, D.: Predictive maintenance 4.0 as next evolution step in industrial maintenance development. In: 2019 International Research Conference on Smart Computing and Systems Engineering, University of Kelaniya, Sri Lanka (2019)
22. Azid, N.A.A., Shamsudin, S.N.A., Yusoff, M.S., et al.: Conceptual analysis and survey of total productive maintenance (TPM) and reliability centered maintenance (RCM) relationship. IOP Conf. Ser. Mater. Sci. Eng. **530**, 1–13 (2019)
23. Yssaad, B., Khiat, M., Chaker, A.: Reliability centered maintenance optimization for power distribution systems. Int. J. Electr. Power Energy Syst. **55**, 108–115 (2014)
24. Martins, A.P.Q.: Maintenance management of a production line-a case study in a furniture industry. Master thesis, Faculdade de Engenharia da Universedade do Porto (2019)
25. Ali, A.Y.: Application of total productive maintenance in service organization. Int. J. Res. Ind. Eng. **8**(2), 176–186 (2019)
26. Eti, M.C., Ogaji, S.O.T., Probert, S.D.: Implementing Total productive maintenance in Nigerian manufacturing industries. Appl. Energy **79**(4), 385–401 (2004)
27. Joochim, O., Meekaew, J.: Applying total productive maintenance in aluminium conductor stranding process. J. Ind. Eng. Manage. Sci. **1**, 1–24 (2016)
28. Raut, S., Raut, N.: Implementation of TPM to enhance OEE in a medium scale industry. Int. Res. J. Eng. Technol. **4**(5), 1035–1041 (2017)

29. Altmannshoffer, R.: Industrielles FM, Der Facility Manager in German (2006)
30. Ylipää, T., Skoogh, A., Bokrantz, J., et al.: Identification of maintenance improvement potential using OEE assessment. Int. J. Product. Perform. Manag. 66(1), 126–143 (2017)
31. Muchiri, P., Pintelon, L., Gelders, L., et al.: Development of maintenance function performance measurement framework and indicators. Int. J. Prod. Econ. 131(1), 295–302 (2011)
32. Baglee, D., Knowles, M.: Maintenance strategy development within SMEs: the development of an integrated approach. Control Cybern. 39(1), 275–303 (2010)
33. Nallusamy, S.: Enhancement of productivity and efficiency of CNC machines in a small scale industry using total productive maintenance. Int. J. Eng. Res. Afr. 25, 119–126 (2016)
34. Ahuja, I.P.S., Khamba, J.S.: Total productive maintenance: literature review and directions. Int. J. Qual. Reliab. Manage. 25(7), 709–756 (2008)
35. Sufian, F., Habibullah, M.S.: The impact of forced mergers and acquisitions on banks' total factor productivity: empirical evidence from Malaysia. J. Asia Pac. Econ. 19(1), 151–185 (2014)
36. Nakajima, S.: Introduction to TPM: Total Productive Maintenance. Productivity Press, Cambridge (1998)
37. Japan Institute of Plant Maintenance (JIPM), TPM Implementation Documents, JIPM Solution
38. Bajaj, A., Kumar, V.: TPM implementation in small scale agriculture industry: a case study. Int. J. Eng. Res. Technol. 2(5), 1122–1135 (2013)
39. Tewari, A., Rawat, E.: Total productive maintenance-a review. Int. J. Res. Appl. Sci. Eng. Technol. 5(4), 406–410 (2017)
40. Venkatesh, J.: An introduction to total productive maintenance (TPM) (2007)
41. Prabowo, H.A., Suprapto, Y.B., Farida, F.: The evaluation of eight pillars total productive maintenance (TPM) implementation and their impact on overall equipment effectiveness (OEE) and waste. Sinergi 22(1), 13–18 (2018)
42. Jha, M.C.K., Singh, A.: Study of total productive maintenance: a case study of OEE improvement in automobile industry, benefits and barriers in TPM implementation. Int. J. Technol. Res. Eng. 3(9), 2400–2406 (2016)
43. Randhawa, J.S., Ahuja, I.S.: 5S - a quality improvement tool for sustainable performance: literature review and directions. Int. J. Qual. Reliab. Manage. 34(3), 334–361 (2017)
44. Filip, F.C., Marascu-Klein, V.: The 5S lean method as a tool of industrial management performances. IOP Conf. Ser. Mater. Sci. Eng. 95, 1–6 (2015)
45. Gurel, D.A.: A conceptual evaluation of 5S model in hotels. Afr. J. Bus. Manage. 7(30), 3035–3042 (2013)
46. Veres, C., Marian, L., Moica, S., et al.: Case study concerning 5S method impact in an automotive company. Procedia Manuf. 22, 900–905 (2018)
47. Shinde, D.D., Prasad, R.: Application of AHP for ranking of total productive maintenance pillars. Wireless Pers. Commun. 100(2), 449–462 (2017). https://doi.org/10.1007/s11277-017-5084-4
48. Parikh, Y., Mahamun, P.: Total productive maintenance: need and framework. Int. J. Innovative Res. Adv. Eng. 2(2), 126–130 (2015)
49. Okpala, C.C., Anozie, S.C., Ezeanyim, O.C.: The application of tools and techniques of total productive maintenance in manufacturing. Int. J. Seman. Comput. 8(6), 18115–18121 (2018)
50. Mugwindiri, K., Mbohwa, C.: Availability performance improvement by using autonomous maintenance–the case of a developing country, Zimbabwe. In: Proceedings of the World Congress on Engineering 2013, 3–5 July, London, U.K. (2013)
51. Farnsworth, M., Bell, C., Khan, S., Tomiyama, T.: Autonomous maintenance for through-life engineering. In: Redding, L., Roy, R. (eds.) Through-life Engineering Services. DE, pp. 395–419. Springer, Cham (2015). https://doi.org/10.1007/978-3-319-12111-6_23

52. Rajput, H.S., Jayaswal, P.: A total productive maintenance (TPM) approach to improve overall equipment efficiency. Int. J. Mod. Eng. Res. **2**(6), 4383–4386 (2012)
53. Graisa, M.M.: An investigation into the need and implementation of total productive maintenance (TPM) in Libyan cement industry. Doctor thesis, The Nottingham Trent University (2011)
54. Singh, R., Gohil, A.M., Shah, D.B., et al.: Total productive maintenance (TPM) implementation in a machine shop: a case study. Procedia Eng. **51**, 592–599 (2013)
55. Ahuja, I.P.S., Khamba, J.S.: An evaluation of TPM initiatives in Indian industry for enhanced manufacturing performance. Int. J. Qual. Reliab. Manage. **25**(2), 147–172 (2008)
56. Adesta, E.Y.T., Prabowo, H.A., Agusman, D.: Evaluating 8 pillars of total productive maintenance (TPM) implementation and their contribution to manufacturing performance. In: IOP Conference Series: Materials Science and Engineering (2018)
57. Tokutaro, S.: Overview of TPM in process industry. In: Tokutaro, S. (ed.) TPM- in Process Industries. Productivity Press, New York (1994)
58. Sethia, C.S., Shende, P.N., Dange, S.S.: A case study on total productive maintenance in rolling mill. J. Emerg. Technol. Innovative Res. **1**(3), 60–66 (2016)
59. Agustiady, T.K., Cudney, E.A.: Total productive maintenance. Total Qual. Manage. Bus. Excellence. 1–8 (2018). https://doi.org/10.1080/14783363.2018.1438843
60. Ahuja, I.S.: Total productive maintenance practices in manufacturing organisations: literature review. Int. J. Technol. Policy Manage. **11**(2), 117–138 (2011)
61. Winatie, A., Maharani, B.P., Rimawan, E.: Productivity Analysis to increase overall equipment effectiveness (OEE) by implementing total productive maintenance. Int. J. Innovative Sci. Res. Technol. **3**(2), 433–439 (2019)
62. Jain, A., Bhatti, R., Singh, H., et al.: Implementation of TPM for enhancing OEE of small scale industry. Int. J. IT. Eng. Appl. Sci. Res. **1**(1), 125–136 (2012)
63. Bhagwat, R., Sharma, M.K.: information system architecture: a framework for a cluster of small- and medium-sized enterprises (SMEs). Prod. Plann. Control **18**(4), 283–296 (2007)
64. Thakkar, J., Kanda, A., Deshmukh, S.G.: Supply chain issues in SMEs: select insights from cases of Indian origin. Prod. Plann. Control **24**(1), 47–71 (2013)
65. Baker, H.K., Kumar, S., Singh, H.P.: Working capital management: evidence from Indian SMEs. Small Enterp. Res. **26**(2), 143–163 (2019)
66. Achanga, P., Shehab, E., Roy, R., et al.: Critical success factors for lean implementation within SMEs. J. Manuf. Technol. Manag. **17**(4), 460–471 (2006)
67. Rymaszewska, D.A.: The challenges of lean manufacturing implementation in SMEs. Benchmarking An Int. J. **21**(6), 987–1002 (2014)
68. Tam, S., Gray, D.E.: The practice of employee learning in SME workplaces. J. Small Bus. Enterp. Dev. **23**(3), 671–690 (2016)
69. Mc Cartan-Quinn, D., Carson, D.: Issues which impact upon marketing in the small firm. Small Bus. Econ. **21**(2), 201–213 (2003)
70. Manihalla, P.P., Gopal, R.C., Rao, S.T.R., et al.: A survey approach to study the influence of management factor in implementing TPM in selected SMEs. Int. Conf. Emerg. Trends Mech. Eng. **2236**, 1–7 (2020)
71. Poduval, P.S., Pramod, V.R., Jagathy Raj, V.P.: Interpretive structural modelling (ISM) and its application in implementation off Total productive maintenance (TPM). Int. J. Qual. Reliab. Manage. **32**(3), 308–331 (2015)
72. Gupta, A., Khanna, I.K.: An analysis of barriers and enablers for effective implementation of total productive maintenance (TPM) in small and medium enterprises (SMEs) in India: literature review. Int. J. Mod. Eng. Manage. Res. **7**(4), 41–61 (2019)
73. Prashanth Pai, M., Ramachandra, C.G., Srinivas, T.R., et al.: A survey approach to study the influence of finance factor and workforce skills in implementing TPM in selected SMEs. Int. J. Prod. Eng. **5**(1), 5–12 (2019)

74. AlManei, M., Salonitis, K., Xu, Y.: Lean implementation frameworks: the challenges for SMEs. Procedia CIRP **63**, 750–755 (2017)
75. Joshi, K.M., Bhatt, D.V.: A modified TPM framework for Indian SMEs. Int. J. Adv. Res. Eng. Technol. **9**(6), 1–14 (2018)
76. Eugen, P.: Implementation and results of total productive maintenance in a SMEs. Fascicle Manage. Technol. Eng. **9**(19), 167–172 (2010)
77. Sharma, R.K., Sharma, R.G.: Integrating six sigma culture and TPM framework to improve manufacturing performance in SMEs. Qual. Reliab. Eng. Int. **30**(5), 745–765 (2013)
78. Chukwutoo, C.I., Nkemakonam, C.I.: Total productive maintenance (TPM) as a business strategy in manufacturing small and medium enterprises in Nigeria. Adv. Res. **15**(5), 1–9 (2018)
79. Nallusamy, S., Kumar, V., Yadav, V., et al.: Implementation of total productive maintenance to enhance the overall equipment effectiveness in medium scale industries. Int. J. Mech. Prod. Eng. Res. Dev. **8**(1), 1027–1038 (2018)

Spark Plasma Sintering of Ceramic Matrix Composite of TiC: Microstructure, Densification, and Mechanical Properties: A Review

Samson Dare Oguntuyi[1]([✉]), Oluwagbenga Johnson[2,3],
and Mxolisi Brendon Shongwe[1]

[1] Institute for Nano Engineering Research, Department of Chemical,
Metallurgical and Materials Engineering, Tshwane University of Technology,
Private Bag X680, Pretoria 0001, South Africa
ShongweMB@tut.ac.za
[2] Department of Mining and Metallurgical Engineering, University of Namibia,
Private Bag, Ongwediva 13301, Namibia
Ojohnson@unam.na
[3] Department of Metallurgy, School of Mining, Metallurgy and Chemical Engineering,
Faculty of Engineering and the Built Environment, University of Johannesburg, PO Box 524,
Johannesburg, South Africa

Abstract. The application of monolithic/un-doped/single-phase ceramics has been limited due to their difficulty in sintering and low fracture toughness. Ceramic matrix composites have gained predominant attention in the past decades in comparison to monolithic/un-doped/single phase ceramics, this is as a result of the high fracture toughness, good wear resistance, and high hardness that they (ceramic matrix composite) possess. Also, the use of sintering additives in collaboration with the application of modern consolidation viz spark plasma sintering (SPS) has gained high prominence to nullify these challenges faced by ceramics. Although, previous review has highlighted the use of diverse techniques (hot press, hot isostatic, pressureless sintering, and SPS) on the consolidation of ceramics and its composites. Amidst all these techniques, SPS has stood to be an effective powder metallurgy route for achieving good microstructure and excellent mechanical properties. This review takes a research on the effects of nitrides based sintering additives on the microstructure, densification, and mechanical properties of titanium carbides ceramic matrix by SPS. The review finally concludes on the potential research importance on the types of sintering additives inclusion that should be in further research processes for improvement in material properties of titanium carbides.

Keywords: SPS · TiC · Microstructure · Densification · Mechanical properties

1 Introduction

Titanium carbide usually demonstrates metallic and ceramic-like features, with a typical crystal structure as depicted in Fig. 1. It has lately grown high interest as a result of its

M. Awang and S. S. Emamian (Eds.): *Advances in Material Science and Engineering*, LNME, pp. 93–101, 2021.
https://doi.org/10.1007/978-981-16-3641-7_13

combined unique properties viz significant elastic modulus (~400 GPa), high melting point (3,160 °C), high hardness, high oxidation resistance, excellent wear-resistance, low thermal expansion, high electrical conductivity, and considerable chemical stability [1–5]. These outstanding properties have made Titanium carbide a high potential material for elevated temperature applications including wear-resistance coatings, corrosion resistance parts, impact- barrier armors, ceramic cutting tools, crucibles, etc., [4–6]. Although monolithically, the sinterability of titanium carbide is challenging in achieving the desired results. These challenges of titanium carbide are due to the solid covalent bond, low self-diffusion coefficient, and oxide layers which are mostly B_2O_3 and TiO_2 [4, 5, 7, 8]. Therefore, consolidation of monolithic TiC with enhanced densification required an elevated sintering temperature greater than 2000 °C with high pressure. These sintering parameters usually resulted in uncontrolled grain growth, poor microstructure, and inefficient mechanical properties [6, 8].

To nullify the aforementioned challenges, sintering additive and/or sintering aid are usually applied to lower the sintering temperature [9, 10]. B_4C, $TiSi_2$, Si_3Ni_4, TiC, $MoSi_2$, WC, TaC, AlN, SiC are examples of non-metallic sintering aids/additives that are being added to ceramic matrix composite to inhibit grain growth and reduce consolidation temperature [11–13]. Additionally, applying metallic additives viz Co, Ni, Fe, Mo also enhances the sinterability and fracture toughness of the manufactured ceramic materials this is attributed to the toughening stimulation mechanism and the creation of liquid phase [12–15]. The improvement of combined fracture toughness, hardness, modulus strength with enhanced performance and densification are prompted by the use of ceramic matrix composites (CMC). CMC also ensures that sintering temperatures are lowered compared to undoped ceramics which usually involves the use of high temperatures for sintering [16–18]. As a substitute route, spark plasma sintering (SPS) is an important method for the consolidation of ceramics, SPS enables the production of fine microstructure which consequently improves mechanical properties, these attributes were as a result of its fast heating and short holding time in comparison with conventional sintering viz, hot pressing, hot isostatic pressing, flash sintering, etc. [19–21]. In the SPS technique, the ceramics materials are introduced in the graphite die, then using a pulsed electric current under an externally applied pressure, the sintering process is accomplished [23, 24]. This article gives a critical review on TiC reinforced with sintering additives consolidated by SPS. An observation will be carried out on how nitride-based material additives have had an influence on the densification, microstructure, and mechanical properties of a TiC ceramic-based matrix.

1.1 Limitations and Challenges of TiC

Difficulty in densification as a result of poor sinterability and high covalent nature of TiC has created some challenges in sintering it, Monolithic application of TiC is limited owning to poor fracture toughness, brittle-like nature, and poor thermal shock resistance [25–30]. Hence, the use of sintering additive in addition to the use of modern techniques of sintering has been observed to minimize these challenges [30–33].

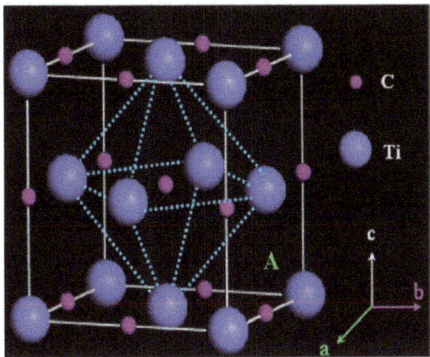

Fig. 1. A crystal structure of TiC [34].

2 Effects of Sintering Additives

Various sintering additives have different influences on the sinterability, microstructure, densification, and mechanical properties of various ceramic materials. The types, quantity, and proper dissipation of these sintering additives go a long way in achieving enhanced properties of the ceramic matrix composites. More also, the individual properties of the sintering additives which are reinforced in the ceramic matrix contribute to the whole properties of the sintered ceramic matrix composites. Some sintering additives have depreciating or enhancing effects on the overall properties of c ceramic matrix composite which are largely attributed to the properties of the reinforcing additives [34].

2.1 Spark Plasma Sintering of TiC Matrix Composites Using Nitrides Based Material as Sintering Additive

Nitrides-based additives such as AlN, TiN, etc., reduced the hardness of TiC ceramic matrix to some percentage but enhances the fracture toughness which has been a challenge for ceramics generally. More also, the reduction in the flexural strength of some TiC ceramic composites was as a result of higher hardness of TiC than the nitrides based sintering additives, therefore the percentage increment of these nitrides based additives reduces the hardness and strength of the TiC ceramics composites but consequently enhances fracture toughness [35, 36].

Pazhouhanfar et al. [37], observed the effects of 5wt.% TiN on the microstructural and mechanical properties of TiC composites. The composites were consolidated at 1900 °C for 10 min under 40 MPa. Densification of 97% was reported for the doped TiC which was 1.6% greater than the relative density of the monolithic (95.5%). Figure 2(a) shows the graphical representation of the relative density of these samples. But densification of 98% and 99% was achieved for a monolithic TiC when sintered by SPS at 1600 °C and 1900 °C respectively, the achievement of the later densification was attributed to the use of fine size particle powder (<2 μm) [30, 38]. The introduction of TiN in the composites as a secondary phase was reported to inhibit grain growth.

The addition of 5wt% TiN to the monolithic TiC reduced the Vickers hardness by 12% compared to the undoped TiC Fig. 2(b) showed this graphically. The formation

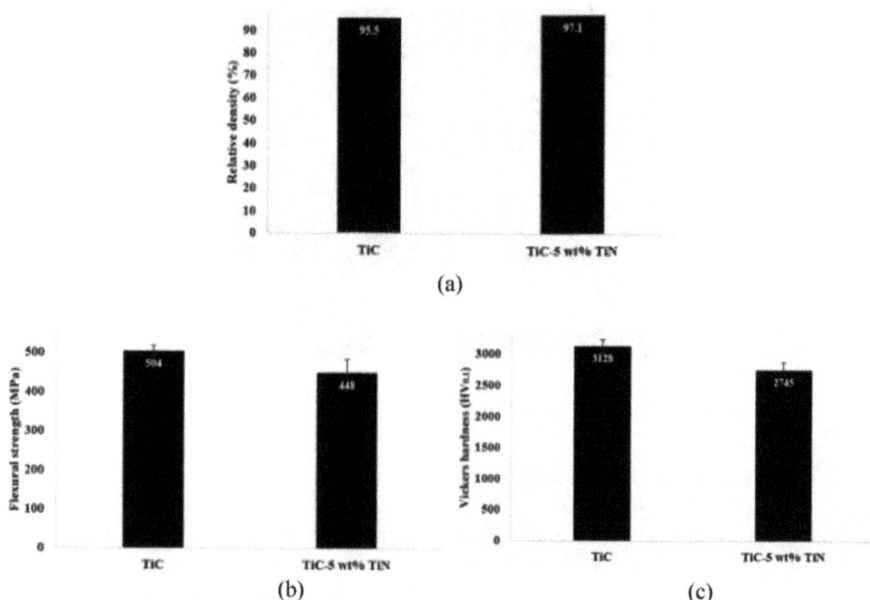

Fig. 2. Shows the densification for undoped TiC and doped TiC with 5wt% of TiN [37]. (b) and (c) showing the flexural strength and Vickers hardness of monolithic TiC and doped TiC with 5wt% TiN respectively [37].

of solid solution phase of Ti(C, N) in the absence of bonding phase contributed to the reduction of the hardness of the doped TiC, similar observations were made in previous works [39, 40]. The flexural strength of the undoped TiC was reported to be greater than the doped TiC, as seen in Fig. 2(c), the existence of the in-situ brittle phase of Ti(C, N) formed was said to contribute to the reduced flexural strength of the doped TiC.

Russias et al. [36] studied the effects of TiN in TiC cermets. It was reported that the addition of TiN to the cermets resulted in grain growth inhibition and transforms the repartition of diverse phases. The existence of TiN reduces the cermet hardness but consequently enhances the cermet's toughness. The hardness reduction was as a result of the lower hardness of TiN as a reinforcement in the cermet which at the same proportion promoted the fracture toughness. This outcome has mostly been observed that the corresponding improvement in hardness and fracture toughness is hard to achieve when sintering additives are being added to a ceramic matrix, as the increase in one leads to the decrease of the other and vice versa [35, 41].

Fattahi et al. [42], reported improved densification and flexural strength when TiC based composites were doped with 5wt% AlN at a sintering temperature of 1900 °C for 10 min under 40 MPa. The addition of AlN and the in-situ Ti_3 Al was reported to influence the full densification of the composites, but its Vickers hardness reduced by 2% in comparison to the monolithic TiC due to the phases of AlN and Ti_3 Al present in the composites whose hardness are lower than TiC, [43, 44], Fig. 3(b) the light- gray color in the micrograph depicted phases of the TiC matrix, while the dark-gray-color were the secondary in-situ formed phases or AlN [43, 44]. The reported flexural strength for the monolithic and doped TiC was 504 MPa and 688 MPa grain size. The enhanced relative density of the doped sample and the formation of the in-situ phase created a clean interface between the secondary phases and the matrix all quantified to the improved flexural strength. The microstructural observation as shown in (Fig. 3a), depicted that the monolithic TiC contained some pores, suggesting inadequate sintering temperature to fully densify the material, while the doped sample with 5 wt% AlN showed highly full densification without any visible porosity in the microstructure (Fig. 3b).

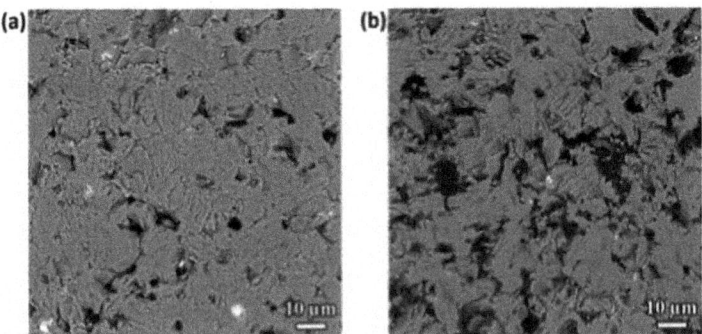

Fig. 3. A SEM graphs of the sintered (a) undoped TiC and (b) doped TiC with 5 wt% AlN [42].

Shaddel et al. [41] reported contrasting densification and mechanical properties of TiC composites when it was doped 5 wt% BN in comparison to Fattahi et al. experiment [42] under the same sintering condition. The addition of BN did not influence the densification of the samples, that both the doped and undoped TiC achieved similar results of approximately 95%. More also, the sintering additive had a depreciating effect on the mechanical properties such that the flexural strength and Vickers hardness reduced at around 15% and 7%, respectively, in comparison to the acquired values for the monolithic samples. The content and texture composites and the in-situ carbonic phases formed when TiC was doped with BN were said to be the cause for the drop in hardness [45], more also, the non-provision by the remaining BN particles in cleaning the interface with TiC had a poor impact on the flexural strength of the doped samples compared with the monolithic sample.

3   Comparison of Different Nitrides in Terms of Properties (Densification and Mechanical Properties)

Different nitrides based have been studied on the microstructure, densification, and mechanical property of TiC (as depicted in graph 1), it was observed that AlN provided improvement in achieving good densification and combined excellent mechanical properties (Table 1). The Fig. 4, majorly showed the influences of sintering additives on the densification and mechanical properties of TiC compared to undoped TiC.

Table 1. Showing the effects of different nitrides sintering additives with TiC using Spark Plasma Sintering for consolidation.

Material composition	Processing condition	Sintered density	Hardness (GPa)	Fracture toughness (MPa m$^{1/2}$)	Flexural Strength (MPa)	References
TiC–5wt% TiN	1900 °C, 40 MPa, 10 min	97	274.5 (HV0.1)	–	450	[37]
Monolithic TiC	1900 °C, 40 MPa, 10 min	99.9	25.7	–	–	[38]
TiC-5wt% AlN	1900 °C, 40 MPa, 10 min	101.27	3050 (HV100)	–	688	[42]
TiC-5wt% BN	1900 °C 40 MPa, 7 min	95	2914 (HV100)	–	429	[41]
Monolithic TiC	1650 °C, 100 MPa, 5 min	97.9	28	5.9	–	[46]

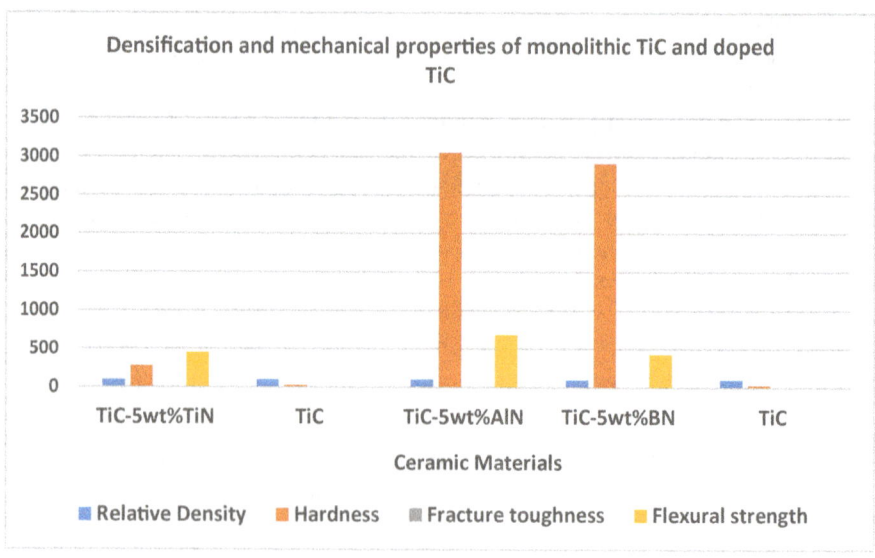

Fig. 4. Relative density and mechanical properties of TiC ceramic materials.

4 Conclusion

The addition of nitrides based on titanium carbides ceramic composites has been observed to produce some contrasting results in the densification and mechanical properties of TiC matrix. It can be inferred that not all nitrides based additives yielded good improvement on the properties of ceramics based matrix, as some of the (nitrides additives) have a depreciating effect while others produce enhancement in the properties of TiC based matrix composites. Therefore, more works should undertake more in adding non-metallic ceramics together with nitrides additive whose hardness is not far less than the parent composites, with this concept an improved combined mechanical properties can be attained without any depreciating effect in any of the desired properties.

Acknowledgments. Appreciation goes to all the author for their contributions to the success of this review.

References

1. Ahmadi, Z., Nayebi, B., Asl, M.S., Farahbakhsh, I., Balak, Z.: Densification improvement of spark plasma sintered TiB2-based composites with micron-, submicron-and nano-sized SiC particulates. Ceram. Int. **44**, 11431–11437 (2018)
2. Ebrahimi, A., Esfahani, H., Fattah-alhosseini, A., Imantalab, O.: In-vitro electrochemical study of TiB/TiB2 composite coating on titanium in Ringer's solution. J. Alloy. Compd. **765**, 826–834 (2018)
3. Balcı, Ö., Ağaoğulları, D., Gökçe, H., Duman, I., Öveçoğlu, M.L.: Influence of TiB2 particle size on the microstructure and properties of Al matrix composites prepared via mechanical alloying and pressureless sintering. J. Alloy. Compd. **586**, S78–S84 (2014)
4. Ji, W., Zhang, J., Wang, W., Wang, H., Zhang, F., Wang, Y., et al.: Fabrication and properties of TiB2-based cermets by spark plasma sintering with CoCrFeNiTiAl high-entropy alloy as sintering aid. J. Eur. Ceram. Soc. **35**, 879–886 (2015)
5. Basu, B., Vleugels, J., Van der Biest, O.: Fretting wear behavior of TiB2-based materials against bearing steel under water and oil lubrication. Wear **250**, 631–641 (2001)
6. Rabiezadeh, A., Ataie, A., Hadian, A.M.: Sintering of Al2O3–TiB2 nano-composite derived from milling assisted sol–gel method. Int. J. Refract. Metal Hard Mater. **33**, 58–64 (2012)
7. Asl, M.S., Ahmadi, Z., Parvizi, S., Balak, Z., Farahbakhsh, I.: Contribution of SiC particle size and spark plasma sintering conditions on grain growth and hardness of TiB2 composites. Ceram. Int. **43**, 13924–13931 (2017)
8. Namini, A.S., Gogani, S.N.S., Asl, M.S., Farhadi, K., Kakroudi, M.G., Mohammadzadeh, A.: Microstructural development and mechanical properties of hot pressed SiC reinforced TiB2 based composite. Int. J. Refract. Metal Hard Mater. **51**, 169–179 (2015)
9. Fu, Z., Koc, R.: Pressureless sintering of TiB2 with low concentration of Co binder to achieve enhanced mechanical properties. Mater. Sci. Eng. A **721**, 22–27 (2018)
10. Farahbakhsh, I., Ahmadi, Z., Asl, M.S.: Densification, microstructure and mechanical properties of hot pressed ZrB2–SiC ceramic doped with nano-sized carbon black. Ceram. Int. **43**, 8411–8417 (2017)
11. Demirskyi, D., Borodianska, H., Sakka, Y., Vasylkiv, O.: Ultra-high elevated temperature strength of TiB2-based ceramics consolidated by spark plasma sintering. J. Eur. Ceram. Soc. **37**, 393–397 (2017)

12. Lin, J., Yang, Y., Zhang, H., Wu, Z., Huang, Y.: Effect of sintering temperature on the mechanical properties and microstructure of carbon nanotubes toughened TiB2 ceramics densified by spark plasma sintering. Mater. Lett. **166**, 280–283 (2016)
13. Germi, M.D., Mahaseni, Z.H., Ahmadi, Z., Asl, M.S.: Phase evolution during spark plasma sintering of novel Si3N4-doped TiB2–SiC composite. Mater. Charact. **145**, 225–232 (2018)
14. Li, B.: Effect of ZrB2 and SiC addition on TiB2-based ceramic composites prepared by spark plasma sintering. Int. J. Refract. Metal Hard Mater. **46**, 84–89 (2014)
15. Cymerman, K., Oleszak, D., Rosinski, M., Michalski, A.: Structure and mechanical properties of TiB2/TiC–Ni composites fabricated by pulse plasma sintering method. Adv. Powder Technol. **29**, 1795–1803 (2018)
16. Black, J.T., Kohser, R.A.: DeGarmo's Materials and Processes in Manufacturing. Wiley, Hoboken (2017)
17. Binner, J., Porter, M., Baker, B., Zou, J., Venkatachalam, V., Diaz, V.R., et al.: Selection, processing, properties and applications of ultra-high temperature ceramic matrix composites, UHTCMCs–a review. Int. Mater. Rev. **65**, 389–444 (2019)
18. Xue, J.-X., Liu, J.-X., Zhang, G.-J., Zhang, H.-B., Liu, T., Zhou, X.-S., et al.: Improvement in mechanical/physical properties of TiC-based ceramics sintered at 1500 C for inert matrix fuels. Scripta Mater. **114**, 5–8 (2016)
19. Asl, M.S., Namini, A.S., Motallebzadeh, A., Azadbeh, M.: Effects of sintering temperature on microstructure and mechanical properties of spark plasma sintered titanium. Mater. Chem. Phys. **203**, 266–273 (2018)
20. Azizian-Kalandaragh, Y., Namini, A.S., Ahmadi, Z., Asl, M.S.: Reinforcing effects of SiC whiskers and carbon nanoparticles in spark plasma sintered ZrB2 matrix composites. Ceram. Int. **44**, 19932–19938 (2018)
21. Oguntuyi, S.D., Johnson, O.T., Shongwe, M.B.: Spark Plasma Sintering of Ceramic Matrix Composite of ZrB 2 and TiB 2: Microstructure, Densification, and Mechanical Properties—A Review, Metals and Materials International, pp. 1–14 (2020)
22. Sabahi Namini, A., Azadbeh, M., Shahedi Asl, M.: Effects of in-situ formed TiB whiskers on microstructure and mechanical properties of spark plasma sintered Ti–B4C and Ti–TiB2 composites. Scientia Iranica **25**, 762–771 (2018)
23. Balak, Z., Asl, M.S., Azizieh, M., Kafashan, H., Hayati, R.: Effect of different additives and open porosity on fracture toughness of ZrB2–SiC-based composites prepared by SPS. Ceram. Int. **43**, 2209–2220 (2017)
24. Nikzad, L., Licheri, R., Ebadzadeh, T., Orru, R., Cao, G.: Effect of ball milling on reactive spark plasma sintering of B4C–TiB2 composites. Ceram. Int. **38**, 6469–6480 (2012)
25. Wang, L., Jiang, W., Chen, L.: Rapidly sintering nanosized SiC particle reinforced TiC composites by the spark plasma sintering (SPS) technique. J. Mater. Sci. **39**, 4515–4519 (2004)
26. Song, G.-M., Guo, Y.-K., Zhou, Y., Li, Q.: Preparation and mechanical properties of carbon fiber reinforced-TiC matrix composites. J. Mater. Sci. Lett. **20**, 2157–2160 (2001)
27. Sribalaji, M., Mukherjee, B., Bakshi, S.R., Arunkumar, P., Babu, K.S., Keshri, A.K.: In-situ formed graphene nanoribbon induced toughening and thermal shock resistance of spark plasma sintered carbon nanotube reinforced titanium carbide composite. Compos. Part B Eng. **123**, 227–240 (2017)
28. Chae, K.W., Niihara, K., Kim, D.Y.: Improvements in the mechanical properties of TiC by the dispersion of fine SiC particles. J. Mater. Sci. Lett. **14**, 1332–1334 (1995)
29. Locci, A.M., Orru, R., Cao, G., Munir, Z.A.: Effect of ball milling on simultaneous spark plasma synthesis and densification of TiC–TiB2 composites. Mater. Sci. Eng. A **434**, 23–29 (2006)
30. Cheng, L., Xie, Z., Liu, G.: Spark plasma sintering of TiC ceramic with tungsten carbide as a sintering additive. J. Eur. Ceram. Soc. **33**, 2971–2977 (2013)

31. Fattahi, M., Babapoor, A., Delbari, S.A., Ahmadi, Z., Namini, A.S., Asl, M.S.: Strengthening of TiC ceramics sintered by spark plasma via nano-graphite addition. Ceram. Int. **46**, 12400–12408 (2020)

32. Fattahi, M., Delbari, S.A., Babapoor, A., Namini, A.S., Mohammadi, M., Asl, M.S.: Triplet carbide composites of TiC, WC, and SiC. Ceram. Int. **46**, 9070–9078 (2020)

33. Chen, J., Li, W., Jiang, W.: Characterization of sintered TiC–SiC composites. Ceram. Int. **35**, 3125–3129 (2009)

34. Nie, J., Wu, Y., Li, P., Li, H., Liu, X.: Morphological evolution of TiC from octahedron to cube induced by elemental nickel. CrystEngComm **14**, 2213–2221 (2012)

35. Liu, N., Chao, S., Huang, X.: Effects of TiC/TiN addition on the microstructure and mechanical properties of ultra-fine grade Ti (C, N)–Ni cermets. J. Eur. Ceram. Soc. **26**, 3861–3870 (2006)

36. Russias, J., Cardinal, S., Aguni, Y., Fantozzi, G., Bienvenu, K., Fontaine, J.: Influence of titanium nitride addition on the microstructure and mechanical properties of TiC-based cermets. Int. J. Refract. Metal Hard Mater. **23**, 358–362 (2005)

37. Pazhouhanfar, Y., Namini, A.S., Delbari, S.A., Nguyen, T.P., Van Le, Q., Shaddel, S., et al.: Microstructural and mechanical characterization of spark plasma sintered TiC ceramics with TiN additive. Ceram. Int. **46**, 18924–18932 (2020)

38. Babapoor, A., Asl, M.S., Ahmadi, Z., Namini, A.S.: Effects of spark plasma sintering temperature on densification, hardness and thermal conductivity of titanium carbide. Ceram. Int. **44**, 14541–14546 (2018)

39. Jiang, C.C., Goto, T., Hirai, T.: Microhardness of non-stoichiometric TiCx, plates prepared by chemical vapour deposition. J. Less Common Metals **163**, 339–346 (1990)

40. Zhang, Z., Geng, C., Ke, Y., Li, C., Jiao, X., Zhao, Y., et al.: Processing and mechanical properties of nonstoichiometric TiCx (0.3 ≤ x ≤ 0.5). Ceram. Int. **44**, 18996–19001 (2018)

41. Shaddel, S., Namini, A.S., Pazhouhanfar, Y., Delbari, S.A., Fattahi, M., Asl, M.S.: A microstructural approach to the chemical reactions during the spark plasma sintering of novel TiC–BN ceramics. Ceram. Int. **46**, 15982–15990 (2020)

42. Fattahi, M., Pazhouhanfar, Y., Delbari, S.A., Shaddel, S., Namini, A.S., Asl, M.S.: Strengthening of novel TiC–AlN ceramic with in-situ synthesized Ti3Al intermetallic compound. Ceram. Int. **46**, 14105–14113 (2020)

43. Yonenaga, I.: Hardness of bulk single-crystal GaN and AlN. Mater. Res. Soc. Internet J. Nitride Semiconductor Res. **7**, 1–4 (2002)

44. Chen, B., Xiong, H., Sun, B., Tang, S., Du, B., Li, N.: Microstructures and mechanical properties of Ti3Al/Ni-based superalloy joints arc welded with Ti–Nb and Ti–Ni–Nb filler alloys. Prog. Nat. Sci. Mater. Int. **24**, 313–320 (2014)

45. Asl, M.S., Ahmadi, Z., Namini, A.S., Babapoor, A., Motallebzadeh, A.: Spark plasma sintering of TiC–SiCw ceramics. Ceram. Int. **45**, 19808–19821 (2019)

46. Teber, A., Schoenstein, F., Têtard, F., Abdellaoui, M., Jouini, N.: Effect of SPS process sintering on the microstructure and mechanical properties of nanocrystalline TiC for tools application. Int. J. Refract. Metal Hard Mater. **30**, 64–70 (2012)

Thermal Performance Modelling of a Flat Bare Tube Bundle Under Deluging Cooling Conditions

Ester Angula[1]([✉]), Paul Chisale[2], and Fillemon N. Nangolo[1]

[1] Department of Mechanical and Industrial Engineering, University of Namibia, Ongwediva, Namibia
eangula@unam.na

[2] Department of Mechanical and Marine Engineering, Namibia University of Science and Technology, Windhoek, Namibia

Abstract. This work focuses on developing and improving a one-dimensional analytical, thermal model to be employed in predicting and optimizing of the performance of a flat bare tube bundle, under deluging cooling condition, commonly found in the literatures. The flat bare tube bundle to be incorporated into the second stage of an induced draft Hybrid (Dry/Wet) Dephlegmator (HDWD) for a Direct Air-Cooled Steam Condenser (ACSC). The developed model is validated theoretically and evaluated analytically by using three approaches, which are: Poppe, Merkel, and heat and mass transfer analogy. The present model was also validated against the models for flat bare tube bundles from the literatures. The geometric orientation of the tubes (flat and round ends sections) was taken into account when modelling was performed. For the flat section of the tube, the governing differential equations were conducted in Cartesian coordinates. The cylindrical coordinates were employed, to derive the governing differential equations, for round ends tube. Heat transfer rate obtained by Poppe method is found to be 2.8% and 9.2% higher than that of Merkel, and heat and mass transfer analogy methods, respectively. By comparing the present model to that of flat bare tube bundle in the literatures, the heat transfer rate and air-side pressure drop were found to be 11% and 88% higher, respectively for both methods. Furthermore, the heat transfer rate for the model of delugeable round bare tube bundle (DRBTB) was found to be higher than that of the present model at both smaller and larger tube pitch, while the air-side pressure drop for the model of DRBTB was found to be lower and higher than that of the present model at smaller and larger tube pitch, respectively.

1 Introduction

As an enhancement technique of improving the performance and availability of the ACSC during the hot periods, the HDWD is found to be appropriate, cost effective and uses around 20% less water than for the pre-cooling technology [2]. Therefore, the prediction of the thermal performance of the flat bare tube bundle to be incorporated into the second stage of induced draft HDWD is essential. The induced draft HDWD is shown in Fig. 1, and its second stage operates in the wet mode as evaporative condenser

M. Awang and S. S. Emamian (Eds.): *Advances in Material Science and Engineering*, LNME, pp. 102–110, 2021.
https://doi.org/10.1007/978-981-16-3641-7_14

during hot and peak periods. Deluge water sprayed on the surface of the flat bare tube bundle during the wet operating mode is collected in collecting troughs under the tube bundle, while the drift eliminator above the tube bundle traps the water droplets blown up by the air.

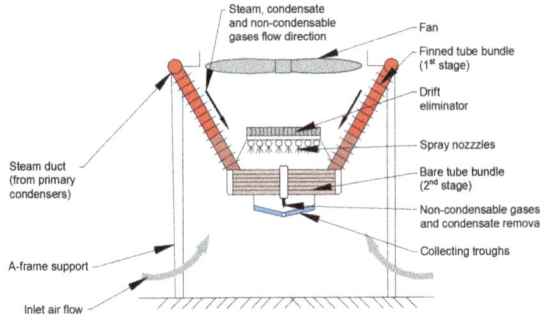

Fig. 1. Schematic diagrams of the Induced draft HDWD

The counter-current flow configuration of air and water over the surface of the flat bare tube bundle is considered. The steam flowing inside the tubes is condensed by the deluge water and cooling air on the exterior of the tubes. A condensation film formed inside the tubes runs down and along the tube wall due to gravity. The Film-wise condensation is taken into account, as the inside tube surface is assumed to be clean and uncontaminated or coated with a substance that inhibits wetting and promote the dropwise condensation [3].

There are numerous recently studies in which the best configuration of the HDWD second stage's tube bundles are identified and modelled. [4] and [5] investigated the performance of a delugeable bundle of 38 mm and 19 mm diameter round tubes for the HDWD, respectively by using a one-dimensional model which was analysed by means of Merkel approach. When the HDWD operates in wet operating mode, [5] reported that the HDWD performance is two to three times of the traditional convectional dephlegmator. [6] and [7] modelled and evaluated the performance characteristics of a round bare tube air-cooled heat exchanger bundle for the second stage of a HDWD during wet and dry operation. [7] determined the thermal performance, and heat and mass transfer coefficients from the experimental data, as well as from a one-dimensional effectiveness-Number of Transfer Units (NTU) model, whereby the Merkel method was employed during wet operation. [8] investigated the performance of a hybrid (dry/wet) cooling system (HDWCS) by using the correlations of [6] and [9]. At different air relative humidity and an ambient dry bulb temperature of 32 °C, the HDWCS' performance was found to be between 35% and 140% relative to a conventional cooling system. The performance a delugeable flat bare tube bundle for the second stage for induced draft HDWD was studied by [1] and [10], in which both one-dimensional analytical and two-dimensional numerical models were employed. The analytical model was based on Merkel, Poppe, and heat and mass transfer analogy methods, while the numerical model was based on heat and mass transfer method. In most of the above-mentioned studies, the round bare tube bundles were employed, except in [1] and [10], where flat bare tube bundle was employed. However, in [1] and [10]'s work the entire tube was modelled as flat, and the round end sections of the tube were not taken into consideration. Therefore, in this study, the round end sections of the tube are modelled separately from the flat section of the tube.

2 Methods

2.1 Model Description

Since the tubes of flat bare tube bundle is considered to be arranged in the symmetrical manner, only the section shown in Fig. 2 is considered, and the differential control volumes are taken at the locations indicated in the same Figure, which are shown in detail in Fig. 3.

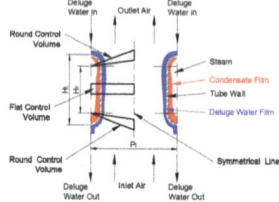

Fig. 2. Physical model of the flat bare tube bundle

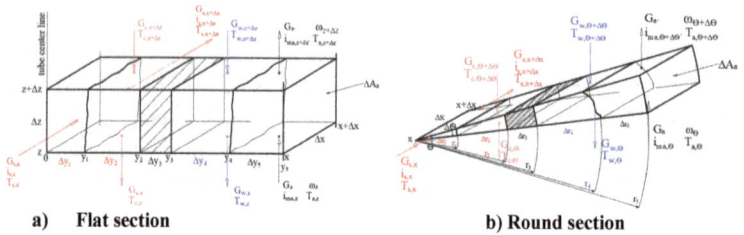

a)   **Flat section**   b) **Round section**

Fig. 3. Elementary control volume of the flat bare tube bundle

To analyse the heat and mass transfer through the end round sections of the tube, the round control volume (Fig. 3(b)) is used, whereby the governing differential equations are derived using cylindrical coordinates. The flat control volume (Fig. 3(a)) is employed, in Cartesian coordinates, for heat and mass transfer analysis in the flat section of the tube. For both control volumes, the mass and energy conservation principles and the following assumptions such as steady operation, uniform deluge water distribution on surface of the tubes and negligible steam velocity effects on condensate film were applied.

2.2 Governing Equations

The one-dimensional analytical model is developed based on Poppe, Merkel as well as heat and mass transfer analogy methods and the approach presented in [11]. Based on Poppe method, the heat transfer processes taking place within a one-dimensional flat control volume of the flat bare tube bundle defined by Eq. (1), (3), (5), and (7), while in round control volume of the flat bare tube bundle the heat transfer processes are described by Eq. (2), (4), (6), and (7). For Merkel's simplified approach, the Lewis factor $(h_a/c_{pma}h_d)$ is taken to be equal to one, the deluge water evaporation (dG_w/dz) is considered to be equal to zero, and the deluge water temperature is assumed to be constant (dT_w/dz). The change in deluge water mass:

$$\frac{dG_w}{dz} = G_a\frac{d\omega_a}{dz} = h_d a_w(w_{sw} - w) \tag{1}$$

$$\frac{dG_w}{d\theta} = \frac{r_4}{r_5}G_a\frac{d\omega_a}{d\theta} = h_d a_w(w_{sw} - w)i_v \tag{2}$$

where a_w is volumetric area of deluge water-side. The change in deluge water temperature:

$$\frac{dT_w}{dz} = \frac{1}{G_{w,z}c_{pw}}\left(G_s\frac{di_s}{dx} + G_a\frac{di_{ma}}{dz} - \frac{dG_w}{dz}c_{pw}T_{w,z}\right) \tag{3}$$

$$\frac{dT_w}{d\theta} = \frac{r_4}{G_{w,\theta}c_{pw}}\left(-\frac{1}{r_4}\frac{dG_w}{d\theta}c_{pw}T_{w,\theta} + G_{s,x}\frac{di_s}{dx} + \frac{1}{r_5}G_{a,\theta}\frac{di_{ma}}{d\theta}\right) \tag{4}$$

The air-water interface vapour mixture enthalpy:

$$\frac{di_{ma}}{dz} = \frac{h_d a_w}{G_a}\left[\frac{h_a}{c_{pma}h_d}(i_{masw} - i_{ma}) + \left(1 - \frac{h_a}{c_{pma}h_d}\right)i_v(w_{sw} - w)\right] \tag{5}$$

$$\frac{di_{ma}}{d\theta} = \frac{h_d a_w}{G_{a,\theta}}r_5\left[\frac{h_a}{c_{pma}h_d}(i_{masw} - i_{ma}) + \left(1 - \frac{h_a}{c_{pma}h_d}\right)i_v(w_{sw} - w)\right] \tag{6}$$

Therefore, the changes in the steam enthalpy can be expressed as:

$$\frac{di_s}{dx} = \frac{U_a a_a}{G_s}(T_s - T_{ws}) \tag{7}$$

2.3 Solution Methods of Governing Equations

For the flat and round control volume, the differential equations are integrated over the vertical flat height (H_f) and angle (θ) respectively. The obtained integrated governing equations are solved analytically using MathCAD software. The one-dimensional analytical model performance is studied based on Poppe, Merkel, and heat and mass transfer analogy methods. For both methods, the outlet air temperature and humidity ratio, as well as the deluge water inlet and outlet temperature are attained through Gauss-Seidel iteration method. The output performance data for the down round end section are taken to be the input performance data of the vertical flat section, while the input performance data of the upper round end section are equal to output performance data of the vertical flat section.

For Poppe method the mass transfer coefficient at the air-water interface is obtained from the relationship between heat and mass transfer coefficient through a Lewis factor which defined by using [12] equation as:

$$Le_f = 0.866^{0.667}\left(\frac{w_{sw}+0.622}{w+0.622}-1\right)/ln\left(\frac{w+0.622}{w+0.622}\right) \tag{8}$$

In Merkel analysis method, the mass transfer coefficient at the air-water interface is obtained in the same manner as in Poppe method. However, in Merkel approach, the Lewis factor is assumed to be equal to unit. For the heat and mass transfer analogy method, the mass transfer convection coefficient is determined from the analogy between heat and mass transfer at the air-water interface by using correlations accessible in literatures.

3   Validation of Model

The model's validation is carried out based on the following typical designing and operating conditions of a delugeable tube bundle as shown in Table 1 [1, 10].

Table 1. Typical designing and operating conditions of a delugeable tube bundle

Description	Symbol	Unit	Value
Steam temperature	T_s	°C	45
Atmospheric pressure	p_a	Pa	101325
Ambient inlet air temperature	T_a	°C	15
Relative humidity	ϕ	%	60
Air velocity	v_a	m/s	2
Deluge water mass flow rate	m_w	kg/s	$10\Delta m_w$

The flow on the air-side was considered as shown in Fig. 4. The height for the entrance region (H_{er}) at which two air-side boundary layers meet was calculated, and the flow within it was considered to be a developing flow. Therefore, the heat transfer rate and air-side pressure drop within entrance region were determined by employing the external flow theories. The flow in the rest of the air flow channel was considered to be a fully developed flow, and therefore the internal flow theories were used. The critical height at which the flow would changes from laminar to turbulent was calculated to be 1020 mm, which greater than the considered tube height ($H_t = 600$ mm), therefore, the laminar flow theories were employed in both entrance and fully developed regions.

In Table 2, the comparisons of overall solutions as per method are presented. For further validation, the model's solutions are compared to that of the model presented in [1] as depicted in Table 3. In [1], the performance of the similar tube bundle is analysed, by employing only a vertical flat control volume and the derived differential governing equations were in Cartesian coordinates, and are integrated over the entire tube height including the bottom and upper round end tube sections.

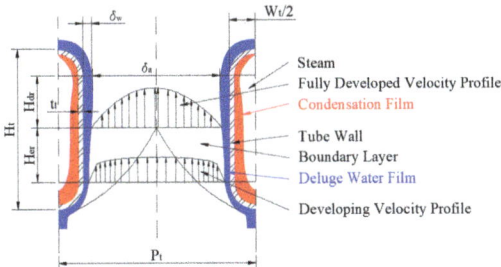

Fig. 4. Tube bundle section, illustrating the air-side flow between two adjacent tubes

Table 2. Comparison of one-dimensional model results as per approaches

Approach	Deluge water evaporation rate	Heat transfer rate	Air-side pressure drop
	Δm_w	Q_a	ΔP_a
	kg/s	W	Pa
Merkel	0.001711	5358.24	278.62
Pope	0.001782	5511.11	279.14
Heat and mass transfer analogy	0.001590	5004.48	275.23

Table 3. Comparison of present and [1] model

Approach	Model	Deluge water evaporation rate	Heat transfer rate	Air-side pressure drop
		Δm_w	Q_a	ΔP_a
		kg/s	W	Pa
Merkel	Present	0.001711	5358.24	278.62 (29.54)
	Angula (2018a)	0.001420	4767.16	31.001
Pope	Present	0.001782	5511.11	279.14 (29.10)
	Angula (2018a)	0.001585	4903.93	30.915
Heat and mass transfer analogy	Present	0.001590	5004.48	275.23 (29.36)
	Angula (2018a)	0.001416	4463.33	30.555

Further validation was performed by employing the performance data for the model of the delugeable round bare tube bundle (DRBTB) represented in [6]. For this, the tube bundle's geometric parameters and operating conditions for DRBTB model were incorporated into the present model. The DRBTB is a multi-row which is 2.87 m wide, 1 m high and 2.5 m long. The DRBTB performance data are 5.126 MW and 22.703 Pa, for the heat transfer rate and air-side pressure drop respectively. During the analysis, the frontal area, steam flow area and fan power, which are 7.172 m², 0.368 m² and 476.58 W respectively, were kept constant and equivalent to that of DRBTB. Since the outer size of the tube bundle was fixed, the pitch between the tubes was varied, in order to determine the effect of tube pitch on heat transfer rate and air-side pressure. The tube pitches 28, 30 and 38 mm were considered. The heat transfer rate ratio (Q_r/Q_f) and air-side pressure drop ratio $(\Delta P_r/\Delta P_f)$, for the DRBTB and present model were computed as depicted in Fig. 5 and 6 respectively.

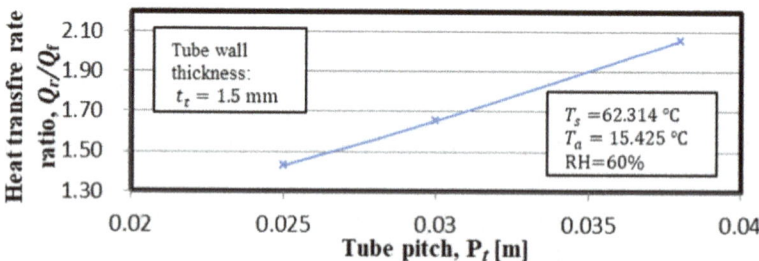

Fig. 5. Tube pitch effect on the heat transfer rate ratio

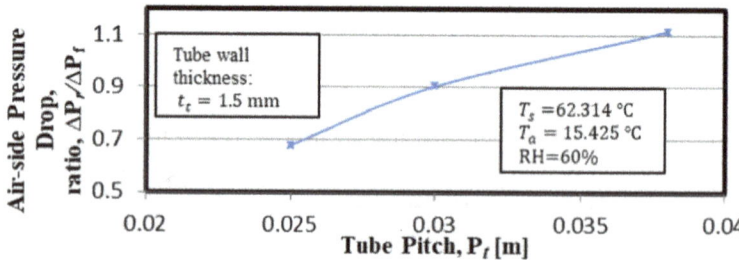

Fig. 6. Tube pitch effect on the air-side pressure drop ratio

4 Results Discussion

The solutions from all the methods are found to be comparable. By comparing the Merkel and heat and mass transfer analogy solutions to Poppe solution, a 2.8% and 9.2% differences in heat transfer rate were achieved, respectively. The similar differences were yielded in [1]. By comparing the present model to that of [1], the increases of around 11%, in heat transfer rates are achieved for both methods. This shows that, modelling of the round end and flat sections of the tube separate, improves and predicts thermal performance of the tube bundle better than when both round and flat sections of the tube are modelled together. Moreover, the air-side pressure drop for present model is found to be 88% higher than for [1] model for all methods, because of the higher drag force that exerted on the down round end tube section by the air, which was not considered in

[1]. Therefore, for the present model only the air-side pressure drop in the flat section (entrance and fully developed region) is comparable to [1] model as indicated in the bracket in Table 3. From Figs. 5 and 6, it is clear that both heat transfer rate and air-side pressure drop ratios increased with tube pitch. This indicates that, the present model yields better performance at smaller tube pitches. However, this comes at the cost of high air-side pressure drop. The heat transfer rate and air-side pressure drop for [6] model were found to be 1.53 and 0.68 times respectively, of that of the present model at smallest tube pitch (25 mm), and 2.06 and 1.12 times respectively, at larger tube pitch (38 mm). The variations in the models results are mainly due to tube shape and the deluge water temperature, which were different.

5 Conclusion

Modelling of the round ends and flat sections of the tube separately, improves and depicts the thermal performance of the flat bare tube bundle, under deluging cooling condition more accurately. Furthermore, the present model was found to be in good agreement with other model's results in the literatures. The increase of 11.8% and 88% in heat transfer rate and air-side pressure drop respectively for both methods was found when the present model was compared with [1] model results. Heat transfer rate obtained by Poppe method is found to be 2.8% and 9.2% higher than that of Merkel, and heat and mass transfer analogy methods, respectively. Moreover, the heat transfer rate for the model of DRBTB was found to be higher than that of the present model at both smaller and larger tube pitch, while the air-side pressure drop for the model of DRBTB was found to be lower and higher than that of the present model at smaller and larger tube pitch, respectively.

Nomenclature

A	m^2	Area	Subscripts	
a	m^2/m^3	Volumetric area	a	Air
c_p	[J/kg K]	Specific heat at constant pressure	v	Air and water mixture (vapour)
H	[m]	Height	s	Sensible
h	[W/mK]	Heat transfer coefficient	l	Latent
h_d	[kg/m²s]	Mass transfer coefficient	er	Entrance region
i	[J/kg]	Enthalpy	w	Deluge water
i_{fg}	[J/kg]	Latent heat	ws	Deluge water surface

(*continued*)

(*continued*)

A	m^2	Area	Subscripts	
k	[W/mK]	Thermal conductivity	f	Flat
L	[m]	Length	dr	Developed region
G	[kg.m/m^{3}s]	Mass flux	g	Gravity
P	[m]	Pitch	ma	Moist air
Q	[W]	Heat transfer rate		
t	[m]	Thickness		
T	[°C]	Temperature		
w	[kg(H$_2$O)/kg dry air]	Humidity ratio		
x, y, z	[m]	Co-ordinate or distance		
y_1, y_2, y_3, y_4, y_5	[m]	Distance		
r_1, r_2, r_3, r_4, r_5	[m]	Radius		
θ	[degree]	Angle		

References

1. Angula, E.: Modelling of a delugeable flat bare tube bundle for an air-cooled steam condenser. EPH – Inter. J. Sci. Eng. 4(4), 28–41 (2018)
2. Heyns, J.A., Kröger, D.G.: Performance characteristics of an air-cooled steam condenser with a hybrid dephlegmator. R & D J. South Afr. Inst. Mech. Eng. 28, 31–36 (2012)
3. Yunus, A.C., Afshin, J.G.: Heat and Mass Transfer, Fundamentals and Applications, 4th edn. McGraw – Hill, New York (2011)
4. Heyns, J.A.: Performance characteristics of an air-cooled steam condenser incorporating a hybrid (dry/wet) dephlegmator. Master thesis, University of Stellenbosch, RSA (2008)
5. Owen, M.: Air-cooled condenser steam flow distribution and related dephlegmator design considerations. Doctor thesis, Stellenbosch University, RSA (2013)
6. Anderson, N.R.: Evaluation of the performance characteristics of a hybrid (dry/wet) induced draft dephlegmator. Master thesis, University of Stellenbosch, RSA (2014)
7. Reuter, H., Anderson, N.: Performance evaluation of a bare tube air-cooled heat exchanger bundle in wet and dry mode. Appl. Ther. Eng. 105, 1030–1040 (2016)
8. Graaff, H.: Performance evaluation of a hybrid (dry/wet) cooling system. Master of Engineering (Mechanical), Stellenbosch University, RSA (2017)
9. Mizushina, T., Ito, R., Miyashita, H.: Experimental study of an evaporative cooler. Int. Chem. Eng. 7(4), 727–732 (1967)
10. Angula, E.: Two-dimensional model of a delugeable flat bare tube air-cooled steam condenser bundle. Am. J. Eng. Res. (AJER) 7(6), 71–86 (2018)
11. Kröger, D.G.: Air-cooled Heat exchangers and Cooling Towers: Thermal-Flow Performance Evaluation and Design. PennWell Corporation, Tulsa, Oklahoma, USA (2004)
12. Bosjnakovic, F.: Technical Thermodynamics, pp. 326–331. Holt Reinhart and Winston, New York (1965)

Investigation of the Mechanical and Microstructural Properties of TIG Welded Ti6Al4V Alloy

P. O. Omoniyi[1,2(✉)], R. M. Mahamood[1,3], N. Arthur[4], S. Pityana[4], S. A. Akinlabi[5], S. Hassan[5], Y. Okamoto[6], M. R. Maina[7], and E. T. Akinlabi[8]

[1] Department of Mechanical Engineering Science, University of Johannesburg, P. O. Box 524, Johannesburg, South Africa
[2] Department of Mechanical Engineering, University of Ilorin, P. M. B. 1515, Ilorin, Nigeria
omoniyi.po@unilorin.edu.ng
[3] Department of Materials and Metallurgical Engineering, University of Ilorin, P. M. B. 1515, Ilorin, Nigeria
[4] CSIR National Laser Centre, P. O. Box 395, Pretoria, South Africa
[5] Department of Mechanical Engineering, Butterworth Campus, Walter Sisulu University, Butterworth, South Africa
[6] Okayama University, Okayama, Japan
[7] Jomo Kenyatta University of Agriculture and Technology, Nairobi 62000-00200, Kenya
[8] Pan Africa University for Life and Earth Sciences Institute, Ibadan, Nigeria

Abstract. The joint integrity of 1 mm thick sheets of Ti6Al4V alloy welded autogenously using TIG welding was investigated in this article. The current and gas flow rate were varied and their effects on the mechanical properties and microstructure of the weld were analyzed. Results show that the microstructure within the weld zone consists of α'martensitic phase and are coarse, which results in higher microhardness within the weld zone compared to the base metal. The samples with a higher gas flow rate were observed to also improve the tensile strength, while samples with a lower gas flow rate resulted in tensile strength below that of the base metal.

1 Introduction

Titanium is one of the most used metals in industries such as aerospace, ship and chemical, due to its unique mechanical properties of good strength to weight ratio and high corrosion resistivity [1]. Joining titanium alloys could pose adverse effects on the mechanical properties if not properly welded in a controlled environment since titanium alloy is adversely affected in atmospheric gases such as nitrogen, oxygen at a temperature of 350 °C [2]. Several welding techniques have been employed in joining of Ti6Al4V alloy sheets, some of which are; gas tungsten arc welding (GTAW) or tungsten inert gas welding (TIG), gas metal arc welding (GMAW), electron beam welding (EBW), friction stir welding (FSW), laser beam welding (LBW) [3, 4]. Each of these techniques have their unique advantages and disadvantages. The EBW is laborious due to the vacuum needed to be created within the working chamber, LBW is expensive to set up, FSW has its

© The Author(s), under exclusive license to Springer Nature Singapore Pte Ltd. 2021
M. Awang and S. S. Emamian (Eds.): *Advances in Material Science and Engineering*, LNME, pp. 111–118, 2021.
https://doi.org/10.1007/978-981-16-3641-7_15

limitation to specific applications and GMAW creates a wider heat affected zone (HAZ) due to the enormous heat input and it is associated with spatter formation. Therefore TIG welding which uses inert gases as a shield from environmental contamination is preferred compared to other welding techniques due to its arc stability, better economy and easier applicability [5–8].

Analysis of research works shows that quite many researches have been done on TIG welding of Ti6Al4V alloy and some advantages and disadvantages have been highlighted. One of the major problems with titanium alloy highlighted by Balasubramanian et al., [6] was the coarsening of grains within the HAZ and fusion zone (FZ) during TIG welding, which is as a result of the heat input and could cause embrittlement of the material. Babu and Rahman [9] also attributed this embrittlement to decreasing size of the α colonies, which controls the maximum dislocation slip length. The decrease in the α colonies are caused by thermomechanical activities during welding. The ductility within the weld zone (WZ) is also attributed to the acicular and martensitic intragranular microstructure and a large prior β grain size. Therefore, controlling the input current and gas flow rate will improve the mechanical properties of the WZ as reported by some works of literature where pulsed current TIG [9, 10] have been used to control the heat input during TIG welding. Even though, it is of less economic advantage and no change in mechanical properties, compared to the manual TIG welding as reported in [11]. Columnar grains have also been reported to be observed in both pulsed and unpulsed TIG welding [12]. Some researchers have identified current, welding speed and voltage as the most important parameters in TIG welding [13, 14]. Thus, it is crucial to control the heat input during welding and the level of gas shielding so as to avoid contamination of the weld.

Therefore, this article investigates the joint integrity of TI6Al4V alloy using TIG welding. It focuses on the effect of welding parameters such as current, voltage and gas flow rate on the microstructure, tensile and microhardness properties of Ti6Al4V alloy.

2 Methods

Ti6Al4V grade 5 of 100 × 60 × 1 mm was joined autogenously using TIG welding technique with argon gas (99.999% purity) as a shield to prevent contamination and oxidation of weld. The parameters used for welding were chosen based on the range of optimum parameters given in [15]. Before the welding process, the samples were cleaned with acetone to remove oxide films and welding was done along the 100 mm length. The elemental composition and mechanical properties of the alloy in accordance with ASTM B265 [16] are presented, Table 1 and 2 respectively. Four welds were carried out by varying current and gas flow rate Table 3.

Table 1. Chemical composition of Ti6Al4V alloy

Element	Ti	Al	V	Fe	C	N	H	O	Others
Weight (%)	Remainder	6.10	4.0	0.15	0.03	0.018	0.002	0.13	Each < 0.10

Table 2. Mechanical properties of Ti6Al4V (grade 5) alloy

Parameters	Tensile strength (MPa)	Yield strength (MPa)	Elongation (%)	Microhardness (HV)
Values	895	825	10	362

Table 3. Experimental process parameters

Sample no	Current (A)	Gas flow rate (L/min)
T11	40	9
T12	30	9
T13	30	7
T14	40	7

The welded plates were cut into 25 × 10 × 1 mm for microstructure analysis, each sample was gradually grinded using grinding papers from 320 to 1200 grit size, they were then polished and etched using Kroll's Reagent. The microstructure of the samples was taken using Olympus DP 25 optical microscope at 50x magnification. Each sample was analyzed at the FZ, HAZ and base material/metal (BM).

Microhardness testing was done using the Indentec digital Vickers microhardness tester. A force of 4.9 N was applied with a dwell period of 15 s. A minimum of twenty indentations was done on each sample across the sample at a distance of 1 mm.

Tensile test samples were cut using the ASTM E8 standard [17] Fig. 1. The tensile samples were pulled using universal testing machine Zwick Roell 2250.

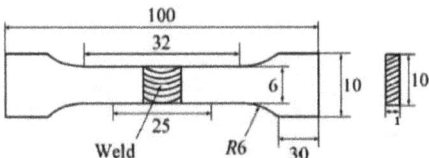

Fig. 1. Tensile sample

3 Results and Discussion

3.1 Microstructure

The microstructure of welded metals is classified into three zones, the BM, HAZ and FZ. Figure 2(a) shows the microstructure of the BM, which is made of the white α and black β phase, as observed in [18]. The HAZ Fig. 2(b) is characterized by the equiaxed α, transformed β and elongated α grains. The FZ Fig. 2(c), which undergoes melting and solidification is observed to have the widmanstatten structure, martensite (α' phase) and recrystallized β phase together with secondary acicular α phase and β phase, which was similar to what was observed in [19].

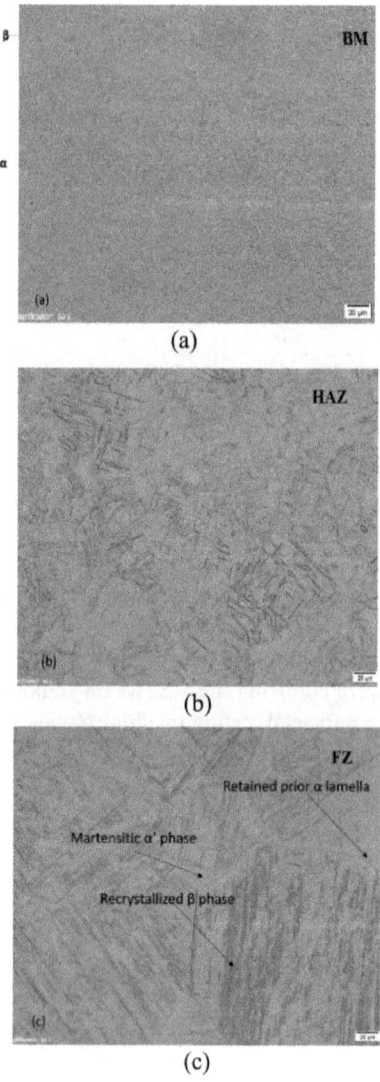

(a)

(b)

(c)

Fig. 2. (a) Optical micrograph of BM (b) Optical micrograph of HAZ (c) Optical micrograph of FZ.

3.2 Microhardness

The microhardness profile is shown in Fig. 3 and it was carried out in accordance with ASTM E384 [20]. The hardness is observed to generally be highest at the FZ for all samples, due to the increment of β grains and extensiveness of the grains within the FZ. The BM was observed to exhibit hardness of the range of 362 ± 4 HV, the HAZ and FZ has an average hardness of 366.4 ± 5 HV and 378.86 ± 5 HV respectively. There is a relationship between the hardness profile and the microstructure, the primary α is lower in hardness and higher in ductility than the β phase, while the α' martensitic shows a higher hardness than the β phase as also reported in [19] and [21]. The higher hardness within the WZ can also be attributed to oxygen pickup during welding, which causes a hardening effect as observed in [22] hence, the FZ has a higher hardness than the BM.

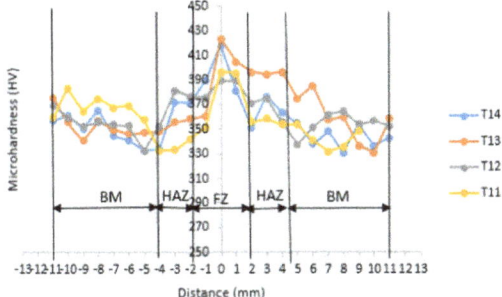

Fig. 3. Microhardness profile of TIG welded samples

3.3 Tensile Strength

Tensile strength was conducted in accordance with ASTM E8 [17]. Figure 4(a) shows the tensile failure points of each sample, while Fig. 4(b) shows the tensile strength of each sample. The maximum tensile stress of 913.67 MPa was observed in sample with current and gas flow rate combination of 30 A and 9 L/min respectively, compared with the tensile strength of the base metal 895 MPa in Table 2. The increase in tensile strength is attributed to the ductility nature of the α' martensitic microstructure observed in the FZ as observed in [23], even though the failure occurred at the HAZ, It could further be attributed to the higher gas flow rate which provided a sustainable inert environment during the cooling of the weldments. Samples with lower tensile strengths below the values of the BM, which were welded with gas flow rate of 7 L/min were observed to fracture at the FZ, irrespective of the current used. This can be attributed to contaminations such as oxygen picked up during welding as a result of lower gas flow rate. Table 4 shows the joint efficiency of the welds, all welds shows a joint efficiency above the standard 70% for butt joint [24], except welds with parameter combination of 30 A and 7 L/min current and gas flow rate respectively, which is weaker as a result of lower heat input and gas flow rate.

Table 4. Joint efficiency of TIG welded samples

Sample no	Joint efficiency (%)
T11	98
T12	102
T13	58
T14	77

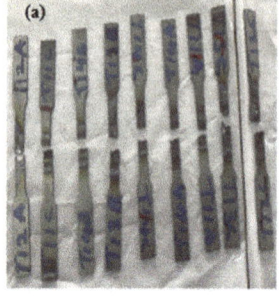
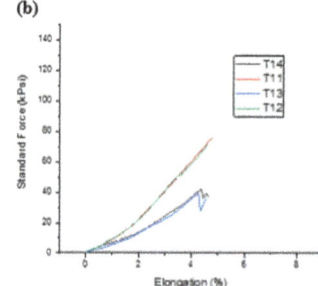

Fig. 4. (a) Joints of welded samples (b) Tensile properties of TIG welded samples

4   Conclusions

TIG welding of Ti6Al4V alloy autogenously has been carried out and the following conclusions are drawn from the results of the characterizations done.

1. The microstructure within the HAZ and FZ is made up of α′ martensitic phase which results in higher hardness with the WZ.
2. The microhardness was observed to increase from the BM towards the FZ.
3. The gas flow rate, has a significant effect on the tensile strength of welds, samples with 9 L/min gas flow rate exhibited an improved tensile strength.
4. Irrespective of the welding current, gas flow rate of 7 L/min exhibited lower tensile strength.

Acknowledgments. The authors will like to acknowledge the funding of the National Research Foundation (NRF) and Japan Society for the Promotion of Science (JSPS) for providing financial assistance. Also the Pan African University for Life and Earth Sciences Ibadan Institute (PAULESI) for the article processing fee payment.

References

1. Vaithiyanathan, V., Balasubramanian, V., Malarvizhi, S., et al.: Gas tungsten constricted arc welding (GTCAW) parameters optimization to attain maximum tensile strength in Ti–6Al–4V alloy sheets used in aero-engine components. Multiscale Multidiscip. Model. Exp. Des. **2**, 291 (2019)

2. Turichin, G., Tsibulsky, I., Somonov, V., et al.: Laser-TIG welding of titanium alloys. IOP Conf. Ser. Mater. Sci. Eng. (2016). https://doi.org/10.1088/1757-899X/142/1/012009

3. Sun, Z., Lv, Y., Xu, B., Liu, Y., Lin, J., Wang, K.: Investigation of droplet transfer behaviours in cold metal transfer (CMT) process on welding Ti-6Al-4V alloy. Int. J. Adv. Manuf. Technol. **80**(9–12), 2007–2014 (2015). https://doi.org/10.1007/s00170-015-7197-9

4. Reda, R., Magdy, M., Rady, M.: Ti–6Al–4V TIG weld analysis using FEM simulation and experimental characterization. Iran. J. Sci. Technol. Trans. Mech. Eng. **44**(3), 765–782 (2019). https://doi.org/10.1007/s40997-019-00287-y

5. Hoye, N., Li, H., Norrish, J., et al.: Post-weld atmospheric contamination of gas tungsten arc deposited welds in commercially pure and Ti-6Al–4V titanium alloys. In: Ti 2011 – Proceedings of the 12th World Conference Titanium, vol. 2, pp. 1629–1633 (2012)

6. Balasubramanian, M., Jayabalan, V., Balasubramanian, V.: Optimizing the pulsed current gas tungsten arc welding parameters. J. Mater. Sci. Technol. **22**, 821–825 (2006)

7. Gnedenkov, A.S., Sinebryukhov, S.L., Mashtalyar, D.V., et al.: Effect of microstructure on the corrosion resistance of TIG welded 1579 alloy. Materials (Basel) (2019). https://doi.org/10.3390/ma12162615

8. Chen, C., Fan, C., Cai, X., et al.: Investigation of formation and microstructure of Ti-6Al-4V weld bead during pulse ultrasound assisted TIG welding. J. Manuf. Process. **46**, 241–247 (2019)

9. Babu, N.K., Raman, S.G.S.: Influence of current pulsing on microstructure and mechanical properties of Ti-6Al-4V TIG weldments. Sci. Technol. Weld. Join. **11**, 442–447 (2006)

10. Mishra, D., Manjunath, A., Parthiban, K.: Interpulse tig welding of titanium alloy (TI-6Al-4V). Indian Weld. J. **50**, 56 (2017)

11. Becker, D.W., Adams Jr, C.M.: The role of pulsed GTA welding variables in solidification and grain refinement. Weld. Res. **58**, 143–152 (1979)

12. Mohandas, T., Madhusudan Reddy, G.: Effect of frequency of pulsing in gas tungsten arc welding on the microstructure and mechanical properties of titanium alloy welds: a technical note. J. Mater. Sci. Lett. **15**, 626–628 (1996)

13. Reddy, V.S., Brahma, R.K., Venkata, S.K.: Optimization of welding parameters of Ti 6al 4v cruciform shape weld joint to improve weld strength based on Taguchi method. Mater. Today Proc. **5**, 4948–4957 (2018)

14. Zaid, A.I.: Investigation into the TIG welded joint of titanium G-5 alloy sheet Investigation into the TIG welded joint of titanium G-5 alloy sheet (2018). https://doi.org/10.1088/1757-899X/377/1/012114

15. Muncaster, P.W.: Practical TIG (GTA) welding. 131 (1991)

16. Metals, R.: Standard specification for titanium and titanium alloy strip, sheet, and plate 1. Annu. B ASTM Stand. **03**, 1–9 (2010)

17. ASTM E8. Standard Test Methods for Tension Testing of Metallic Materials 1 (2016). https://doi.org/10.1520/E0008

18. Gope, D.K., Kumar, U., Chattopadhyaya, S., Mandal, S.: Experimental investigation of pug cutter embedded TIG welding of Ti-6Al-4V titanium alloy. J. Mech. Sci. Technol. **32**(6), 2715–2721 (2018). https://doi.org/10.1007/s12206-018-0528-7

19. Yan, G., Tan, M.J., Crivoi, A., et al.: Improving the mechanical properties of TIG welding Ti-6Al-4V by post weld heat treatment. Procedia Eng. **207**, 633–638 (2017)

20. ASTM. ASTM E384–2016: Standard Test Method for Knoop and Vickers Hardness of Materials. ASTM Stand i:1–43 (2016)
21. Kishore, B.N., Ganesh, S.R.S., Mythili, R., et al.: Correlation of microstructure with mechanical properties of TIG weldments of Ti-6Al-4V made with and without current pulsing. Mater. Charact. **58**, 581–587 (2007)
22. Mehdi, B., Badji, R., Ji, V., et al.: Microstructure and residual stresses in Ti-6Al-4V alloy pulsed and unpulsed TIG welds. J. Mater. Process. Technol. **231**, 441–448 (2016)
23. Beris, B.: Effects of Gas Shielding Flow Rate on Weld Quality of TIG Weldind in Ti6Al4V Alloy. Instabul Technical University (2012)
24. ASME. ASME Boiler and Pressure Vessel Code VIII: An International Code. Stand. No. Div. I UW-20 (2004)

Prediction of the Temperature Behaviour During Friction Stir Welding (FSW) Using Hyperworks®

Bahman Meyghani[1], Mokhtar B. Awang[2(✉)], and Reza Teimouri[3]

[1] Institute of Materials Joining, Shandong University, 17923, Jingshi Road, Jinan 250061, China
[2] Department of Mechanical Engineering, Faculty of Engineering, Universiti Teknologi PETRONAS (UTP), 32610 Bandar Seri Iskandar, Perak Darul Ridzuan, Malaysia
Mokhtar_awang@utp.edu.my
[3] Faculty of Mechanical Engineering, University of Kashan, Kashan, Iran

Abstract. Fundamentally, heat is needed during friction stir welding (FSW) for joining materials together. This heat is usually produced by friction force and material deformation. Therefore, friction coefficient is a significant factor and in order to increase the efficiency and the quality of the simulated model, there is a need to explore it accurately. However, previous studies used constant values of the friction coefficient resulting in inaccuracy of the model. This paper proposes, a mathematical formulation for predicting temperature dependent values of the friction coefficient using coulomb friction and von Mises yield laws. Then, the friction coefficient values are used to simulate a finite element model. HyperMesh® and HyperView® solvers have been employed from Altair Hyperworks® to simulate the process. The results of the model showed that, the temperature at the shoulder surface is always higher than the pin area, thus the heat generated by shoulder is found to be higher around 60 °C. Furthermore, higher temperature at the advancing side (around 30 °C) was obtained. Finally, the model is verified to show the accuracy of the predicted friction coefficient values and the results of the finite element model.

Keywords: Friction stir welding · Heat · Friction force · Material deformation · Friction coefficient · Finite element model

1 Introduction

There is an increasing need to reduce structures weight, especially for the aircraft panels and aerospace industries. Advanced welding techniques play a significant role in manufacturing of these lightweight structures. Friction stir welding (FSW) is one of the most important and relatively new solid state welding methods which was invented and patented in the last decades [1–3]. FSW was developed in the automotive industry as an alternative method for welding of aluminium sheets in 2001 [4, 5]. Nowadays, FSW can be used to join a variety of materials without the use of filler material and the process currently is using in a lot of industrial applications, such as railway, marine,

© The Author(s), under exclusive license to Springer Nature Singapore Pte Ltd. 2021
M. Awang and S. S. Emamian (Eds.): *Advances in Material Science and Engineering*, LNME, pp. 119–130, 2021.
https://doi.org/10.1007/978-981-16-3641-7_16

land transportation and aerospace [6–8]. As can be seen in Fig. 1, through this welding method, a rotational cylindrical tool, which consists of a pin and a shoulder plunges into the workpieces, then moves along the welding seam [9–11]. Some of the key benefits of FSW are the elimination of solidification cracking, liquation cracking and porosity, and the better-quality for mechanical properties of the join [12–14]. On the other hand, FSW is a complicated procedure, because the process involves highly nonlinear (coupled) physical phenomenon, such as severe plastic deformation, complex flow of the material, complicated interactions behaviour between the tool and the workpiece, and complex thermal behaviour [15–17]. These multiple parameters highly affect the joint quality and the welding efficiency [18, 19]. Consequently, FSW has been studied by a considerable amount of studies [20–25].

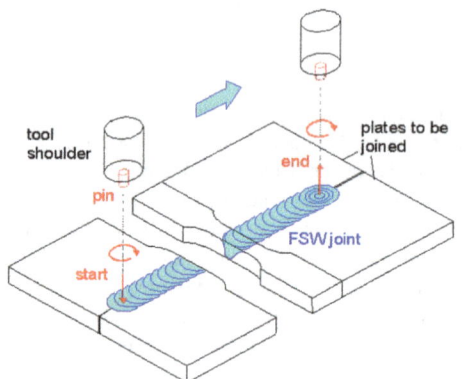

Fig. 1. Friction stir welding schematic view

In the meantime, without finite element modelling techniques, FSW can only be studied in experiments [26–29] which is time consuming and expensive [30]. As mentioned earlier, the welding process has a very complicated behaviour, therefore the investigation of the FSW process such as predicting the temperature evaluation or finding the material behaviour during the weld is very difficult [31–34], while finite element methods (FEMs) are able to solve complicated governing equations by providing an effective way for analysing the joint formation and the behaviour of the weld [19, 35].

Basically, FE is a technique in which approximate solutions of complex equations can be solved by using boundary conditions, because the calculations of variations in FE minimize the error and produce a stable condition. To illustrate, in FEMs many tiny lines will produce a large domain, because the method uses various simple elements for connecting all of the small sub domains (named finite elements) in order to find approximate shape over a larger domain. In the meantime, modelling of FSW is a very complicated process that involves a highly nonlinear thermomechanical behaviour. For example, some nonlinear behaviours are heat transfer mechanics and extremely severe plastic deformations near the pin (in the stirring zone). Consequently, by using FEMs the material behaviour and the thermal history during different process parameters and different tool geometries can be investigated. Hence, numerical simulations can be appropriate for investigating significant features of the process [36–38].

The contact interactions contain a pair of interactions which defines the contact between the workpiece and the tool. There are two different surfaces in the interaction, first one that is the tool surface has to be defined as a master surface, because it is harder in comparison with the workpiece which needs to be defined as a slave surface. In this regard, this "balanced master slave" arrangement is significant for getting more uniform pressure of the contact in the welding surfaces and also avoiding the hourglass effects. It is also combined with a softened contact interaction property to promote a sharing of the local contact pressure among nodes on both sides of the interface. One of the most important facts in modelling of the FSW process is the choice of the contact model. Many researchers [39, 40] have chosen the classical Coulomb friction law and considered that the Coulomb friction model allows realistic results. Two contact conditions were used to investigate the heat generation which compared the classical and the modified Coulomb law. The results showed that, the temperature and friction force are controlled by the contact pressure. Since in the research, there was a comparison between Norton and Coulomb laws, it was determined that the coefficient in the Norton model affected by forces, but these forces should be computed experimentally [41]. Two important parameters in the contact condition are the friction coefficient and the slip rate. Several studies [42, 43] have estimated these parameters by measuring the tool torque and the axial pressure. In the meantime, full sticking, full sliding or partial sliding/sticking conditions have been used. For example, full sliding condition was employed in a study [44], but the pin depth was neglected and the contact geometry was not realistic. Another study assumed [45] an uniform shear stress in the tool workpiece surface by using a machine power input. Meanwhile, in order to discover the mechanical and the thermal behaviours of FSW, a thermal model was proposed in a full sliding contact condition [46]. In addition, some researchers [47, 48] measured the plastic deformation to investigate the contact behaviour. Two conditions, full sliding and full sticking were considered [49], but unfortunately, both models were uncoupled due to the complexity of the process. Furthermore, Gerlish et al. [50] and Schmidt et al. [51] results showed that, the contact condition has a partial sliding/sticking condition.

Meanwhile, due to the simplifications, the friction coefficient values were taken from other machining processes or were considered as a hypothetical amount that were taken from the literature. Based on the discussion above, friction coefficient has a significant role in increasing the accuracy of the results. In this paper, in order to increase the accuracy and development of the modelling of the FSW process, a mathematical formulation is proposed to resolve the governing equations of the FSW contact condition. Then, FEMs are used to predict the temperature behaviour during the process.

2 Material and Methodology

2.1 Sliding, Sticking and Partial Sliding/Sticking Conditions

True estimation of the friction coefficient by using straightforward methods is still undetermined [52], because the friction coefficient values depends on many different parameters such as material properties and tribological behaviours [53]. A linear sliding rate is presented when the intensity of the shear stress on the workpiece segments is available in the model. In the Coulomb law, the shear stress of the contacting interface is written as follows,

$$\tau_{fric} = \mu P_0 \tag{1}$$

where τ_{fric} is the shear stress, μ is the friction coefficient and P_0 is the pressure.

When the contact shear stress is more than the yield stress, the workpiece surfaces stick to the tool segments and with the increase of the temperature, von Mises shear stress criterion controls the material behavior. The shear stress, based on von Mises theory is calculated as follows [54],

$$\tau_y = \frac{\sigma_y}{\sqrt{3}} \tag{2}$$

In Johnson-cook model the equivalent yield stress is defined as a function of temperature and strain rate which is written as follows,

$$\sigma_y = \left[A + B(\varepsilon_P)^n \right] \left[1 + C \left[\frac{\dot{\varepsilon}_P}{\dot{\varepsilon}_0} \right] \right] \left[1 - \left[\frac{T_{FSW} - T_{room}}{T_{melt} - T_{room}} \right]^m \right] \tag{3}$$

where the values are the temperature (T_{FSW}), yield stress (A), strain factor (B), strain rate factor (C), strain exponent (n), temperature exponent (m), material melting point (T_{melt}) and room temperature (T_{room}). A mixed state of the sliding and the sticking conditions can be established when the full sliding ($\delta = 0$) and the full sticking ($\delta = 1$) conditions are explained as below,

$$\tau_{fric} = \mu P_0 \quad when \quad \delta = 0 \tag{4}$$

$$\tau_{fric} = \tau_y \quad when \quad \delta = 1 \tag{5}$$

In addition, the pressure at the shoulder and the pin bottom can be calculated as,

$$P_0 = \frac{F_N}{\pi R_S^2} \tag{6}$$

where F_N is the normal force and R_S is the shoulder radius.

In a partial sliding/sticking condition, the shear stress at the shoulder and pin bottom [55] is written as fallows,

$$\tau_0 = \delta \tau_y + (1 - \delta)\mu P_0 \tag{7}$$

where δ is the slip rate. For the pin side, the shear stress is defined as,

$$\tau_1 = \mu P_0 sin\alpha \quad when \quad \delta = 0 \tag{8}$$

$$\tau_1 = \tau_y \quad when \quad \delta = 1 \tag{9}$$

Hence, in the partial sliding/sticking condition the shear stress for pin side area can be explained as,

$$\tau_1 = \delta \tau_y + (1 - \delta)\mu P_0 sin\alpha \tag{10}$$

where α is the cone angle.

By using Eq. 7 and 10 the amount of the friction coefficient can be calculated as,

$$\mu = \frac{\tau_0 - \tau_1}{(1 - \delta)P_0(1 - \sin\alpha)} \tag{11}$$

And by solving Eq. 7 or 10, the slip rate is written as follows,

$$\delta = \frac{\tau_1 - \tau_0\sin\alpha}{(1 - \sin\alpha)\tau_y} \tag{12}$$

2.2 Finite Element Model Descriptions

During FSW, the interface temperature reaches up to 80–90% of the base material melting point [56], thus the temperature dependent material properties for aluminium 6061-T6 are selected for the workpiece material. C3D8RT element in dynamic explicit is used for the mesh. Near the welding seam, small mesh size is selected, while outside of the welding area larger size for the mesh is set. The rotational speed of 800 rpm and the transverse speed of 40 mm/min and the tilt angle of 2° are applied to the model. To decrease the computational time, mass scaling is set, also the tool is considered as a rigid body, because in this situation there is no need to calculate the outcomes for the rigid body. Therefore, the computational time of the simulation would be decreased.

3 Results and Discussion

According to the Eq. 2, when the temperature increases, material shear stress decreases, due to the direct proportion between the material shear stress and the material yield stress. It should be noted that, according to Johnson cook law, when the temperature rises, the material becomes softer, then the yield stress decreases. Furthermore, when the temperature increases, the material becomes weaker and whereby the welding force and the friction coefficient decrease. This is because the coefficient is highly influenced by the welding temperature. Figure 2 shows that, the friction coefficient has dropped with the rise of the temperature. In the term of the friction coefficient, the value of the 0 means there is no friction between surfaces and the value of 1 means that the frictional force is equal to the normal force (applied by the machine). It needs to be explained that, the coefficient depends on the contact geometry, forces, material property, and welding parameters.

The results of the literature [9] showed that, since the heat generation depends on the friction and the deformation, by the growth of the heat the coefficient would reduce. This issue happens, because of the growth of the stirring phenomenon at higher temperatures. To illustrate more, higher frictional force and higher deformation result in lower resistance of the material, and this issue increases the peak temperature. Consequently, higher frictional forces and deformations decrease the welding forces and increase the temperature [43, 57].

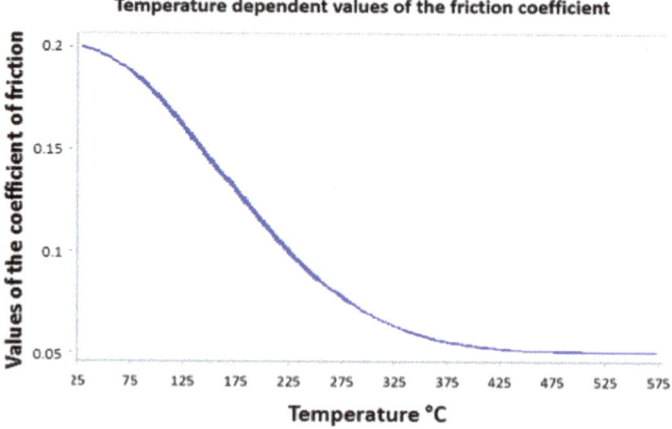

Fig. 2. The calculated values of the friction coefficient

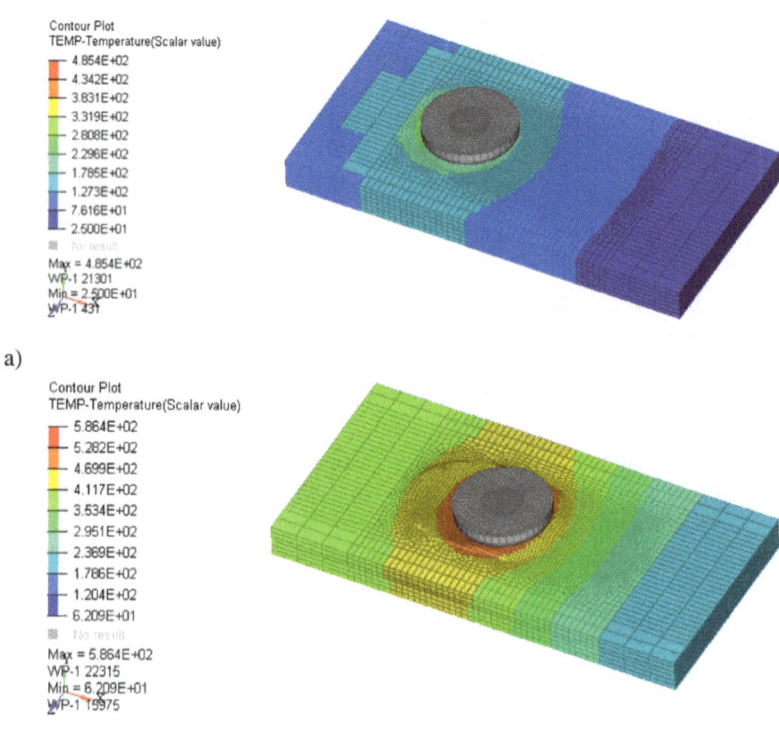

a)

b)

Fig. 3. Three dimensional view of the workpiece a) plunging and b) welding steps

The measured values showed that, the temperature varies from the room temperature to the material melting point. As can be seen in Fig. 3, the maximum temperature at the plunging step is around 485 °C and in the welding step is about 586 °C. This issue confirms by the literature who claimed that the welding temperature usually is in a range of 80% to 90% of the base material melting point [56]. Therefore, the achieved values for the temperature have a good agreement with the reported values in the literature [12, 58].

Figure 4 indicates the distribution of the temperature at the plunging and welding steps. As can be seen, the deformation pattern for the mesh is regular. At the plunging step, below the material moves up due to the plunging force of the tool, however in the welding step, the material which is located in the outside of the shoulder, formed an "U" shape patter. This results shows that the majority of the temperature is produced by the shoulder.

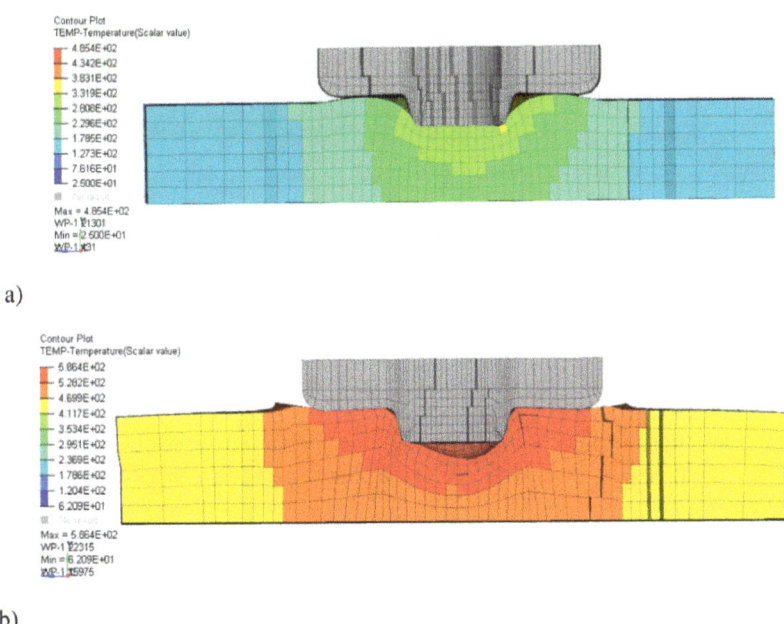

a)

b)

Fig. 4. Cross section of the workpiece at the a) plunging and b) welding steps

Figure 5 indicates the top view of the workpiece at the welding stage with and without the tool. It can be observed that, the temperature has a confined pattern at the back side of the welding, while at the front side the distance between the temperature circles is wider. The most important reason of this issue is the presence of the tilt angle which causes an additional forging force at the back side of the welding. It needs to be mentioned that, this additional forging force enhances the mixing of the material, whereby improves the quality of the welding.

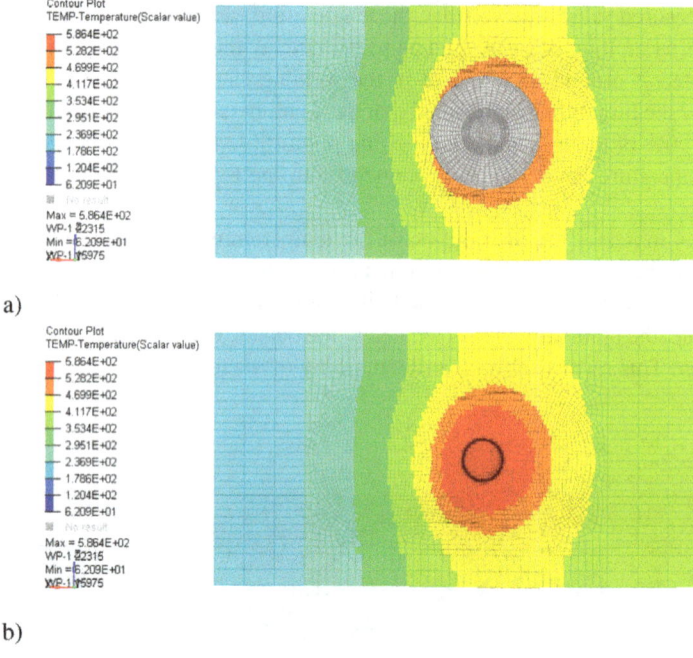

a)

b)

Fig. 5. Top view of the workpiece a) with tool and b) without tool

4 Conclusions

Based on the discussed results, below points are concluded:

- Friction coefficient values were declined as the temperature of the welding increased.
- The maximum temperature of 485 °C at the plunging and 586 °C at the welding stages were obtained.
- At the plunging step the material which is located under the shoulder pushed upward, however at the welding step the material which is located outside of the shoulder area pushed upward.
- Compared to the pin side and pin bottom surfaces, the shoulder bottom surface generates more heat, thus it had higher temperature.
- Confined pattern for the temperature at the front side of the welding were observed, while at the welding back side, a wider distance between the temperature contour circles were achieved.

Acknowledgement. The authors would like to acknowledge the fellowship of the government of China, Shandong University from the International Postdoctoral Exchange Program and the Universiti Teknologi PETRONAS (UTP), Malaysia for the financial support from YUTP-FRG grant cost center 0153AA-H18. Moreover, the authors would like to thank Altair Engineering Sdn Bhd, Malaysia and Professor Wallace Kaufman for their endless support and collaboration.

References

1. Thomas, W.: Friction stir butt welding. International Patent Application No. PCT/GB92/0220 (1991)
2. Emamian, S.S., Awang, M., Yusof, F., Sheikholeslam, M., Mehrpouya, M.: Improving the friction stir welding tool life for joining the metal matrix composites. Int. J. Adv. Manuf. Technol. **106**(7–8), 3217–3227 (2020). https://doi.org/10.1007/s00170-019-04837-1
3. Meyghani, B., Awang, M., Wu, C.: Finite element modelling of friction stir welding (FSW) on a complex curved plate. J. Adv. Joining Process. **1**, 100007 (2020)
4. Hancock, R.: Friction welding of aluminum cuts energy costs by 99%. Weld. J.-New York **83**(2), 40–43 (2004)
5. Meyghani, B., Wu, C.: Progress in thermomechanical analysis of friction stir welding. Chin. J. Mech. Eng. **33**(1), 12 (2020)
6. Meyghani, B., Awang, M.B., Emamian, S.S., Mohd Nor, M.K.B., Pedapati, S.R.: A comparison of different finite element methods in the thermal analysis of friction stir welding (FSW). Metals **7**(10), 450 (2017)
7. Meyghani, B., Awang, M.: A comparison between the flat and the curved friction stir welding (FSW) thermomechanical behaviour. Arch. Comput. Methods Eng. **27**(2), 563–576 (2019). https://doi.org/10.1007/s11831-019-09319-x
8. Meyghani, B., Awang, M.: A novel tool path strategy for modelling complicated perpendicular curved movements. Key Eng. Mater. **796**, 164–174 (2019)
9. Meyghani, B., Awang, M., Emamian, S.: A mathematical formulation for calculating temperature dependent friction coefficient values: application in friction stir welding (FSW). Defect Diffus. Forum **379**, 73–82 (2017)
10. Meyghani, B., Awang, M., Emamian, S., Akinlabi, E.: A comparison between temperature dependent and constant Young's modulus values in investigating the effect of the process parameters on thermal behaviour during friction stir welding: Vergleich zwischen den temperaturabhängigen und konstanten Elastizitätsmodulwerten in der Untersuchung der Prozessparameter auf die Wärmewirkung beim Rührreibschweißen. Materialwiss. Werkstofftech. **49**(4), 427–434 (2018)
11. Emamian, S., Awang, M., Hussai, P., Meyghani, B., Zafar, A.: Influences of tool pin profile on the friction stir welding of AA6061. ARPN J. Eng. Appl. Sci. **11**(20), 12258–12261 (2016)
12. Meyghani, B., Awang, M., Emamian, S., Khalid, N.: Developing a finite element model for thermal analysis of friction stir welding by calculating temperature dependent friction coefficient. In: Awang, M. (ed.) 2nd International Conference on Mechanical, Manufacturing and Process Plant Engineering. LNME, pp. 107–126. Springer, Singapore (2017). https://doi.org/10.1007/978-981-10-4232-4_9
13. Meyghani, B., Awang, M., Emamian, S., Mohd Nor, M.K.B.: Thermal modelling of friction stir welding (FSW) using calculated young's modulus values. In: Awang, M. (ed.) The Advances in Joining Technology. LNME, pp. 1–13. Springer, Singapore (2019). https://doi.org/10.1007/978-981-10-9041-7_1
14. Meyghani, B., Awang, M.: Developing a finite element model for thermal analysis of friction stir welding (FSW) using hyperworks. In: Awang, M., Emamian, S.S., Yusof, F. (eds.) Advances in Material Sciences and Engineering. LNME, pp. 619–628. Springer, Singapore (2020). https://doi.org/10.1007/978-981-13-8297-0_64
15. Emamian, S., Awang, M., Yusof, F., Hussain, P., Meyghani, B., Zafar, A.: The effect of pin profiles and process parameters on temperature and tensile strength in friction stir welding of AL6061 alloy. In: Awang, M. (ed.) The Advances in Joining Technology. LNME, pp. 15–37. Springer, Singapore (2019). https://doi.org/10.1007/978-981-10-9041-7_2

16. Emamian, S., et al.: A review of friction stir welding pin profile. In: Awang, M. (ed.) 2nd International Conference on Mechanical, Manufacturing and Process Plant Engineering. LNME, pp. 1–18. Springer, Singapore (2017). https://doi.org/10.1007/978-981-10-4232-4_1
17. Meyghani, B., Awang, M., Wu, C.: Thermal analysis of friction stir welding with a complex curved welding seam. Int. J. Eng. **32**(10), 1480–1484 (2019)
18. Meyghani, B., Awang, M.B., Poshteh, R.G.M., Momeni, M., Kakooei, S., Hamdi, Z.: The effect of friction coefficient in thermal analysis of friction stir welding (FSW). IOP Conf. Ser. Mater. Sci. Eng. **495**(1), 012102 (2019)
19. Su, Y., Li, W., Patel, V., Vairis, A., Wang, F.: Formability of an AA5083 aluminum alloy T-joint using SSFSW on both corners. Mater. Manuf. Process. **34**(15), 1737–1744 (2019)
20. Meyghani, B., Awang, M.B.: Prediction of the temperature distribution during friction stir welding (FSW) with a complex curved welding seam: application in the automotive industry. MATEC Web Conf. **225**, 01001 (2018)
21. Meyghani, B., Awang, M.B., Momeni, M., Rynkovskaya, M.: Development of a finite element model for thermal analysis of friction stir welding (FSW). IOP Conf. Ser. Mater. Sci. Eng. **495**(1), 012101 (2019)
22. Ansari, M.A., Samanta, A., Behnagh, R.A., Ding, H.: An efficient coupled Eulerian-Lagrangian finite element model for friction stir processing. Int. J. Adv. Manuf. Technol. **101**(5–8), 1495–1508 (2018). https://doi.org/10.1007/s00170-018-3000-z
23. Meyghani, B., Awang, M., Wu, C., Emamian, S.: Temperature distribution investigation during friction stir welding (FSW) using smoothed-particle hydrodynamics (SPH). In: Emamian, S.S., Awang, M., Yusof, F. (eds.) Advances in Manufacturing Engineering. LNME, pp. 749–761. Springer, Singapore (2020). https://doi.org/10.1007/978-981-15-5753-8_70
24. Meyghani, B., Awang, M., Wu, C.: Thermal analysis of friction stir processing (FSP) using arbitrary Lagrangian-Eulerian (ALE) and smoothed particle hydrodynamics (SPH) meshing techniques. Materialwiss. Werkstofftech. **51**(5), 550–557 (2020)
25. Meyghani, B., Awang, M., Wu, C.: Finite element modeling of friction stir welding (FSW) on a complex curved plate. J. Adv. Joining Process. **1**, 100007 (2020)
26. Zafar, A., Awang, M., Khan, S., Emamian, S.: Investigating friction stir welding on thick nylon 6 plates. Weld. J. **95**(6), 210S-218S (2016)
27. Sulaiman, S., Emamian, S.: Optimum speed of friction stir welding on 304L stainless steel by finite element method (2014)
28. Emamian, S., et al.: Comparison of carbon-based nanomaterials characteristics on H13 tool steel: Vergleich der Eigenschaften Kohlenstoff-basierter Nanomaterialien auf H13 Werkzeugstahl. Materialwiss. Werkstofftech. **48**(3–4), 198–204 (2017)
29. Kumar, S., Wu, C., Padhy, G., Ding, W.: Application of ultrasonic vibrations in welding and metal processing: a status review. J. Manuf. Process. **26**, 295–322 (2017)
30. Meyghani, B.: Thermomechanical analysis of friction stir welding (FSW) on curved plates by adapting calculated temperature dependent properties. Universiti Teknologi PETRONAS (2018)
31. Dialami, N., Chiumenti, M., Cervera, M., Segatori, A., Osikowicz, W.: Enhanced friction model for Friction Stir Welding (FSW) analysis: simulation and experimental validation. Int. J. Mech. Sci. **133**, 555–567 (2017)
32. Sun, Z., Wu, C., Kumar, S.: Determination of heat generation by correlating the interfacial friction stress with temperature in friction stir welding. J. Manuf. Process. **31**, 801–811 (2018)
33. Miles, M., Nelson, T., Gunter, C., Liu, F., Fourment, L., Mathis, T.: Predicting recrystallized grain size in friction stir processed 304L stainless steel. J. Mater. Sci. Technol. **35**(4), 491–498 (2019)
34. Salloomi, K.N., Hussein, F.I., Al-Sumaidae, S.N.: Temperature and stress evaluation during three different phases of friction stir welding of AA 7075–T651 alloy. Model. Simul. Eng. **2020**, 1–11 (2020)

35. Pan, X., Wu, C.T., Hu, W., Wu, Y.: A momentum-consistent stabilization algorithm for Lagrangian particle methods in the thermo-mechanical friction drilling analysis. Comput. Mech. **64**(3), 625–644 (2019). https://doi.org/10.1007/s00466-019-01673-8
36. Bakhtiari Argesi, F., Shamsipur, A., Mirsalehi, S.E.: Dissimilar joining of pure copper to aluminum alloy via friction stir welding. Acta Metallurgica Sinica (English Lett.), 1–14 (2018). https://doi.org/10.1007/s40195-018-0741-5
37. Leon, J.S., Jayakumar, V.: Effect of thermal boundary conditions in friction stir welding using polygonal tool pins
38. Ansari, M.A., Behnagh, R.A.: Numerical study of friction stir welding plunging phase using smoothed particle hydrodynamics. Model. Simul. Mater. Sci. Eng. **27**(5), 055006 (2019)
39. Schmidt, H., Hattel, J.: A local model for the thermomechanical conditions in friction stir welding. Model. Simul. Mater. Sci. Eng. **13**(1), 77 (2004)
40. Zhang, Z., Liu, Y., Chen, J.: Effect of shoulder size on the temperature rise and the material deformation in friction stir welding. Int. J. Adv. Manuf. Technol. **45**(9), 889–895 (2009)
41. He, X., Gu, F., Ball, A.: A review of numerical analysis of friction stir welding. Prog. Mater. Sci. **65**, 1–66 (2014)
42. Schmidt, H.B., Hattel, J.H.: Thermal modelling of friction stir welding. Scripta Mater. **58**(5), 332–337 (2008)
43. Chao, Y.J., Qi, X., Tang, W.: Heat transfer in friction stir welding—experimental and numerical studies. J. Manuf. Sci. Eng. **125**(1), 138–145 (2003)
44. Song, M., Kovacevic, R.: Numerical and experimental study of the heat transfer process in friction stir welding. Proc. Inst. Mech. Eng. Part B J. Eng. Manuf. **217**(1), 73–85 (2003)
45. Khandkar, M., Khan, J.A., Reynolds, A.P.: Prediction of temperature distribution and thermal history during friction stir welding: input torque based model. Sci. Technol. Weld. Joining **8**(3), 165–174 (2003)
46. Hamilton, C., Sommers, A., Dymek, S.: A thermal model of friction stir welding applied to Sc-modified Al–Zn–Mg–Cu alloy extrusions. Int. J. Mach. Tools Manuf. **49**(3), 230–238 (2009)
47. Ulysse, P.: Three-dimensional modeling of the friction stir-welding process. Int. J. Mach. Tools Manuf. **42**(14), 1549–1557 (2002)
48. Colegrove, P., Shercliff, H.: CFD modelling of friction stir welding of thick plate 7449 aluminium alloy. Sci. Technol. Weld. Joining **11**(4), 429–441 (2006)
49. Heurtier, P., Jones, M., Desrayaud, C., Driver, J.H., Montheillet, F., Allehaux, D.: Mechanical and thermal modelling of friction stir welding. J. Mater. Process. Technol. **171**(3), 348–357 (2006)
50. Gerlich, A., Yamamoto, M., North, T.: Strain rates and grain growth in Al 5754 and Al 6061 friction stir spot welds. Metall. Mater. Trans. A. **38**(6), 1291–1302 (2007)
51. Schmidt, H.N.B., Dickerson, T., Hattel, J.H.: Material flow in butt friction stir welds in AA2024-T3. Acta Mater. **54**(4), 1199–1209 (2006)
52. Nandan, R., DebRoy, T., Bhadeshia, H.: Recent advances in friction-stir welding–process, weldment structure and properties. Prog. Mater Sci. **53**(6), 980–1023 (2008)
53. Zahmatkesh, B., Enayati, M., Karimzadeh, F.: Tribological and microstructural evaluation of friction stir processed Al2024 alloy. Mater. Des. **31**(10), 4891–4896 (2010)
54. Schmidt, H., Hattel, J., Wert, J.: An analytical model for the heat generation in friction stir welding. Model. Simul. Mater. Sci. Eng. **12**(1), 143 (2003)
55. Su, H., Wu, C.S., Bachmann, M., Rethmeier, M.: Numerical modeling for the effect of pin profiles on thermal and material flow characteristics in friction stir welding. Mater. Des. **77**, 114–125 (2015)
56. Tang, W., Guo, X., McClure, J., Murr, L., Nunes, A.: Heat input and temperature distribution in friction stir welding. J. Mater. Process. Manuf. Sci. **7**, 163–172 (1998)

57. Padmanaban, R., Kishore, V.R., Balusamy, V.: Numerical simulation of temperature distribution and material flow during friction stir welding of dissimilar aluminum alloys. Procedia Eng. **97**, 854–863 (2014)
58. Dialami, N., Chiumenti, M., Cervera, M., Agelet de Saracibar, C., Ponthot, J.P., Bussetta, P.: Numerical simulation and visualization of material flow in friction stir welding via particle tracing. In: Idelsohn, S.R. (ed.) Numerical Simulations of Coupled Problems in Engineering. CMAS, vol. 33, pp. 157–169. Springer, Cham (2014). https://doi.org/10.1007/978-3-319-061 36-8_7

Investigation on Hybrid Geometric Modelling Construction for Core and Cavity of Injection Mould via CAD System

Noor Atikah Abdul Malek and Mohd Salman Abu Mansor[✉]

School of Mechanical Engineering, Engineering Campus, Universiti Sains Malaysia, Seri Ampangan, 14300 Nibong Tebal, Seberang Perai Selatan, Pulau Pinang, Malaysia
mesalman@usm.my

Abstract. The use of injection moulding process is becoming more popular recently due to its profitable returns in plastic manufacturing industry. Its flexibility in changing the product type have made injection-moulding process widely used. Time consuming and complexity in designing the injection mould has led to developing a hybrid geometric modelling construction that can be used in designing a mould with CAD system for injection moulding process. A system that consist of a combination of two representation schemes will produce a mould design from the product part model. The algorithm for the two representation schemes are integrated to produce a hybrid scheme that later can generate mould design without a complex way. This research aims to study the existing methods in creating the core and cavity for the mould by developing algorithm. At the end of this study, the developed algorithm able to detect the topology for the part model. It able to recognise the characteristic of the model and interpret it in form of topology data. Then, it can develop a design for pair of core and cavity for injection mould.

1 Introduction

Injection moulding process is usually used in manufacturing industry especially in plastic product manufacturing industry since it can produce large quantity of product with low cost involved within a short time of process. In injection moulding process, the flexibility in changing the product design is the main criteria in using this process. The product design is determined by the mould design, which is consist of two separate parts that are core and cavity. Therefore, design of the mould is the most important factor to be taken note since it will affect the product outlook. In order to achieve a flexible production process by producing various types of product design, designing process for injection mould should be easier and not time consuming.

Designing the injection mould can be within minutes for a simple product design or few days for complex product design. In line with the current technology trend, which is the Industry 4.0, designing process can have so much improvement [18]. Advances in technology nowadays have given a big impact on manufacturing industry. Moving from Industry 1.0 to Industry 4.0, there are many changes and improvements have been

M. Awang and S. S. Emamian (Eds.): *Advances in Material Science and Engineering*, LNME, pp. 131–139, 2021.
https://doi.org/10.1007/978-981-16-3641-7_17

done to the technologies and approaches used in the current industry [11, 12]. Industry 4.0 is the current trend of technologies used in manufacturing industry. It is about the automated technologies and computerised data controlling that enable to improve the quality of the industry [1].

One of the approach to achieve the Industry 4.0 stage is by implement a flexible programmable tool that will help in improving the industry performance such as use of computer software [22, 23]. Computer Aided Design (CAD), Computer Aided Manufacturing (CAM), Computer Aided Engineering (CAE), and other computer software are the examples of technologies that used in current industry. These computer software help in facilitate the parts designing processes since it includes constructing, combining, modification and even analysis processes. Work of designing can be easier with aid of these tools.

In CAD system, geometric modelling construction is one of its functional programme [2, 26]. It is use to display, generate and analyse the part model of the product by applying mathematical description as its functions. Even though geometric modelling is not the final goals in engineering, however it is the basic requirement in running an analysis and simulation for the product. Geometric modelling will present the complete part or product by including the topological and geometrical data. Topology is about the link of the part entities [3]. It specifies the information that relates between the two object entities, which is shape of edges while the geometry will determine the shape and dimension of the part [9].

In latest technology, many things have improved. Process of generating a mould are no longer complicated [18]. Design of mould that is used in injection moulding process that includes two separate parts, can be generated at the same time without complex process [10]. In previous time, design for these two core and cavity are generated separately. From the part model, the geometric of the part is determined then, the design for core and the cavity are generated by constructing them separately. With nowadays technology [6], there are computer software that can straight away generate the mould design for injection moulding process by using the part model design.

Based on research done by Chan et al. [6], injection moulding is a process that used in industry for producing products or parts that usually made from plastic polymer material. It is about 30% of plastic products are being manufactured by injection moulding process [13, 16]. This process involves injecting the material into the mould which made of core and cavity [7, 20]. The most crucial part in this overall process is in designing the mould. This part will determine the geometry, functionality and the final looks for the product or part produced. In nowadays industry trend, injection-moulding process is broadly used in manufacturing industry since it can produce large amount of products, involves low cost and flexible in producing variety of product types.

Injection moulding is one of the option because it can coop both of quality and of productivity. Quality and productivity should move parallel in order to capture demand from market. Quality that injection-moulding offers are producing complex-shape plastic products and at the same time, it can produce product with good dimensional accuracy within short cycle [14]. Quality and productivity can be optimize by applying the right and suitable parameters during injection moulding process. The parameters that involve during the injection moulding process are temperature, time, injection pressure and it might involves amount of gas release during the process [15].

There are three geometric modelling methods that available in industry, which are wireframe, surface and solid modellers [8]. These three geometric modelling systems have their own advantages and disadvantages. Among these three systems, wireframe modeller is the simplest. In this model, basic features such as straight line, conics or simple spline curve are used to generate the boundaries of the part model. However, because of its simple features, the surface information is not very clear, this can cause some difficulty in interpreting the image, and even impossible to manufacture a 3D object [24].

The second and the third geometric modelling systems, which are surface and solid models. These two models are more sophisticated compared to wireframe model [21]. These two models can overcome the weakness that occur in first model. A more complex shape can be done by using these two techniques. Overall, there are two basic approaches for these three techniques. There are transfinite interpolation and discrete approximation with interpolation. In first approach that is transfinite interpolation, the surface is construct as if it would pass through the given collection of curves. While for the discrete approximation, it construct the surface that interpolates with a given set of data points.

Moving to the more advanced world that equipped with various types of latest technologies, previous method and conventional way have been integrated in order to achieve something new or some sort of improvement in current achievement. A study done by Alexander [5], issue discussed is about the hybrid approach in geometric modelling construction. There are several options in developing hybrid method in geometric modelling. Since geometric modelling is built by more than one field. In geometric modelling, hybrid method can be apply either in the representation scheme, techniques, programming languages, operating systems, computer systems, modern graphic hardware or software-hardware algorithm implementation. These different work fields involve different expertise.

One of the important element in solid modelling is the representation scheme. Representation scheme is the symbolic structure that used to build the object [5]. The scheme is built by mathematical expression, which carries a particular function for every mathematic equation. Different representation schemes perform different function and give different outcome too [19]. Several schemes usually used in developing 3D model or system. However, the most popular scheme that widely used is the Boundary representation scheme (B-rep). B-rep is about describing the boundary of the solids that usually sufficient for basic visualisation. Started to be used in the middle of 1970s, B-rep represents objects in term of the surfaces, edges and vertices [27]. Vertices use to bound the edges, edges use to bound the face and faces are meet at certain edge to produce an object. The basic B-rep perform faster than other representation scheme.

Other than B-rep, there is another representation that used to describe geometric object, which is Functional representation scheme (F-rep). F-rep use real continuous function in representing the geometric of objects [25]. Other than real continuous function, the signed distance function also have been used in interpreting the function. This method has advantage in fast developing visualisation algorithm. However, this function is more restrictive. Another advantage of this F-rep is easy to implement the nonlinear transformations and other complicated operation in representing. F-rep has two classes of visualisation algorithm, first is polygonization based which transform F-rep into B-rep and the second class is ray tracing based which it straight away visualise from observers' viewpoint without convert.

The next representation scheme is parametrized primitive instancing. This scheme is based on assumption of family objects [5]. Every member of the family is different from each other by a few parameters. The object family is named generic primitive. For the individual object in the family, it is named primitive instances. However, by combining them with other scheme, it is favourable to describe the basic primitive. When a mechanism for constructing more complex object is added, it will remove the main weakness of this scheme. The other existed representation scheme is Constructive Solid Geometry (CSG). It used Boolean construction or other combination primitive through regularized set in presenting the rigid solid. In current trend, the use of CSG and boundary representation is the most popular scheme to represent the solid. CSG is unambiguous however, this scheme is not unique. It depends on the half space, which it uses the set of primitive solid and the combination of operator. In current developing process, combination this scheme with GSC is possible.

The investigation on hybrid geometric modelling construction for injection mould is one way to overcome the current weakness that occur in each approach [4]. Therefore, by combining two geometric modelling techniques, limitation in each techniques probably can be solved and lead to current technology improvement [5]. In addition, this approach is one of the achievement that can lead to industry's trend nowadays that is Industry 4.0. This work is conducted to study an automatic design recognition for the design of the mould from the part model by implementing the hybrid between two techniques of geometric modelling construction. The approach of hybrid techniques is relevant since it is possible to improve weaknesses in existing techniques.

2   Methodology

In this investigation, the combination of two representation schemes was used for developing the design for injection mould. The schemes used were Boundary Representation (B-rep) and Construction Solid Geometry (CSG). The B-rep scheme was used as the presenting scheme for the solid geometry for the injection mould. As the method for creating the injection mould, CSG scheme was used in this investigation. The ability to merge, subtract and intersection operation was used to create pattern on injection mould and splitting injection mould into core and cavity.

SolidWorks software was used in preparing the 3D CAD model for the product model. The part option was chosen for creating the product. The part model was created according to the demand or the form that the product required. As for this investigation, the product chose was a connecting rod. For the file to work in ACIS 3D modeller, the part model had to be saved in format of acis with .SAT file extension. The part model in .SAT file that opened with software ACIS 3D. The .SAT file also has been used in the algorithm since the results needed should be opened by using ACIS 3D modeler.

The main algorithm that used to generate the parting line, parting surface, core and cavity was developed from a series of combination of programming code that had been done by Md Yusof and Abu Mansor [20, 28]. In the Fig. 1, the algorithm flowchart of parent process was started by inserting the input part model as in .SAT file format. The algorithm extracted out the data from the input file. The data that have been extracted was about the topology and geometry information.

Two sub-processes were involved in the parent process. The sub-process A was generated at the third step and the sub-process B was generated at the last step. Sub-process A was developed to build the bounding box for the input model. While the sub-process B was developed to generate the core and cavity.

In the sub-process A, the algorithm developed was for building the bounding box around the part model. During the process, a box was created to enclosed the part model inside it. The flowchart for the algorithm in Sub-process A as shown in Fig. 2. For the sub-process B, it involved splitting the box into two parts which are core and cavity. In this investigation, the core and cavity for injection mould were constructed using hybrid of two representation schemes. The presentation schemes are Boundary representation (B-rep) and Constructive Solid Geometry (CSG). In this process, the B-rep was used in presenting the geometry of the part model while the CSG used in creating the core and the cavity.

The algorithm was started with inserting the part model which in .SAT file format into the algorithm. The topological information of the part model were analysed to determine the type of faces that built the part model. The number of face was used in determining the parting direction. After that, pattern of the core and cavity were created inside the box. Undercut feature such as considered by another researchers [17] was not considered for this method. Without the presents of undercut feature, the parting line, core and cavity were generated and the process was end.

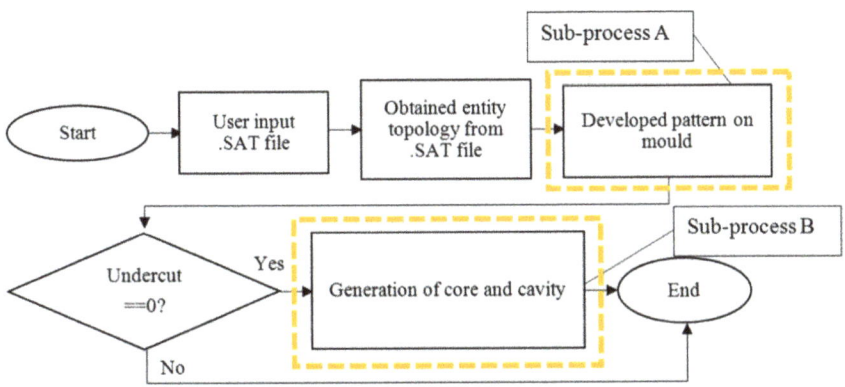

Fig. 1. The flowchart for parent process

In the process of developing pattern on mould, the solid geometry for the mould was created. It is called a bounding box. This bounding box is a geometry that will be a mould without splitting line yet. At first, determined the position of the part model by setting of one of its point with coordinate x = 0, y = 0, z = 0. This step was done to ensure the part model has finite position and to ensure the bounding box bounded the part model at the centre. Function of api_make_box was used to developed the bounding box. By using this function of bounding box, a box that completely enclosed the part model was created. Options for minimum point and maximum point for geometry of bounding box was declared as (0, 0, 0). Then the derivative value for the mould to extrude was determined. Boolean function used to intersect the pattern of the part model with the mould box. This function allowed the pattern of the part model to be translated into the box.

For the sub-process B, the flowchart for the algorithm flow is shown in Fig. 3. This step functions to separate the mould into two parts, which are the core and the cavity. In the sub-process A, the pattern of the part were successfully translated into the centre of the mould box. After analysed the absent of undercut in the model part, the parting direction was determined. However, this solid box still in the form of one solid geometry. In this step, another box that half size of the mould box was created. This box used to separate the mould box into half part by using Boolean subtraction operation. This process was repeated for another half of mould. The half part that has pattern of outer surface of part model was determined as core and mould that has pattern of inner surface was determined as the cavity.

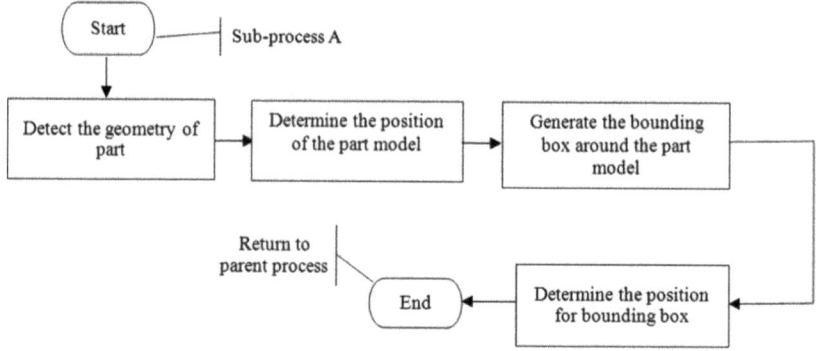

Fig. 2. The flowchart for sub-process A

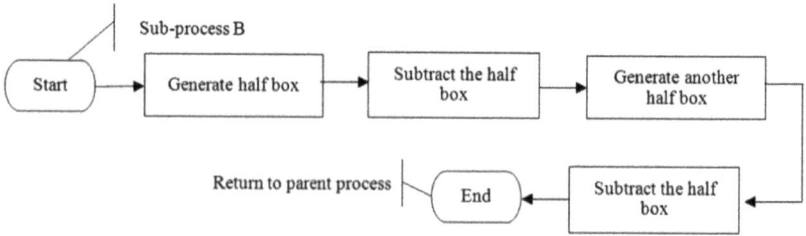

Fig. 3. The flowchart for sub-process B

3   Result and Discussion

The developed algorithm had generate the core and cavity that is the pair of injection mould. The injection mould consist of the core and cavity as shown in Fig. 4 and Fig. 5 respectively. The generated pattern was results from Boolean subtraction operation between the part model and the generated bounding box. In this bounding box, the part model was subtracted from the bounding box. This caused the pattern of the part model generated in the bounding box. During this stage, the injection mould was still in the form of one unit. Therefore, further process was done to split this bounding box into two parts.

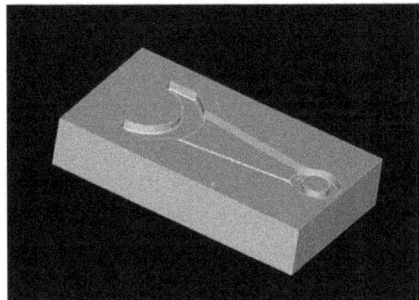
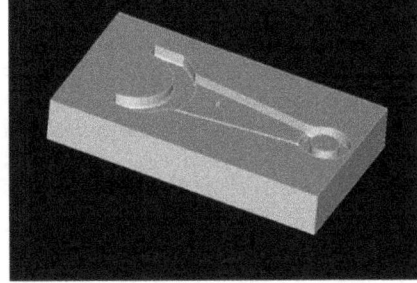

Fig. 4. Core of the injection mould   **Fig. 5.** Cavity of the injection moulds

These two parts were results from the Boolean subtraction operation once again. This subtraction operation was done between the bounding box and another box that had been generated with the half size of the bounding box. The surface of the half box must be defined at the middle of the part model. This would act as the parting line for the bounding box to split into two parts and produced half of the pattern of part model at the core and cavity.

4   Conclusions

With regards to the investigation of hybrid geometric modelling construction for the core and cavity of injection mould, the basic concept in designing the plastic injection mould is studied. The hybrid representation scheme to build the geometry is investigated. All the elemental information that relates to core and cavity of injection mould with CAD system are used. The hybrid between boundary representation (B-rep) and Constructive Solid Geometry (CSG) schemes are studied in order to develop the algorithm from these two schemes to produce a design for core and cavity of injection mould. B-rep scheme is used in presenting the geometry of the part model. While for the CSG, it use in developing the core and the cavity for the injection mould. This is due to complexity of the object in CSG used Boolean operation in the representation schemes. In future, more algorithms that can combines all the possible series of geometric modelling construction can be investigated for designing more sophisticated core and cavity of injection mould.

Acknowledgments. This research is supported by the Universiti Sains Malaysia and Ministry of Higher Education Malaysia under the Research University Grant (Reference No. 8011114) and Fundamental Research Grant Scheme (FRGS) (Reference No. 6071369).

References

1. MacDougall, W.: Industries 4.0: Smart Manufacturing for the Future. Germany Trade and Invest, Berlin (2014)
2. Bedworth, D., Henderson, M., Wolfe, P.: Computer-Integrated and Manufacturing. McGraw-Hill, New York (1991)
3. Agarwal, S.C., Waggenspack, W.N.: Decomposition method for extracting face topologies from wireframe models. Comput. Aided Des. **24**(3), 123–140 (1992). https://doi.org/10.1016/0010-4485(92)90032-6
4. Wypysinski, R.: Hybrid modelling in CAD. Adv. Technol. Mech. **2**(1(2)), 15–22 (2015). https://doi.org/10.17814/atim.20151(2).14
5. Penev, A.: Computer graphic and geometric modelling - a hybrid approach. Int. J. Pure Appl. Math. **85**(4), 781–811 (2013). https://doi.org/10.12732/ijpam.v85i4.14
6. Chan, W.M., Yan, l., Xiang, W., Cheok, B.T.: A 3D CAD knowledge-based assisted injection mould design system. Int. J. Adv. Manuf. Technol. **22**(5–6), 387–395 (2003). https://doi.org/10.1007/s00170-002-1514-9
7. Md Yusof, M., Abu Mansor, M.S.: Core and cavity generation using slicing and boolean-based algorithm. Appl. Mech. Mater. **761**, 175–179 (2015). https://doi.org/10.4028/www.scientific.net/AMM.761.175
8. Dimas, E., Briassoulis, D.: 3D Geometric modelling based on NURBS: a review. Adv. Eng. Softw. **30**(9), 741–751 (1999). https://doi.org/10.1016/S0965-9978(98)00110-0
9. Rao, P.N.: CAD/CAM: Principles and Applications. McGraw Hill Education Private Limited, Noida (2014)
10. Fu, M.W., Fuh, J.Y.H., Nee, A.Y.C.: Core and cavity generation method in injection mould design. Int. J. Prod. Res. **39**(1), 121–138 (2001). https://doi.org/10.1080/00207540010002379
11. Rubmann, M., et al.: Industry 4.0: the future of productivity and growth in manufacturing industry (2015)
12. Gaub, H.: Customization of mass-produced parts by combining injection molding and additive manufacturing with Industry 4.0 technologies. Reinf. Plast. **60**(6), 401–404 (2016). https://doi.org/10.1016/j.repl.2015.09.004
13. Mathivanan, D., Nouby, M., Vidhya, R: Minimization of sink mark defects in injection molding process–Taguchi approach. Int. J. Eng. Sci. Technol. **2**(2), 13–22 (2010). https://doi.org/10.4314/ijest.v2i2.59133
14. Singh, G., Verma, A.: A brief review on injection moulding manufacturing process. Mater. Today Proc. **4**(2, Part A), 1423–14330 (2017). https://doi.org/10.1016/j.matpr.2017.01.164
15. Guo, W., Mao, H., Li, B., Guo, X.: Influence of processing parameters on molding process in microcellular injection molding. Procedia Eng. **81**, 670–675 (2014). https://doi.org/10.1016/j.proeng.2014.10.058
16. Ogorodnyk, O., Martinsen, K.: Monitoring and control for thermoplastics injection molding: a review. Procedia CIRP **67**, 380–385 (2018). https://doi.org/10.1016/j.procir.2017.12.229
17. Md Yusof, M., Abu Mansor, M.S.: Undercut feature recognition for core and cavity generation. IOP Conf. Ser. Mater. Sci. Eng. **290**(1), 012070 (2018). https://doi.org/10.1088/1757-899X/290/1/012070
18. Kong, L., et al.: A Windows-native 3D plastic injection mold design system. J. Mater. Process. Technol. **139**(1), 81–89 (2003). https://doi.org/10.1016/S0924-0136(03)00186-9

19. Requicha, A.: Representation for rigid solids: theory, methods and systems. ACM Comput. Surv. **12**(4), (1980). https://doi.org/10.1145/356827.356833
20. Md Yusof, M., Abu Mansor, M.S.: Alternative method to determine parting direction automatically for generating core and cavity of two-plate mold using B-rep of visibility map. Int. J. Adv. Manuf. Technol. **96**(9–12), 3109–3126 (2018). https://doi.org/10.1007/s00170-018-1695-5
21. Varley, P.A.C., Company, P.P.: A new algorithm for finding faces in wireframes. Comput. Aided Des. **42**(4), 279–309 (2010). https://doi.org/10.1016/j.cad.2009.11.008
22. Qiu, Z.M., et al.: Geometric model simplification for distributed CAD. Comput. Aided Des. **36**(9), 809–819 (2004). https://doi.org/10.1016/j.cad.2003.09.007
23. Zhou, J., Li, L., Hu, Y., Yang, J., Cheng, K.: Plastic mold design of top-cover of out-shell of mouse based on CAE. Procedia Eng. **15**, 4441–4445 (2011). https://doi.org/10.1016/j.proeng.2011.08.834
24. Zhengxu, Z., Ghosh, S.K., Link, D.: Recognition of machined surfaces for manufacturing based on wireframe models. J. Mater. Process. **24**, 137–145 (1990)
25. Pasko, A., Adzhiev, V., Sourin, A., Savchenko, V.: Function representation in geometric modeling: concepts, implementation and applications. Vis. Comput. **11**(8), 429–446 (1995). https://doi.org/10.1007/bf02464333
26. Tan, J.X., Abu Mansor, M.S.: Construction of a hybrid geometric model for an injection mould using CAD/CAM system. In: Emamian, S.S., Awang, M., Yusof, F. (eds.) Advances in Manufacturing Engineering. LNME, pp. 323–333. Springer, Singapore (2020). https://doi.org/10.1007/978-981-15-5753-8_30
27. Stroud, I.: Boundary Representation Modelling Techniques. Springer, London (2006). https://doi.org/10.1007/978-1-84628-616-2
28. Md Yusof, M., Abu Mansor, M.S.: Automatic core and cavity generation for 3D CAD model using normal vector and scanning ray approaches. ARPN J. Eng. Appl. Sci. **12**(14), 4250–4254 (2017). http://www.arpnjournals.com/jeas/volume_14_2017.htm

Cost of Equipment Failure Modelling as a Tool for Maintenance Strategy

George Decruz(✉), Habibah Norehan Haron, and Khairur Rijal Jamaludin

Universiti Teknologi Malaysia, Johor Bahru, Malaysia

Abstract. The cost of equipment failure has been recognized as a significant tool for the assessment of maintenance performance. Sound strategic maintenance decision making based on maintenance problem identification leading to planning and improvement is central to the improved operational performance of manufacturing plants. Researchers have developed models based on the fundamental premise that poor maintenance has a cost. This article illustrates how this tool is developed practically to ensure a structured approach in developing specific strategies with respect to maintenance for the kiln of a cement plant based on equipment failures over a ten-year period. The cost of failure elements included in the analysis are the opportunity losses, direct incremental fixed element costs and the impact of equipment failure on variable costs. In examining the last, a detailed model development and analysis of the energy efficiency losses related to equipment failure was carried out. The resulting cost of equipment failure model was used in a simulation exercise for a five-year period from 2013 to 2017. The results of this simulation indicate that while opportunity losses are the most significant and make up more than 85% of the losses, efficiency losses are also important, comprising close to ten percent of losses. The variable cost losses, which are independent of selling price assume added relevance in driving efficiencies when competitive market conditions are at play.

1 Introduction

Maintenance as defined in a holistic manner examines actions directed towards ensuring that an equipment can achieve its desired state of functionality [1]. The study at the cement plant was focused on the clinker production line at one of the kilns with the objective of improving maintenance performance. Clinker, a semi-finished product, is the principal input material for the manufacture of cement, which requires inter-grinding of clinker with gypsum. The manufacture of clinker from raw materials requires the crushing of limestone and shale and the subsequent grinding and burning of raw meal, which is a proportioned mix of limestone, shale, silica sand and ferrum containing laterite. The grinding process takes place in a mill whilst the burning process is carried out in a pyro-processing unit comprising the preheater, kiln, and cooler. The pyro-processing unit utilizes coal as the principal fuel whilst diesel is used for start-ups. The general process flow of clinker production and the energy inputs is illustrated in Fig. 1, [2].

© The Author(s), under exclusive license to Springer Nature Singapore Pte Ltd. 2021
M. Awang and S. S. Emamian (Eds.): *Advances in Material Science and Engineering*, LNME, pp. 140–149, 2021.
https://doi.org/10.1007/978-981-16-3641-7_18

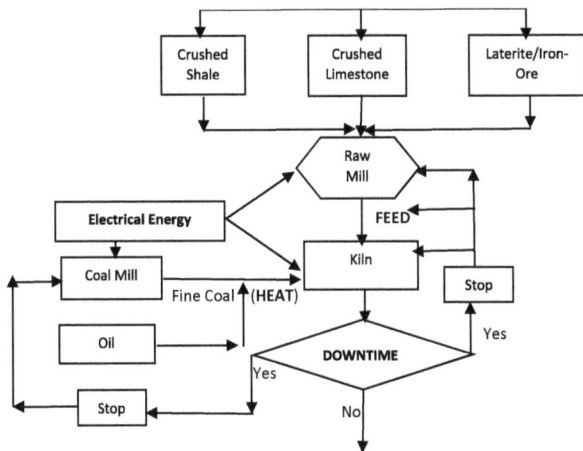

Fig. 1. Process flow for clinker with energy inputs

A review of maintenance performance assessment criteria identified that maintenance cost was the most frequently used tool, utilized in 15% of articles reviewed [3]. Early work on maintenance cost assessment focused on the costs incurred to return a system to its previous level of operability after a failure, [4]. This approach was limited in that it only looked at maintenance costs related to cost of maintenance repair items as well cost of labor to execute the repairs. Subsequent research sought to examine maintenance cost more rigorously and understand it from the perspective of improved revenue through the avoidance of maintenance failure [5]. In this approach, based on the Prevention-Appraisal-Failure Model, maintenance failure costs are the summation of the costs related to the repair costs incurred to return the equipment to normal operation, opportunity losses related to the revenue loss as well other losses.

Alternative approaches in disassembling maintenance failure costs were also developed. The "Overall Equipment Cost Loss" model summates the integrated cost losses related to reductions in availability, performance efficiency and quality of identified critical equipment [7]. This concept of performance efficiency is of special relevance to energy intensive industries where energy consumption forms a significant portion of variable costs and where the re-starting of the plants after an interruption consumes elevated energy consumption rates. The assessment of maintenance costs from the perspective of activity-based costing as well as life cycle costing suggests that maintenance activities encompass more than just the costs incurred in equipment or component repair [8, 9]. The activities leading up to the repair as well as the post repair activities such as cooling down and re-heating which are executed solely because of repair work are an integral part of maintenance and should form part of the cost assessment of maintenance. The failure to assign cost drivers correctly can provide an inaccurate assessment of product costing [10]. This is the case when energy losses associated with equipment failure is integrated as part of production energy costs.

The various incremental heat energy losses can be identified from energy balances of the kiln [11]. Research on electrical energy losses associated with plant stoppages has also been assessed and evaluated in one study to be 0.25 kilowatt-hour/ton cement [12]. The utilization of costs associated with both Mean Time Before Failure (MTBF) and Mean Time To Repair (MTTR) have been used to identify maintenance strategy for equipment [13]. The utilization of maintenance decision models that focus on improved equipment reliability and reduced costs has been linked to manufacturing performance [14].

2 Method

2.1 General Model

The approach in determining the cost of equipment failure was based on previous research as outlined above,

i)   Incremental increase in variable costs or efficiency losses resulting from equipment failure, $C_{effL,\,i}$.
ii)   Opportunity losses resulting from contribution margin losses and losses arising out of impact of production loss on unit fixed costs, $C_{oppL,\,i}$.
iii) The incremental increase in fixed costs because of equipment failure, $\Delta C_{FMD,i}$.

This is expressed as follows in Eq. 1, for the cost of failure, C_{MFi}, of event, "i".

$$C_{MF,i} = C_{effL,i} + C_{oppL,i} + \Delta C_{FMD,i}, \tag{1}$$

The measure of equipment failure is the kiln downtime which is defined as the time from when the feeding to the kiln is stopped until it is resumed. The research process is detailed in Fig. 2.

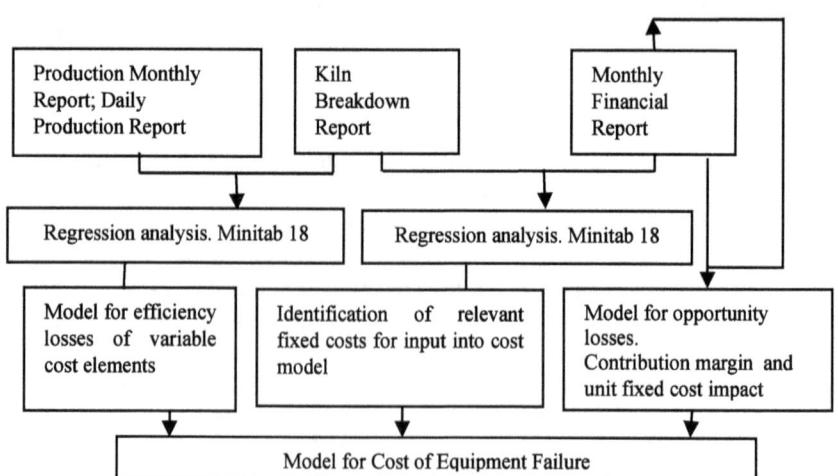

Fig. 2. Development of model for cost of equipment failure

2.2 Efficiency Losses

The following variable cost elements were identified based on preliminary investigation.

i)   Specific heat consumed in calories per ton clinker in the production of clinker, HC_{sp}
ii) Specific consumption of electricity in kilowatt-hour per ton clinker in the kiln, EC_{spOVK}
iii) Specific consumption of electricity in kilowatt-hour per ton clinker for raw meal production, EC_{spOVRM}
iv) Specific consumption of electricity in kilowatt-hour per ton clinker for production of pulverized coal for clinker burning. EC_{spOVCM}.·

HC the total heat consumed during the month, derived from the product of HC_{sp} and the total clinker production for the month, is the sum of HC_{prod}, the heat consumed solely for production, HC_{prev}, the total consumed for preventive maintenance outages and HC_{effL}, the total heat loss contributed by efficiency losses resulting from equipment failure.

$$HC = HC_{prod} + HC_{prev} + HC_{effL} \tag{2}$$

Equation 2 can be further simplified to Eq. 3 by excluding the months when scheduled preventive maintenance was carried out.

$$HC = HC_{prod} + HC_{effL} \tag{3}$$

HC_{prod} and HC_{effL} are determined as follows,

i)   Identification of days with full 24-h operation and total heat consumed in calories, HC_{24}, and total clinker production in tons, $Prod_{24}$
ii) Computation of unit calories for production, $HC_{24}/Prod_{24}$.
iii) Total production of clinker, $Prod_T$.
iv) Utilizing ii) and iii), Eq. 4 gives the overall heat required for production, HC_{prod},

$$HC_{prod} = (HC_{24} \times Prod_T)/Prod_{24} \tag{4}$$

v) HC_{effL} for the month is the difference between HC and HC_{prod}.

There are other operational factors that influence heat consumption of production, such as the chemistry of raw meal, quality of coal, leakages of air into the system [12, 15]. However, Eq. 4 is developed based on the assumption that these other parameters remain constant for the duration of each individual month.

The monthly heat efficiency losses, HC_{effL} determined above were regressed against several predictors such as kiln output, PR_K, calorific value of coal, CV_{CO}, number of failure events in the kiln, N_K, and kiln downtime with all values exceeding 24 h capped to 24 h, D_{K24}, utilizing Minitab18. In a general form, this is expressed in Eq. 5.

$$HC_{effL} = f[D_{K24}, PR_K, CV_{CO}, N_K] \tag{5}$$

The capping of kiln downtime is based on the operational norms where the heating up period for any downtime exceeding 24 h or more is the same. The cost incurred by the heat efficiency losses for an equipment failure event, "i", $HC_{effL,i,cost}$ is obtained as a product of the heat efficiency loss, HC_{effLi}, computed for the event, composite price of fuel for the month when the failure event occurred, $P_{compfuel}$, and the price index of the fuel relative to December 2017, $P_{index,fuel}$. This is expressed in Eq. 6.

$$HC_{effL,i,cost} = HC_{effL,i} \times P_{comp.fuel} \times P_{index,fuel} \tag{6}$$

As only monthly power meter readings are available, a different approach was taken to compute electrical efficiency losses. The methodology utilized was to develop a relationship between the specific electricity consumed in kilowatt-hour per ton clinker at each of the operating units and identified predictors including production rates and kiln downtime and utilizing Minitab 18 to conduct a regression analysis. In the case of kiln downtime, alternative approaches were tested utilizing different values for the capping of downtime. The specific efficiency loss related to an equipment failure can then be expressed as a function of the kiln downtime if kiln downtime is assessed to be a significant predictor of specific energy consumption for each of the cases. The various sets of response variables and predictor variables are shown in the following Eqs. 7 and 8.

$$EC_{spOVK} = f[EC_{spK}, EC_{speffLK}] \tag{7}$$

$$EC_{spOVK} = f[PR_K, D_K/D_{K24}/D_{96}, N_K] \tag{8}$$

where EC_{spOVK} is the overall monthly specific electrical energy consumed per ton clinker in the kiln measured in kilowatt-hour per ton clinker, EC_{spK} is the specific electrical energy consumed per ton clinker for production of clinker in kilowatt-hour per ton clinker, $EC_{speffLK}$ is the specific efficiency losses in kilowatt-hour per ton clinker, PR_K is the production rate of clinker measured in tons per hour, D_K or D_{K24}, that is the kiln downtime either taken as recorded or capped to 24 h, N_K is the number of kiln failures for the month. A similar approach is taken to identify the electrical energy losses associated with the raw mill, EC_{effLRM}, and coal mill, EC_{effLCM}, due to equipment failure leading to kiln downtime.

The cost incurred by electrical efficiency losses for an equipment failure event, "i", $EC_{effL,i,cost}$ is obtained as a product of the sum of the electrical efficiency losses, $EC_{effLK,i}$, $EC_{effLRM,i}$, $EC_{effLCM,i}$ computed for the event, "i", unit price of electricity for the month when the failure event occurred, $P_{elect,i}$, and the price index of electricity cost relative to December 2017, $P_{index,elect}$. This is expressed in Eq. 9.

$$EC_{effL,i,cost} = (EC_{speffLK,i} + EC_{speffLRM,i} + EC_{speffLCM,i}) \times (PR_K) \times (P_{elect}) \times (P_{index,elect}) \tag{9}$$

2.3 Fixed Cost Elements Impacted By Equipment Failure

This is determined by carrying out a regression analysis of the maintenance cost, C_M, manpower cost, C_{MP}, and overtime cost, C_{OT}, against kiln equipment failure downtime and tested for significance at a p-value of less than 0.05. The incremental costs for the significant fixed cost resulting from the failure of event, "i", expressed as $\Delta C_{M,i}$, $\Delta C_{MP,i}$ and $\Delta C_{OT,i}$ shall then be incorporated into the cost of equipment failure model.

2.4 Opportunity Losses

The computation of opportunity losses related to the failure of event, "i", $C_{oppL,i}$, is determined for both the contribution margin component, $C_{margin,i}$, and the impact of increase in unit fixed cost component, $FC_{impact,i}$.

$$C_{oppL,i} = C_{margin,i} + FC_{impact,i} \tag{10}$$

i.   The contribution margin loss is expressed in Eq. 11,

$$C_{margin,i} = \left(C_{contmargin,20XX} \times D_{Ki} \times PR_{K,i,mon,20XX}\right) \times \frac{PPI_{DEC2017}}{PPI_{mon,i,20XX}} \tag{11}$$

where $C_{margin,i}$ is the contribution margin loss resulting from the equipment failure, "i", $C_{contmargin20XX}$, is the contribution margin for the year when the event "i" occurred, D_{Ki}, is the downtime caused by the failure event, "i", $PR_{K,mon,i,20XX}$, is the clinker production rate for the month when the failure event, "i" occurred, $PPI_{DEC2017}$ and $PPI_{mon,i,20XX}$ are the production price index for December 2017 and the month when the failure event, "i" occurred respectively as sourced from the website of the Department of Statistics, Malaysia.

ii. The impact of increase in unit fixed cost, $FC_{impact,i}$, is derived from Eqs. 12 and 13.

$$FC_{chan20XX} = \{[FC_{20XX} \div CP_{K20XX}] \\ - [(FC_{20XX}) \div (CP_{K20XX} + \langle PR_{K20XX} \times D_{K20XX}\rangle)]\} \tag{12}$$

$$FC_{impact.i} = FC_{chan20XX} \times D_{K,i} \times PR_{K,i,mon,20XX} \times \frac{PPI_{DEC2017}}{PPI_{mon,i,20XX}} \tag{13}$$

Equation 12 computes the change in unit fixed cost for the year in which the failure event, "i" occurred caused by total equipment failure downtime where FC_{20XX} is the fixed cost for the year under consideration, CP_{K20XX} is the clinker production for the year under consideration, PR_{K20XX} is the clinker production rate in tons per hour for the year and D_{K20XX} is the total equipment failure downtime for the year. The change in unit fixed cost $FC_{chan20XX}$ obtained from Eq. 12 is utilized together with, $D_{K,i}$, the kiln downtime for the equipment failure event, "i", $PR_{K,i,mon,20XX}$, the clinker production rate for the month in the year when the failure event, "i" occurred and the ratio of production price index to obtain the impact of the event "i" on fixed costs. The total opportunity losses, $C_{oppL,i}$, is thus expressed in Eq. 14.

$$C_{oppL,i} = \left(C_{contmargin,i} + FC_{chanY}\right) \times D_{K,i} \times PR_{K,i,mon,20XX} \times \left(\frac{PPI_{DEC2017}}{PPI_{mon,i,20XX}}\right) \tag{14}$$

3   Results and Discussion

The models utilized data from Plant Monthly Reports, Kiln Breakdown Reports, Plant, Daily Production Reports and Monthly and Yearly Financial Reports for a 10-year period from 2008 to 2017. The results of the analysis were then utilized to carry out a simulation exercise over the period from 2013 to 2017 to identify equipment criticality and develop maintenance strategy. The models for efficiency losses are as shown in Table 1.

Table 1. Regression models for efficiency losses

Variable cost element	Regression model for determining efficiency losses
$HC = HC_{prod} + HC_{effL}$ Megacalories	$\log_{10}HC_{effL,i} = 3.917 + 2.460(\log_{10}D_{K24,i})-(1.248(\log_{10}D_{K24,i})^2) + (0.3107(\log_{10}D_{K24,i})^3)$
EC_{spOVK}, Kilowatt hour per ton clinker	$EC_{spOVK} = 81.46 - 0.29PR_K + 0.027D_{K96}$ $EC_{speffLK,i} = 0.027D_{K96,i}{}^*$, *downtime hours for any failure capped to 96 h
EC_{spOVRM}, Kilowatt hour per ton clinker	$ECspOVRM = 44.82 - 0.056PRRM + 0.018DK24 + 0.025NSRM$ $EC_{speffLRM,i} = 0.018D_{K24,i}$,
EC_{spOVCM}, Kilowatt hour per ton clinker	$EC_{spOVCM} = 13.17 - 0.046PR_K + 0.0036D_{K24}$ $EC_{speffLK,i} = 0.0036D_{K24,i}{}^*$,

Table 2. Regression analysis of fixed cost elements against kiln downtime.

Fixed cost elements regressed against downtime, DK	R-squared, %	p-value	Remarks
Maintenance, CM	10.05	<0.002	
Manpower, CMP	2.52	<0.117	Absence of a significant relationship is attributed to the maintenance of a constant workforce; additional maintenance labour costs are covered under maintenance costs
Overtime Hours, COT	0.0	<0.987	The overtime is rostered and independent of downtime hours

The opportunity costs are expressed in a general form for the period 2013 to 2017 in Table 3 with contribution margin expressed relative to the 2017 value stated as M as this value is confidential.

Table 3. Contribution margin losses and impact of downtime on unit fixed costs.

Year	Contribution margin per ton, $C_{contmargin}/ton$	Impact on unit fixed costs per ton clinker, $FCchan20XX$ $(FC_{20XX}/CP_{K20XX})-(FC_{20XX}/\{CP_{K20XX}+[D_{K20XX} \times PR_{K20XX}]\}$; RM/ton
2013	1.58M	0.72
2014	1.74M	0.92
2015	1.76M	1.90
2016	1.88M	2.21
2017	1M	6.87

The overall equation to determine the cost of failure related to a failure event, "i", is illustrated in the equation below for the year 2017. The subscripts, i, month and 2017 define the failure event during the specific month and year, respectively.

$$
\begin{aligned}
C_{MFi} = \\
\sum_{mon=JAN}^{mon=DEC} \left(\left[\left\{ 10^{\begin{subarray}{l}\{3.917+2.460(\log_{10}D_{K24,i,mon,2017})-(1.248(\log_{10}D_{K24,i,mon,2017})^2)+\\ (0.3107(\log_{10}D_{K24,i,mon,2017})^3)\}\end{subarray}} \\ \times P_{compfuel,i,mon,2017} \times P_{index,fuel,i,mon,2017} \right\} \right] + \right. \\
\left[\left\{ \begin{array}{l} [0.027 \times D_{K96,i,mon,2017}] + \\ [0.018 \times D_{K24,i,mon,2017}] + [0.0030 \times D_{K24,i,mon,2017}] \\ \times PR_{K,i,mon,2017} \times P_{elect,i,mon,2017} \times P_{index,elect,i,mon,2017} \end{array} \right\} \right] + [\Delta C_{M,i,2017}] + \\
\left. \left\{ \begin{array}{l} (1M \times D_{K,i,mon,2017} \times PR_{K,i,mon,2017}) \\ +(6.87 \times PR_{K,i,mon,2017} \times D_{K,i,mon,2017}) \end{array} \right\} \times \langle PPI_{DEC2017}/PPI_{mon,i,2017} \rangle \right)
\end{aligned}
$$

$$(15)$$

Contribution margin losses make up 84.3% of the total losses. This supports earlier literature which found it to make up more than 80% of total failure costs [6]. The total efficiency losses comprise almost 10% of total costs and is almost equally divided between heat efficiency losses and electrical efficiency losses. The results of the computation of the cost of failure by equipment and failure type was ranked according to total cost impact using a standardization process based on RM/rated output. The resultant ranking of equipment failure based on overall cost impact over 5 years as well as single year maximum impact. together with strategies identified by the Plant operations and maintenance team are tabulated in Tables 4 and 5.

Table 4. Overall cost of equipment failure by percentage for the period 2013–2017.

Equipment failure cost component	Percentage (%)
Heat efficiency losses, HCeffLcost	5,24
Electrical efficiency losses, ECeffL	4.61
Contribution margin losses	84.33
Fixed cost impact losses	2.23
Maintenance repair costs	3.59
Total	100.00

Table 5. Identification of major equipment failure and strategies.

Failure type	5-year analysis/1-year maximum	Percentage of total	Strategic plan
Kiln refractory failure	3.48/6.72	17.43	Refractory selection in critical areas according to economic value of returns and past performance Kiln alignment at 2-year intervals
Clinker deep bucket breakdown	2.75/11.89	13.78	Five-year overhaul plan
Cyclone blockages	1.94/2.66	9.72	Capital expenditure to address operational complexity. Redundant detection system for cyclone
Cooler grate plate failure	1.06/2.7	5.31	Cooler alignment at 5-year intervals
Kiln main drive tripping**	0.62/2.26	3.10	**Not applicable Previous improvement in equipment design in 2015 to variable frequency drive motor has reduced failure impact

4 Conclusions

The utilization of the cost model as shown in Eq. 15 provides a composite evaluation based on failure frequency and severity, where severity is measured based on the cost impact of failure. Despite the significantly smaller contribution of the efficiency losses, it should be recognized that as contribution margins drop in a competitive marketplace, survival may hinge on the reduction and elimination of such efficiency losses. The simulation exercise showed that the top five major failures contributed close to fifty percent of the total cost of equipment failure and opens the window of opportunity for strategic solutions.

References

1. BSI. Glossary of Maintenance Terms in Terotechnology. London British Standard Institution, (BSI), BS3811 (1984)
2. George, D., Habibah @ Norehan, H.: The impact of kiln downtime on the variable cost elements of clinker production-a case study. J. Adv. Res. Bus. Manag. Stud. **18**(1), 1–6 (2020)
3. Simões, J.M., Gomes, C.F., Yasin, M.M.: A literature review of maintenance performance measurement a conceptual framework and directions for future research. J. Qual. Maintenance Eng. **17**(2), 22 (2011)
4. Bovaird, R.: Characteristics of optimal maintenance policies. Manag. Sci. **7**(3), 16 (1961)
5. Salonen, A., Deleryd, M.: Cost of poor maintenance. J. Qual. Maintenance Eng. **17**(1), 63–73 (2011). https://doi.org/10.1108/13552511111116259
6. Peimbert-Garcia, R.E., Limon-Robles, J., Beruvides, M.G.: Cost of quality modeling for maintenance employing opportunity and infant mortality costs: an analysis of an electric utility. Eng. Econ. **61**(2), 112–127 (2016). https://doi.org/10.1080/0013791x.2016.1152619
7. Wudhikarn, R.: Implementation of the overall equipment cost loss (OECL) methodology for comparison with overall equipment effectiveness. J. Qual. Maintenance Eng. **22**(1), 13 (2016)
8. Aoudia, M.: Towards the design of a new framework for maintenance costing. Contemp. Eng. Sci. **8**, 1475–1483 (2017). https://doi.org/10.12988/ces.2015.59264
9. Mirghani, M.A.: Application and implementation issues of a framework for costing planned maintenance. J. Qual. Maintenance Eng. **9**(4), 436–449 (2003). https://doi.org/10.1108/135 52510310503268
10. Haroun, A.E.: Maintenance cost estimation: application of activity-based costing as a fair estimate method. J. Qual. Maintenance Eng. **21**(3), 258–270 (2015). https://doi.org/10.1108/jqme-04-2015-0015
11. Atmaca, A., Yumrutas, R.: Analysis of the parameters affecting energy consumption of a rotary kiln. Appl. Therm. Eng. **66**, 10 (2014)
12. Virendra, R., Kumar, D.B.S.P., Babu, J.S., Kant, D.R.: Detailed energy audit and conservation in a cement plant. Int. Res. J. Eng. Technol. (IRJET) **2**(1), 8 (2015)
13. Bevilacqua, M., Braglia, M.: The analytic hierarchy process applied to maintenance strategy selection. Reliab. Eng. Syst. Saf. **70**, 13 (2000)
14. de Almeida, A.T., Bohoris, G.A.: Decision theory in maintenance decision making. J. Qual. Maintenance Eng. **1**(1), 7 (1995)
15. Radwan, A.M.: Different possible ways for saving energy in the cement production. Adv. Appl. Sci. Res. (2012)

Evaluation of Fracture Energy of Aluminium Alloy 1050-F and Carbon Steel EN - 3 (~1015) 0.15% C at Different Temperatures Gradient

O. M. Ikumapayi[1]([✉]), E. T. Akinlabi[2], S. O. Fatoba[3], R. A. Kazeem[4], S. O. Afolabi[1], A. O. M. Adeoye[1], and S. A. Akinlabi[5]

[1] Department of Mechanical and Mechatronics Engineering, Afe Babalola University, Ado Ekiti, Nigeria
ikumapayi.omolayo@abuad.edu.ng
[2] Directorate, Pan African University for Life and Earth Sciences Institute (PAULESI), Ibadan, Nigeria
[3] Department of Mechanical Engineering Science, University of Johannesburg, Auckland Park Kingsway Campus, Johannesburg 2006, South Africa
[4] Department of Mechanical Engineering, University of Ibadan, Ibadan, Nigeria
[5] Department of Mechanical Engineering, Faculty of Engineering and Technology, Walter Sisulu University, Butterworth Campus, Butterworth, South Africa

Abstract. The present study examined the mechanical properties of pure aluminium as well as carbon steel by evaluating the effects of temperature gradient on the impact strength of carbon steel as well as an aluminium alloy following ASTM A370 standard. Pure aluminium (1050-F) and mild steel of 0.15% C (EN-3(~1015) were the specific materials used. The impact test is used in this study to evaluate the material toughness and its notch sensitivity. The test determines the toughness or impact strength of a material in the presence of a notch and rapid loading condition per ASTM E23 standard. The test is vital for the investigation of material's mechanical properties, materials such as ceramics, polymers, composites, and metals can be tested with this method. The level of the fracture determines the quantity of energy the material store during the fracturing process. Hence, during this present test materials or specimens used were subjected to various temperatures gradient. The testing temperatures for both specimens were from –50 °C to 125 °C with a step of 25 °C. To determine the impact strength of each material at defined temperatures, Liquid Nitrogen, Methanol, and boiling water at 100 °C were mixed and used to get the specimens to below and above 0 °C. The specimens were dipped into the mixture for 2 min and then removed with care and taken to the testing machine. The observations were documented in a tabular form. It was revealed that Aluminium is more malleable than steel while steel is more ductile than aluminium. The ductile to brittle transition temperature was found to be 36.33 °C and 49.34 °C for aluminium and EN3 specimen, respectively. The fractured surfaces were captured with the scanning electron microscope (SEM) at –50 °C, 0 °C and 125 °C and the morphological structure revealed were documented.

Keywords: Aluminium · Brittleness · Carbon steel · Ductility · Fracture · Transition

© The Author(s), under exclusive license to Springer Nature Singapore Pte Ltd. 2021
M. Awang and S. S. Emamian (Eds.): *Advances in Material Science and Engineering*, LNME, pp. 150–168, 2021.
https://doi.org/10.1007/978-981-16-3641-7_19

1 Introduction

As engineers, we are required to know the properties of the materials we use before designing and building the desired components for use. This way fatalities and injuries, as well as the loss of money, can be mitigated through the appropriate use of materials. One way of determining a material's property is to use the Charpy Impact Test. The test was named after the French engineer Georges Charpy (1865 – 1945) [1]. The testing of materials goes way back to the 1800s where scientists and engineers from around the world formed the International Association for Testing Materials. The members would meet after 2 to 3 years in different cities around the world to discuss and share ideas on the progress of material testing. By the early 1900s the association consisted of 2682 members and amongst those were Brinell (hardness test), Heyn (grain size), Martens (martensite), Le Chatelier as well as Charpy which is the main man behind the invention of Charpy impact test. Georges Augustin Albert Charpy was a French-born from 1865 to 1945. Charpy obtained his engineering degree that majored in marine artillery from École Polytechnique in 1887 and later became a metallurgical engineer [1]. Because of a lot of failure with the boilers and steam engines in that time, Charpy became more concerned and interested in what caused these premature failures in these structures. That is when he started experimenting on the impact properties of steel because a lot of engineering structures then were built using steel as it was the strongest material in those days. During his experiments, he discovered that using a notch on test pierce improved the sensitivity of the measurement. He tested the specimen with the use of a hammer attached to a pendulum [2].

Properties such as material strength and material toughness can be determined through this test. The physical properties that determine its behaviour under the action of loading can be attributed to mechanical properties. Mechanical properties of materials are few in numbers, among which are strength, hardness and ductility. These properties depend on many factors among which are types of materials and operating conditions such as temperature. It is very important to know the properties of materials to make a proper material selection. The Charpy impact test is one of the techniques that aid to determine the mechanical properties of materials. This test is performed on the small sample extracted from the material, it determines how the mechanical properties of materials change under the influence of impact loading and temperature [3]. A Charpy test can be described as the scientific standard test that is used to measure the impact test of materials. The test simply displays the characteristics of the tested material during the fracture. Charpy Impact Test is a comparative dynamic test which evaluates the resistance of a material to fracture when subjected to an impact loading. From the test, the impact energy indicates the properties of the material (characteristics), toughness determined relative to the amount of impact energy and from that, a specimen can be determined to either be brittle or ductile depending on the fracture surface. The Charpy test machine measures the amount of energy absorbed on impact. This is generally a fast and economical test and mostly used on quality control applications [1–3]. The Charpy impact test is usually a method employed to determine the temperature dependence of the measured impact energy to be able to evaluate the transition from the ductile-brittle transition in metals. It can also be used to test the strength of composites, polymers, as well as ceramics. The Charpy test specimen has a U-notched or V-notched segment in the midsection and it is the supported at each end on the machine stand is impacted by a single blow of a free pendulum [4, 5]. The Charpy impact test is one of the numerous ways employed in the engineering field to determine the integrity and the failure mode of engineering materials. The test can be easily used by anyone because it is not complicated and the other advantage is that it costs less, this is the reason why the Charpy Impact Test is very versatile. The other purpose of the impact test is also to determine whether a specimen of material is brittle or ductile and to understand the effect of temperature of the specimen and the impact loading. This plays a huge role in engineering applications because it helps engineers to wisely select materials that are suitable for the surroundings where that materials are to be used and to also predict the mode of failure possible for that material [6–10].

Steel has an alloy of iron usually contains some elements of carbon to enhancing strength and resistance to fracture. Steel is a very important part of engineering materials because, unlike all other structural applications used in various fields, it has the greatest useful applications. Steels are commonly used as construction members in ship buildings, aerospace, military equipment, automobile etc. because they possess various desirable characteristics including good wear and corrosion resistance, low cost, good magnetic attributes as well as high strength behaviour. While rolling and casting are some of the methods traditionally used for steel production, in recent years the method of powder metallurgy has become preferred. Steels are categorized into low, medium, and high carbon steels dependent on their carbon content [11]. Samples of carbon steel grade AISI 1015 were exposed to heat rolling (700° C) by Medrea et al. (2014) [12] with 36.4% deformation. The materials' microstructure and mechanical characteristics were then evaluated by comparing to normalized specimen. The experiments have shown that warm rolling creates a very fine microstructure with comparable mechanical properties to those acquired via normalization. It is shown by considering technological and economic aspects that warm rolling offers important advantages compared to hot and cold working [12]. Zakir Hossain et al. (2019) [13] investigated effect of Cinnamaldehyde as a Green Inhibitor on the carbon steel (AISI 1015). In this study, 10% HCl was used as a medium for corrosion mitigation, it was established that the optimum dose of cinnamaldehyde and the performance of resistance were 200 ppm and 95.36% respectively. It was thought that the inhibition was primarily due to the presence of cinnamaldehyde on the surface of the metal that accompanied the isotherm Langmuir. The effect of temperature ranges from (25–85 °C), indicate that cinnamaldehyde's binding affinity on the surface of the metal is high. Chemical adsorption was associated with the slight increase in inhibition effectiveness. The value of the activation energy (Ea) acquired was observed to be reduced than that acquired in the HCl solution free of inhibitors. Another experiment was carried out by Boztepe and Bayramoglu, (2016) [14], In this report, AISI 1050 steel samples were taken to boronizing mechanism through using Ekabor 2 powder inside the bond jar in stainless steel. The experiments were conducted for 3, 6 and 9 h at temperatures of 800 °C, 850 °C and 900 °C to analyze the influence of various parameters on the sensitivity of boronized specimens to wear. Wear checking for pin-on-disk is used to describe the wear behaviour of boronized materials. Wear tests were conducted under static loading of 30 N in dry conditions, utilising abrasive paper of 220 mesh size Al2O3. For each of the test specimens, various rotational speeds of the pin on disk were selected as 300, 600, 900, 1200, 1500 revolutions. Weight losses of the specimens were measured after the abrasive study to confirm the abrasive wear resistance of boronized samples. The findings were also matched respectively with unboronized and conventionally hardened AISI 1050 steel samples. Similarly, in the research work of Aztekin and Ateş (2015) [15], it was noted that in compliance with ISO 3685, cutting variables were chosen as cutting depth (2,5 mm), cutting speed (100 m/min), and three feed rates (0,24, 0,32 and 0,40 mm/rev). AISI 1050 steel was machined under the same cutting conditions of Unsaturated Polyester Composite materials in order to allow a contrast with a material whose machining attributes are quite well established. Tool wear testing took more than 15 min. Tool wear was investigated using weight loss procedure and an electron scanning microscope. It is reported that the cutting forces produced during the operation of AISI 1050 (between 270 and 1383 N) are substantially lower (between 2 and 40 N) as compared to the cutting forces produced during UPC service.

In the same vein, this contribution of aluminium alloy AA1050 cannot be neglected in this research. The work of Mhedhbi et al. (2015) [16] aims at assessing the performance of the homogenized AA1050 alloy, and the various metallurgical factors like, annealing, cold-rolling as well as cold-rolling. Therefore, microhardness, tensile tests as well as well microstructures were examined. From the most important results, the optical micrographs show that the equiaxed grains are evidently elongated along the rolling direction, with increasing cold-rolling reduction rate. The concentration of rolling reduction enhances the hardening impact of the material, which is a good deal with improving strength and low plasticity. It was noted that, the elongation is increased to 36%, but the tensile strength is reduced to 86 MPa after annealing for 1 h at 350 °C. An article on the mechanical performance of aluminum metal matrix composites has been established by Singh and Kumar (2015) [17]. The following measurement were used 1050/10 wt percent Al_2O_3 alloy and 1050/10 wt percent SiC composite alloy. To produce these composites, the casting route was followed by stir. Specific experiments were carried out to study the physical behavior on these two MMC's and base alloy. In addition, the effects of improved particle size distribution on the composites' microstructure were explored using SEM. From this analysis it appears that heavy interfacial bonding of Al 1050 to SiC & Alumina enhances the compressive strength. Through this analysis it appears that the magnitude of UTS and yield strength increases with the SiC and Al_2O_3 weight percentage in the matrix being 10%. An experimental study was performed concerning the variability of the mechanical properties of aluminum alloy sheets 1050A due to heat treatment [18]. For this purpose, the laboratory prepared end sheets and foils from continuous casting strip (CCS) and hot rolling strip (HRS). Originally CCS and HRS were 7.5 mm thick and 3.0 mm thick, respectively. The final thickening reported for both cases was 0.20 mm. Samples of various thicknesses were taken during the rolling process to get curves for CCS and HRS sheets and foils to harden. Oven processes were performed for drawing softening curves of foils. CCS foil hardening is caused by precipitation of Al_3Fe particles at a temperature range of 160 °C to 220 °C. Despite Fe's low over-saturation, HRS foil has lower hardening in the same way.

2 Materials and Methods

2.1 Materials

The materials used in this study were aluminium alloy AA1050-F and the carbon steel EN - 3 (~1015) of 0.15% C. A spark test was employed on both samples to determine the chemical composition of the materials used. The chemical composition of the AA1050-F aluminium material used is presented in Table 1 while that of carbon steel (EN 3) is presented in Table 2.

Table 1. Spectrometer analysis result for Pure Aluminium 1050–F

Element	Cu	V	Fe	Ti	Mg	Zn	Si	Mn	Al
Wt. %	0.05	0.05	0.40	0.03	0.05	0.05	0.25	0.05	99.50

Table 2. Spectrometer analysis result for mild steel EN–3 (~1015)

Element	C	P	Mn	Cu	Si	Al	Fe
Wt. %	0.15	0.025	0.18	0.09	0.168	0.07	Balance

2.2 Experimental Procedures

In this experiment, aluminium alloy 1050-F and carbon steel 1015 were tested for impact loading. Impact samples of each metal were fabricated, cut and v-notch at an angle of 45 degrees. Each of the specimen was subjected to heat treatment before impact testing. To achieve this, a cryogenic flask containing liquid nitrogen was used alongside with methanol as well as boiled water at 100 °C. Boiling water was poured in a mixture of nitrogen and methanol which serves a stabilizing agent. So, once the boiled water is poured into the beaker, there will be no violent reaction. At this juncture, the temperature reading was measured and recorded. We must ensure that that the sample is first place inside the sample holder before dipping it inside the mixture. Each Specimen was soaked into liquid nitrogen which is at approximately –90 °C held by tongs in the cryogenic flask and as the hot water at 100 °C is poured until the desired temperature is reached, and the infrared thermometer is used in the process to measure the temperature attained during the process. We must ensure that we keep pouring boiled water to raise the temperature until the temperature of the mixture is below –50 °C. This is to ensure that the infrared thermometer has the capacity to read such lower temperatures.

It is pertinent to state that, Liquid nitrogen was poured into the cryogenic flask, due to the low temperature of the nitrogen, methanol was added to increase the temperature of the contents in the flask to the desired temperature of –50 °C the two sample pieces were then placed into the flask so they can come up to the same temperature as the content in the flask, the sample pieces were left for a few seconds and then they were taken out refer to Fig. 1. Each of the specimens was then placed onto the Charpy impact machine to test the impact energy. The same procedures were repeated for different temperatures which are from –50 °C to 125 °C with a step of 25 °C. For the 25 °C, the water was assumed to be at room temperature and for the 50 °C, the boiled water in the kettle was used to bring up the temperature of the mixture.

Fig. 1. Specimen after removal from liquid nitrogen

2.3 Charpy Experimental Test

The Charpy impact tester is strictly designed according to the standards ASM E23; ISO 148 and IS 1598:1997. The specimen that is used for the Charpy test is a bar containing a notch. The source of the impact is a swinging hammer. This swinging hammer is suspended at an angle of 45 ° from the horizontal surface. When the arm is released during the test, it hammers the specimen directly on the opposite side of the notch, causing the specimen to break. The impact energy from the hammer can give a clear and visual indication of the material property such as toughness. The specimen which is being tested decelerates the movement of the swinging hummer by directly absorbent of the energy of the impact. A specimen with higher toughness will have a great absorption of the energy during the impact, causing the hummer to swing at a very low height after the impact [19]. The Charpy impact test was conducted with a V-notch specimen, the temperature for Charpy impact test for this study was from –50 °C to 125 °C with a step of 25 °C. The test was conducted by ASTM A370 and ASTM E23 standard. The schematic illustration of the test samples before and after the 45° included angle V-notch with 2 mm depth at the centre is presented in Fig. 2a. Each of the steel and aluminium bars is in the dimension of 55 mm by 10 mm as shown in Fig. 2a. The prepared samples for aluminium and steel with the notch is presented in Fig. 2(b and c) respectively. The fracture surfaces for aluminium and steel are shown in Fig. 3(d and e) respectively.

During the experiment, the hammer is being released from a predetermined height of h, it strikes the specimen directly at the notch, causing it to break into two separate

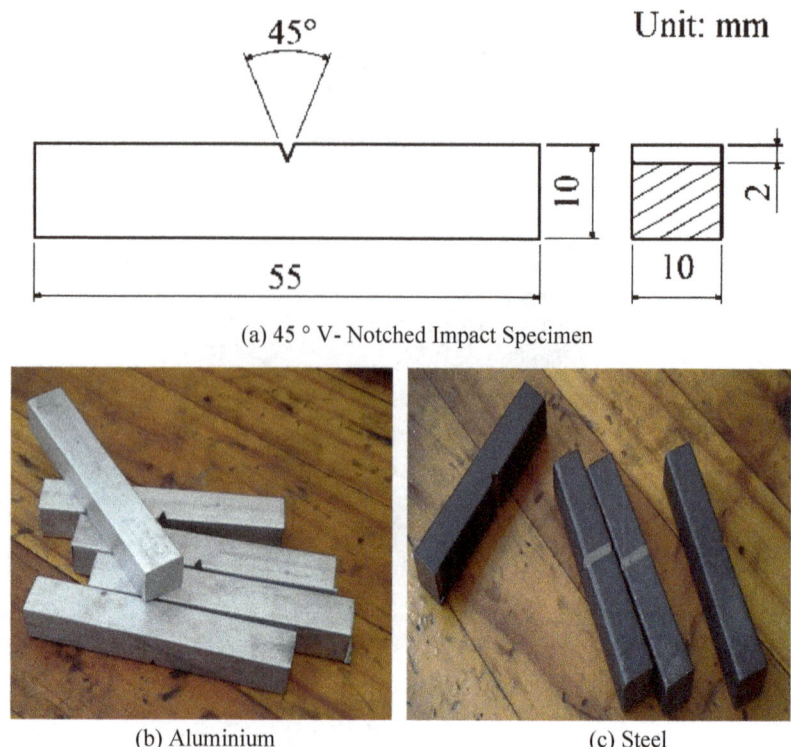

(a) 45 ° V- Notched Impact Specimen

(b) Aluminium   (c) Steel

Fig. 2. Notched specimen

pieces or fractured. The pendulum will carry on swinging rising a maximum height of h^i while the energy absorbed (E) at the fracture can be obtained by calculating the potential energy difference of the hammer which oscillates like the pendulum before and after the test [4]. The equation for the potential difference is shown below:

$$E = m * g(h - h^{\cdot}) \tag{1}$$

Where m stands for the mass of the pendulum and g stands for acceleration due to gravity [4].

2.4 Digital IR Infrared Thermometer

The digital IR Infrared Thermometer is shown in Fig. 4. It was used to read and record the temperature of the liquid mixture employed during the Charpy Impact Test. In the case of non-sensitivity of the IR infrared, there is a connection at the top which is comprising of contact and penetration probe.

3   Results and Discussions

The results obtained from the experiment presented in Tables 3 and 4. The material A is for aluminium result and material B is for carbon steel material (EN-3). Each table is

(a) Fractured EN-3 specimen   (b) Fractured Aluminium specimen

Fig. 3. Fractured specimens

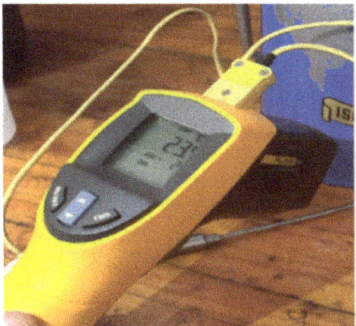

Fig. 4. Infrared thermometer

accompanied by graphs which displays the behaviour of the specimen during the test. Table 3 shows the results obtained during the Charpy Impact Test of Aluminium specimen. Table 4 shows the results obtained during the Charpy Impact Test of EN3 specimen. Each experiment at each temperature was taken in triplicate to ensure consistency and accuracy. The average value of the three readings was then taken and recorded. The graph in Fig. 5 shows the impact of energy vs temperature of aluminium alloy. Three specimens were tested at each temperature for reproducibility.

The Charpy test was conducted and the experiment reading was recorded in tabular form, as shown above, Steel and aluminium absorbed different impact energy during fracture. Table 3 (Aluminium reading) and Table 4 (Steel reading) above shows reading of impact energy recorded with an initial temperature of −50 °C with an increment of 25 °C up to a maximum temperature of 125 °C. The absorbed energy at the point of fracture was well recorded and for one reading of temperature, there are three impact energy readings added together to get the average impact energy. The readings recorded

Table 3. Impact test results for aluminium specimen

Material A	Impact energy (J)			
Temperature (˚C)	A_1	A_2	A_3	Average impact energy (J)
−50	2.5	2	2	2.1
−25	2.5	2	2	2.1
0	5	3	2.5	3.5
25	10	7	4	7
50	14	12	8	11.33
75	16	16	13	15
100	12	16	16	14.67
125	16	16	16	16

Table 4. Impact test results for EN3 specimen

Material B	Impact energy (J)			
Temperature (˚C)	B_1	B_2	B_3	Average impact energy (J)
−50	8.5	8.5	8.5	8.5
−25	9	8.5	8.5	8.67
0	9	9	9	9
25	10	10	9	9.67
50	14	12	10	12
75	16	16	13	15
100	16	16	18	16.67
125	16	16	16	16

in both Tables 3 and 4 show that when the temperature increases also impact energy increases and this implies a direct proportionality relationship. The higher the energy the greater the impact energy. The absorbed impact energy readings of steel from −50 °C to 0 °C were greater than 5J while for aluminium were less than 5J. Steel absorbed a lot of energy compared to the impact energy absorbed by the aluminium. It was revealed that aluminium specimen experienced an increase in impact energy up to 75 °C, it then declined and rose again up until 125 °C. A similar trend was noticed in carbon steel specimen, the impact energy increases with increase in temperature until 100 °C and a reduction in impact energy was experienced at 125 °C resulting to absorbed energy of 16J. The maximum impact energy absorbed by both material specimen at maximum temperature OF 125 °C was found to be 16 J but there exists higher impact energy for steel at 100 °C resulting to 16.67 J. For average impact energy of aluminium and steel, the two material only had two similar values of impact energy, which was 15 J at 75 °C and 16 J at 125 °C (see Tables 3 and 4 above). For the aluminium the impact energy distribution over the temperature used is less than that of the steel, this shows that the aluminium is more malleable and will absorb less energy in different conditions compared to the steel. The ductility of steel/ the more energy absorbed by the steel is proportional to the test temperature, hence the higher the temperature the more impact energy and this enhances its ductility. The graph in Fig. 6 shows the impact of energy vs the temperature for individually tested steel. Again, three specimens were tested at a particular temperature to ascertain the consistency.

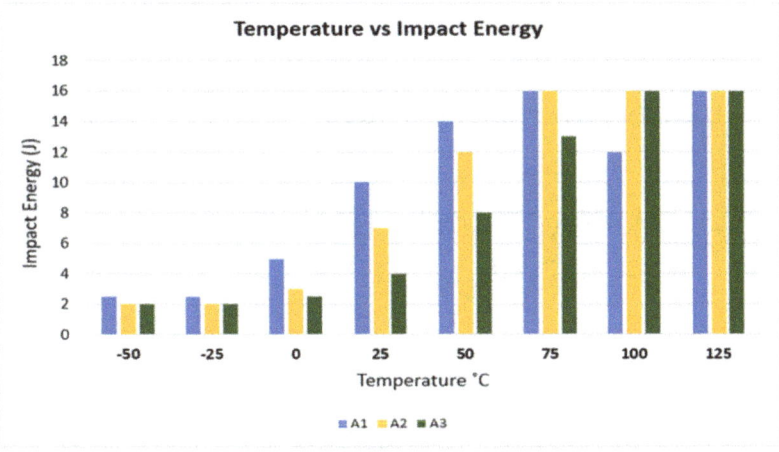

Fig. 5. Temperature versus impact energy for individual tested aluminium

Figure 7 shows graphical results obtained during the Impact Test Experiment of Aluminium specimen. The point where ductile to brittle transition occurs is shown by the red arrow. While Fig. 8 shows graphical results obtained during the Impact Test Experiment of Aluminium specimen. The point where ductile to brittle transition occurs is shown by the red arrow.

For steel, the transition temperature is between 40 and 50 °C which gives approximately an average of 49.3 °C which becomes the transition temperature for steel. For Aluminium, the transition temperature is approximately 36.3 °C. The ductile to brittle transition is dependent on the composition of the steel. The metal with high carbon content EN3 (0.15%wt carbon) has a higher transition temperature compared to aluminium. The EN-3 has a wider transition zone compared to the one for aluminium. Fracture toughness of the material is related to the impact strength of the metal and form the graph it can be seen on both graphs that EN-3 steel has higher fracture toughness than aluminium because the mean impact energy is higher. At low temperatures, the aluminium is more brittle because less impact strength [20] is seen on the graph and, at

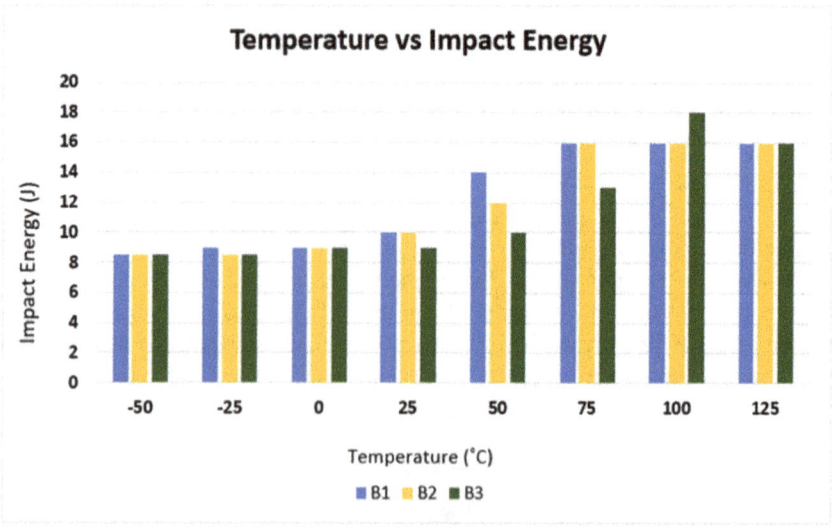

Fig. 6. Temperature versus impact energy for the individual tested steel

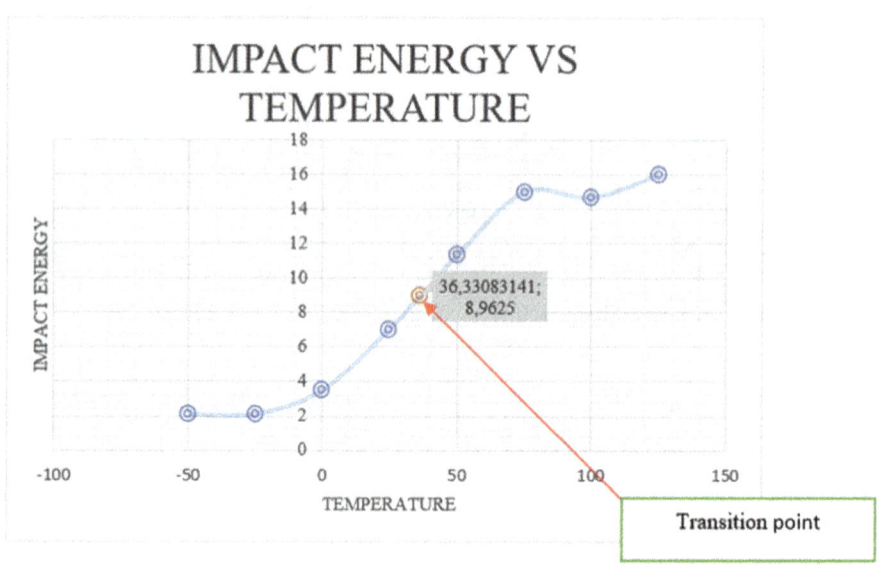

Fig. 7. Graphical results for aluminium specimen

intermediate temperatures, the aluminium is more ductile than EN-3 steel. After all, the impact strength is increasing at a faster rate and at higher temperatures the aluminium is more ductile because impact strength is higher than that of EN-3 steel. The metal with higher carbon content is supposed to be more brittle. The carbon content seems to be having no effect on impact strength at lower temperatures because the EN-3 steel is brittle compared to the aluminium at higher temperatures.

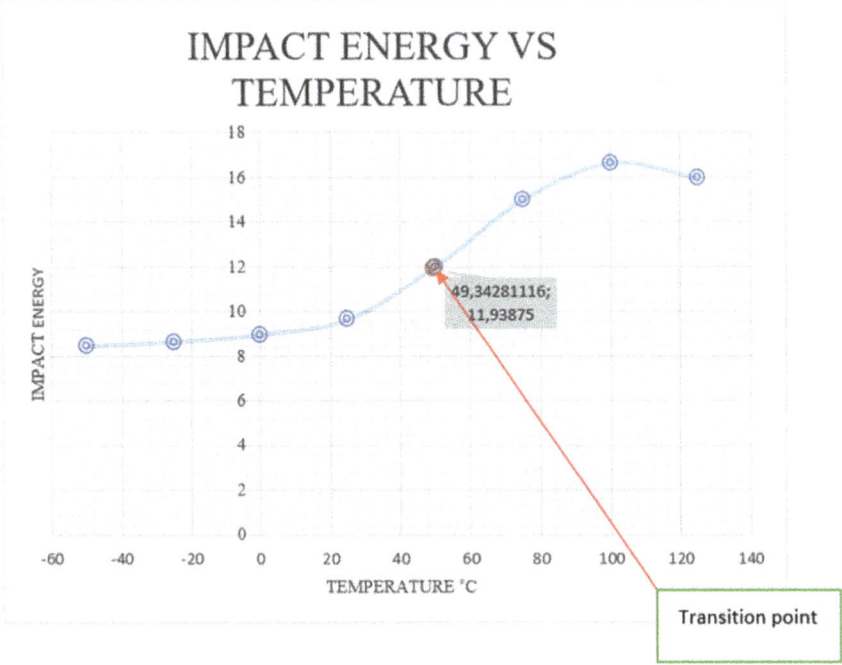

Fig. 8. Graphical results for EN3 specimen

Carbon content affects the impact energy because an increase in carbon content makes the specimen harder. As the hardness increase, the specimen becomes more brittle and thus decreases the impact energy. The opposite is also true, a decrease in carbon content, decreases the hardness of material making it more ductile and as a result increases the impact energy. Carbon content affects the ductile-to-brittle transition temperature. Increasing the carbon content shifts the curve toward the brittle end of the curve, decreasing impact energy. Decreasing carbon content shifts the curve to the left i.e. to the ductile side, increasing impact energy. Brittle and ductile fractures can be distinguished by the textures of the fractured materials (T, 2019). The surface of the specimens was smooth, shiny and more granular at low temperatures, as opposed to the cup and cone surfaces with a fibrous texture, showing characteristics of brittle fracture. As the temperature increased the surfaces became less levelled and had a more fibrous texture which showed that the materials became more ductile as the temperature increased. Lateral deformation refers to the necking of a material as it deforms. Increased carbon content makes the material harder and more brittle [7, 8]. As the material becomes more brittle, the necking or lateral deformation becomes less, instead, the material reaches a point whereby it fractures without necking. Thus, the more carbon content the material has, the lesser it necks. The more the temperature increases the more ductile the material becomes. The more it deforms laterally before it fractures.

Three specimens were tested at each temperature, starting from –50 °C to 125 °C by an increment of 25 °C. From Table 1 and 2, it is evident that the increase in temperature

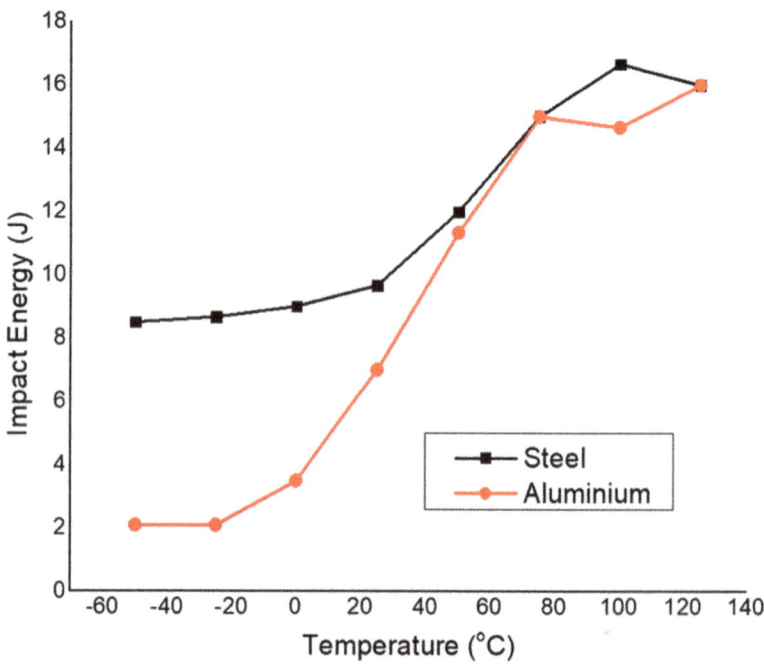

Fig. 9. Comparison of impact energy for EN3 and AA1050-F specimen

leads to an increase in impact energy, and it should be also noted that the impact energy at the same temperature is not always the same. The cause of that might be that the specimens were not subjected to an exact loading. The pendulum might have been given a small, unintentional and unnoticeable push when it was released. This minor error produces unfair results, hence, to mitigate the error we take the average of the energies as reflected in Tables 5 and 6. There is a direct relationship between the temperature and the impact energy, i.e. the increase in one parameter leads to the increase in the other. With the increase in temperature, the material becomes more ductile, and thus absorbs more energy. This is in line with the theory [21]. Recall that a ductile fracture is characterized by absorbing a high amount of energy before fracture. In contrast, a brittle fracture is distinguished by absorbing a low amount of energy before fracture. Hence, it can be established that the lower the temperature, the more brittle the material becomes and the higher the temperature, the ductile the material becomes. Two types of specimens were examined in this experiment, i.e. steel and aluminium. These two specimens were tested over the same temperatures and loading. This was done to examine and compare their behaviour and to obtain fair and reliable results. Three specimens were tested at each temperature, to make the experiment even fairer. Upon testing the two specimens, it was found that the Aluminum specimen absorbed less energy before it ruptured than the Steel specimen. This suggests that the aluminium specimen is more malleable than the steel while steel is more ductile than aluminium specimen under the same environmental conditions. The graph of temperature vs average in Fig. 9 shows that the steel curve is always above the aluminium curve, reinforcing that Aluminum is more ductile than steel.

3.1   Calculating Impact Strength

To compute the impact strength of each material, considering the details in Fig. 10, The details are used to evaluate the impact strength of each specimen under investigation.
   The impact strength is given by:

$$\text{Impactstrength} = \frac{E}{A} \tag{2}$$

Where:
   E = Energy needed to fracture the specimen
   A = ligament cross sectional area of the specimen in m^2
   h = is the width of the specimen excluding the notch width in (mm)
   d = is the width of the specimen in (mm)

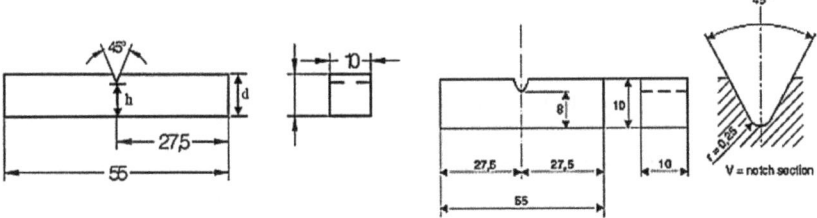

Fig. 10. Impact specimen showing computing details for impact strength

The specimen has the following dimensions:

$$d = 10mm$$

$$h = 8\ mm$$

$$V\ notch\ angle = 45°$$

$$V\ notch\ depth = 2\ mm$$

Therefore, the ligament cross-sectional area is given by:

$$A = \frac{10\ x\ h}{10^6} \tag{3}$$

$$= \frac{10\ x\ 8}{10^6} = 8\ x\ 10^{-6}$$

   Table 6 below shows the rest of the results of impact energy for EN-3 specimen. The calculations were computed according, and the results are displayed in Table 6.
   It was observed that an increase in temperature also leads to an increase in impact strength as well as impact energy. The impact strength for both materials is presented

Table 5. Impact strength results for aluminium specimen

Temperature (°C)	Average impact energy, E (J)	Area (m²) ×10⁻⁶	Impact strength $= (\frac{E}{A}) (\frac{KJ}{m^2})$
−50	2.1	8	262.5
−25	2.1	8	262.5
0	3.5	8	437.5
25	7.0	8	875.0
50	11.33	8	1416.25
75	15.0	8	1875
100	14.67	8	1833.75
125	16.0	8	2000

Table 6. Impact strength for EN3 specimen

Temperature (°C)	Average impact energy, E (J)	Area (m²) ×10⁻⁶	Impact strength $= (\frac{E}{A}) (\frac{KJ}{m^2})$
−50	8.5	8	1062.5
−25	8.67	8	1083.75
0	9.0	8	1125
25	9.67	8	1208.75
50	12.0	8	1500
75	15.0	8	1875
100	16.67	8	2083.75
125	16.0	8	2000

in Tables 5 and 6. The ductile to brittle transition for all specimen used in this experiment were found to be 36.3 for aluminium specimen and 49.34 for EN-3 specimen. Carbon content makes the material hard and increase its brittleness. The ductile to brittle transition temperature decreases with an increase in carbon content. Furthermore, the results for EN-3 specimen evidenced that carbon content increases impact test and hence impact strength also increases. The deformation of material also increases with an increase in temperature. The morphology of the EN3 shows that EN3 is more ductile than steel, while the morphology for Aluminium specimen shows that Aluminium is less ductile compare to steel but more malleable. For Aluminium, the impact energy remains constant at low temperature and increases at high temperature. The maximum impact energy of aluminium occurs at Temperature of 85 °C while for EN-3 occurs at a temperature of 100 °C.

From the obtained results of the experiment, each specimen fractured differently, as the brittle fracture and ductile fracture were observed. The brittle fracture was observed where there was no deformation and where the impact energy was very low, while the ductile fracture was observed where there was too much deformation of the specimen at high impact energy. The steel easily fractured at low temperature, but as the temperature increased, the more it was difficult for the crack to initiate and propagate. The carbon steel changed the phase from ferrite at low temperatures to austenite at high temperatures. The steel was brittle at low temperatures and too ductile as the temperature increased, meaning at high temperatures the steel did not break into two parts after the impact, thus more impact energy was absorbed. From theory, we know that carbon steel has low carbon content therefore it becomes brittle at low temperatures. The carbon steel shows the ductile to brittle transition temperature between –25 °C to 70 °C resulting in 49.34 °C. Aluminium has a face-centred cubic crystal structure, meaning it does not obey the behaviour of ductile to brittle transition. The addition of carbon to steel makes it to be stronger but it does not change the ductile to brittle transition behaviour. Adding alloys such as chrome, vanadium and nickel to steel can decrease the ductile to the brittle transition temperature.

3.2 Examination of Fractured Surfaces

The fracture surfaces were studied using a scanning electron microscope and the fractography at the beginning of the test which is –50 °C, at 0 °C as well as 125 °C were studied to know the effects of temperature at fracture. It was revealed in Fig. 11A that the fracture mechanism of aluminium alloy has large fibroid structures with lots of thread-like structures and large dimples which may be due to the ductility property of the sample. Figure 11B shows more revealing fibroid structures with onion rings structures and smaller dimples which indicated that is more ductile than the sample in Fig. 11A. The fracture surfaces also influenced by the number of carbon contents in each tested sample. Carbon content affects the impact energy because an increase in carbon content makes the specimen harder. As the hardness increase, the specimen becomes more brittle and thus decreases the impact energy. The opposite is also true, a decrease in carbon content, decreases the hardness of material making it more ductile and as a result increases the impact energy. Carbon content affects the ductile-to-brittle transition temperature. Increasing the carbon content shifts the curve toward the brittle end of the curve, decreasing impact energy. Decreasing carbon content shifts the curve to the left i.e. to the ductile side, increasing impact energy. Brittle and ductile fractures can be distinguished by the textures of the fractured materials [5, 22, 23]. The surface of the specimens was smooth, shiny and more granular at low temperatures, as opposed to the cup and cone surfaces with a fibrous texture, showing characteristics of brittle fracture. As the temperature increased the surfaces became less levelled and had a more fibrous texture which showed that the materials became more ductile as the temperature increased. Lateral deformation refers to the necking of a material as it deforms. Increased carbon content makes the material harder and more brittle. As the material becomes more brittle, the necking or lateral deformation becomes less, instead, the material reaches a point whereby it fractures without necking. Thus, the more carbon content the material has, the lesser it necks. The more the temperature increases the more ductile the material becomes. The more it deforms laterally before it fractures [24–26].

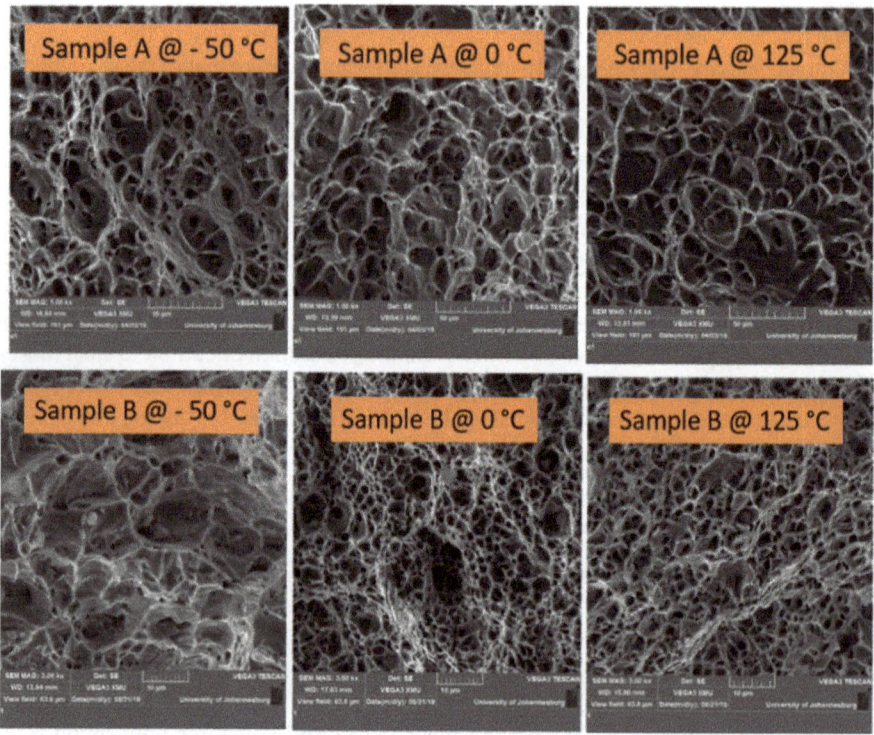

Fig. 11. Fractography of samples A and B at the lowest, zero and highest temperatures.

4 Conclusion

The experiment was conducted successfully, with the results recorded, the effect of temperature on material properties was shown. As the temperature increased, the impact energy absorbed increased. Steel and aluminium absorb more impact energy at higher temperatures compared to a lower temperature. From the results obtained the two materials absorbed different impact energy at different temperatures. Steel absorbed a lot of energy at the lowest temperature compared to that of aluminium and it became stronger but remained ductile. The results obtained were consistent with the theoretical background behaviour of steel. The steel at room temperature fractured in a ductile manner, when the temperature has lowered the point where the behaviour changed to brittle was reached and it was the ductile-brittle transition temperature. It is evident from the graphs, the captured of the fracture surfaces and the table of results that material becomes more ductile with an increase in temperature. From the specimen, it can be observed that the fracture surfaces become dull as temperature increases.

From this experiment, it can be concluded that steel is more ductile than aluminium, but aluminium is more malleable than steel. This was proven and tested under different temperatures. Hence in design, when selecting the material, one should have clear information about what temperatures the design will be operating to select the suitable material. The experiment supports the theory and impact energy does increase with an increase in temperature. The experiment has also shown that the content of carbon and temperature play a huge role in the properties of a material.

The ductile to brittle transition temperature also increases with an increase in temperature. It is also further noticed that an increase in carbon content increases the hardness and brittleness of material and hence decrease the impact energy. The Charpy impact test has successfully experimented, and it helped to determine the relationship between temperature and impact energy, which allows engineers in real-life situations to know how to select materials for specific functions.

Acknowledgements. The authors wish to acknowledge the financial support offered by Pan African University for Life and Earth Sciences Institute (PAULESI), Ibadan, Nigeria for the payment of article publication charges (APC).

References

1. Hrabe, N., White, R., Lucon, E.: Effects of internal porosity and crystallographic texture on Charpy absorbed energy of electron beam melting titanium alloy (Ti-6Al-4V). Mater. Sci. Eng. **742**, 269–277 (2019)
2. Sorrentino, L.: Comparison of falling dart and Charpy impacts performances. Compos. B **165**, 102–108 (2019)
3. Veronica, S.C.: Toughness of polyester matrix composites reinforced with sugarcane bagasse fibers evaluated by Charpy impact tests. Mater. Res. Technol. **6**(4), 334–338 (2017)
4. Chang, H.T.: Microstructural characteristerizer of Charpy-impact tested nanostructured Bainite. Mater. Characteristics **107**, 63–69 (2015)
5. Groove, M.P.: Fundamentals of Modern Manufacturing: Materials, Processes and Systems, 4th edn., pp. 693–1190. John Wiley & Sons, Inc., Hoboken (2010)
6. Han, Y., Lach, R., Grellmann, W.: The Charpy impact fracture behaviour in ABS materials. Die Angewandte Makromolekulare Chemie **270**(1), 13–21 (1999)
7. Ikumapayi, O.M., Okokpujie, I.P., Afolalu, S.A., Ajayi, O.O., Akilabi, E.T., Bodunde, O.P.: Effects of quenchants on impact strength of single-vee butt welded joint of mild steel. In: IOP Conference Series: Materials Science and Engineering, vol. 391, p. 012007 (2018)
8. Ikumapayi, O.M., Akinlabi, E.T.: Composition, characteristics and socioeconomic benefits of palm kernel shell exploitation-an overview. J. Environ. Sci. Technol. **11**, 220–232 (2018)
9. Muhammad Said, N.B., Ali, M.B., Zakaria, K.A., Daud, M.A.M.: Comparison of impact duration between experiment and theory from Charpy impact test. MATEC Web Conf. **7**, 01054 (2016)
10. Ikumapayi, O.M., Akinlabi, E.T., Majumdar, J.D., Akinlabi, S.: Characterization of high strength aluminium – based surface matrix composite reinforced with low-cost PKSA Fabricated by friction stir processing. Mater. Res. Express **6**, 1–27 (2019)

11. Abdulkareem, S., Ogedengbe, T.S., Aweda, J.O., Ajiboye, T.K., Khan, A.A., Babatunde, M.A.: Investigation on effect of material compositions on machinability of carbon steels. In: Journal of Physics: Conference Series, vol. 1378, p. 022046 (2019)
12. Medrea, C., Negrea, G., Domsa, S.: Study on mechanical properties of warm-rolled AISI 1015 carbon steel. Z. Met. **95**(3), 176–178 (2014)
13. Zakir Hossain, S.M., et al.: Cinnamaldehyde as a Green Inhibitor in Mitigating AISI 1015 Carbon Steel Corrosion in HCl. Arab. J. Sci. Eng. **44**, 1–12 (2019). https://doi.org/10.1007/s13369-019-03793-y
14. Boztepe, M.H., Bayramoglu, M.: The effect of process parameters on the abrasive wear resistance of boronized AISI 1050 steel. In: Proceedings of the ASME 2016 International Mechanical Engineering Congress and Exposition, IMECE2016, pp. 1–5 (2016)
15. Aztekin, K., Ateş, E.: Machinability of unsaturated polyester composite and AISI 1050 in turning. J. Multi. Eng. Sci. Technol. **2**(11), 3048–3053 (2015)
16. Mhedhbi, M., Khlif, M., Bradai, C.: Investigations of microstructural and mechanical properties evolution of AA1050 alloy sheets deformed by cold-rolling process and heat treatment annealing. J. Mater. Environ. Sci. **8**(8), 2967–2974 (2015)
17. Singh, K., Kumar, M.: Analysis on mechanical behaviour of AA 1050/SiC and AA 1050/Al2O3 composites. Int. J. Sci. Res. **6**(6), 2025–2030 (2015)
18. Pérez-Ilzarbe, J., Fernández Carrasquilla, J., Luis Pérez, C.: A study of the mechanical properties and recrystallization of two types of foil of aluminium 1050 A. Key Eng. Mater. **423**, 137–145 (2010)
19. Anderson, T.L.: Fracture Mechanics: Fundamentals and Applications, 3rd ed, pp. 299–344. Taylor & Francis Group, Florida (2005)
20. Callister, W., Rethwisch, D.: Materials Science and Engineering, 8th edn., pp. 235–274.John Wiley & Sons Inc., Massachusetts (2010)
21. Kawata, H., Umezawa, O.: Two-step ductile to brittle transition behavior on ferrite +pearlite structure steel sheet. ISIJ Int. **57**(7), 1–7 (2017)
22. Adeoti, O.M., Dahunsi, O.A., Awopetu, O.O., Oladosu, K.O., Ikumapayi, O.M.: Optimization of clay-bonded graphite crucible using d-optimal design under mixture methodology. Int. J. Sci. Technol. Res. **8**(7), 444–461 (2019)
23. Azeez, T.M., Ikumapayi, O.M., Bodunde, O.P., Babalola, S.A., Ogundayomi, M.O.: Measurement of surface roughness on a transmission shaft using CNC and conventional lathes machining. Int. J. Sci. Technol. Res. **8**(10), 1626–1633 (2019)
24. Afolalu, S.A., et al.: Experimental analysis of the wear properties of carburized HSS (ASTM A600) cutting tool. Int. J. Appl. Eng. Res. **12**(19), 8995–9003 (2017)
25. Okokpujie, I.P., et al.: Modeling and optimization of surface roughness in end milling of aluminium using least square approximation method and response surface methodology. Int. J. Mech. Eng. Technol. **9**(1), 587–600 (2018)
26. Oyinbo, S.T., Ikumapayi, O.M., Jen, T.C., Ismail, S.O.: Experimental and numerical prediction of extrusion load at different lubricating conditions of aluminium 6063 alloy in backward cup extrusion. Eng. Solid Mech. **8**(2), 119–130 (2020)

The Optimization of the Surface Roughness of Milled Polypropylene + 60wt.% Quarry Dust Composite Using the Taguchi Technique

Harrison Shagwira[1], F. M. Mwema[1,2(✉)], J. O. Obiko[3], T. O. Mbuya[4], and E. T. Akinlabi[5]

[1] Department of Mechanical Engineering, Dedan Kimathi University of Technology, Nyeri, Kenya
`fredrick.mwema@dkut.ac.ke`
[2] Department of Mechanical Engineering Science, University of Johannesburg, Auckland Park Kingsway, South Africa
[3] Department of Mining, Materials and Petroleum Engineering, Jomo Kenyatta University, Nairobi, Kenya
[4] Department of Mechanical and Manufacturing Engineering, University of Nairobi, Nairobi, Kenya
[5] Pan African University for Life and Earth Sciences Institute (PAULESI), Ibadan, Nigeria

Abstract. This study is based on the optimization of the parameters that influence the computer numerical control (CNC) milling operation during the machining of polypropylene+60wt.% quarry dust composite. The input parameters studied are the cutting speed, the feed rate and the depth of cut. These input parameters were optimized using the Taguchi optimization technique with the output response taken into consideration was the surface roughness. An L_9 orthogonal array (OA) was selected and formulated in a commercial software Minitab 19 based on three factors and three levels combination. The signal-to-noise (S/N) ratio was analysed to give a combination of values of the input parameters that produced optimum results for surface roughness. The analysis of variance (ANOVA) was then conducted to determine the significance and percentage contribution of each parameter. From the results, the optimum values obtained were cutting speed of 1000 rpm, feeding rate of 120 mm/min and depth of cut of either 0.5 mm or 0.8 mm. The cutting speed had the highest contribution towards the surface roughness at 81.98%, followed by the depth of cut at 7.43% and the feed rate having the least contribution at 3.69%.

1 Introduction

The use of quarry dust in the fabrication of plastic composites is a novel idea. Only a few publications exist for the production of plastic-sand composites [1, 2]. The interaction between the quarry dust and the plastic matrix, therefore, has not been fully explored. Owing to the hardness property of quarry dust, this composite can find its application in the construction industry. However, to achieve dimensional accuracy and avoiding

M. Awang and S. S. Emamian (Eds.): *Advances in Material Science and Engineering*, LNME, pp. 169–174, 2021.
https://doi.org/10.1007/978-981-16-3641-7_20

fitting problems, it is necessary to carry out a machining operation on the fabricated composites.

A lot of research has been carried out on the machinability of fibre reinforced plastic (FRP) composites [3]. The machining operations such as turning, shaping, milling, drilling, to mention but a few can be performed on plastic-quarry dust composite. Due to inhomogeneity and the anisotropic nature of fibre reinforced plastic composites, the machining process of these composites is different from that of metals and tends to be complex. Plastic-quarry dust composite is not an exception. The orientation and the hard nature of the quarry dust may affect the machining of the composite. The cutting mechanism applied to the composed may cause shearing, deformation and rupturing of the quarry dust particles in the plastic matrix. This will tend to affect the surface roughness of the machined surface of the plastic-quarry dust composite. The guidelines of machining FRP are available [4] but none exists for plastic-quarry dust composite.

The existing optimization methods such as the Taguchi optimization are required to find optimum input machining parameters with the theoretical models used for prediction purposes to achieve good surface quality and better dimensional tolerances. The Taguchi optimization technique can be used for such purposes conveniently. Sreenivasula [5] utilized the Taguchi optimization technique in optimizing the input parameter used in end milling to achieve a better surface roughness. Several researchers have carried out the optimization of input parameters towards the surface roughness on plastics composite materials [6–8].

This study aims at performing Taguchi optimization on the surface roughness of CNC milled PP+60wt.% quarry dust. An L_9 orthogonal array was used to create a combination of the input parameters to be used in the study. The input parameter investigated were the cutting speed, the feed rate and the depth of cut. The signal to noise (S/N) ratio and the analysis of variance (ANOVA) were used in the analysis to find the optimal parameters.

2   Experimental Method

The material used in this study was PP+60wt.% quarry dust composite produced by authors of this research thorough compression moulding [9]. The machine used was a computer numerical control (CNC) machine model (EMX 2000). A high-speed steel tool was used during the milling process.

The design of experiments (DoE) was based on the Taguchi technique which has proved to be a powerful tool in optimization. A commercial software Minitab 19, was used to formulate the optimization. Three factors (cutting speed feed rate and depth of cut) and three levels as shown in Table 1 were used to come up with an L_9 orthogonal array as shown in Table 2. A series of nine experiments were performed and the surface roughness of the machined surface was taken as the response. Each sample measured 40 mm × 10 mm × 4 mm. A handheld roughness tester (TR200, Netherlands) was used to measure the surface roughness as the response.

A signal-to-noise (S/N) ratio was then carried out to determine the optimum parameters that affected the surface roughness. A small-is-better quality characteristic was chosen because the surface roughness is desired to be as small as possible, approaching zero. The ANOVA was then performed on a 5% significant level to determine the significance and percentage contribution of each factor towards the surface roughness.

Table 1. The three factors and three levels used in the Taguchi Optimization of surface roughness of PP+60wt.% quarry dust.

Factor	Notation	Level of factors		
		L_1	L_2	L_3
Speed of cut (rpm)	A	300	600	1000
Feed rate (mm/min)	B	30	120	200
Depth of cut (mm)	C	0.3	0.5	0.8

Table 2. The Taguchi L9 Orthogonal array having the experimental values of the factor levels.

Experiment number	Milling parameters levels		
	Speed of cut (rpm)	Feed rate (mm/min)	Depth of cut (mm)
1	300	30	0.3
2	300	120	0.5
3	300	200	0.8
4	600	30	0.5
5	600	120	0.8
6	600	200	0.3
7	1000	30	0.8
8	1000	120	0.3
9	1000	200	0.5

3   Results and Discussion

3.1   The Taguchi Optimization of the Surface Roughness

Table 3 shows the various surface roughness values obtained on the milled surface of the PP+60wt.% quarry dust composite. The smaller-is-better criterion was selected since the lowest values of surface roughness were desirable. For each of the nine experiments, three roughness values were obtained and the average value was recorded to minimize the errors that might arise during the measurements.

The signal-to-noise ratios were then obtained. When carrying out the Taguchi technique, the term 'signal' means the output characteristic's desirable values i.e. the mean, while the term 'noise' means the output characteristic's undesirable value i.e. the standard deviation. Therefore, the signal-to-noise ratio represents the deviation of the output quality characteristic from the mean value.

Table 3. Experimental parameter with corresponding results for surface roughness of PP+60wt.% quarry dust composite

No.	Speed of cut (rpm)	Feed rate (mm/min)	Depth of cut (mm)	Surface roughness, Ra (µm)
1	300	30	0.3	2.87767
2	300	120	0.5	2.67767
3	300	200	0.8	2.59625
4	600	30	0.5	1.08000
5	600	120	0.8	0.91233
6	600	200	0.3	2.15433
7	1000	30	0.8	1.23900
8	1000	120	0.3	1.07250
9	1000	200	0.5	1.00867

Table 4 shows the response table for the signal-to-noise ratios for the surface roughness of the surface of the machined PP+60wt.% quarry dust composite. The speed of cut is observed to have the biggest influence on the surface roughness as compared to all other parameters, with the depth of cut having a moderate influence while the feed rate has the least influence on surface roughness during end milling of PP+60wt.% quarry dust composite.

Table 4. Response table for signal to noise ratios for surface roughness based on Smaller is better quality characteristic

Level	Cutting speed (rpm)	Feed rate (mm/min)	Depth of cut (mm)
1	−8.6743	−3.9036	−5.4850
2	−2.1793	−2.7887	−3.0995
3	−0.8481	−5.0094	−3.1171
Delta	7.8262	2.2207	2.3855
Rank	1	3	2

Figure 1 shows the main effects plot for signal-to-noise ratios for surface roughness of PP+60wt.% quarry dust composite. It is observed that the optimum combination input parameters that can produce quality and a better surface finish are $A_3B_2C_2$ or $A_3B_2C_3$ which corresponds to the largest values obtained for the S/N ratio of the input parameters. The real optimum values obtained are cutting speed of 1000 rpm, feeding rate of 120 mm/min and depth of cut of either 0.5 mm or 0.8 mm.

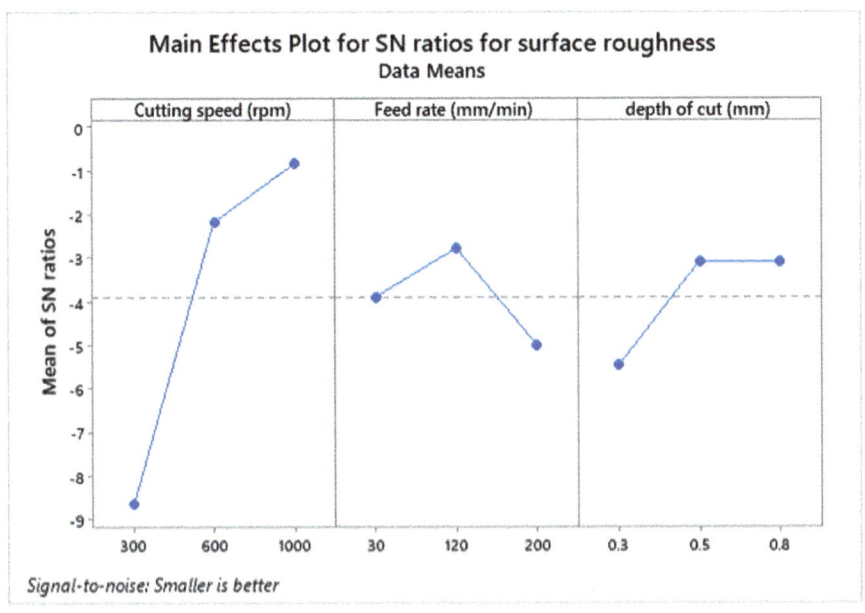

Fig. 1. Main effects plot for S/N ratios of surface roughness for PP+60wt.% quarry dust composite

3.2 The ANOVA for Surface Roughness

Table 5 shows the ANOVA results obtained after statistically analyzing the percentage contribution and the relative significance of the input parameters towards the surface roughness. It is observed that the cutting speed had the highest contribution towards the surface roughness at 81.98%, followed by the depth of cut at 7.43% and the feed rate having the least contribution at 3.69%. However, the p-values for all the factors were greater than 0.05 level of significance making them insignificant.

Table 5. The ANOVA for surface roughness of the PP+60wt.% quarry dust composite

Source	DF	Adj SS	Adj MS	F-value	P-value	% contribution
Cutting speed (rpm)	2	4.4517	2.2258	11.89	0.078	81.98
Feed rate (mm/min)	2	0.2005	0.1003	0.54	0.651	3.69
Depth of cut (mm)	2	0.4036	0.2018	1.08	0.481	7.43
Error	2	0.3744	0.1872			6.89
Total	8	5.4302				

4 Conclusions

The Taguchi optimization technique was successfully carried out and the optimum CNC milling parameters for PP+60wt.% quarry dust composite were obtained. The optimum values for the milling factors obtained were cutting speed of 1000 rpm, feeding rate of 120 mm/min and depth of cut of either 0.5 mm or 0.8 mm. It is also be concluded that the cutting speed has the highest contribution towards the surface roughness when milling PP+60wt.% quarry dust composite, at 81.98%, followed by the depth of cut at 7.43% and the feed rate having the least contribution at 3.69%.

References

1. Slieptsova, I., Savchenko, B., Sova, N., et al.: Polymer sand composites based on the mixed and heavily contaminated thermoplastic waste. IOP Conf. Ser.: Mater. Sci. Eng. **111**, 12027 (2016). https://doi.org/10.1088/1757-899X/111/1/012027
2. Ndibi Mbozo'o, M.P., Tchuisseu Mikla, N.: Physical Characterization of Composite Materials Based on Plastic Waste (Polyethylene Terephthalate) and Sand (2017). https://doi.org/10.5281/zenodo.1036278
3. König, W., Wulf, C., Graß, P., et al.: Machining of fibre reinforced plastics. CIRP Ann. **34**(2), 537–548 (1985). https://doi.org/10.1016/S0007-8506(07)60186-3
4. Eriksen, E.: Influence from production parameters on the surface roughness of a machined short fibre reinforced thermoplastic. Int. J. Mach. Tools Manuf **39**(10), 1611–1618 (1999)
5. Sreenivasulu, R.: Optimization of surface roughness and delamination damage of GFRP composite material in end milling using Taguchi design method and artificial neural network. Procedia Eng. **64**, 785–794 (2013). https://doi.org/10.1016/j.proeng.2013.09.154
6. Hussain, S.A., Pandurangadu, V., Kumar, K.P.: Optimization of surface roughness in turning of GFRP composites using genetic algorithm. Int. J. Eng. Sci. Tech **6**(1), 49 (2014). https://doi.org/10.4314/ijest.v6i1.6
7. Palanikumar, K.: Cutting parameters optimization for surface roughness in machining of GFRP composites using Taguchi's method. J. Reinf. Plast. Compos. **25**(16), 1739–1751 (2006). https://doi.org/10.1177/0731684406068445
8. Parida, A.K., Routara, B.C., Bhuyan, R.K.: Surface roughness model and parametric optimization in machining of GFRP composite: Taguchi and Response surface methodology approach. Mater. Today Proc. **2**(4–5), 3065–3074 (2015). https://doi.org/10.1016/j.matpr.2015.07.247
9. Shagwira, H., Mwema, F., Mbuya, T., et al.: Dataset on impact strength, flammability test and water absorption test for innovative polymer-quarry dust composite. Data Brief **29**, 105384 (2020). https://doi.org/10.1016/j.dib.2020.105384

Balancing of the Production Line Process in the Manufacturing of the Hand Grinder

Jean Luc Habiyaremye$^{(\boxtimes)}$, Fillemon Nangolo, Ester Angula, Erasmus Shaanika, Sam Shaanika, Mutiu Erinosho, Ignatius Shahonya, and Nikanor Shikomba

Department of Mechanical and Industrial Engineering, University of Namibia, P.O. Box 3624, Ongwediva, Namibia
{jhabiyaremye,fnangolo,eangula,shaanikae,sshaanika,merinosho, ishahonya,nshikomba}@unam.na

Abstract. The mechanical and manual operations in the production of a hand grinder present many problems. There is a wide range of issues such as massive inventory of work in progress (WIP), irrational allocation of work procedures, disorderly assembly line layout, and poorly balanced assembly line. The visualization of the working area layout and process charts of the production line is achieved using the Value Stream Map (VSM). This study produces the future value flow diagram to optimize the hand grinder production line. In this survey, industrial engineering (IE) methodologies helped improve the manufacturing sequences and layout and identify all workflow bottlenecks. Also, when IE methods are combined with the genetic algorithm (GA), they made a useful tool to balance successfully and eliminate all sequential manufacturing problems. The improved production line successfully shortens the production lead time (PLT), cuts down the inventory activities, removes unnecessary movements of people, and balances the flow-line production system and layout.

1 Introduction

As Africa starts investing in the industrialization process, factories face many production problems that need to be solved to increase production efficiency. So many production factories in Africa are still using manual production methods. A few of them have introduced a semi-automated production process, wasting lots of time trying to balance the production line [1]. As a result, it is essential to investigate the production line balance problem and optimize the production line layout. Also, there is an urgent need to cut down the production cycles and rationalize procedural distribution for achieving the production goal to shorten the PLT, save capital, and reduce production expenses [2].

It is essential to remove the waste and non-value-added activities throughout the flow-line production system The VSM approach is used to build value flow and future flow diagrams for technical evaluation, whereas the IE approach addresses the optimization problem and assessment of the production line. This study will use a mathematical model to balance the flow-line production system [3].

M. Awang and S. S. Emamian (Eds.): *Advances in Material Science and Engineering*, LNME, pp. 175–180, 2021.
https://doi.org/10.1007/978-981-16-3641-7_21

2 Methodologies

2.1 Business Process Review

The research team had a chance to evaluate the methods used in the existing assembly line of a hand grinder (Ken2450W-230mm) before executing any lean manufacturing (LM) approaches. The crucial elements of information were collected from the book of best operation practices (BOP) guides, face-to-face interviews with operators and senior engineers, and individual observations of the field (go to Gemba). The evaluation process focused on the daily operational activities, time to complete a production process, production plant layout, and manufacturing challenges [4].

The VSM visualizes the movement of materials and the actual LM processes. Also, it shows the current and future causes of waste, which, once corrected, can have a positive impact on resource control and management, cash flows, and customer satisfaction [5].

The compiled data in Table 1 show that the availability of resources amounts to 58.89%. It means there is a substantial amount of waste that could be eliminated to increase the production plant's performance and meet customer demand. Steps to mitigate production efficiency are defined as follows:

a) Techniques of managing inventory (supermarket method) will be introduced to ensure a continuous flow, b) a Kanban method must be used to avoid any inventory pile-ups, c) optimization of the production plant layout. Accurate planning and eliminating any unnecessary movement of workers and assembly parts, d) eliminate bottlenecks in a chain of processes, e) use all available resources effectively to improve production efficiency.

2.2 Value Flow Diagram

The daily consumer demand is per shift per day is 823 units. The Talk time is equal to productive time per shift per day, divided by customer demand. [(Talk time = 8h × 3600/823 units) = 35 s]. The value flow diagram of the production line of this grinder is explained in Fig. 1. Procedures and continual flow of materials are successive with aligned work-in-progress inventories. This study thoroughly investigated the VSM diagram and summarized it professionally and quantitatively.

After a thorough investigation of all items of information and data collected in various ways, this paper identified several problems in the grinder, the flow-line production system, and the layout that needs to be solved.

i) There is a lack of synchronized ideas, inadequate procedural arrangement, and low balance efficiency, ii) The production floor layout is irrational and could lead to unnecessary movement of people and materials that cause massive waste, iii) Some work stations were overstaffed; thus, it affected the entire production line's efficiency.

3 The Improvement of the Production System

The continuous improvement process is meant to use LM methods to eliminate the waste, identify Gemba operations, and provide an improved VSM diagram (see Fig. 2).

Table 1. Process data

Manufacturing sections		Operating time (s)	Number of staff	Process cycle time (s)	Inventory	Availability %
Assembly I		116.26	7	16.61	300 items	51.01
Assembly II		145.57	9	16.17	510 cpts	65.82
Inspection	Functional test	21.7	1	21.7	381 units	56.6
	High-pressure test	20.1	1	20.1	504 units	57.89
	Noise test	20.93	1	20.93	480 units	54.98
Packaging		54.44	4	13.61	205 units	67.04

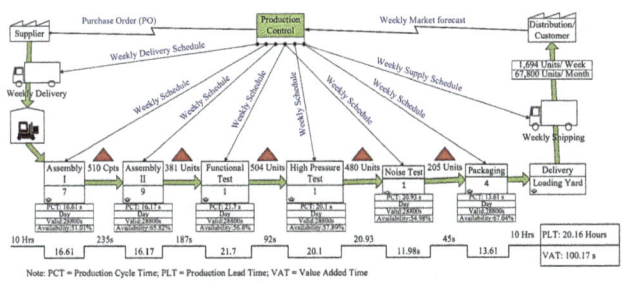

Fig. 1. Value flow diagram

This study found it useful to maintain production control using the pulling system [6]. When a continuous flow cannot be established directly to the next production step, this paper found it essential to use supermarkets for internal customers to address the internal supply issues [7].

This survey achieved the production floor layout's effective optimization and reduced handling space for improved production floor layout. The production floor was reduced by 6.46 m in length and 5.94 m in width translating into 3.88 × 102 m² reduction. Before implementing the improved flow-line production system, the workers could only depend on a common-sense judgment. It could eventually incur procedural complexity in a manufacturing procedure, failure to manage time properly, frequent interruption of process flow, etc. [8].

3.1 Genetic Algorithm and Investigation on Balancing Production Line

3.1.1 Mathematical Model Development

The talk time of this factory's flow-line production system was computed earlier before. During tight working conditions, it is essential to balance the workload of each section. The production line balance rate is developed through the reduction of the number of

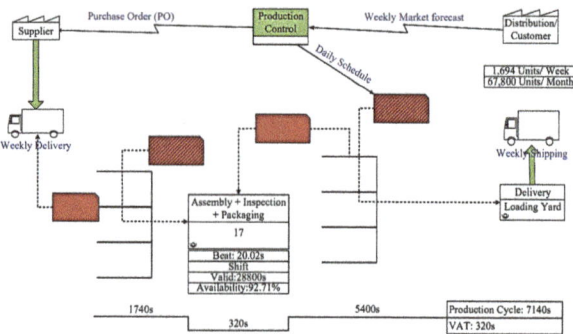

Fig. 2. Future process flow diagram

segments.

$$Max\ \eta\ (\%) = \frac{100 * \sum T_i}{PT * m} \tag{1}$$

$$\varphi = \omega_1 * m + \omega_2 * SI \tag{2}$$

$$SI = \sqrt[2]{\frac{\sum_{i=1}^{m}(PT - T_i)^2}{m}} \tag{3}$$

In these expressions, η stands for production line balance rate in %, T_i stands for the standard operation time element, PT stands for Takt time, m stands for a given number of work stations, SI stands for section index, $\omega 1$ and $\omega 2$ stand for weight constants for optimization and $\omega 1 + \omega 2 = 1$. This study has established constraints to the flow-line production system.

a) Assigning every operating unit of the production line to the matching workstations and it is expressed as follows:

$$H = \cup_{j=1}^{m} X_j (j = 1, 2, 3, 4 \ldots .) \tag{4}$$

b) Assigning one operation unit to one workstation is mandatory, and it is obtained by.

$$X_i \cap X_j = \Psi; \tag{5}$$

c) Knowing that x ≤ y, it is a must to ensure the operation unit's compliance with the priority condition matrix (PM) as follows:

$$PM = \left(PM_{ij}\right)_{n \bullet n} if\ PM_{ij} = 1, PM_{ij} = 1, j \in X_y \tag{6}$$

d) Each section's working time is strictly limited to the talk time T(Xj) ≤ PT.

H stands for operation units, Xi and Xj stand for ith and jth operation units, respectively. The following expression gives the fitness function:

$$F(\varphi) = 1 - \frac{\varphi}{\Gamma} \tag{7}$$

In which φ is the objective function, Γ is higher than the value of φ. The following expression leads to the process of choosing the operation to be done.

$$C_i = \frac{F(\varphi)}{\sum_{i=1}^{n} F(\varphi)} \tag{8}$$

In this expression, 'n' stands for the number of the population. And Ci is the chance for the chromosome 'i' to be chosen.

3.2 Data Simulation

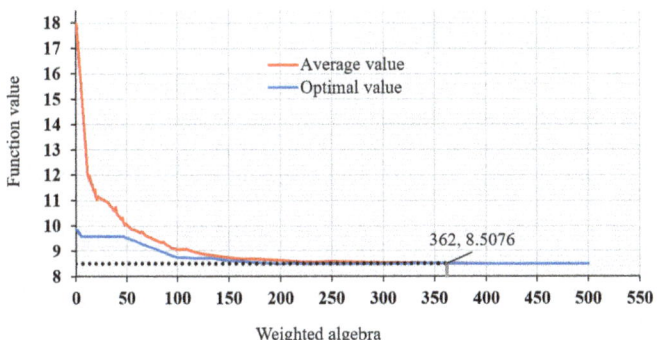

Fig. 3. Simulation results

The simulated results with the weight constants for optimization $\omega1 = 0.6$ and $\omega2 = 0.4$ are displayed in Fig. 3. The minimum number of sections was 12 and which diminished by 6, after performing 362 cycles. This survey registered an improvement of 11% when measured against the unimproved one. A summary of the average optimized time for all workstations shows that the staff-hour utilization amounted to 85.27%, the average actual and basic time count for 21.13 s, and 18.97 s, respectively.

These results show that for the improved manufacturing sequences, the number of staff was cut from 23 to 17, the production line time was shortened by 1090.6 min, and the production balance rate grew from 60.16% to 94.02%. The appreciation rate was improved from 0.14% to 4.48%. These results confirm that the improvement strategies implemented in this factory are efficient and workable.

4 Conclusion

This study has investigated an existing production line of an angle grinder manufacturing and used an LM approach to improve the production process to ensure the continuous production process. This paper used a pull system method, and when combined with the IE method, it resulted in the effective optimization of the entire flow-line production system and layout. This paper developed a successful genetic algorithm approach and an adequate mathematical model to address the balance problem that is often found in the advanced manufacturing industries. This study identified and removed all production line bottlenecks, produced a reasonable flow-line production layout and improved the assembly line. Furthermore, this study removed non-value-added activities, shortened the PLT, and improved the production line efficiency.

References

1. Gong, Q., Yang, Y., Wang, S.: Information and decision-making delays in MRP, KANBAN, and CONWIP. Int. J. Prod. Econ. **156**, 208–213 (2014)
2. Bhat, M., Bhandarkar, V.: Investigating the impact of lean philosophy for identification and reduction of delays associated with performance of production line (2020)
3. Pedell, S.: Picture scenarios: an extended scenario-based method for mobile appliance design. Ozchi 2004. Wollongong, Australia (2004)
4. Lee, S.-H., et al.: Identifying waste: applications of construction process analysis. In: Proceedings of the Seventh Annual Conference of the International Group for Lean Construction (1999)
5. Grout, J.R.: Mistake proofing: changing designs to reduce error. BMJ Qual. Saf. **15**(suppl 1), i44–i49 (2006)
6. Jiménez, E., et al.: Applicability of lean production with VSM to the Rioja wine sector. Int. J. Prod. Res. **50**(7), 1890–1904 (2012)
7. Rahman, N.A.A., Sharif, S.M., Esa, M.M.: Lean manufacturing case study with Kanban system implementation. Procedia Economics and Finance **7**, 174–180 (2013)
8. Braglia, M., Carmignani, G., Zammori, F.: A new value stream mapping approach for complex production systems. Int. J. Prod. Res. **44**(18–19), 3929–3952 (2006)

Evaluation on the Performance of Logistics Companies in Malaysia with TOPSIS Model

Lam Weng Hoe[1,2], Lam Weng Siew[1,2(✉)], Mohd Azam Bin Din[1,2], and Liew Kah Fai[1,2]

[1] Department of Physical and Mathematical Science, Faculty of Science, Universiti Tunku Abdul Rahman, Kampar Campus, Jalan Universiti, Bandar Barat, 31900 Kampar, Perak, Malaysia
lamws@utar.edu.my

[2] Centre for Business and Management, Universiti Tunku Abdul Rahman, Kampar Campus, Jalan Universiti, Bandar Barat, 31900 Kampar, Perak, Malaysia

Abstract. An efficient and effective logistics management is important to the logistics companies as it can increase the company's revenue, improve customer service, reduce the overall transportation costs, and improve the operating cost structure. Therefore, many logistics companies are striving to enhance their logistics management from all the aspects from time to time. Logistics plays a key role in the development of economy because it supports the movement and flow of many economic transactions. The objective of this study is to propose a conceptual framework in evaluating and ranking the financial performance of the logistics companies in Malaysia with Technique for Order Preference by Similarity to Ideal Solution (TOPSIS) model. The findings of the paper indicate that COMPLET gives the highest performance, followed by FREIGHT, SEEHUP, HARBOUR, and lastly TASCO. The significance of this paper is to measure the financial performance of the logistics companies in Malaysia with the proposed conceptual framework using TOPSIS model.

1 Introduction

Logistics management is very important in operating the supply chain. A good logistics management can help logistics companies to reduce costs and improve efficiency. Moreover, good logistics management can deliver better service to consumers. Thus, it will attract more business and generate greater revenue. Logistics is a kind of activity that involved with transfer of related information, services, and goods [1]. Logistics is responsible to manage the transfer of services and products from the point of origin to the end-users [2]. Logistics is crucial in supply chain management (SCM). SCM is the process associated with coordination in the supply chain network [3]. The SCM is essential to transform the raw materials into finished goods. After that, the finished goods will be moved to the end-users through effective and efficient SCM. In order to achieve a better goal, supply chain integration should be well established by linking the humanitarian logistics idea [4–6].

© The Author(s), under exclusive license to Springer Nature Singapore Pte Ltd. 2021
M. Awang and S. S. Emamian (Eds.): *Advances in Material Science and Engineering*, LNME, pp. 181–188, 2021.
https://doi.org/10.1007/978-981-16-3641-7_22

TOPSIS is a multi-criteria decision-making (MCDM) model and has been broadly applied to choose the best decision alternative based on multiple criteria [7]. The concept of TOPSIS model is that the alternative with the best performance should have the closest separation distance to the positive ideal solution (PIS) and the longest separation distance from the negative ideal solution (NIS) [8, 9]. In TOPSIS model, the NIS consists of all worst values of the criteria, while PIS consists of all best values of the criteria [10, 11]. The decision alternatives are evaluated and ranked based on the relative similarity to the ideal solution (C_i^*). In this paper, the ranking of the logistics companies in Malaysia is identified by TOPSIS model.

In Malaysia, no comprehensive study has been done on the evaluation of logistics companies' financial performance using TOPSIS model. Thus, this study aims to measure the logistics companies' financial performance in Malaysia using TOPSIS model. The remainder of this paper is organized as follows. The data and methodology of the study is discussed in Sect. 2. The next section presents the main findings obtained in this study. The conclusion of this study is drawn in the last section of the paper.

2 Data and Methodology

2.1 Data

In this study, the listed logistics companies, namely COMPLET, FREIGHT, HARBOUR, SEEHUP, and TAS are evaluated in terms of financial performance from year 2017 to 2019 [12]. The data are gathered from the logistics companies' financial annual report. Table 1 presents the proposed conceptual framework.

Table 1. Proposed conceptual framework.

Level	
Level 1 (Main aim)	Evaluation of logistics companies
Level 2 (Decision criteria)	Current ratio (CR)
	Debt to assets ratio (DAR)
	Debt to equity ratio (DER)
	Earnings per share (EPS)
	Return on asset (ROA)
	Return on equity (ROE)
Level 3 (Decision alternative)	COMPLET
	FREIGHT
	HARBOUR
	SEEHUP
	TASCO

Based on the past study, crucial financial ratios are taken into consideration in this paper. The financial ratios such as CR, DAR, DER, EPS, ROA, and ROE are used to assess the companies' financial performance. Hence, these financial ratios are treated as the decision criteria utilized to assess the logistics companies' financial performance. In this paper, the decision criteria that need to be minimized are DAR and DER. On the other hand, CR, EPS, ROA, and ROE are the decision criteria that need to be maximized [13–16].

2.2 TOPSIS

TOPSIS model is used to identify the decision alternative with the best performance among multiple alternatives based on multiple decision criteria [7]. Moreover, TOPSIS model is able to rank the decision alternatives according to C_i^* [17]. In TOPSIS model, the alternative with the best performance will have the closest separation distance to the PIS and the longest separation distance from the NIS [8]. The NIS is identified to maximize the cost criterion and minimize the benefit criterion [18, 19]. In contrast, the PIS is determined to minimize the cost criterion and maximize the benefit criterion [18, 19]. In this paper, the financial ratios are equally important in determining the companies' financial performance [20–23]. TOPSIS model comprised of 6 steps as presented below [24]:

Step 1: Establish the decision matrix (x_{ij}).
The decision matrix with a size of $m \times n$ is established where it composed of m decision alternatives and n decision criteria.

$$x_{ij} = \begin{bmatrix} x_{11} & x_{12} & \cdots & x_{1n} \\ x_{21} & x_{22} & \cdots & x_{2n} \\ \vdots & \vdots & \vdots & \vdots \\ x_{m1} & x_{m2} & \cdots & x_{mn} \end{bmatrix} \tag{1}$$

Step 2: Form the normalized decision matrix (R).

$$r_{ij} = \frac{x_{ij}}{\sqrt{\sum_{i=1}^{m} x_{ij}^2}}, \quad i = 1, 2, ..., m; j = 1, 2, ..., n \tag{2}$$

$$R_{ij} = \begin{bmatrix} r_{11} & \cdots & r_{1n} \\ \vdots & \ddots & \vdots \\ r_{m1} & \cdots & r_{mn} \end{bmatrix} \tag{3}$$

Step 3: Form the weighted normalized decision matrix (V).

$$W = (w_1, w_2, ..., w_n) \text{ where } \sum_{j=1}^{n} w_j = 1 \tag{4}$$

To establish the weighted normalized decision matrix, each of the elements in the rows of R matrix is multiplied by w_j. Equation (5) demonstrates the weighted normalized decision matrix (V).

$$V_{ij} = \begin{bmatrix} w_1 r_{11} & \cdots & w_n r_{1n} \\ \vdots & \ddots & \vdots \\ w_1 r_{m1} & \cdots & w_n r_{mn} \end{bmatrix} \qquad (5)$$

Step 4: Identify the PIS (A^+) and NIS (A^-).

$$A^+ = \left\{ \left(\max V_{ij} | j \in J \right) \left(\min V_{ij} | j \in J' \right) \right\} = \{v_1^+, v_2^+, ..., v_n^+\} \qquad (6)$$

$$A^- = \left\{ \left(\min V_{ij} | j \in J \right) \left(\max V_{ij} | j \in J' \right) \right\} = \{v_1^-, v_2^-, ..., v_n^-\} \qquad (7)$$

Step 5: Compute the separation distance from PIS (d_i^+) and NIS (d_i^-) for each decision alternative.

$$d_i^+ = \sqrt{\sum_{j=1}^{n} (v_{ij} - v_j^+)^2}, i = 1, 2, ..., m \qquad (8)$$

$$d_i^- = \sqrt{\sum_{j=1}^{n} (v_{ij} - v_j^-)^2}, i = 1, 2, ..., m \qquad (9)$$

Step 6: Compute the relative closeness of the alternative to the ideal solution (C_i^*).

$$C_i^* = \frac{d_i^-}{d_i^- + d_i^+} \text{ where } C_i^* \in [0, 1], i = 1, \ldots, m \qquad (10)$$

Step 7: Determine the ranking of the decision alternatives by according to the relative closeness to the ideal solution (C_i^*).
The best decision alternative with the largest value of C_i^* is identified.

3   Results and Discussion

Table 2 presents the decision matrix. After that, the normalized decision matrix is determined and displayed in Table 3.

Table 2. Decision matrix.

Logistics company	CR	DAR	DER	EPS	ROA	ROE
COMPLET	18.6560	0.0949	0.1104	0.0568	5.9664	5.9901
FREIGHT	6.0253	0.0638	0.0682	0.0344	6.5313	6.9563
HARBOUR	0.3676	0.1423	0.1675	0.0238	3.6991	4.1814
SEEHUP	33.9001	0.0649	0.0717	0.0163	1.2295	1.3462
TASCO	2.1034	0.5067	1.0919	0.0630	2.3193	3.9940

Table 3. Normalized decision matrix.

Logistics company	CR	DAR	DER	EPS	ROA	ROE
COMPLET	0.4757	0.1750	0.0990	0.5921	0.6002	0.5479
FREIGHT	0.1536	0.1176	0.0612	0.3587	0.6570	0.6363
HARBOUR	0.0094	0.2622	0.1503	0.2476	0.3721	0.3825
SEEHUP	0.8644	0.1197	0.0643	0.1696	0.1237	0.1231
TASCO	0.0536	0.9341	0.9797	0.6563	0.2333	0.3653

Table 4. Weighted normalized decision matrix.

Logistics company	CR	DAR	DER	EPS	ROA	ROE
COMPLET	0.0793	0.0292	0.0165	0.0987	0.1000	0.0913
FREIGHT	0.0256	0.0196	0.0102	0.0598	0.1095	0.1060
HARBOUR	0.0016	0.0437	0.0250	0.0413	0.0620	0.0637
SEEHUP	0.1441	0.0199	0.0107	0.0283	0.0206	0.0205
TASCO	0.0089	0.1557	0.1633	0.1094	0.0389	0.0609

Table 5. NIS (A^-) and PIS (A^+).

	CR	DAR	DER	EPS	ROA	ROE
NIS (A^-)	0.0016	0.1557	0.1633	0.0283	0.0206	0.0205
PIS (A^+)	0.1441	0.0196	0.0102	0.1094	0.1095	0.1060

Table 4 demonstrates the weighted normalized decision matrix.

Table 5 displays the NIS (A^-) and PIS (A^+).

Figure 1 and 2 present the separation distance of the logistics companies from the NIS and PIS, respectively.

As presented in Fig. 1, SEEHUP (0.2490) gives the longest separation distance from the NIS among the logistics companies, followed by COMPLET (0.2447), FREIGHT (0.2424), HARBOUR (0.1881), and finally TASCO (0.0927). The results demonstrate that SEEHUP is the farthest from the NIS. In contrast, TASCO is very closest to the NIS.

According to Fig. 2, COMPLET (0.0689) has the shortest separation distance to the PIS, followed by FREIGHT (0.1284), SEEHUP (0.1476), HARBOUR (0.1726), and TASCO (0.2593). This indicates that TASCO is the farthest from the PIS whereas COMPLET is the closest to the PIS.

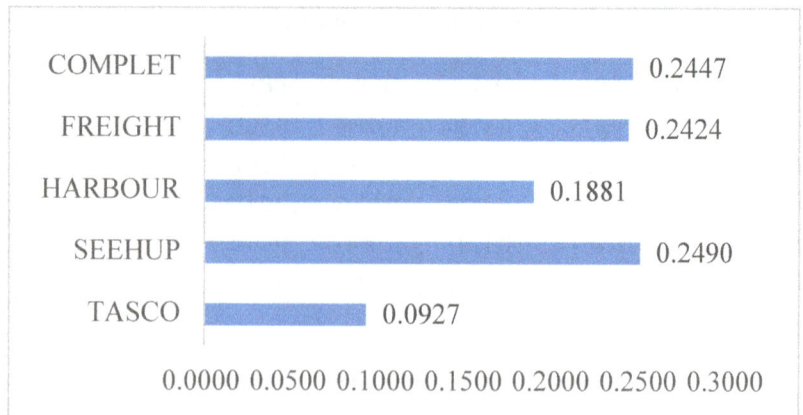

Fig. 1. Separation distance from the NIS.

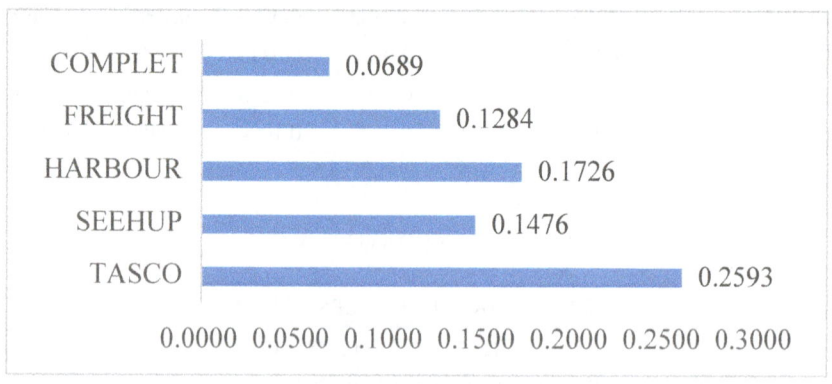

Fig. 2. Separation distance from the PIS.

Table 6 displays the relative closeness and ranking of logistics companies.

Table 6. Relative closeness and ranking of companies

Logistics company	C_i^*	Ranking
COMPLET	0.7802	1
FREIGHT	0.6536	2
HARBOUR	0.5215	4
SEEHUP	0.6278	3
TASCO	0.2634	5

Based on Table 6, COMPLET achieves the highest C_i^* value of 0.7802. Therefore, COMPLET obtains the first ranking in the evaluation of financial performance. This indicates that COMPLET gives the best financial performance over the study period. Moreover, it also implies that COMPLET has the closest separation distance to the PIS and the farthest separation distance from the NIS. FREIGHT, SEEHUP, and HARBOUR obtain C_i^* value of 0.6536, 0.6278, and 0.5215, respectively. As a result, the second, third, and fourth ranking are achieved by FREIGHT, SEEHUP, and HARBOUR, respectively. Lastly, TASCO obtains the last ranking in this study. TASCO obtains C_i^* value of 0.2634, which is the lowest among the logistics companies. This means that the financial performance of TASCO is farthest from the PIS and closest to the NIS. In summary, COMPLET achieves the best financial performance based on crucial financial ratios with TOPSIS model.

4 Conclusions

The logistics companies' financial performance is measured using TOPSIS model. Based on the results, COMPLET obtains the best financial performance among the logistics companies, followed by FREIGHT, SEEHUP, HARBOUR, and TASCO. The results reveal that COMPLET is very far from the NIS and the closest to the PIS. This paper is significant as it assists to assess the logistics companies' financial performance using TOPSIS model.

Acknowledgments. The study is supported by Universiti Tunku Abdul Rahman (UTAR), Malaysia.

References

1. Bowersox, D.J., Closs, D.J., Cooper, M.B.: Supply Chain Logistics Management. McGraw-Hill, Boston (2007)
2. Azmi, I., Hamid, N.A., Hussin, M.N.M., et al.: Logistics and supply chain management: the importance of integration for business processes. J. Emerg. Econ. Islamic Res. **5**, 73–80 (2017)
3. Jain, J., Dangayach, G.S., Agarwal, G., et al.: Supply chain management: literature review and some issues. J. Stud. Manuf. **1**, 11–25 (2010)
4. Van Wassenhove, L.N.: Humanitarian aid logistics: supply chain management in high gear. J. Oper. Res. Soc. **57**, 475–489 (2006)
5. Maon, F., Lindgreen, A., Vanhamme, J.: Developing supply chains in disaster relief operation through cross-sector socially oriented collaborations: theoretical model. Surg. Endosc. Other Interv. Tech. **14**, 149–164 (2009)
6. Balcik, B., Beamon, B.M., Krejci, C.C., et al.: Coordination in humanitarian relief chains: practices, challenges and opportunities. Int. J. Prod. Econ. **126**, 22–34 (2010)
7. Hwang, C.L., Yoon, K.: Multi-attribute Decision Making. Springer, Heidelberg (1981). https://doi.org/10.1007/978-3-642-48318-9
8. Benitez, J.M., Martin, J.C., Roman, C.: Using fuzzy number for measuring quality of service in the hotel industry. Tour. Manage. **28**, 544–555 (2007)

9. Zhang, H., Gu, C.L., Gu, L.W., et al.: The evaluation of tourism destination competitiveness by TOPSIS & information entropy – A case in the Yangtze River Delta of China. Tour. Manage. **32**, 443–451 (2011)
10. Stank, T.P., Daugherty, P.J., Ellinger, A.E.: Pulling customers closer through logistics service. Bus. Horiz. **41**, 74–80 (1998)
11. Ertugrul, D., Karakasoglu, N.: Performance evaluation of Turkish cement firms with fuzzy analytic hierarchy process and TOPSIS methods. Expert Syst. Appl. **36**, 702–715 (2009)
12. Bursa Malaysia. Company Announcements | Bursa Malaysia Market. Online available from http://www.bursamalaysia.com/market/listed-companies/company-announcements/#/?category=all (n.d.)
13. Liew, K.F., Lam, W.S., Lam, W.H.: Financial analysis on the company performance in Malaysia with multi-criteria decision making model. Syst. Sci. Appl. Math. **1**, 1–7 (2016)
14. Lam, W.S., Liew, K.F., Lam, W.H.: Evaluation on the financial performance of the Malaysian banks with TOPSIS model. Am. J. Serv. Sci. Manage. **4**, 11–16 (2017)
15. Lam, W.S., Lam, W.H., Liew, K.F.: Data analysis on the performance of technology sector in Malaysia with entropy-TOPSIS model. Commun. Comput. Inf. Sci. **886**, 194–203 (2018)
16. Lam, W.S., Liew, K.F., Lam, W.H.: Investigation on the performance of construction companies in Malaysia with Entropy-TOPSIS model. In: IOP Conference Series: Earth and Environmental Science, vol. 385, p. 012006. IOP Publishing (2019)
17. Kabir, G.: Third party logistic service provider selection using fuzzy AHP and TOPSIS method. Int. J. Qual. Res. **6**, 71–79 (2012)
18. Wang, Y.M., Elhag, T.M.S.: Fuzzy TOPSIS method based on alpha level sets with an application to bridge risk assessment. Expert Syst. Appl. **31**, 309–319 (2006)
19. Wang, Y.J., Lee, H.S.: Generalizing TOPSIS for fuzzy multiple-criteria group decision-making. Comput. Math. Appl. **53**, 1762–1772 (2007)
20. Bakirci, F., Shiraz, S.E., Sattary, A.: Financial performance analysis of iron, steel metal industry sector companies in the Borsa İstanbul: DEA super efficiency and TOPSIS methods. Ege Acad. Rev. **14**, 9–19 (2014)
21. Bulgurcu, B.K.: Application of TOPSIS technique for financial performance evaluation of technology firms in Istanbul stock exchange market. Procedia Soc. Behav. Sci. **62**, 1033–1040 (2012)
22. Gundogdu, A.: Measurement of financial performance using TOPSIS method for foreign banks of established in Turkey between 2003–2013 years. Int. J. Bus. Soc. Sci. **6**, 139–151 (2015)
23. Lam, W.H., Din, M.A., Lam, W.S., et al.: Evaluation on the performance of suppliers in Malaysia with TOPSIS model. J. Fundam. Appl. Sci. **10**, 406–415 (2018)
24. Lam, W.H., Lam, W.S., Liew, K.F.: Performance analysis on telecommunication companies in Malaysia with TOPSIS model. Indonesian J. Electr. Eng. Comput. Sci. **13**, 744–751 (2019)

TRIZ Application: An Innovative Approach in Redesigning an Ergonomics Car Interior for Limbs Disabled

Salami Bahariah Suliano[✉], Siti Azfanizam Ahmad, Azizan As'arry,
and Faieza Abdul Aziz

Department of Mechanical and Manufacturing, Universiti Putra Malaysia, Serdang, Malaysia

Abstract. Being physically varieties gives an additional impact to limbs disabled in driving either normal car or modified car. Despite modifying car based on references of the modified personal car can improve limbs disabled in mobility, it is poor in ergonomics and does not match the needs of limbs disabled. Therefore, the aim of this study was to provide efficient solutions in redesigning car by applying the Theory of Inventive Problem Solving (TRIZ) on an existing normal car and transpire a more innovative design capacity. In this paper, the TRIZ contradiction matrix and 40 principle solution tools were applied for the problem's solution generation. The workflow contains several steps based on TRIZ solution map including function analysis, cause and effect chain analysis and trimming. Presenting seven interior parts (handle, egress, upholstery back, upholstery bottom, pedals, gear knob, and steering) after seven solved inventive problems. Based on these result, there are seven innovative solutions to the car interior in contemplation of limbs disabled.

1 Introduction

Living as a disabled reflect to live in a different way of lifestyle as compared to normal people. The ability of a person with disabilities towards using the vehicle was found to be very important as one of the survival factors, community integration [1], and employment opportunity. Therefore, it is an important measure for disabled to have access to every single equipment or facilities from door to door to mobile. The term door to door illustrating an independent moves of disabled start from exiting house door to till closing car door. Then after, they can start mobile and sequence repeated along the way throughout their daily journey. In the process, multiple problems might be faced by disabled as such complexity of transferring wheelchair into the car (arm and reach) and accessibility to the car interior [2].

A medical condition in which accompanied by physical impairments and relative individual functional performance (varies depending on personal characteristics) is often being reasons that causing a challenge in assessing driving abilities of disabled [3]. However, independent driving is an important aspect particularly for a person with disabilities as significant activities of daily life [4]. For improvement of quality of life, the majority of disabled is driving an adapted or modified car as the most important means

M. Awang and S. S. Emamian (Eds.): *Advances in Material Science and Engineering*, LNME, pp. 189–199, 2021.
https://doi.org/10.1007/978-981-16-3641-7_23

of independency [5] on access to personal vehicular transportation [6]. Modification or assistive driving has led to improvement for limbs disabilities driver [4]. As an effort to reduce the problem faced by limbs disabled in driving, modification has been made based on what has been offered in the market or reference of senior users. Disabled tend to adapting with available modification besides considering their actual needs. Based on past research work by Suliano et al. an analysis of the needs of limbs disabled, covering the driver's area of car interior design has been made [7]. The results concluded that certain major needs to be taken into consideration to redesign the interior part of the driver's area to achieve ergonomic friendly car for limb disabled driver. These include handle at the door and surrounded area, pedals, egress, upholstery (back) and upholstery (bottom). One research by Suliano et al. (2020), a comparison of modification and assistive technology in market was presented. These include the importance of five main aspects of ergonomics namely safety, comfort, ease of use, productivity and performance and aesthetic to be implied in this in designing or redesigning car to support disabled drivers [8].

During the redesigns, organizations are interested in increasing the "usefulness" of their products. The term usefulness includes all valuable results of the product's function such as specific features, ergonomics, capacity, aerodynamic and stability [9]. It is also desirable to reduce the harmful effects of a product. Using TRIZ methodology, the idea of increasing useful effects and decreasing harmful effects is captured within the law of ideality [10]. Therefore, this study intends to taking into consideration of the above needs of limbs disabled raised problems and adapting TRIZ as a solution mechanism.

2   Theory of Inventive Problem Solving (TRIZ)

TRIZ, a Russian acronym meaning theory of inventive problem solving, can serve as one of the promising candidates to meet this requirement. Based on analysing numerous patents, TRIZ theory suggests that design problems can be solved in predictable ways and 95% of the inventive problems in any particular field have already been solved in some other fields [11].

Pioneered by Althuller in 1946, the TRIZ method offers an extensive series of tools to help designers to be more innovative [12], avoid a trial-and-error procedure in the design process and to more easily solve inventive problems [13]. The TRIZ method includes tools for problem analysis and knowledgebase tools for system change.

In TRIZ, inventive solutions come after inventive problems as mapped in Fig. 1. Technically, an inventive situation results from inability to technical system to fulfil current functional requirements. For the formulation of Inventive problems, it should be sufficient enough to combine the description of the situation, effects and goal to be achieved [14]. With clear inventive problems, any tools in TRIZ are applicable to most handy and appropriate. One of the major key discoveries of TRIZ is the 40 inventive principles, which are summarized over hundreds of thousands of patents starting from 1946 [15–17]. This is the basic concept of TRIZ that have been widely applied to generate ideas to solve many engineering problems.

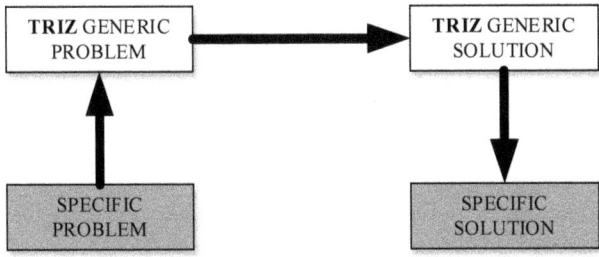

Fig. 1. General model of TRIZ [17])

2.1 Contradiction Matrix

The main theories of TRIZ include contradiction analysis, substance-field method, ideal final result, ARIZ (algorithm of inventive problem solving), etc. Among them, Contradiction Matrix is the most famous one in application. Contradiction Matrix is an answer to an elimination of lists of contradiction to be eliminated. Contradiction matrix laid of 39×99 matrix where the x-axis is the parameter that worsen, while the y-axis is the parameter that improves and in the middle is intersection of numbers that made up of inventive principles [18]. There are two method to apply the inventive principle, as below:

Method 1: Utilize the contradiction matrix in order to get a set of recommended inventive principle to solve the engineering contradiction. If no good solution at the recommended inventive principle, look at the remaining 40 Inventive Principle.

Method 2: Familiarize with all 40 Inventive Principle and apply each or a combination to solve the Engineering Contradiction.

With respect to contradiction table, its resolve technical contradictions. The clear understanding of each system parameters is very much needed to ensure a wise selection. The parameter to a problem is taken from statement of "If, then and but" in which this statement hold improving and worsening parameters that contradicting with each other. This statement helps to crystallize very clear what the problem is about and what parameter is a contradiction to each other [19]. This statements is well known as Engineering Contradiction (EC). An EC arises in an Engineering system if an improvement in one characteristics results in degradation of another characteristics [20].

The approach of using the contradiction matrix mainly depends on ways of user's working and the problem with the parameter and principles remain fixed [15]. A Good definition of contradiction will bring to effective TRIZ. There are five steps to apply contradiction matrix and principles [21] which included as follows:

Step 1: Define problem
Define the elements of the design that must be improved.

Step 2: Analyse the problem
Map these elements with the 39 parameters of contradiction matrix.

Step 3:   Select parameters

Identify the solution directions that will help to narrow down the problem. Identify the related parameters in these solution directions.

Step 4:   Find invention principles

Map these elements into the 39 parameters of the contradiction matrix to get pairs of improving features. Then, according to the pair, find the corresponding invention principles (from 40 inventive principle).

Step 5:   Develop and evaluate the feasible solutions

Develop alternatives according to the corresponding invention principles and evaluate the feasibility.

Consider the above steps, elements of design must be defined (step 1) prior problem identification (step 2). In order for the user to solve a problem, the user needs to identify a parameter that he needs to improve to solve the problem and due to that improvement, another the parameter that will become worsen (step 3). Based on the matching identified improving and worsening parameters, the engineering contradiction tool will provide a list of recommended inventive principles (out of the 40 inventive principles mentioned earlier) that may be applied to solve that particular problem [13] based on the improving parameter and the worsening parameter (step 4). The corresponding grid intersected by both contradiction parameters is the solution principle recommended [22].

In applying TRIZ engineering contradiction to solve a problem, not all the recommended inventive principles must be used to solve a particular problem. Sometimes none of the recommended inventive principles can be applied to solve a particular problem, In fact, quite a few of the matching identified improving and worsening parameter has no recommended inventive principles (there are quite a number of blanks in the contradiction matrix which means there is no recommended inventive principle, but, all 40 inventive principles yet to be diagnoses to find the best match). These inventive principles provide good ideas or hints to the user to think of solutions for a particular problem (step 5).

2.2 Idealized Design Theory

The concept of idealized design development develop by Genrich Altshuller emphasis that any technical system is always evolved to a simplified, effective, idealized system within its life cycle [22]. Supplying a systematic method to help a designer to correctively define questions, the idealized design theory can generate innovative concepts for idealized design. The law of ideality states that any technical system tends to reduce costs, to reduce energy wastes, to reduce space and dimensional requirements, to become more effective, more reliable, and simpler. Any technical system, during its lifetime, tends to become more ideal [23]. The ideal system is a non-existent one in which all of its functions are still executed. Restated, function is ideally performed by existing resources. Nevertheless, an actual systems approach ideally increases its beneficial functions and eliminates harmful factors [24]. Increase its beneficial functions or in other words improves its usefulness mean [25]:

- System performs more useful function (useful work)

- Work is carried out effectively and efficiently

  Meanwhile, eliminates harmful factors is including all payment factors:

- Cost of time and facilities
- Harmful effects, cost of energy, substance, components, space, etc.

Besides, an ideal substance is no substance but its function are performed. When substance is more ideal, it's created more useful functions, reduces cost, and reduces undesired effect generated. The evolvements of ideality should have a final result as show in.

2.3 Method

Based on conducted study involved 30 Malaysian independent upper limb or lower limb or combined limb disabled drivers as respondents covering the driver's interior area only, there eight elements are crucial for redesign upon their needs as listed below (Suliano et al. 2020):

(1) Handle (Door)
(2) Handle (Headliner)
(3) Ingress/egress
(4) Upholstery (Back)
(5) Upholstery (Bottom)
(6) Pedals
(7) Steering
(8) Gear Knob

The problem of each part above was defined (step 1) and improving and worsening parameter was identified from available 39 system parameter (step 2). Next, the selected parameter was mapped in the contradiction matrix (step 3) to find suggested solution corresponding to 40 Inventive Principles (step 4) and the feasible solution was developed respectively to the problem highlighted (step 5).

3 Results and Discussions

3.1 Contradiction Matrix's Analysis

A total of eight elements as listed before undergoing the analysis. Each element has more than one worsening and improving factor. However, the author only considers one or two from each as to focus on the most applicable solution to tackle solution for limbs disabled driver's problems. Below steps shows analysis (by step mention above) of one of the listed elements which is ingress/egress (door).

Step 1: Determining problem arises related to ingress/egress. One if the problem is door opening. Some of independent disabled are using wheelchair. Therefore, they need to open the door widest possible to be able to exit or enter the car, and to lift up the wheelchair. But, once the door open maximum, they have difficulties to reach the door to close. Therefore, an additional forces or stress needed to reach and close the door. The engineering contradiction (EC) for this problem is:

If <u>door open widest</u> **(a)**
Then <u>easier for handling and moving</u> **(b)**
But <u>difficult to close door</u> **(c)**

Step 2: Analyse the problem using EC. Each EC consist of a manipulative variable (from **If** statement), and two responding variables (one from **Then** statement and another one from **But** statement). Relation of statement from Step 1 and Step 2 is as below:

(a) <u>Manipulative variable</u>;

    To keep

(b) <u>Responding variable 1</u>;

    Good → improving factor (IF)

(c) <u>Responding variable 2</u>;

    Bad → worsening factor (WF)

Step 3: Once the parameter are clearly stated in EC. Therefore, the IF & WF need to be mapped with 39 system parameters.

IF: <u>easier for handling and moving</u>
WF: <u>difficult to reach to close</u>

Using 39 system parameters, most appropriate system parameter as below:

IF: Ease of operation (33)
WF: Force (10)

Step 4: Refer to contradiction matrix, capture the recommended inventive principle. Table 1 shows simplified contradiction matrix with respect to Step 3 IF and WF.

Step 5: From the recommended solution given, the inventive principle is evaluated to solve problem.

Table 2 illustrating an overall outcome of the contradiction matrix application towards 8 elements mentioned.

Table 1. Contradiction matrix analysis

Worsening Factor Improving Factor	Force (10)
Ease of operation (33)	28, 13, 35

Therefore, the recommended solutions is:

#28 Mechanics substitution
#13 The other way around
#35 Parameter change

3.2 Idealized Design Theory's Analysis

From above analysis of door, an idealized design theory are:

1. Finalized targets for design:
   Improvement for door. Since a manual door give a bad impact to disabled. A modification to a door need to take place.
2. Idealized solution:
   As mention in introduction of ideal system in Sect. 2.2, an ideal system is no system. Taking an example of current situation, if the manual door is system, an ideal system is no manual door. Means, an additional trendy automated/robotic parts might be a replacement. But an ideal system need to minimize the harmful effect, in which cost are one of harmful effect.
3. Obstacles for idealized solution:
   Complexity of designing the parts and high cost are expected.
4. Results with obstacles occurred:
   Employment of creation, modifying and maintaining automated/robotic parts.
5. Conditions without obstacles occurred:
   To design improvement of the door that can have an automatic function replacing the manual function that give stress and pressure to the disabled.
6. Usable resources for creation of these conditions:
   An automatic button to eliminate a manual car door opener.

Table 2. Contradiction matrix analysis

No.	Elements	Improving factor	Worsening factor	Inventive principle	Solution
1.	Handle (Door)	**#5** Area or moving object	**#11** Stress/pressure	**#10** Prior action **#15 Dynamization** **#28** Mechanics substitution **#36** Phase transition	To dynamize static door handle to movable and changeable direction
2.	Handle (Headliner)	**#39** Productivity	**#11** Stress/pressure	**#11 Do in advance** **#37** Thermal expansion **#14 Curvature**	To make the headliners curvy in a way it is hold able
3.	Egress	**#33** Ease of operation	**#10** Force	**#28 Mechanics substitution** **#13** The other way around **#35** Parameter change	To equipped the door with an auto open & closed button
4.	Upholstery (back)	**#32** Ease of manufacture	**#11** Stress/pressure	**#35** Parameter change **#19** Periodic action **#1 Segmentation** **#37** Thermal expansion	To segmentize the back upholstery to few segmentation to support limbs disabled back cushioning
5.	Upholstery (bottom)	**#4** Length of stationary object	**#13** Stability of object composition	**#39** Inert environment **#37** Thermal expansion **#35 Parameter change**	To change bottom seat cushion to height changeable
6.	Steering	**#39** Productivity	**#5** Length of moving object	**#10 Preliminary action** **#20** Copying **#34** Discarding & recovering **#32** Color change	To preliminary create a button to make the steering adjustable

(continued)

Table 2. (*continued*)

No.	Elements	Improving factor	Worsening factor	Inventive principle	Solution
7.	Pedals	#35 Adaptability and versatility	#30 Object generated harmful factors	#35 **Parameter change** #11 Beforehand cushioning #32 Color change #31 Porous material	To combined brake pedal and accelerator pedal to one same cross section pedals plates
		#35 Adaptability and versatility	#36 Device complexity	#29 Pneumatic & hydraulic #15 **Dynamization** #28 Mechanics substitution #37 Thermal expansion	To dynamize plate of pedal from static to movable to fit limbs disable needs
8.	Gear knob	#39 Productivity	#4 Length (Angle) of stationary object	#30 Flexible shells/Thin films #7 **Nested doll** #14 Curvature #26 Copying and models	To change the gear shift from fixed height and length to extendable using nested doll concept
		#39 Productivity	#11 Stress/pressure	#10 **Preliminary action** #37 Thermal expansion #14 Curvature	To preliminary make the gear nob adjustable alike suggested for steering

4 Conclusions

In conclusion, the projected selection of solutions from arises problem and needs of limbs disabled was presented using contradiction matrix. The results concluded that each crucial elements of car interior can be improved by taking into consideration suggested inventive principles in order to achieve ergonomic friendly car for limb disabled driver. These include Dynamization, Curvature, Mechanics substitution, Segmentation, Parameter change, Preliminary action, and nested doll. All inventive principle holding its own strength and solution is always available outside and very much applicable to solve the respective problem. Although few others improving and worsening factor applicable to each element of car interior parts were not highlighted, they still need can be used as a measure in developing the solution in future regards to the needs of limbs disabled. By considering the findings of the study, a solution will be developed to redesign a disabled-friendly car that is more ergonomic. As this is an ongoing project, the final outcome is expected to be able to improve the driving experience, in particular to limbs disabled drivers.

Acknowledgments. The authors would like to acknowledge MyTRIZ Innovation Association group who provided insight, expertise and documentation that greatly guided the research.

References

1. Prasad, R.S., Hunter, J., Hanley, J.: Driving experiences of disabled drivers. Clin. Rehabil. **20**, 445–450 (2006)
2. Branowski, B., Pohl, P., Rychlik, M., Zablocki, M.: Integral model of the area of reaches and forces of a disabled person with dysfunction of lower limbs as a tool in virtual assessment of manipulation possibilities in selected work environments. In: Stephanidis, C. (ed.) Universal Access in Human-Computer Interaction. Users Diversity. LNCS, vol. 6766, pp. 12–21. Springer, Heidelberg (2011). https://doi.org/10.1007/978-3-642-21663-3_2
3. Greve, J.M.D.A., Santos, L., Alonso, A.C., Tate, D.G.: Driving evaluation methods for able-bodied persons and individuals with lower extremity disabilities: a review of assessment modalities. Clinics **70**, 638–647 (2015)
4. Dahuri, M.K.A.M., Hussain, M.N., Yusof, N.F.M., Jalil, M.K.A.: Factors, effects, and preferences on vehicle driving modification for the Malaysia independent disabled. J. Soc. Automot. Eng. Malaysia. **1**(1), 103–110 (2017)
5. Peters, B., Ostlund, J.: Joystick controlled driving for drivers with disabilities (2005)
6. Roosmalen, L., Paquin, G.J., Steinfeld, A.M.: Quality of life technology: the state of personal transportation. Phys. Med. Rehabil. Clin. **21**, 111–125 (2017)
7. Suliano, S.B., Ahmad, S.A., As'arry, A., Aziz, F.A., Aziz, A.R.A., Shokshk, A.A.: Limbs disabled needs for an ergonomics assistive technologies and car modification. In: Awang, M., Emamian, S.S., Yusof, F. (eds.) Advances in Material Sciences and Engineering. LNME, pp. 67–73. Springer, Singapore (2020). https://doi.org/10.1007/978-981-13-8297-0_9
8. Suliano, S.B., Ahmad, S.A., As'arry, A., Aziz, F.A.: Review on ergonomics application on car modification for limbs disabled drivers. In: Emamian, S.S., Awang, M., Yusof, F. (eds.) Advances in Manufacturing Engineering. LNME, pp. 575–589. Springer, Singapore (2020). https://doi.org/10.1007/978-981-15-5753-8_53
9. Sapuan, S.M., et al.: Design of composite racing car body for student based competition. Sci. Res. Essays **4**, 1151–1162 (2009)
10. Fitzgerald, D., Herrmann, J., Schmidt, L.: Improving environmental design using TRIZ inventive principles. In: 16th Proceedings of the 16th CIRP International Design Seminar, pp. 96–100 (2006)
11. Yang, C., Kao, C., Liu, T.: An innovative product design approach based on TRIZ's inventive principles. J. Syst. Innov. **2**, 1–8 (2012)
12. Ng, K.W., Ang, M.C., Cher, D.T., Ahmad, S.A., Abdul Wahab, A.N.: Combining ARIZ with shape grammars to support designers. In: Badioze Zaman, H., et al. (eds.) Advances in Visual Informatics. LNCS, vol. 11870, pp. 305–317. Springer, Cham (2019). https://doi.org/10.1007/978-3-030-34032-2_28
13. Altshuller, G.: And Suddenly the Inventor Appeared. Technical Innovation Center Inc, Worcester (2004)
14. Guin, A.A., Kudryavtsev, A.V., Boubentsov, V.Y., Seredinsky, A.: Level 1 Study Guide: Theory of Inventive Problem Solving. First Fruit Sdn. Bhd., Selangor, Malaysia (2015)

15. Wang, C.-N., Huang, Y.-F., Le, T.-N., Ta, T.-T.: An innovative approach to enhancing the sustainable development of Japanese automobile suppliers. Sustain. **8**, 420 (2016)
16. Wang, C.N., Lin, M.H., Huang, C.J., Huang, C.C., Liao, R.Y.: Using TRIZ to improve the procurement process of spare parts in the Taiwan Navy. Sustain **9**, 1–12 (2017)
17. Feniser, C., Burz, G., Mocan, M., Ivascu, L., Gherhes, V., Otel, C.C.: The evaluation and application of the TRIZ method for increasing eco-innovative levels in SMEs. Sustainability **9**, 1–19 (2017)
18. GEN3. TRIZ group training manual (Level 1 Practitioner) (2006)
19. GEN3. TRIZ group training manual (Level 3 Practitioner) (2017)
20. Yeoh, T.S. Yeoh, T.J., Song, C.L.: TRIZ: Systematic Innovation in Manufacturing, 10th Print. Selangor, Malaysia: First Fruit Sdn. Bhd. (2015)
21. Mann, D.: Hands-on systematic innovation (2010)
22. Manohar, N., Kalla, P.: Innovative conceptual design on car using TRIZ method for optimum parking space. IOSR J. Eng. (IOSRJEN) **2**(8), 52–57 (2012)
23. Navas, H.V.G.: TRIZ: design problem solving with systematic innovation. In: Tech (2013)
24. Yang, C.J., Chen, J.L.: Accelerating preliminary eco-innovation design for products that integrates case-based reasoning and TRIZ method. J. Clean. Prod. **19**(9–10), 998–1006 (2011)
25. Petrov, V.: The Complete Book of Classical TRIZ. First Fruit Sdn. Bhd. (2018)

Evaluation on the Customer Satisfaction in Intercity Bus Transportation Using Analytic Hierarchy Process Model

Lam Weng Hoe[1,2], Lam Weng Siew[1,2(✉)], Yeoh Hong Beng[1,3], Fong Choy Yan[4], and Yeap I-Xin[4]

[1] Department of Physical and Mathematical Science, Faculty of Science,
Universiti Tunku Abdul Rahman, Kampar Campus, Jalan Universiti, Bandar Barat, 31900
Kampar, Perak, Malaysia
lamws@utar.edu.my

[2] Centre for Business and Management, Universiti Tunku Abdul Rahman,
Kampar Campus, Jalan Universiti, Bandar Barat, 31900 Kampar, Perak, Malaysia

[3] Centre for Learning and Teaching, Universiti Tunku Abdul Rahman,
Kampar Campus, Jalan Universiti, Bandar Barat, 31900 Kampar, Perak, Malaysia

[4] Faculty of Science, Universiti Tunku Abdul Rahman, Kampar Campus, Jalan Universiti,
Bandar Barat, 31900 Kampar, Perak, Malaysia

Abstract. Intercity bus transportation involves bus services that serve the general public and operates on a regular schedule to different destinations. It makes important connection between two or more areas that are not in close distance. Due to stiff competition from other transportation modes, bus service providers need to improve their services by examining and understanding their customers' preference. The objective of this paper is to propose a conceptual framework with Analytic Hierarchy Process (AHP) model to identify the priority of decision criteria in the selection of intercity bus transportation companies among the passengers. AHP model provides solution to this Multi Criteria Decision Making (MCDM) problem by decomposing the problem into a hierarchical structure. The proposed AHP model is illustrated with the case study among the passengers in Malaysia who have experienced the route from Kampar to Penang in Malaysia. The decision criteria studied in this paper are service provided, accessibility, time, environment and availability. Besides, this study also aims to determine the most preferred intercity bus transportation company among Transnational Express, Kesatuan Express and Konsortium Express with AHP Model. The results of this study show that Transnational Express is the most preferred intercity bus transportation company followed by Kesatuan Express and Konsortium Express. Service provided and accessibility are ranked as the top two decision criteria by the passengers. This study provides recommendations to intercity bus transportation companies to improve the service quality and customers' satisfaction level among the passengers in different parts of Malaysia.

Keywords: Customer satisfaction · Service quality · Intercity bus transportation · Analytic hierarchy process · Multi-criteria decision making

© The Author(s), under exclusive license to Springer Nature Singapore Pte Ltd. 2021
M. Awang and S. S. Emamian (Eds.): *Advances in Material Science and Engineering*, LNME, pp. 200–209, 2021.
https://doi.org/10.1007/978-981-16-3641-7_24

1 Introduction

Intercity transportation involves three main modes, namely bus, rail, and air. Intercity bus transportation is defined as the bus services that serve the general public and operates on a regular schedule to different destinations [1]. It makes important connection between two or more areas that are not in close distance. Most of the intercity travellers will prefer bus transportation due to its flexibility. According to Islam, et al. [2], customer satisfaction is an experience of fulfilment of the customer toward an expected outcome. The level of satisfaction with a program or facility is affected by the customers' expectation and perception toward the service quality provided. However, in recent years, intercity bus transportations are facing stiff competition from other transportation modes. Wrong perception in customers' needs by the bus transportation companies is one of the factors affecting the usage of bus transportation in long distance travelling. When a service is provided, customers have the chance to evaluate it based on their level of satisfaction on the service. Based on research carried out by Mokhlis [3], the management often assumes the needs of customers are the same yet in fact are not in alliance with the current demand of customers. As a result, bus service might lose its attractiveness to customers due to inefficient service and low customers' satisfaction level. Although past studies by Goh et al. [4], Chimba et al. [5] and others reviewed that bus transportation is far more safety than other modes of transportation, the crash rate for buses and the injury rate for bus passengers are relatively high compared to other transport modes in Malaysia. Hence, it is important to identify the customers' satisfaction in bus transportation in order to attract more users to choose intercity express bus as their first choice in long distance travelling and improve constantly low customers' satisfaction in bus transportation. The objective of this paper is to propose a conceptual framework using Analytic Hierarchy Process (AHP) model to identify the priority of decision criteria in the selection of intercity bus transportation companies among the passengers. AHP model solves Multi Criteria Decision Making (MCDM) problem by decomposing the problem into a hierarchical structure. AHP model has been applied by researchers to solve MCDM problem in different fields [6–14].

According to Murambi and Bwisa [15], customer satisfaction is the most important consideration after a product or service is provided. If the customers are dissatisfied with the company, they will find other service provider to replace the company. Putra et al. [16] stated that the customer satisfaction is associated with the service quality. The overall performance of the service providers will affect customer satisfaction. The level of customer satisfaction depends on the quality of service experienced by the customers.

Islam et al. [2] concluded that the quality of service provided will affect the overall customer satisfaction in bus transportation services. Service provided includes the behavior of the workers, price of the ticket, physical environment of the bus, layout of the bus stops and route safety. In relation to it, Freitas [17] found out that the behavior of the personnel is one of the most important factors that will affect the customer satisfaction. According to Islam et al. [2], accessibility is one of the decision criteria that affects customer satisfaction. Besides, Le-Klahn et al. [18] found out that accessibility is one of the significant elements in providing sustainable and higher quality services. Accessibility consists of the access in the bus ticket as well as the access to bus stop.

Kostakis and Pandelis [19] discovered that some of the passengers who consume bus transport for professional reason will consider departure and arrival times as their main concern. The possible delay of bus's departure and arrival creates bad impression for customers and causes their dissatisfaction. In relation to it, Minhans et al. [20] identified that the inconsistent travel time is due to the poor boarding and longer waiting time for the customers. On top of it, Kamaruddin et al. [21] concluded that environment factor will have an impact on customer satisfaction. Environment factor refers to customer perception towards the bus pollution and environmental impacts. As mentioned by Mouwen [22], the customer satisfaction level is lower in highly urbanized environment compared to low urbanized environment. This indicates that there is growing number of customers who look forward to a less polluted environment. Islam et al. [2] stated that availability factor comprises the frequency of the service, the coverage of network and availability of service at bus terminal. Similarly, Tyrinopoulos and Antoniou [23] showed that route frequency and coverage of network are significant aspects for customer satisfaction in intercity bus transportation. Customers will appreciate the efficient and effective transportation system with high service frequency, convenient schedule and larger coverage of network.

2 Methods

This study examines the customer satisfaction on intercity bus transportation using the proposed conceptual framework with AHP model. It aims to identify the ranking of the three selected bus transportation companies and the five decision criteria. The bus transportation companies selected are companies that serve specifically route from Kampar to Penang, namely Konsortium Express, Transnational Express and Kesatuan Express. The decision criteria for this study are service provided, accessibility, time, environment and availability. For this research, all selected target respondents are required to have experience to use all these studied bus transportation companies. A total of 87 respondents participated in this research fieldwork.

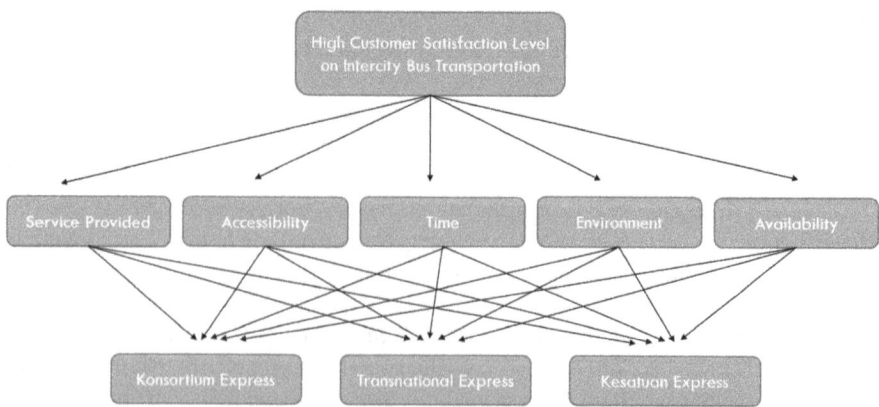

Fig. 1. Proposed conceptual framework

AHP model is designed to solve MCDM problem. MCDM problem involves ranking the decision alternatives based on multiple decision criteria [24–33]. The proposed conceptual framework is shown in Fig. 1.

The basic procedure to carry out the AHP consists of the following steps.

Step 1: Structure a decision problem and selection of criteria. Then, arrange all the components in a hierarchy that provides an overview of the complex relationships to help the decision maker to assess whether the elements in each level are of the same magnitude of importance as shown in Table 1.

Table 1. Hierarchy structure for the selection of intercity bus transportation companies.

Top level : main objective	High customer satisfaction level on intercity bus transportation
Middle level: decision criteria	1. Service provided
	2. Accessibility
	3. Time
	4. Environment
	5. Availability
Third level: decision alternatives	1. Konsortium express
	2. Transnational express
	3. Kesatuan express

Step 2: Priority setting of the criteria by pairwise comparison and weightings are then normalized and averaged in order to obtain an average weight for each criterion. The ratio scale used for pairwise comparison is summarized in Table 2 [18, 34].

Table 2. Ratio scale used for pairwise comparison.

Scale	Definition
1	Equal importance
3	Moderate importance
5	Strong importance
7	Very strong importance
9	Absolute importance
2, 4, 6, 8	Intermediate values

Step 3: Make a pairwise comparison of options on each criterion. Then, ratings are normalized and averaged.

Step 4: Normalization method is used by adding the matrices in on each column and dividing all elements in the column by the columns total and repeated for all pairwise matrices. Therefore, the overall weighted score for each decision criteria which reflects the level of influence of decision alternative (*FDW*) is obtained by multiplying criteria (*Q*) with decision criteria (w^T). The higher the weight is, the higher the level of influence is. This will cause higher in ranking of the alternatives.

Step 5: Examine the consistency of Consistency Ratio.

CR = CI/RI which CI is the consistency index and RI is the random index.

It reflects that reliable results with relatively high consistency is formulated if the CR ≤ 0.10 [9, 34].

3   Results and Discussion

Figure 2 shows the weights and rankings of all decision criteria in the selection of bus companies.

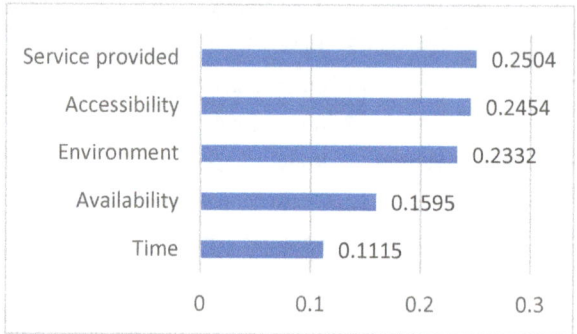

Fig. 2. Priority of decision criteria in the selection of bus companies.

As shown in Fig. 2, the priority of decision criteria in the selection of bus companies is the service provided (0.2504) followed by accessibility (0.2454), environment (0.2332), availability (0.1595) and time (01115). Service provided has the highest priority among all the decision criteria. Figures 3, 4, 5, 6, 7 shows the preference of bus companies based on each decision criterion. As shown in Figs. 3, 4, 5, 6, 7, Transnational Express has the top ranking for all decision criteria. This means that Transnational Express is the most favourable bus companies among the others in terms of service provided, accessibility, time, environment and availability. Kesatuan Express and Konsortium Express are the second and third ranked respectively under all the decision criteria.

Figure 8 presents the overall weights and priority in the selection of bus companies in this study. As shown in Fig. 8, the result shows that Transnational Express (0.4095) is the most preferred bus company among the passengers with respect to all decision criteria which are service provided, accessibility, time, environment and availability. The preference of the bus companies is followed by Kesatuan Express (0.3413) and Konsortium Express (0.2492). The overall consistency ratio is 0.0235 which is well below 0.1000. Thus, the result of this study is consistent and acceptable.

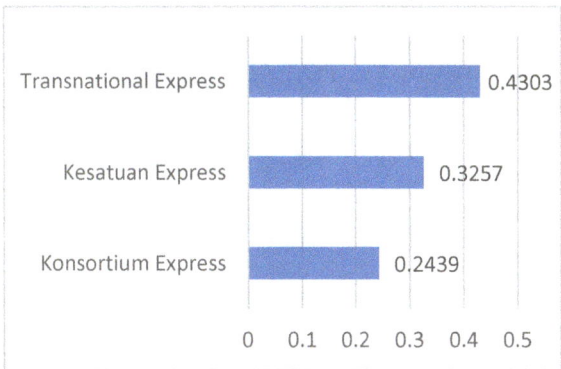

Fig. 3. Preference of bus companies based on service provided.

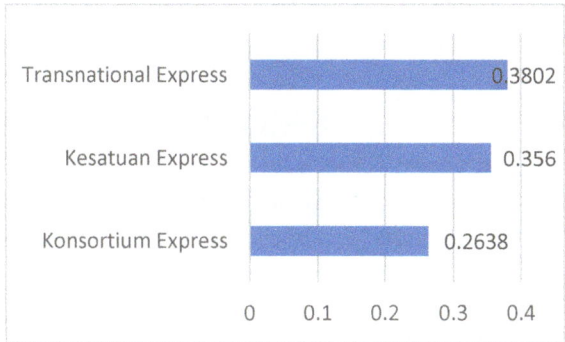

Fig. 4. Preference of bus companies based on accessibility.

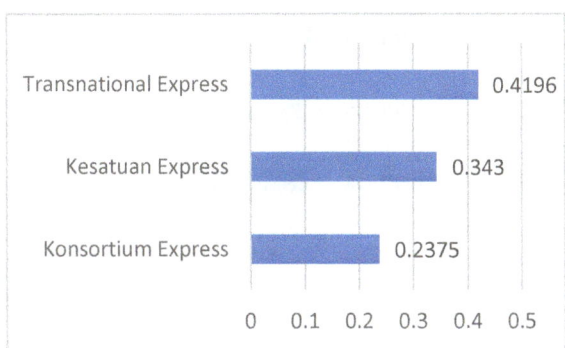

Fig. 5. Preference of bus companies based on time.

Based on the results of this study, there are recommendations for the bus companies as the guidance to improve the service quality and customer satisfaction level among the passengers. It is suggested to the bus companies to take into account service provided

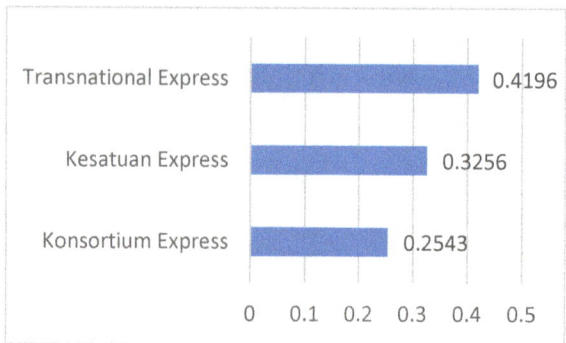

Fig. 6. Preference of bus companies based on environment.

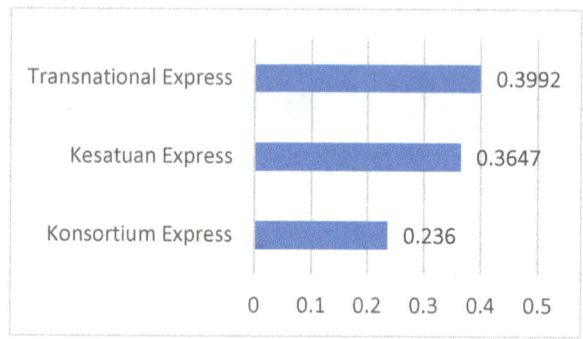

Fig. 7. Preference of bus companies based on availability.

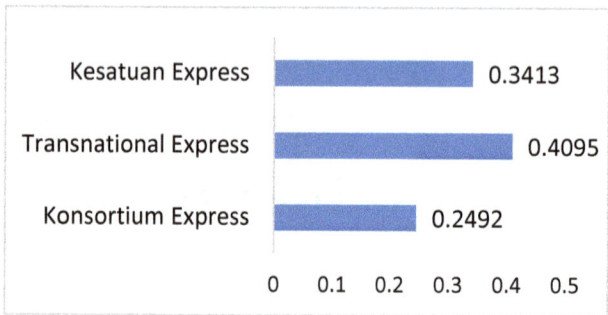

Fig. 8. Overall weights in the selection of bus companies.

first as it is the primary concern of the customer satisfaction. Correspondingly, accessibility of a bus company can be enhanced by increasing more ticketing channels such as promotions through social media that is part and parcel of our daily life and mobile application ticket platform to allow more customers to access to the tickets easily.

Environmental problem is being concerned by the public in recent years. People start to realize the significance and consequence of it. Therefore, bus companies are recommended to develop and maintain a systematic approach to environmental management that contributes toward less pollution and comply with current legislative requirements and national standards such as the amount of greenhouse gas emission.

Availability of the bus company is to be maintained by increasing its coverage network. Although time is the least considerate decision criteria among all, it does not mean that bus companies are to neglect this area because lead time issue may incur high cost and loss to the company in terms of its financial income and reputation. Thus, communication is the key to maintain a good lead time as bus coordinator needs to ensure bus driver is clear with the route as well as times of arrival and departure of the bus trip.

4 Conclusions

The main objective of this research is to study the customer satisfaction on intercity bus transportation by using the proposed conceptual framework with AHP model. Among the five decision criteria of service provided, accessibility, time, environment and availability, service provided is ranked as the most important factor that affects customer satisfaction on intercity bus transportation. This reflects that passengers are very concern about the services rendered by the bus company. It is directly reflected the customer's impression towards the particular bus company, right from services of front desk such as price of the ticket, the behavior of the personnel until the personal experience of bus journey physical conditions of the bus, layout of the bus stops and route safety. To certain extent, it is about the level of efficiency and effectiveness of the services that are provided to the passengers by the bus company.

The second and third preferred criteria for the selection of intercity bus transportation are accessibility and environment. The priority setting is then followed by availability and time. Besides, this research aims to identify the most preferred bus company among passengers whenever it comes to the journey specifically from Kampar to Penang. Transnational Express overtakes Konsortium Express and Kesatuan Express to be the most preferred bus company among the passengers. Above all, this research concludes that the primary determinant affecting customer satisfaction is service provided and Transnational Express is the most favourable bus company among passengers from Kampar to Penang. It is recommended that the service quality shall be further improved to enhance customer satisfaction level among the passengers in the intercity bus transportation.

Acknowledgments. The study is supported by Universiti Tunku Abdul Rahman (UTAR), Malaysia.

208   L. W. Hoe et al.

References

1. Woldeamanuel, M.: Evaluating the competitiveness of intercity buses in terms of sustainability indicators. J. Public Transp. **15**(3), 77–96 (2012)
2. Islam, R., Chowdhury, M.S., Sarker, M.S., Ahmed, S.: Measuring customer's satisfaction on bus transportation. Am. J. Econ. Bus. Admin. **6**(1), 34–41 (2014)
3. Mokhlis, S.: Passenger satisfaction and loyalty: a case of inter-city coach travel in Malaysia. Jurnal Manajemen dan Bisnis **11**(2), 1–20 (2016)
4. Goh, K., Currie, G., Sarvi, M., Logan, D.: Factors affecting the probability of bus drivers being at-fault in bus-involved accidents. Accid. Anal. Prev. **66**, 20–26 (2014)
5. Chimba, D., Sando, T., Kwigizile, V.: Effect of bus size and operation to crash occurrences. Accid. Anal. Prev. **42**(6), 2063–2067 (2010)
6. Harshan, R.K., Chen, X., Shi, B.: Analytic hierarchy process (AHP) based model for assessing performance quality of library websites. Inf. Technol. J. **16**, 35–43 (2017)
7. Lam, W.S., Bishan, R.S., Lam, W.H.: An empirical study on the mold machine-tool selection in semiconductor industry with analytic hierarchy process model. Adv. Sci. Lett. **23**(9), 8286–8289 (2017)
8. Lam, W.S., Chen, J.W., Lam, W.H.: An empirical study on the selection of fast food restaurants among undergraduates with AHP model. Am. J. Inf. Sci. Comput. Eng. **2**(3), 15–21 (2016)
9. Lam, W.S., Leong, W.B., Lam, W.H.: Selection of mobile network operator based on multi-criteria decision making model using analytic hierarchy process. Math. Statist. J. **1**(1), 12–18 (2015)
10. Lam, W.S., Lee, W.K., Lam, W.H.: Multi-criteria decision making in job selection problem using analytic hierarchy process model. Math. Statist. J. **1**(2), 3–7 (2015)
11. Lam, W.S., Chen, J.W., Lam, W.H.: Analysis on the preference of fast food restaurants with analytic hierarchy process model. Int. J. Psychol. Cogn. Sci. **3**(6), 72–76 (2017)
12. Lam, W.S., Chen, J.W., Lam, W.H.: An Empirical study on the preference of laptop in Malaysia with analytic hierarchy process model. SCIREA J. Comput. Sci. Technol. **1**(2), 127–141 (2016)
13. Lam, W.S., Bakar, M.A., Lam, W.H., Liew, K.F.: Multi-criteria decision making in the selection of mobile network operators with AHP-TOPSIS model. J. Eng. Appl. Sci. **12**, 6382–6386 (2017)
14. Lam, W.S., Lam, W.H., Liew, K.F., Chen, J.W.: An Empirical study on the preference of fast food restaurants in Malaysia with AHP-TOPSIS model. J. Eng. Appl. Sci. **13**, 3226–3231 (2018)
15. Murambi, D.N., Bwisa, H.M.: Service quality and customer satisfaction in public transport sector of Kenya: a survey of shuttle travelers in Kitale terminus. Int. J. Acad. Res. Bus. Soc. Sci. **4**(9), 402–412 (2014)
16. Putra, A.A., Yamin, J.M., Riyanto, R., Mulyono, A.T.: The satisfaction analysis for the performance of public transport urban areas. Int. Refereed J. Eng. Sci. **3**(8), 38–44 (2014)
17. Freitas, A.L.P.: Assessing the quality of intercity road transportation of passengers: an exploratory study in Brazil. Transp. Res. Part A **49**, 379–392 (2013)
18. Le-Klahn, D.T., Hall, C.M., Gerike, R.: Analysis of visitor satisfaction with public transport in Munich. J. Public Transp. **17**(3), 68–85 (2014)
19. Kostakis, A.P., Pandelis, I.: measuring customer satisfaction in public transportation an empirical study based in urban buses in the city of Larissa (Greece)-the MUSA methodology. MIBES **1**(3), 260–275 (2009)
20. Minhans, A., Shahid, S., Ahmed, I.: An Investigation into qualitative differences between bus users and operators for intercity travel in Malaysia. Jurnal Teknologi (Sci. Eng.) **70**(4), 71–81 (2014)

21. Kamaruddin, R., Osman, I., Che Pei, C.A.: Public transport services in Klang valley: customer expectations and its relationship using SEM. Procedia. Soc. Behav. Sci. **36**, 431–438 (2012)
22. Mouwen, A.: Drivers of customer satisfaction with public transport services. Transp. Res. Part A **78**, 1–20 (2015)
23. Tyrinopoulos, Y., Antoniou, C.: Public transit user satisfaction: variability and policy implications. Transp. Policy **15**(4), 260–272 (2008)
24. Lam, W.S., Liew, K.F., Lam, W.H.: An optimal control on the efficiency of technology companies in Malaysia with data envelopment analysis model. J. Telecommun. Electron. Comput. Eng. **10**(1), 107–111 (2018)
25. Liew, K.F., Lam, W.S., Lam, W.H.: Financial analysis on the company performance in Malaysia with multi-criteria decision making model. Systems Science and Applied Mathematics **1**(1), 1–7 (2016)
26. Lam, W.S., Liew, K.F., Lam, W.H.: Evaluation on the financial performance of the Malaysian banks with TOPSIS model. Am. J. Serv. Sci. Manage. **4**(2), 11–16 (2017)
27. Lam, W.S., Lam, W.H., Liew, K.F.: Data analysis on the performance of technology sector in Malaysia with entropy-TOPSIS model. Commun. Comput. Inf. Sci. **886**, 194–203 (2018)
28. Lam, W.S., Liew, K.F., Lam, W.H.: An empirical comparison on the efficiency of healthcare companies in Malaysia with data envelopment analysis model. Int. J. Serv. Sci. Manage. Eng. **4**(1), 1–5 (2017)
29. Lam, W.S., Liew, K.F., Lam, W.H.: Improvement on the efficiency of technology companies in Malaysia with data envelopment analysis model. Lect. Notes Comput. Sci. **10645**, 19–30 (2017)
30. Lam, W.S., Liew, K.F., Lam, W.H.: Investigation on the efficiency of financial companies in Malaysia with data envelopment analysis model. J. Phys. Conf. Ser. **995**, 012021 (2018)
31. Lam, K.F., Lam, W.S., Lam, W.H.: An empirical evaluation on the efficiency of the companies in Malaysia with data envelopment analysis model. Adv. Sci. Lett. **23**(9), 8264–8267 (2017)
32. Lam, W.S., Chen, J.W., Lam, W.H.: Data driven decision analysis in bank financial management with goal programming model. Lect. Notes Comput. Sci. **10645**, 681–689 (2017)
33. Lam, W.H., Lam, W.S., Liew, K.F.: Performance analysis on telecommunication companies in Malaysia with TOPSIS model. Indonesian J. Electr. Eng. Comput. Sci. **13**(2), 744–751 (2019)
34. Saaty, T.L.: Decision making with the analytic hierarchy process. Int. J. Serv. Sci. **1**, 83–98 (2008)

Synthesis and Characterization of a Cellulose-Based Polymeric Surfactant Towards Applications in Enhanced Oil Recovery

Funsho Afolabi[1(✉)], Syed M. Mahmood[1], Jonathan Johnson[2], and Omolara A. Peters[2]

[1] Department of Petroleum Engineering,
Universiti Teknologi PETRONAS, Seri Iskandar, Malaysia
funsho_18002873@utp.edu.my
[2] Chemical Sciences Department,
Afe Babalola University Ado Ekiti, Ekiti, Nigeria

Abstract. The oil and gas industry is in constant search for new solutions for recovery and production enhancement. This need is triggered by the huge hydrocarbon reserves left in subsurface reservoirs after the primary and secondary recovery schemes. Polymeric surfactants are a potential candidate because of their multifaceted functionality. Cellulose-based polymeric surfactants provide an additional advantage of being cost-effective and environmental-friendly. However, a major shortfall of existing polymeric surfactants is their limited surface activity which is essential to manipulate capillary forces within porous media for favorable displacement. Based on new insights into polymer physics and cellulose chemistry, this study aims to develop a novel cellulose-based surfactant with improved capability to reduce surface and interfacial tension to new levels of low while maintaining its solution viscosifying functions. Here, cellulose sulfate is selected as a suitable anionic cellulose derivative backbone while dodecyl polyoxyethylene acrylate is chosen as the surfactant macromonomer for the hydrophobic modification. Using microwave-assisted free radical polymerization, a new compound with a unique molecular structure is synthesized. The various functional groups anticipated in the novel amphiphilic cellulose derivative are confirmed by FTIR spectra analysis. SEM micrograph showed interesting features that could be valuable for chemical enhanced oil recovery applications. The authors suggested further characterization and tests to ascertain the potency of the new product.

1 Introduction

There is a renewed interest in cellulose utilization lately in several industries and facets of technology because of growing concern over environmental issues associated with synthetic polymeric systems, besides the cost of synthesizing and producing the latter [1]. Studies on polymeric surfactants derived from cellulose and cellulose derivatives were more focused on applications for the detergent industry where the washing power and anti-re-depositing efficiency were compared to the 'mother' cellulose compounds and commercial surfactants like TWEEN20[2–9]. Based on the current environmental

M. Awang and S. S. Emamian (Eds.): *Advances in Material Science and Engineering*, LNME, pp. 210–219, 2022.
https://doi.org/10.1007/978-981-16-3641-7_25

sustainability drive for a paradigm shift to eco-friendlier and more renewable 'green technologies,' there are few cases where cellulose-based biopolymeric surfactants were tested directly for Chemical Enhanced Oil Recovery purposes. Using various synthesis routes, carboxymethyl cellulose has been modified hydrophobically [10], and so was hydroxyethyl cellulose [11–13]. Amphiphilic cellulose-based polymers have been observed to exhibit tendencies for mobility control comparable to conventional polymers [2, 5, 6, 10–14]. They have demonstrated the ability to reduce IFT and emulsify oil [2–4, 7, 10, 14, 15], and even alter the wettability properties of rock surface [11]. Also, their recovery potentials have been tested via coreflood displacement experiments using representative reservoir rock samples [11–13]. However, a major shortcoming is their inability to reduce interfacial tension to as low as possible to perpetrate in situ generations of microemulsions which will aid in actualizing effective displacement of trapped and bypassed oil in realistic porous media of subsurface formations.

The limitations in the surface activity of cellulose derivatives and/or cellulose-based surfactants can be linked to the rigid polymeric backbone which prevents adequate adsorption at interfaces as well as optimized conformations thereof [15]. Here, these features are a factor of molecular structure complexity, the functional groups present, and essentially the size and location of the hydrophobic group(s) on the molecule. In the majority of cellulose-based polymeric surfactants, the hydrophobic units were attached to the chain in a manner that enhances the rheology properties more than surface activity. This is due to the location and availability of the reactive hydroxyl groups for targeted modification. Cao and Li's work [10] showed that if cellulose chain scission can be attained, and replace glycoside units with hydrophobic monomers, then the outcome product can be likened to polysoaps i.e. a polymeric-chain compound made up of repeating units of surfactants that behaves like low-weight monomeric surfactants [16–18]. So these compounds with the aforementioned molecular structure can influence multifunctional properties during displacement in porous media i.e. maintain mobility control by the reason of the improved rheology through increased viscosity and viscoelasticity while reducing IFT significantly and changing rock wettability. The latter surface properties are achieved because individual molecules have more chain flexibility with optimized conformation at interfaces [19].

The main aim of this study is to develop a pre-conceived anionic cellulose-based polymeric surfactant with a novel molecular structure that has certain strategic functional groups for purposed enhanced surface activity while retaining the traditional viscosifying properties of polymers. The synthesis route shall be microwave irradiation-assisted free radical polymerization technique. The novel cellulose derivative with new molecular features will be verified and tendencies for a strong surface activity will be identified via SEM and FTIR characterization methods.

2 Methods

2.1 Materials

The chemicals utilized for synthesis and purification are cellulose sulfate (CS) as the anionic cellulose derivative; dodecyl polyoxyethylene acrylate (DPEA) as the nonionic surfactant macromonomer for the intended hydrophobic modification; distilled water, ethanol, and acetone for preparation of samples and extraction of impurities respectively. All chemicals are of analytical grade. Cellulose sulfate and DPEA are KERMEL products. Acetone was gotten from Guangdang Guanghua chemical factory Co Ltd China while Ethanol from Central Drug House (P) Ltd New Delhi India. All chemicals were supplied through TOPJAY Scientific. The microwave oven utilized is a SHARP R-7G17 model with 850W output.

2.2 Synthesis

0.6% solution of cellulose sulfate and 0.5% DPEA were mixed using a magnetic stirrer with rod for 30 min. Afterward, the reaction mixture was transferred to a microwave oven and allowed a resident time of 5 to 20 min under stirring at low to medium-low radiation intensity. The product was cooled and oven-dried on glass Petri dishes at 60 ° C for 8 to 12 h. The solid product is immersed in a solution mixture of ethanol and water of ratio 70/30% to eliminate unreacted monomers. The sample was now subjected to thorough washing using acetone to remove remaining impurities. The final solid product, which is in the form of white flakes, was oven-dried at 40 °C to a constant weight. The size of flakes was reduced by pulverization before further use.

2.3 Characterization

The Fourier Transform Infra-Red (FT-IR) spectra of the synthesized sample were measured using the KBr pellet technique for the standard range of 4000 cm^{-1} to 400 cm^{-1}. The machine model is a Nicolet iS10 FT-IR Spectrometer. Particle shape, size, structure, and morphology were examined by Field Emission Scanning Electron Microscopy (FE-SEM) using the JOEL-JSM 7600F equipment.

3 Results and Discussion

3.1 Synthesis and Mechanism

It is somewhat difficult for high molecular weight, large molecular size amphiphilic polymers to have the same conformation at interfaces, and micellar morphology in solution as a regular low weighted surfactant [20]. That explains the physics behind the weak surface activity. However, in the case of hydrophobically modified cellulose and their derivatives, there is an indication that the surface activity will tremendously increase if the hydrophobic unit is attached at C_1 and C_4 rather than the traditional C_2, C_3, and C_6 carbon atoms [21]. The challenge here is that the process requires random chain scission of the polymeric system and re-copolymerization using a suitable targeted surfactant monomer. Studies have revealed that this reaction path is difficult to create except using heat from irradiation sources to thermo-mechanically break the chain at unusual points along the cyclic cellulosic chain [22]. Examples of solid-state irradiation techniques that have been used are UV-rays, gamma rays, X-rays, microwave, and ultrasonic rays.

The intention in this study is to use microwave heat to generate cellulose macro radicals in the presence of unsaturated reactive macro surfactants so that the copolymerization reaction can take place spontaneously in solution. The schemes below depict the reaction mechanism and path for the synthesis process.

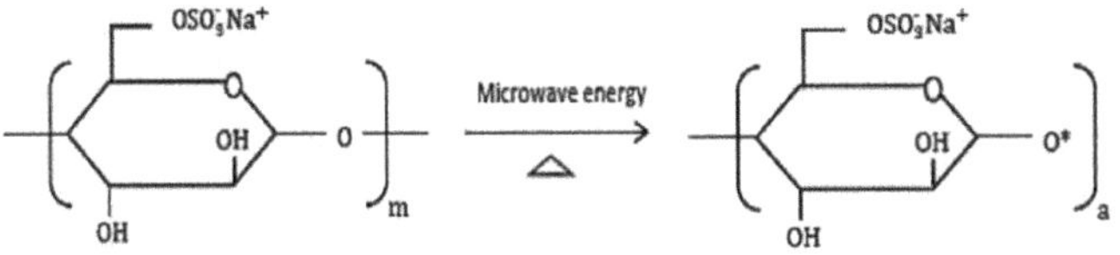

Scheme 1. Initiation step: the creation of cellulose sulfate radicals by exposure to microwave heat

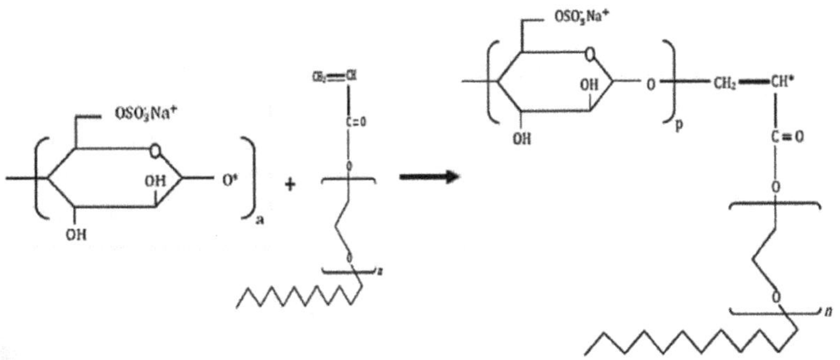

Scheme 2. Propagation step: chain-growth copolymerization reaction between surface-active macromonomers and cellulose sulfate radicals

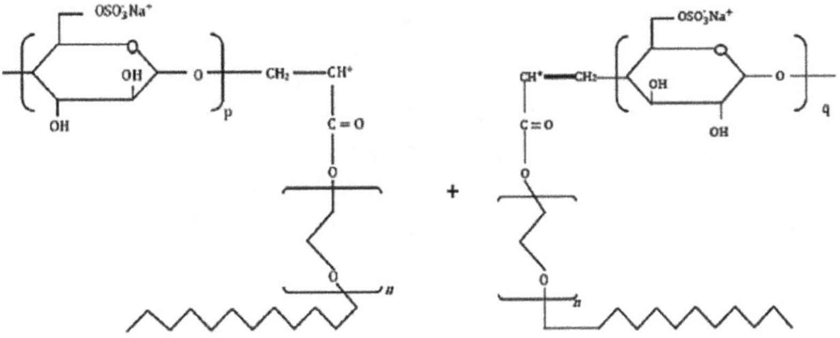

Scheme 3. Termination step: chain-growth reaction is brought to an end by the co-joining of macro-radicals.

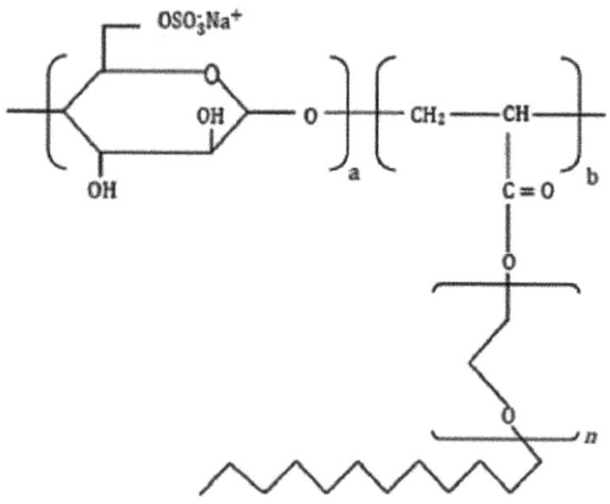

Scheme 4. End product depicting the molecular structure of the proposed novel cellulose-based polymeric surfactant.

The concentrations for the reaction were selected based on stoichiometry according to Cao and Li [10]. The resident time in the reaction oven and intensity of microwave heat were intentionally reduced so that elongated exposure will not lead to the total breakdown and decomposition of the entire cellulose chain. Hence, moderate chain scission will serve as a control for random conjoining of the reactive macro surfactant. Several rounds of washing and filtering using appropriate solvent were to ensure no contaminant remains in the final product which can potentially affect the results, and probably the physicochemical performance in later analysis. The ethanol/water mixture is to take care of polar impurities while acetone removes the remaining unreacted organic components if present.

3.2   FTIR Analyses

The pre-conceived chemical structure of the novel cellulose-based polymeric surfactant was strategically designed and developed to include certain functional groups after intended purposes. Hence the precursors were selected for these purposes. Classical surfactants that reduce IFT to ultra-low levels in EOR applications are commonly known to be of large hydrophilic head groups, preferably anionic if they're to be applied in sandstone reservoirs. While the hydrophobic tail is essentially a long one of a minimum of 8 carbon alkyl chain and can run up to 18 to 20 carbon atoms. Surfactants whose hydrophilic head groups houses poly (ethylene oxide) (PEO) and poly (propylene oxide) (PPO) are known to have improved hydrophilic-lipophilic balance (HLB) in aqueous-oleic solutions because of better adsorptive properties at interfaces. These described features above were intentionally included in the novel amphiphilic cellulose derivative i.e. see Scheme 4.

The FTIR spectra aided in isolating and identifying these anticipated functional groups in the synthesized product. Given in Fig. 1 below is the spectroscopy. Besides the free OH stretching vibration of primary alcohol noticed at 3428.00 cm^{-1} for cellulose sulfate (Fig. 1), the hydrophobically modified derivative gave additional spikes (valleys) at 3694.44 cm^{-1} and 3629.50 cm^{-1} (Fig. 2). The extras are due to inter-molecular and intramolecular hydrogen bonding indicative of possible hydrophobic associating properties in solutions typical of amphiphilic polymers. Stretching vibration noticed at 2928.71 cm^{-1} is due to CH$_2$ located on the cyclic glycoside chain. Spikes due to the carboxylate group are noticed at 1820.35 cm^{-1} and 1637 cm^{-1}, nonconjugated, and conjugated respectively. These confirm the fusion of the surfactant macromonomer unto the cellulose sulfate backbone. Furthermore, the spike at 1025.00 cm^{-1} is due to the C-O stretching vibration of alkyl-substituted ether, supporting the earlier observation. Additionally, the valley at 759.25 cm^{-1} is due to the rocking vibration of CH$_2$ on the alkyl chain. Finally, a spike at 688.29 cm^{-1} confirms the sulfate ion. In all, and as expected, there is a general reduction and slight shift in the intensity of cellulose sulfate peaks as reflected in the hydrophobically modified version. The presence of all these functional groups proves that the reaction took place between the cellulose sulfate radical and the macrosurfactant to give the unique molecular structure presented in Scheme 4.

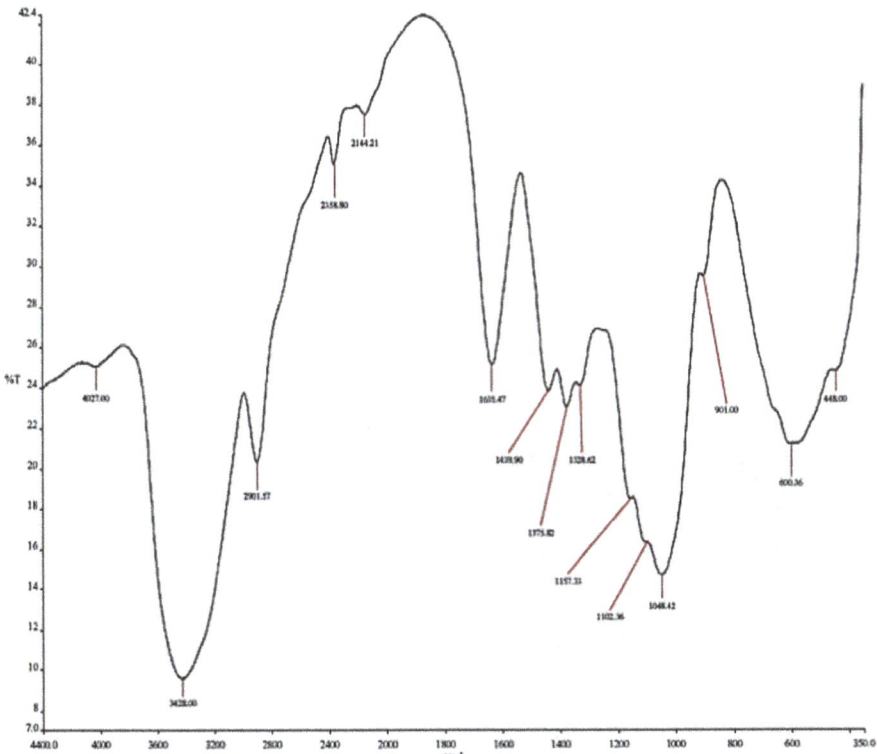

Fig. 1. FTIR Spectra measured for the cellulose sulphate

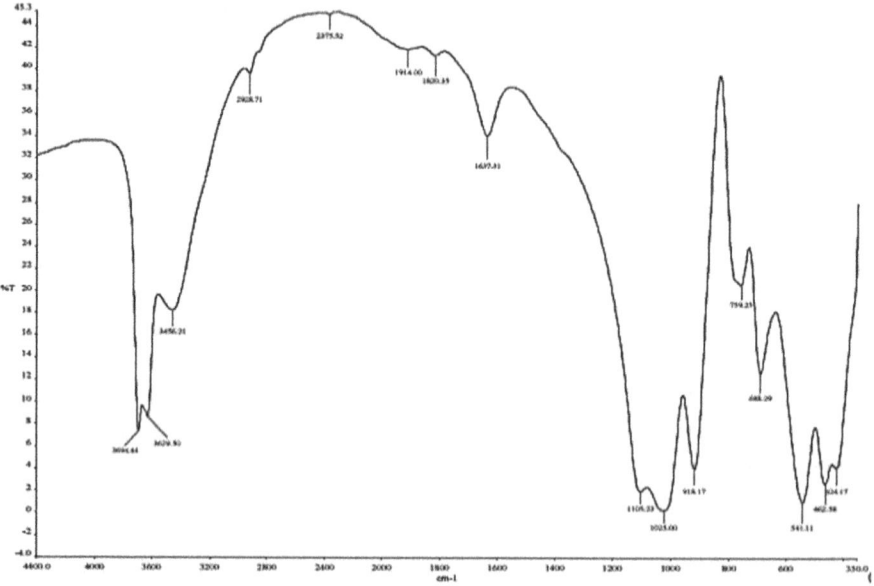

Fig. 2. FTIR Spectra measured for the cellulose-based polymeric surfactant DPEA-Cell-OSO$_3^-$

3.3   SEM Analysis

A scanning electron micrograph was used to study the particle size, structure, shape, texture, and morphology at solid-state. Here, that of cellulose sulfate and the hydrophobically modified derivative is analyzed and compared critically as seen in Fig. 3. The cellulose sulfate arrangement is seen as mesh-like overlays of particles (Fig. 3), and are highly fibrous (Fig. 4) typical of cellulosic materials. The particles are long, thick, and rod-like.

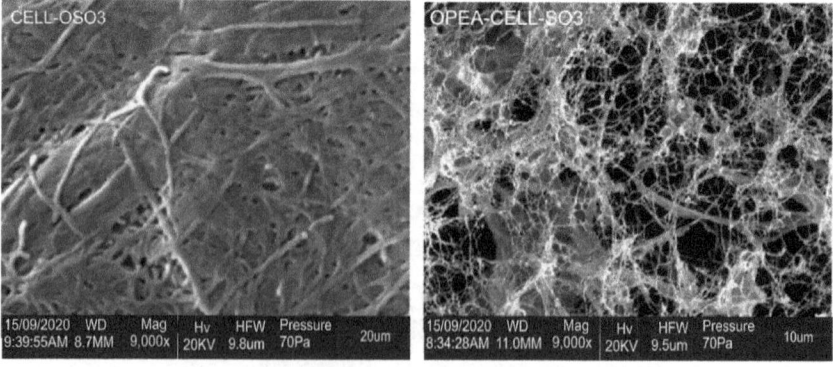

Fig. 3. SEM Micrograph at 9,000x magnification for (a) cellulose sulfate (b) DPEA-Cell-OSO$_3^-$

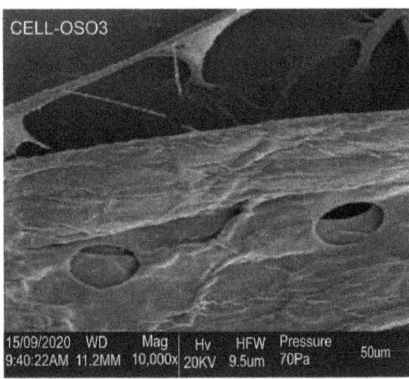 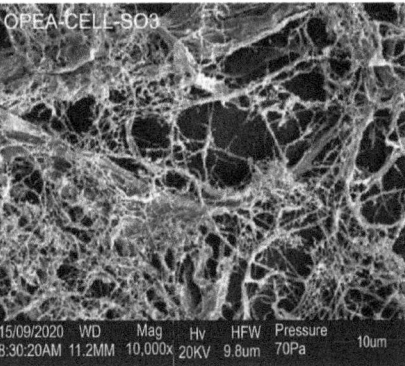

Fig. 4. SEM Micrograph at 10,000x magnification for (c) cellulose sulfate (d) DPEA-Cell-OSO_3^-

On the other hand, the hydrophobically modified cellulose sulfate particle arrangement takes the form of a highly dense dendritic network (Fig. 3), which is normally attributed to hydrophobic communication between molecules. The particles are thinner and spiky, and appear to be more crystalline compared to cellulose sulfate (Fig. 4).

The improved physical appearance of the hydrophobically modified cellulose sulfate is indicative of better physicochemical properties when dispersed in solution. The extensive web-like network aggregation has been pointed out to be an important criterion for strong emulsification in the colloid state [23]. Also, the particle form and structure shows that the novel derivative will have a high tensile strength compared to the unmodified cellulose sulfate [24]. This implies that the modified version will most likely have a better tolerance to salinity, temperature, and other physicochemical harsh conditions that might rather affect the former adversely. In all, the micrograph did not only point out an improvement in particle properties of cellulose compounds but was also indicative of possible better solution properties.

3.4 Conclusions

Polymeric surfactants provide a new realm of opportunity in chemical enhanced oil recovery technology because of their potential multifunctional capabilities. These apply in both microscopic and macroscopic displacement of residual and bypassed oil saturation at secondary and tertiary stages of reservoir development. Furthermore, cellulose-based biopolymeric surfactants can drive down the risks associated with cost and environmental sustainability in oilfields.

In the present study, based on new insights in polymer physics and cellulose chemistry, a novel amphiphilic cellulose derivative was proposed and synthesized using a unique combination of reaction routes. Infra-red spectra were used to identify the strategic functional groups incorporated into the new compound and SEM analysis was used to study the particles and their aggregative properties in solid form. In-line with the targeted end improvement in interfacial and colloidal properties, further molecular structure characterization and solution properties shall be the immediate next line of investigations.

Acknowledgments. The authors would like to acknowledge the center for graduate studies, university Teknologi PETRONAS for the financial support; and the department of chemical sciences, Afe Babalola University Ado Ekiti for providing the facilities for the research.

References

1. Kang, H., Liu, R., Huang, Y.: Graft modification of cellulose: methods, properties, and applications. Polymer **70**, A1–A16 (2015)
2. Tomanova, V., Srokova, I., Ebringerova, A., et. al.: Surface-active and associative properties of ionic polymeric surfactants based on carboxymethylcellulose. Polymer Engineering and Science. Wiley Online Library (2011). https://doi.org/10.1002/pen.22014.
3. Talaba, P., Srokova, I., Ebringerova, A., et al.: Cellulose-based biodegradable polymeric surfactants. J. Carbohydr. Chem. **16**(4 & 5), 573–582 (1997)
4. Huang, X., Liu, H., Shang, S., et al.: The equilibrium and dynamic surface tension of polymeric surfactants based on epoxidized soybean oil grafted hydroxyethyl cellulose. Royal Society of Chemistry Advances (2016). https://doi.org/10.1039/C6RA09769C.
5. Wei, Y., Cheng, F.: Synthesis and aggregates of cellulose-based hydrophobically associating polymer. Carbohyd. Polym. **68**, 734–739 (2007)
6. Hong, P., Fa, C., Wei, Y., et al.: Surface Properties and synthesis of the cellulose-based amphoteric polymeric surfactant. Carbohydrate Polymers (69), pp. 625–630 (2007). https://doi.org/10.1016/j.carbpol.2007.01.021
7. Sun, W., Sun, D., Wei, Y., et al.: Oil-in-water emulsions stabilized by hydrophobically modified hydroxyethyl cellulose: adsorption and thickening effect. J. Colloid Interface Sci. **311**, 228–236 (2007)
8. El-Sakhawy, M., Kamel, S., Salama, A., et al.: Preparation and infrared study of cellulose based amphiphilic materials. Cellulose Chem. Technol. **52**(3–4), 193–200 (2018)
9. Zhong, J., Chai, X., Fu, S.: Homogeneous grafting poly (methyl methacrylate) on cellulose by atom transfer radical polymerization. Carbohyd. Polym. **87**, 1869–1873 (2012)
10. Cao, Y., Li, H.: Interfacial activity of a novel family of polymeric surfactants. Eur. Polymer J. **38**, 1457–1463 (2002)
11. Bai, Y., Shang, X., Wang, Z., et al.: Experimental study on hydrophobically associating hydroxyethyl cellulose flooding system for enhanced oil recovery Energy & Fuels (32), pp. 6713–6725 (2018)
12. Wang, C., Liu, P., Wang, Y., et al.: Experimental study of key effect factors and simulation on oil displacement efficiency for a novel modified polymer BD-HMHEC. Sci. Rep. **8**, 3860 (2018)
13. Liu, P., Mu, Z., Wang, C., et al.: Experimental study of rheological properties and oil displacement efficiency in oilfields for a synthetic hydrophobically modified polymer. Sci. Rep. **7,** 8791 (2017)
14. Tomanova, V., Pielichowski, K., Srokova, I., et al.: Microwave-assisted synthesis of carboxymethyl cellulose-based polymeric surfactants. Polym. Bull. **60**, 15–25 (2008)
15. Huang, X., Liu, H., Shang, S., et al.: Preparation and characterization of polymeric surfactants based on epoxidized soybean oil grafted hydroxyethyl cellulose. J. Agric. Food Chem. **63**, 9062–9068 (2015)

16. Barakat, Y., Basily, I.K., Mohamad, A.I., et al.: Polymeric surfactants for enhanced oil recovery. part iii – interfacial tension features of ethoxylated alkylphenol-formaldehyde nonionic surfactants. Br. Polym. J. **21**, 459–465 (1989)

17. Barakat, Y., Gendy, T.S., Basily, I.K., et al.: Polymeric surfactants for enhanced oil recovery. Part II – The HLB-CMC Relationship of Ethoxylated Alkylphenol-Formaldehyde Polymeric Surfactants. British Polymer J. (21), 451–457 (1989)

18. Barakat, Y., Gendy, T.S., Mohamad, A.I., et al.: Polymeric surfactants for enhanced oil recovery. part i – critical micelle concentration of some ethoxylated alkylphenol-formaldehyde nonionic. British Polymer J. (21), 383–389 (1989)

19. Somasundaran, P., Markovic, B., Krishnakumar, S., et al.: Colloid systems and interfaces stability of dispersions through polymer and surfactant adsorption. In: Handbook of Surface and Colloid Chemistry, pp.559–601. CRC Press (1997)

20. Huang, X., Liu, H., Shang, S., et al.: Synthesis and characterization of castor oil-based polymeric surfactants. Front. Agr. Sci. Eng. **3**(1), 46–54 (2016)

21. Cao, Y., Li, H.: Synthesis of a novel family of polymeric surfactants with low interfacial tension by ultrasonic method. Polym. J. **31**(11–1), 920–923 (1999)

22. Carraher, C.E., Jr.: Polymer Chemistry (Sixth Edition). ISBN: 0–8247–0806–7 Marcel Dekker Inc. (2003)

23. Sun, J., Xu, X., Wang, J., et al.: Synthesis and emulsification properties of an amphiphilic polymer for enhanced oil recovery. J. Dispersion Sci. Technol. **31**(7), 931–935 (2010). https://doi.org/10.1080/01932690903224284

24. Rodriguez, F., Cohen, C., Ober, C.K., et al.: Principles of Polymer Systems – Sixth Edition. CRC Press (2015). ISBN: 978-1-4822-2379-8

Combined H_2O and CO_2 Reforming of CH_4 Over Ca Promoted Ni/Al_2O_3 Catalyst: Enhancement of Ni-CaO Interactions

Ahmad Salam Farooqi[1], Mohammad Yusuf[1,2], Muhammad Afiq Isyraf Ishak[1],
Noor Asmawati Mohd Zabidi[2,3], R. Saidur[4,5], Afrasyab Khan[6],
and Bawadi Abdullah[1,2(✉)]

[1] Chemical Engineering Department, Universiti Teknologi PETRONAS,
32610 Seri Iskandar, Malaysia
bawadi_abdullah@utp.edu.my
[2] Centre of Contaminant Control and Utilization (CenCoU), Institute of Contaminant
Management for Oil and Gas, Universiti Teknologi PETRONAS, 32610 Seri Iskandar, Malaysia
[3] Fundamental and Applied Sciences Department, Universiti Teknologi PETRONAS, 32610
Seri Iskandar, Malaysia
[4] Research Centre for Nano-Materials and Energy Technology (RCNMET), School of Science
and Technology, Sunway University, Jalan University, 47500 Bandar Sunway, Selangor, Darul
Ehsan, Malaysia
[5] Department of Engineering, Lancaster University, Lancaster LA1 4WY, UK
[6] Department of Hydraulics and Hydraulic and Pneumatic Systems, South Ural State University,
Lenin Prospect 76, Chelyabinsk 454080, Russian Federation

Abstract. Methane reforming with the application of combined H_2O and CO_2 is
appropriately known as bi-reforming of methane (BRM). The method is promising because it has potential to deliver an alternative energy source and is effective
in mitigating greenhouse gases (GHGs). In this study, Ni-based catalyst with 5
wt% supported Al_2O_3 and various Ca (2–4 wt%) loadings were prepared by following the wetness impregnation method. Characterization techniques, including
Brunauer-Emmett-Teller (BET) and X-ray diffraction (XRD), were applied to
observe prepared catalysts' structural morphology and physicochemical properties. Based on XRD analysis, the CaO-4wt% catalyst exhibited higher crystallinity
than the CaO-2wt% catalyst, enabling it to withstand greater forces and take longer
to deteriorate. To improve the catalyst activity, CaO was added as a promoter that
altered the contact between Ni and Al_2O_3 and modified the properties of the catalysts for an excellent activity. A fixed-bed continuous reactor with a feed ratio
$CH_4 : CO_2 : H_2O$ of 3:1:2 at 1073 K was used to observe the catalysts performance.
BRM catalytic reaction at 800 °C having Ni-4CaO/Al_2O_3 catalyst, proved greatly
in converting CH_4 and CO_2 with 79.6% and 87.2% respectively during the 6 h
reaction.

Keywords: Bi reforming · Catalyst development · Greenhouse gas · Syngas ·
Impregnation method

© The Author(s), under exclusive license to Springer Nature Singapore Pte Ltd. 2021
M. Awang and S. S. Emamian (Eds.): *Advances in Material Science and Engineering*, LNME, pp. 220–229, 2021.
https://doi.org/10.1007/978-981-16-3641-7_26

1 Introduction

The exponential increase in the demand for energy due to the growing world population is the major contributor to climate change and global warming. The extreme consumption of fossil fuels such as coal, oil, and natural gas releases enormous GHGs, including mainly CO_2 and CH_4 [1, 2]. For this concern, many researchers have placed extensive efforts to develop innovative and economical technologies such as CO_2 utilization to generate synthesis gas (syngas), a useful product [3, 4]. Many primary feedstocks can produce syngas such as natural gas, biomass, and coal, but natural gas is the cleanest and cheapest source [5]. Reforming processes have been proved as the most conventional and cheaper technologies to form syngas from methane in industrial applications [6–8]. Among various available reforming methods on syngas production, following have been found effective; steam reforming (SR), partial oxidation (PO), and dry reforming (DR), also known as CO_2 reforming [9–11]. The salient reaction in these methods are expressed as,

$$SR : CH_4 + H_2O \rightarrow CO + 3H_2 \ (\Delta H^o_{298K} = 206 \ kJ \ mol^{-1}) \tag{1}$$

$$PO : CH_4 + 1/2O_2 \rightarrow CO + 2H_2 \ (\Delta H^o_{298K} = -38 \ kJmol^{-1}) \tag{2}$$

$$DR : CH_4 + CO_2 \rightarrow 2CO + 2H_2 \ (\Delta H^o_{298K} = 248 \ kJ \ mol^{-1}) \tag{3}$$

In practice, the catalytic route being followed in SR reaction is the most common to form syngas. However, SRM faces drawbacks due to the higher H_2:CO ratio ≥ 3 and a vast quantity of CO_2 being formed. Also, since the SRM process is endothermic, there is a need to inject rigorous energy that makes the method excessively expensive [1, 12]. In the DRM case, both unwanted CH_4 and CO_2 are utilized to produce the valuable syngas, but the H_2:CO ratio of less than unity is unsuitable for Fischer Tropsch Process [13, 14]. Even though the POM possesses numerous advantages including considerably short residence time, generation of desirable H_2:CO ratio of 2. However, local hot spots on the catalytic bed due to its exothermicity are the major setbacks, with possible explosion risks, in adopting a safe industrial operation [15]. Recently, BRM, involving both SRM and DRM, gave impetus in raising the scientific and academic community's interest to accrue benefits while mitigating drawbacks associated with each of the three basic routes [16–19]. Bi reforming is an alternate operation to form syngas having a suitable H_2:CO ratio that prevents additional expenses needed to separate oxygen from the air and separate CO_2 from biogas [15]. Thus, it can be applied in various real-life industrial applications since it has economic, technical, environmental, and industrial benefits.

$$BR : 3CH_4 + CO_2 + 2H_2O \rightarrow 4CO + 8H_2 \ (\Delta H^o_{298K} = +712 \ kJ \ mol^{-1}) \tag{4}$$

Due to the high selectivity and catalytic activity as well as high resistance to Carbon formation, possessed by some of the Noble metals including Ru, Pd, Rh and Pt, they were found suitable catalysts for BRM reaction [20–25]. While Ni-based catalysts are enormously used in industrial reforming operations due to its ease in availability and being cheap, but they are liable to coke formation [26–28]. Thus, it is vital to explore a fresh catalyst system for BRM, which must oppose to carbon deposition by the realization of a suitable combination of support and promoter. In the present work, Ni-xCaO/Al$_2$O$_3$ (x = 0.2, 0.4 wt%) catalyst with various CaO loadings was investigated for the BRM reaction to assess the promoter's influence on the catalytic activity in view of conversion of methane and CO$_2$.

2 Methodology

2.1 Catalyst Preparation

The Ni-originated catalysts were synthesized by the wetness impregnation method [29] with a range of variation in materials and reaction conditions. The appropriate amount of Ca (NaO$_3$)$_2$·4H$_2$O powder was weighed accurately and subsequently dissolved in deionized water to obtain a homogenous solution. Afterward, an appropriate quantity of Ni(NO$_3$)$_2$·6H$_2$O was dissolved in deionized water to make Ni aqueous solution. Both Ca solution and Ni solution was added simultaneously in a beaker and magnetically stirred for mixing. Prior to that, Al(NO$_3$)$_3$.9H$_2$O powder was weighed, and the resulting Ca-Ni solution was added dropwise into the powder and stirred by using a magnetic stirrer for 1 h at room temperature. After that, the slurry mixture was placed into an oven for drying overnight at temperature 120 °C. Subsequently, the dried powder was calcined using a PROTHERM furnace at a temperature of 800 °C for 5 h with a 5 °C/min heating rate. Finally, the calcined powder was cooled and crushed using a sieve shaker to set the range of the particle between 50–125 μm for catalytic testing and characterizations. In this study, two kinds of CaO-Ni/Al$_2$O$_3$ catalysts, analogous to Ca loadings of 2 wt% and 4 wt%, and constant Ni loading of 5 wt%, were synthesized.

2.2 Characterization of Catalysts

X-ray diffraction (XRD) was used to obtain the spectra owing to the crystallinity of the prepared catalysts by incorporating X-ray source of Cu Kα radiation with a wavelength of 1.54 Å in the range of diffraction angle, 2θ = 3–80°. The voltage and current utilized were 30 kV and 15 mA, respectively, with a scan rate of 1°min^{-1} and a small stride size of 0.02° to ensure high pattern resolution during scanning. The Brunauer-Emmett-Teller (BET) (Micromeritics ASAP-2010 instrument) was utilized to examine the surface area and porosimetry of CaO-Ni/Al2O3 catalysts by using the measurement of N2 adsorption-desorption analysis acquired at 77K.

2.3 Catalyst Testing

A tubular flow reactor (TFR) was employed to perform the bi reforming of methane (BRM) reactions at atmospheric pressure. For catalytic activity testing, the catalyst with approximately 50–125 μm average particle size was weighed for about 0.1 g and sandwiched between quartz wool in the middle of the reactor. The reduction reaction was then carried out under a flow of 60 ml min^{-1} of 50% H$_2$/N$_2$ mixture at 800 °C for 1 h to activate the catalyst. Individual mass flow controllers were used to precisely control gaseous reactants' flow rates while the syringe pump was used to regulate the flow rate of water. Before entering the tubular flow reactor, gaseous reactants, including CH$_4$ and CO$_2$ being previously diluted with N$_2$, were mixed with steam. BRM reaction was conducted initially at 800 °C and with feed ratio CH$_4$:CO$_2$:H$_2$O of 3:1:2. Catalytic assessment can be calculated based on the conversion of the reactants, XCH$_4$ or XCO$_2$; yield, selectivity, and syngas ratio (H$_2$/CO). The respective equations are shown in Table 1.

Table 1. Assessment parameters with equations

Parameters	Equations
Conversion (X)	$X_{CH4}\,(\%) = \dfrac{F_{CH_4}^{in} - F_{CH_4}^{out}}{F_{CH_4}^{in}} \times 100\%$
	$X_{CO2}\,(\%) = \dfrac{F_{CO_2}^{in} - F_{CO_2}^{out}}{F_{CO_2}^{in}} \times 100\%$
Yield (Y)	$Y_{CO}\,(\%) = \dfrac{F_{CO}^{out}}{F_{CO_2}^{in} - F_{CH_2}^{in}} \times 100\%$
	$Y_{H2}\,(\%) = \dfrac{F_{CO}^{out}}{F_{CO_2}^{in} - F_{CH_4}^{in}} \times 100\%$
Selectivity (S)	$S_{CO}\,(\%) = \dfrac{F_{CO}^{out}}{\sum_{j=H2,CO} F_{CO}^{out}} \times 100\%$
	$S_{H2}\,(\%) = \dfrac{F_{H2}^{out}}{\sum_{j=H2,CO} F_{H2}^{out}} \times 100\%$
Syngas ratio (SR)	H$_2$/CO ratio $= \dfrac{F_{H2}^{out}}{F_{CO}^{out}}$

3   Results and Discussion

3.1 XRD Analysis

The XRD analysis was employed to identify the crystallinity of the prepared catalysts for the present study. Figure 1 shows the Ni/Al$_2$O$_3$ catalyst XRD patterns as a function of different wt% of Ca. The peak of 2-theta value at 46° indicates the crystallinity peak of NiO. These crystallinity peaks were enhanced in proportional to Ca's loading, which has increased the crystallinity of the catalyst [30]. Moreover, it can also be noticed of occurrence of two sharp crystallinity peaks of CaO at 2-theta = 37° and 68°. The intensity of crystallinity peaks of Ca was enhanced due to higher concentration of Ca in both catalysts while the peaks of NiAl$_2$O$_4$ that are located at 2-theta = 39° and 45°, decrease concurrently [29]. Thus, it can be concluded that as the amount of Ca loading increases, the crystallinity of the catalysts increases which makes the catalyst more rigid while the intensity of alumina reduces. Furthermore, the Ni peak shifted slightly to a lower 2-theta value with Ni-CaO-4wt% catalyst compared to the other catalyst. Therefore, it can be stated that the possible interactions between the Ni and Ca in both catalysts, can affect the location and intensity of the peaks.

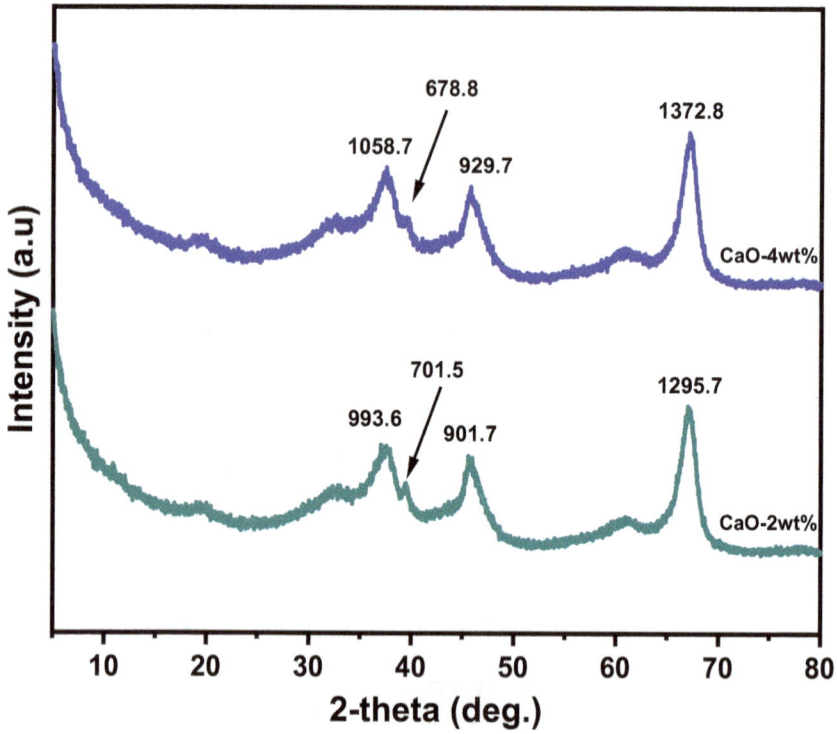

Fig. 1. XRD patterns of synthesized catalysts

3.2 BET Analysis

Figure 2 displays the patterns of N_2 adsorption-desorption isotherms for the prepared catalysts based on BET analysis. Both catalysts demonstrated similar adsorption-desorption patterns where all the resulting isotherms are associated with type IV isotherms, including hysteresis loops of H1 shaped for all the catalysts. Hence, these isotherms are significant indications for characteristics of mesoporous (i.e. 2–50 nm) materials (2–50 nm), which follow the IUPAC [31] norm.

Table 2 presents the values for surface area, pore-volume, and pore size of Ni based catalysts, $2CaO-Ni/Al_2O_3$ and $4CaO-Ni/Al_2O_3$. Out of these two catalysts, 4CaO-Ni/Al_2O_3 demonstrates greater surface area and pore volume than the 2 wt% Ca. However, the pore size of both catalysts remains the same, which is around 8.6 nm. This indicates that the higher Ca loading will result in a greater surface area as well as the pore volume of the catalyst.

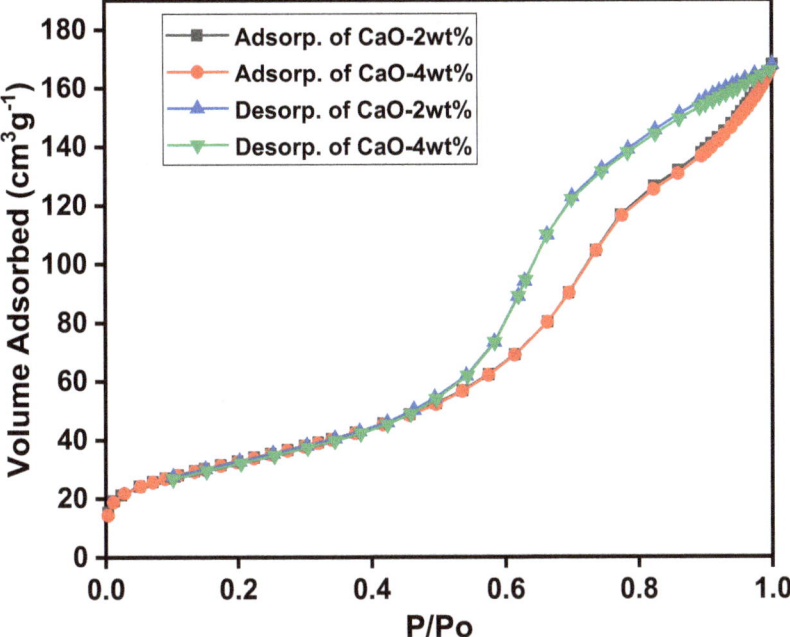

Fig. 2. N$_2$ adsorption-desorption isotherms of CaO-2wt% and CaO-4wt% catalysts

Table 2. Textural properties of the synthesized catalysts

Catalyst name	Surface area (m^2/g)	Pore volume (cm^3/g)	Pore size (nm)
2CaO-Ni/Al$_2$O$_3$	115.4	0.25	8.6
4CaO-Ni/Al$_2$O$_3$	116.2	0.26	8.6

3.3 Catalytic Bi Reforming Reaction Evaluation

The evaluation of the catalyst performance of synthesized catalysts has been carried out for 6 h time on stream in BRM reaction as shown in Fig. 3. From the results obtained, both catalysts showed good stability in the stated test period where the expected conversion of the CH$_4$ and CO$_2$ reactants was achieved around 50–80%. The highest conversion of methane and carbon dioxide is 82% and 74% respectively for Ni-4CaO/Al2O3 catalyst as shown in Figs. 3a and 3b. Whereas the highest conversion of methane and CO$_2$ is 71% and 87% respectively, for Ni-4CaO/Al$_2$O$_3$. Higher CO$_2$ conversions are often recognized to the rise in reverse water gas shift reaction (i.e. CO$_2$ + H$_2$ → CO + H$_2$O) [32]. Sengupta et al. (2015) demonstrated that Ni catalysts' catalytic activity supported on Al$_2$O$_3$ improved with addion of MgO and CaO promoters [33]. As shown in Fig. 3c, the results have classified on the basis of the molar ratio between H$_2$ and CO, whereas the ratio was lower than 2 for Ni-2CaO/Al$_2$O$_3$ (1.72) and Ni-2CaO/Al$_2$O$_3$ (1.81).

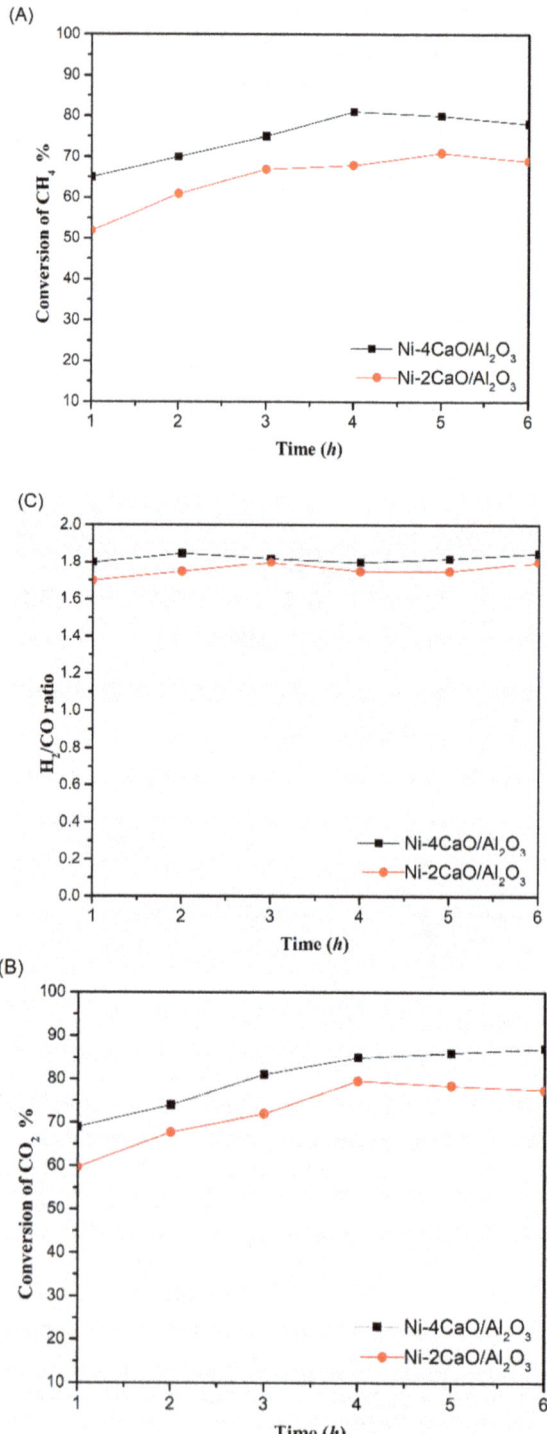

Fig. 3. Performance of the synthesized catalysts, (a) CH_4 conversion, (b) CO_2 conversion and (c) H_2:CO ratio at 800 °C, feed composition CH_4 : CO_2 : $H_2O = 3$:1:2 and feed flow rate of 60 ml/min.

4 Conclusions

The Ni-based catalysts incorporated with various Ca loading were successfully synthesized by application of wetness impregnation technique with reinforcement of Al2O3 as support. From the characterization analysis of XRD and BET, the properties including structural morphology, surface area, and porosimetry of the prepared catalysts were determined. CaO-4wt% catalyst demonstrates better characteristics than the CaO-2wt% catalyst due to the higher loading of Ca. Moreover, by performing a BRM reaction, both respective catalysts have achieved the desired conversion. Ni-4CaO/Al_2O_3 catalyst demonstrated the maximum activity providing favorable CH_4 and CO_2 transformation rates of 79.6% and 87.2%, respectively, at a suitable H_2/CO ratio of 1.81 at the temperature of 1073 K with a ratio CH_4: CO_2: H_2O of 3:1:2. Further benefits include the deposition of the active metal content on the catalyst's surface, which proves to enhance the reducibility and the basicity of the catalyst.

Acknowledgments. The authors would like to thank the Ministry of Education (MOE), Malaysia for providing financial assistance under FRGS/1/2018/TK02/UTP/02/10 and Universiti Teknologi PETRONAS for providing the required facilities to conduct this research work.

References

1. Farooqi, A.S., Al-Swai, B.M., Ruslan, F.H., et al.: Catalytic conversion of greenhouse gases (CO_2 and CH_4) to syngas over Ni-based catalyst: effects of Ce-La promoters. Arab. J. Chem. **13**, 5740–5749 (2020)
2. Liao, C.-H., Horng, R.-F.: Experimental study of syngas production from methane dry reforming with heat recovery strategy. Int. J. Hydrogen Energy **42**, 25213–25224 (2017)
3. Zhang, T., Liu, Z., Zhu, Y.A., et al.: Dry reforming of methane on Ni-Fe-MgO catalysts: Influence of Fe on carbon-resistant property and kinetics. Appl. Catal B Environ. **264**, 18497 (2020)
4. Wei, T., Jia, L., Luo, J.L., et al.: CO_2 dry reforming of CH_4 with Sr and Ni co-doped LaCrO3 perovskite catalysts. Appl. Surf. Sci. **506**, 144699 (2020)
5. Guczi, L., Stefler, G., Geszti, O., et al.: Methane dry reforming with CO_2: a study on surface carbon species. Appl. Catal. A Gen. **375**, 236–246 (2010)
6. Feng, J., Ding, Y., Guo, Y., et al.: Calcination temperature effect on the adsorption and hydrogenated dissociation of CO_2 over the NiO/MgO catalyst. Fuel **109**, 110–115 (2013)
7. Rajaeiyan, A., Bagheri-Mohagheghi, M.M.: Comparison of sol-gel and co-precipitation methods on the structural properties and phase transformation of γ and α-Al_2O_3 nanoparticles. Adv. Manuf. **1**, 176–182 (2013)
8. Bhavani, A.G., Kim, W.Y., Kim, J.Y., et al.: Improved activity and coke resistance by promoters of nanosized trimetallic catalysts for autothermal carbon dioxide reforming of methane. Appl. Catal. A Gen. **450**, 63–72 (2013)
9. Yusuf, M., Farooqi, A.S., Keong, L.K., et al.: Contemporary trends in composite Ni-based catalysts for CO_2 reforming of methane. Chem. Eng. Sci. **229**, 116072 (2021)
10. Selvarajah, K., Phuc, N.H.H., Abdullah, B., et al.: Syngas production from methane dry reforming over Ni/Al_2O_3 catalyst. Res. Chem. Intermed. **42**, 269–288 (2016)
11. Yang, N., Wang, R.: Sustainable technologies for the reclamation of greenhouse gas CO_2. J. Clean Prod. **103**, 784–792 (2015)

12. Usman, M., Wan Daud, W.M.A., Abbas, H.F.: Dry reforming of methane: influence of process parameters-a review. Renew. Sustain. Energy Rev. **45**, 710–744 (2015)

13. Olah, G.A., Goeppert, A., Czaun, M., et al.: Single Step bi-reforming and oxidative bi-reforming of methane (Natural Gas) with steam and carbon dioxide to metgas (CO-2H$_2$) for methanol synthesis: self-sufficient effective and exclusive oxygenation of methane to methanol with oxygen. J. Am. Chem. Soc. **137**, 8720–8729 (2015)

14. Siang, T.J., Pham, T.L.M., Van, C.N., et al.: Combined steam and CO$_2$ reforming of methane for syngas production over carbon-resistant boron-promoted Ni/SBA-15 catalysts. Microporous Mesoporous Mater **262**, 122–132 (2018)

15. Olah, G.A., Goeppert, A., Czaun, M., et al.: Bi-reforming of methane from any source with steam and carbon dioxide exclusively to metgas (CO-2H$_2$) for methanol and hydrocarbon synthesis. J. Am. Chem. Soc. **135**, 648–650 (2013)

16. Álvarez, M.A., Centeno, M.Á., Odriozola, J.A.: Ru-Ni catalyst in the combined dry-steam reforming of methane: the importance in the metal order addition. Top Catal. **59**, 303–313 (2016)

17. Stroud, T., Smith, T.J., Le Saché, E., et al.: Chemical CO$_2$ recycling via dry and bi reforming of methane using Ni-Sn/Al$_2$O$_3$ and Ni-Sn/CeO$_2$-Al$_2$O$_3$ catalysts. Appl. Catal. B Environ. **224**, 125–135 (2018)

18. Zhao, Z., Ren, P., Li, W., et al.: Effect of mineralizers for preparing ZrO$_2$ support on the supported Ni catalyst for steam-CO$_2$ bi-reforming of methane. Int. J. Hydrogen Energy **42**, 6598–6609 (2017)

19. Santos, B.A.V., Loureiro, J.M., Ribeiro, A.M., et al.: Methanol production by bi-reforming. Can. J. Chem. Eng. **93**, 510–526 (2015)

20. Gangadharan, P., Kanchi, K.C., Lou, H.H.: Evaluation of the economic and environmental impact of combining dry reforming with steam reforming of methane. Chem. Eng. Res. Des. **90**, 1956–1968 (2012)

21. Özkara-Aydnolu, Ş., et al.: Thermodynamic equilibrium analysis of combined carbon dioxide reforming with steam reforming of methane to synthesis gas. Int. J. Hydrogen Energy **35**, 12821–12828 (2010)

22. He, S., Wu, H., Yu, W., et al.: Combination of CO$_2$ reforming and partial oxidation of methane to produce syngas over Ni/SiO$_2$ and Ni-Al$_2$O3/SiO$_2$ catalysts with different precursors. Int. J. Hydrogen Energy **34**, 839–843 (2009)

23. Safariamin, M., Tidahy, L.H., Abi-Aad, E., et al.: Dry reforming of methane in the presence of ruthenium-based catalysts. Comptes Rendus. Chim. **12**, 748–753 (2009)

24. Fu, Z.,, Yang, Q., Liu, Z., et al.: Photocatalytic conversion of carbon dioxide: from products to design the catalysts. J. CO$_2$ Util **34**, 63–73 (2019)

25. Mondal, U., Yadav, G.D.: Perspective of dimethyl ether as fuel: Part II- analysis of reactor systems and industrial processes. J. CO$_2$ Util **32**, 321–338 (2019)

26. Guo, J., Lou, H., Zhao, H., et al.: Dry reforming of methane over nickel catalysts supported on magnesium aluminate spinels. Appl Catal A Gen **273**, 75–82 (2004)

27. Montoya, J.A., Romero-Pascual, E., Gimon, C., et al.: Methane reforming with CO$_2$ over Ni/ZrO$_2$-CeO$_2$ catalysts prepared by sol-gel. Catal Today **63**, 71–85 (2000)

28. Hally, W., Bitter, J.H., Seshan, K., et al.: Problem of coke formation on Ni/ZrO$_2$ catalysts during the carbon dioxide reforming of methane. Stud. Surf. Sci. Catal. **88**, 167–173 (1994)

29. Siew, K.W., Lee, H.C., Gimbun, J., et al.: Characterization of La-promoted Ni/Al$_2$O$_3$ catalysts for hydrogen production from glycerol dry reforming. J. Energy Chem. **23**, 15–21 (2014)

30. Quincoces, C.E., Dicundo, S., Alvarez, A.M., et al.: Effect of addition of CaO on Ni/Al$_2$O$_3$ catalysts over CO$_2$ reforming of methane. Mater Lett. **50**, 21–27 (2001)

31. Donohue, M.D., Aranovich, G.L., et al.: Classification of Gibbs adsorption isotherms. Adv. Colloid Interface Sci. **76–77**, 137–152 (1998)

32. Fan, M.-S., Abdullah, A.Z., Bhatia, S.: Catalytic technology for carbon dioxide reforming of methane to synthesis gas. ChemCatChem **1**, 192–208 (2009)
33. Sengupta, S., Deo, G.: Modifying alumina with CaO or MgO in supported Ni and Ni–Co catalysts and its effect on dry reforming of CH_4. J. CO_2 Util **10,** 67–77 (2015)

Investigation on Bending Strength Between Different Designs of Pedicle Screw to Obtain an Optimum Design

Rosdi Daud[1]([✉]), M. Izuddin[1], H. Mas Ayu[2], and A. Shah[3]

[1] Faculty of Mechanical and Automotive Engineering Technology,
Universiti Malaysia Pahang, 26600 Pekan, Pahang, Malaysia
rosdidaud@ump.edu.my
[2] Faculty of Manufacturing and Mechatronic Engineering Technology,
Universiti Malaysia Pahang, 26600 Pekan, Pahang, Malaysia
[3] Faculty of Technical and Vocational,
Universiti Pendidikan Sultan Idris, 35900 Kuala Kubu Bharu, Perak, Malaysia

Abstract. Pedicle screw is made from titanium alloy which gives support in the treatment of specific condition like lumbar, acute and chronic instabilities or deformities of thoracic, sacral spine. Mostly, the screw comes with rods act as connector, nuts, sleeves and plates. As a system, it increases the rigidity and holds the bone in placed while the bone heals itself over time. Over the years, many researchers have studied the pullout strength and the bending strength of pedicle screw. They found out that factors like core design, thread design, geometry and insertion techniques influenced the strength in pullout and bending. Failure of a screw is unavoidable. Sometime breakage happened at the neck or at the radial hole of pedicle screw due to high local stress. Thus, this project focused on the finding the most optimize diameter and position of radial in cylindrical screw to increase its bending strength. The pedicle screw inserted 40 mm into cylindrical block as a replacement for real human bone. It was then validated using Ansys software after gone through several simulation until the error become less than 10%. As a validated model, it was modified into pedicle screw with 2 radial holes varied diameter of 1.0 mm, 1.5 mm and 2.0 mm at 3 different position of radial hole. It was given bending moment of 1.29 Nm for starters and increase 0.3 Nm until the pedicle screw reached the yield stress at 790 Mpa. The head fixed at all directions and bending moment applied at tip of pedicle screw. As a result, high stress mostly distributed around the radial holes. All of the proposed designs able to withstand bending moment of 1.29 Nm–2.79 Nm while maintaining its original shape. Unlike the original pedicle screw, it can only resist bending moment at best 2.03 Nm before it broke. Finally, the most optimized diameter and position of radial hole was 1.0 mm at parallel positioned with improvement of 10.75% and 21.51% better than 1.5 mm and 2.0 mm respectively despite any position of radial hole.

1 Introduction

Osteoporosis is a common disease characterized by loss in bone mass and density which can easily cause to fractures in population of older ages people, and the after effect of

M. Awang and S. S. Emamian (Eds.): *Advances in Material Science and Engineering*, LNME, pp. 230–240, 2021.
https://doi.org/10.1007/978-981-16-3641-7_27

spinal surgeries on osteoporosis-related fractures also increased [1]. The normal bone mineral density is between -1 to -2.5 and below than that will have the risk of osteoporosis. Research has showed that the possibility to get bone fractures has increase by 1.5 to 3 times as the standard deviation of bone mineral density (BMD) decreases [2]. Osteoporosis in elderly patients are liable to progressive spinal deformities and potential neurologic, these conditions is going to be a concern before performing spine surgery [3]. Many orthopaedic surgeons still not prefer to perform surgical treatment to osteoporosis condition. Patients with the condition are hard to treat surgically because of late age and side effects of anaesthesia, making them poor surgical candidates [3, 4]. Furthermore, osteoporosis is a risky spine surgery, especially complications with bone graft fusions and device implementation.

When the lowest mass density locates at the fracture area, normally pedicle screws unable to provide enough fixation power between the bone and screw interface, causing to greater risk of loosening or breakage [5, 6]. The pedicle screw spinal system is commonly applied to treat patient with condition of lumbar instability disorders, though the implementation of the system in cancellous bone are risked to screw loosening. There were several cases that frequently observed on the failure rate of pedicle screw relatable to that instrumentation [7]. In osteoporotic patients, the composition of the bone become abnormal, thus increase the risk of fixation failure. Past researchers reviewed the post operation cases observed in a group total of 38 patients above the age of 65 years old who had undergone five-level fusions [8]. Pedicle fractures and compression fractures present in early complications and the most common late complications were pseudarthrosis with device failure and adjacent-level disc degeneration with herniation. Low fixation strength in the cancellous bone of patients with low bone mass density leads to an increased complications of device failure.

Osteoporosis also lower the opportunity to a successful fusion in these patients. As a result, device that is supposedly to provide temporary support experiences more prolonged stress loading, likely to suffer from delayed fusion and pseudarthrosis. In the case of posterior fixation with screws, device is prone to screw pull-out and loosening. Anterior instrumentation suffers failure to continuous repeated loading, causing to fatigue failure and left broken screw into the osteoporotic bone [9, 10]. There are a lot of innovative techniques and devices have been created to avoid the screw looseness such as increasing the strength of vertebrae, optimizing screw trajectory, balancing the outer and core diameter ratio of threaded shaft [11], thread profile design (12,13), or adding anchoring mechanism of screws [14]. Other methods to prevent loosening, cement-augmentation has been implemented in cannulated pedicle screw by injecting the PMMA into the osteoporotic vertebrae to increase the area of interface between the bone and screw. Cannulated pedicle screw used for solely purpose in cement-augmentation process by secreting poly methyl methacrylate at distal radial holes in pedicle. After poly methyl methacrylate injected through radial holes, it solidified between the interfaces of screw-bone thus integrate between the threaded shaft and the osteoporotic bone in the vertebral body. As a result, cement gives instant boost of strength and stiffness, and most importantly able to increase pullout strength in cancellous bone compared with cancellous bone without the augmentation method. Initial clinical follow up results tremendous achievement of this method. Even Chao et al. [15] has demonstrated on his osteoporotic cadaveric thoracolumbar model that by using this technique the effect on pullout strength of the screw is triple than normal.

Past studies have proved that poly methyl methacrylate augmentation increases average stiffness, strain energy and initial fixation and fatigue strength of pedicle screws [16, 17]. However, after applying the technique, it is difficult to retrieve the broken screw in revision surgery, and the patient with porous bone had a higher risk of cement leakage from the vertebral body when using cannulated pedicle screw. Furthermore, the strength of cement-augmented screw often deteriorates after experiencing the cyclic loading especially for the osteoporotic patient [18]. There was researcher [19] study diameter, quantity and location of radial holes that are perpendicular to threaded shaft. The investigation was informative as researcher able to learn the location of radial holes. By learning to control these variabilities, pedicle screw has room for improvement. Thus, the aim of this research is to investigate the bending strength of cannulated pedicle screw between different designs of pedicle screw to obtain optimum design.

2 Methods

The research starts with construction of reference model of pedicle screw. The model constructed using a SolidWorks version of 2018 software. The model converted into format Parasolid x_t which then transfer into Ansys Workbench version 18.1 as a finite element model. The model simulated to obtain values such as total strain energy, maximum tensile stress, von misses stress (VMS) and maximum deflection. These values were verified by experimental values of past research. The purpose was to gain accuracy of our model compared to the real experiment.

Next a proposed design was constructed. The model was constructed with different diameter of radial hole and simulated using Ansys. The model then compared to reference model in term of their VMS values. The model needs to be reconstructed until VMS value of proposed design better than VMS value of reference model. The design with optimized diameter of radial hole was only accepted if only the VMS value got better. Then, proposed design with different position of radial hole is constructed. The position of radial hole adjusted many times to obtain optimized position. Again, the model simulated with Ansys and obtain the VMS value. After compared to reference model and obtained good result of VMS value, then the proposed design with different position of radial hole was optimized.

Finally, both optimized proposed designs with optimized diameter and position of radial hole combined into one model. The model simulated using Ansys for VMS value. The model was compared to reference model for the last time. The model reconstructed when VMS was not better and accepted when the value got better. This means that both combined designs were better than the reference model. All the data was recorded and prepared to be present. Four models were created for analysis. One model was a reference model, second model was cannulated pedicle screw with radial holes positioned parallel to each other, pedicle screw with radial hole at 90° and pedicle screw with radial holes opposite to each other. All three models were created using SolidWorks 2018. The geometry and dimensions given in Table 1 below.

Table 1. Geometry of pedicle screw

Geometry	Dimension (mm)
Length	45
Inner diameter	4.9
Outer diameter	6.5
Pitch	2.8
Proximal root radius	0.8
Distal root radius	1.2
Proximal half angle	14
Distal half angle	25
Thread width	0.2
Cannulated diameter	2

The diameter of radial hole is 1.0 mm, 1.5 mm and 2.0 mm. The models were designed like in the Fig. 1 below is to assist cement augmentation. The cement will be injected through the cannulated hole and exited through the radial holes. The cement then solidified to hold the pedicle screw into the position. The models were designed for solely purpose on cancellous bone.

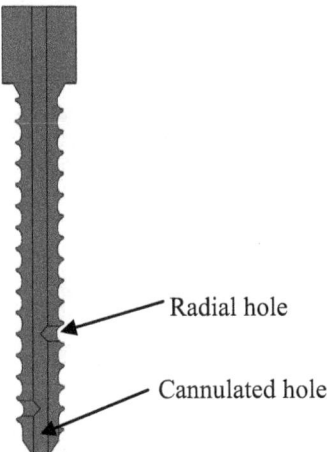

Fig. 1. Cannulated pedicle screw with 2 radial holes

In order to mimic the actual condition, the assemblies consisted of screws and cylinder block to represent bone. When all the models were constructed, the model were ready to assembly. The pedicle screw will mate together with cylinder block to replicate the condition of the real bone. The material properties of the pedicle screw and cylindrical blocks were set to match the commercial screw and bone respectively (Table 2).

The pedicle screw and the block were made from titanium alloy (Ti-Al-4V) and high-molecular-weighted polyethylene, respectively. All material properties were considered to each part in order to perform FEA and calculate the von Misses stress distribution.

Table 2. Mechanical properties of components

Components	Materials	Modulus of elasticity	Poisson's ratio
Pedicle screw	Titanium alloy (Ti-Al-4V)	114 Gpa	0.3
Block	Polyethylene	2.6 Gpa	0.3

In order to evaluate the performance of the pedicle screw, proposed design used in this process. One model was cannulated pedicle screw with radial holes positioned parallel to each other, pedicle screw with radial hole at 90° and pedicle screw with radial holes opposite to each other. The proposed design has similar diameter of radial holes varied from 1.0 mm–2.0 mm. The proposed designs were then compared to a journal from [19]. H.C. Chen et al. has proposed 3 types of commercially available cannulated pedicle screws. Each of pedicle screw has different number of radial holes. The S4: 4 holes, S9: 9 holes, S12: 12 holes (Fig. 2).

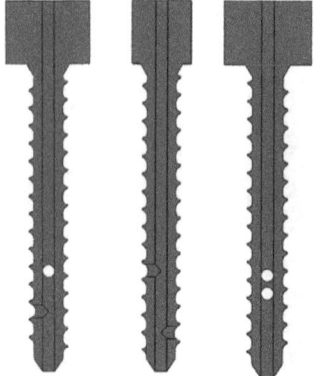

Fig. 2. Positions of radial holes

During the simulation, polyethylene bone was being excluded. Only the pedicle screw was remained. The Fig. 3 below shows the head of the screw fixed in all directions. The tip was given a moment of 1.29 Nm in clockwise direction. In the next simulation, the moment increased 0.3 Nm until the pedicle screw reach the yield stress.

Fig. 3. Bending moment of pedicle screw

3   Results and Discussion

Before any performance comparison has been done, the model must be validated first. The Fig. 4 below shows the stress distribution in a pedicle screw when a load of 220N applied on the surface of cylinder bone. As a result, the maximum stress corresponding to the point of failure was 880.12 Mpa. The point of failure tends to happen around the neck of the pedicle screw. This has been also proved by researcher [20, 21] experimentally and using FEA.

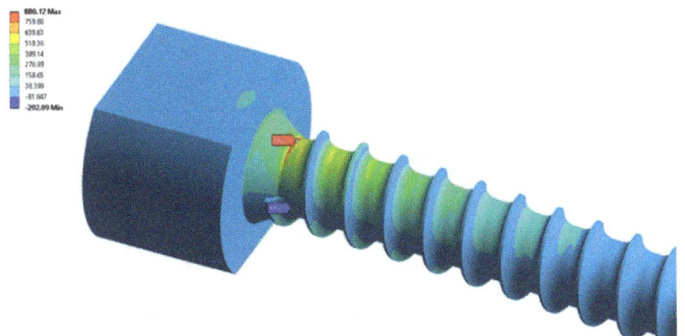

Fig. 4. Principle stress

Based on the journal that have the same geometry as this study, their result shows 891 Mpa of stress at the exact location of failure. The percentage error between the journal's result and this study result is around 1.22%. The calculation was using the formula below:

$$\frac{current\ result - journal\ result}{journal\ result} \ x\ 100\%$$

In order to get a good validated model. The percentage error should be less than 10%. Exceeding the suggested range of percentage error could result inaccurate data. Since the model has 1.22% of error, this model is acceptable to be used for further improvisation or modification on the design of the pedicle screw.

3.1   Finite Element Analysis Result

In the finite element analysis, all variants of 90° radial hole pedicle screw starts with given bending moment of 1.29 Nm at the tip of the screw. The Fig. 5 shows 2.0 mm diameter has the highest stress at the beginning compared to 1.0 mm and 1.5 mm. The colored line represents the diameter of radial hole. When the given moment increased, the stress also increased linearly. Until all of the model reached maximum moment, 1.0 mm can withstand 2.79 Nm of moment with stress of 765.94 Mpa. Meanwhile, 1.5 mm and 2.0 mm can only withstand moment of 2.49 Nm and 1.89 Nm respectively. Eventually all of the model will break once it reached the yield stress of titanium alloy which is 790 Mpa. But the best model is the pedicle with 1.0 mm diameter radial hole as it has the highest resistance to the given moment.

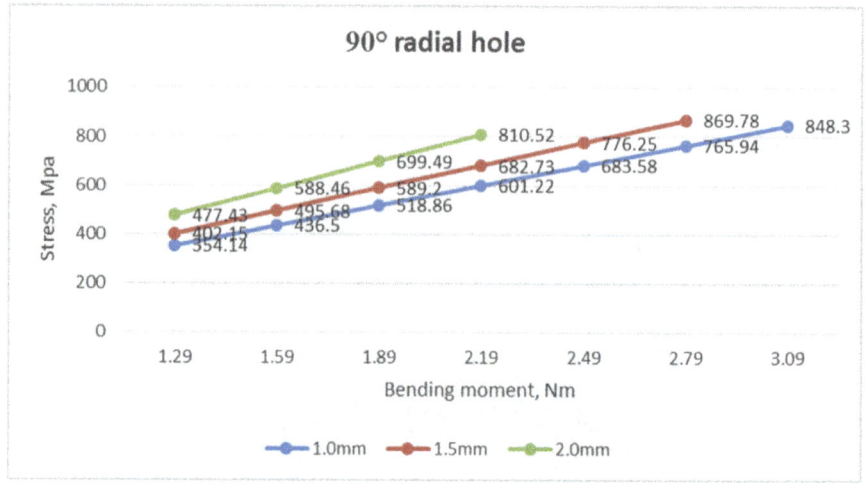

Fig. 5. Line graph for 90° radial hole position

In parallel radial hole model, the line increased rapidly at the beginning and stop after exceeding the yield stress of titanium alloy (Fig. 6). The pattern almost the same as the previous model but the stress is rather not the same. The model with 1.0 mm diameter radial hole achieves 738.29 Mpa stress at 2.79 Nm of bending moment. The 1.5 mm breaks at 2.79 Nm which has exceeded the yield stress by 47.64 Mpa. The 2.0 mm has the lowest bending resistance which break at 2.49 Nm.

Based on the data, the stress in the pedicle screw did not show significant differences when changed the position of radial hole. Nonetheless, the diameter differences show big improvement on the pedicle screw. So, among all the model the optimum model that has the lowest stress distributed on the body was pedicle screw with parallel position and radial hole of 1.0 mm diameter.

3.2 Comparison Between Current Design and Proposed Design of Pedicle Screw Performance

When we compared between pedicle screws from H.C. Chen et al. with the proposed design that has 1.0 mm diameter radial hole at parallel position, we able to evaluate the performance of the pedicle screws. Based on the bar chart above, proposed design took

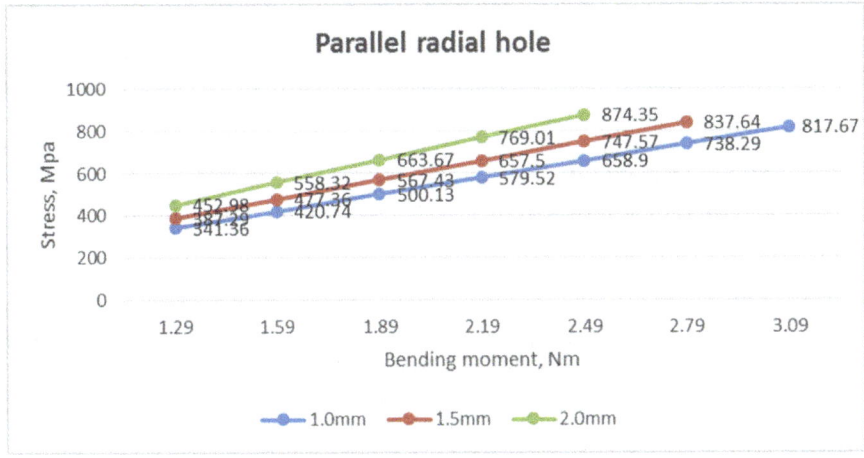

Fig. 6. Line graph for parallel radial hole position

the lead in term of resistance of bending moment. Not only it has the lowest stress but also able to withstand the highest bending moment of 3.09 Nm. Among all of the others, proposed design eventually will breaks, but it can withstand higher than the S4, S9, and S12. The proposed design has more improvement than the S12. The reason was, the S12 reached the limit of 790 Mpa first at 1.64 Nm. Theoretically in real situation, when the S12 were given bending moment, it will break first compared to others. This means the others can withstands bending moment a lot higher before it reach the limit (Fig. 7).

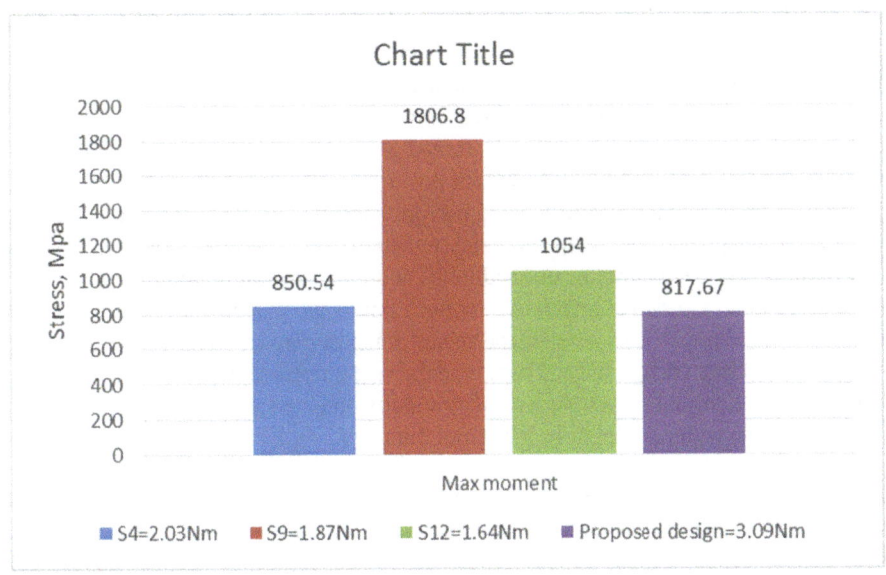

Fig. 7. Comparison of Von Mises Stress result between commercial and proposed design

It can be concluded that the existence of cement augmented screws help increases the pullout strength and may help enhance the bending strength. Even with augmented screw, the risk of cement leakage can still happen. In order to overcome cement leakage, radial holes are placed around the distal 1/3 of the pedicle screw. From the finite element analysis in this study, when Von Misses stress reached 790 Mpa under bending condition, the maximum bending moment of 1.0 mm was 10.75% and 21.51% better than 1.5 mm and 2.0 mm respectively despite any position of radial hole. During the bending condition, screw breakage always happened around the radial holes in all position of radial hole, which correspond to previous journal [19]. All of the pedicle screw can withstand any ranges of bending moment from 1.29 Nm–2.79 Nm but not exceed 3.09 Nm. Nonetheless, there was improvement from the previous journal. The number of radial hole does influenced the failure as high density of radial holes can result high concentration and increase the risk of pedicle screw breakage. Beside this study proof that the proposed design are potentially improve the pedicle screw performance, but several limitations were discovered when using the finite element method:-

- Finite element maybe not all accurate to characterize the model in real situation. It just gives us the expectation result for us to assume the condition in very short of time.
- The real shape of spinal vertebrae was being simplified to cylindrical shape. The vertebrae are not cylindrical actually. Because of the simplification, the result could be not accurate and hard to predict the real condition and biomechanical performances.

4   Conclusions

In conclusion, cannulated pedicle screw with radial holes is specially designed to help patient with osteoporosis condition. Unfortunately, this screw had face failures such as loosening and breakage. A lot of breakage happens around the radial hole. Thus, this study focuses on bending strength on cylindrical cannulated pedicle screw with radial hole. Many design improvements had been introduced by other researchers to increase the bending strength such as core design, thread geometry, and thread root radius. Many of them had achieved their objectives but the position and diameter of radial hole still be a question. In order to improve the bending performance of cylindrical cannulated pedicle screw, three design had been proposed to be studied in detail. The design is pedicle screw with parallel hole, 90 degree positioned holes and opposite positioned radial hole. All of the designs have varied diameter of 1.0 mm, 1.5 mm and 2.0 mm. All of the proposed design will be compared to a reference model for purposes of signification in design changes in bending performance. These models were constructed and analyze using the help of SolidWorks and Ansys. As a result, the study obtained an optimum design which is pedicle screw with parallel hole at 1.0 mm diameter.

Acknowledgments. We would like to thank Universiti Malaysia Pahang through research grant RDU180395 for fully support the facilities and resources for this study. The authors also would like to acknowledge research grant FRGS/1/2018/TK03/UMP/02/5 which provided by Ministry of Higher Education Malaysia for support the resources for this study.

References

1. Hirano, T., Hasegawa, K., Washio, T., Hara, T., Takahashi, H.: Fracture risk during pedicle screw insertion in osteoporotic spine. J. Spinal Disord. **11**(6), 493–497 (1998)
2. Kanis, J.A., Melton, L.J., 3rd., Christiansen, C., Johnston, C.C., Khaltaev, N.: The diagnosis of osteoporosis. J. Bone Miner. Res. **9**(8), 1137–1141 (1994)
3. Ponnusamy, K.E., Iyer, S., Gupta, G., Khanna, A.J.: Instrumentation of the osteoporotic spine: biomechanical and clinical considerations. Spine J. **11**(1), 54–63 (2011)
4. Hadjipavlou, A.G., Katonis, P.G., Tzermiadianos, M.N., Tsoukas, G.M., Sapkas, G.: Principles of management of osteometabolic disorders affecting the aging spine. Eur. Spine J. **12**(Suppl 2), S113–S131 (2003)
5. Chang, M.C., Liu, C.L., Chen, T.H.: Polymethylmethacrylate augmentation of pedicle screw for osteoporotic spinal surgery: a novel technique. Spine (Phila Pa 1976) **33**(10), E317–24 (2008)
6. Halvorson, T.L., Kelley, L.A., Thomas, K.A., Whitecloud, T.S., Cook, S.D.: Effects of bone mineral density on pedicle screw fixation. Spine **19**, 2415–2420 (1994)
7. Esses, S.I., Botsford, D.J., Huler, R.J., Rauschning, W.: Surgical anatomy of the sacrum: a guide for rational screw fixation. Spine (Phila Pa 1976) **16**(6s), S283–288 (1991)
8. De Wald, C.J., Stanley, T.: Instrumentation-related complications of multilevel fusions for adult spinal deformity patients over age 65 surgical considerations and treatment options in patients with poor bone quality. Spine **31**(Suppl 19), S144–S151 (2006)
9. Andersen, T., Christensen, F.B., Niedermann, B., Helmig, P., Høy, K., Hansen, E.S., et al.: Impact of instrumentation in lumbar spinal fusion in elderly patients: 71 patients followed for 2–7 years. Acta Orthop. **80**(4), 445–450 (2009)
10. Butler, T.E., Jr, Asher, M.A., Jayaraman, G., Nunley, P.D., Robinson, R.G.: The strength and stiffness of thoracic implant anchors in osteoporotic spines. Spine (Phila Pa 1976) **19**(17), 1956–1962 (1994)
11. Kim, Y.Y., Choi, W.S., Rhyu, K.W.: Assessment of pedicle screw pullout strength based on various screw designs and bone densities - an ex vivo biomechanical study. Spine J. **12**(2), 164–168 (2012)
12. Amaritsakul, Y., Chao, C.K., Lin, J.: Biomechanical evaluation of bending strength of spinal pedicle screws, including cylindrical, conical, dual core and double dual core designs using numerical simulations and mechanical tests. Med. Eng. Phys. **36**(9), 1218–1223 (2014)
13. Patel, P.S., Shepherd, D.E., Hukins, D.W.: The effect of screw insertion angle and thread type on the pullout strength of bone screws in normal and osteoporotic cancellous bone models. Med. Eng. Phys. **32**(8), 822–828 (2010)
14. Shea, T.M., et al.: Designs and Techniques That Improve the Pullout Strength of Pedicle Screws in Osteoporotic Vertebrae: Current Status. BioMed Research International (2014)
15. Chao, C.K., Lin, J., Putra, S.T., Hsu, C.C.: A neurogenetic approach to a multi objective design optimization of spinal pedicle screws. J. Biomech. Eng-T ASME **132**(9), 091006 (2010)
16. Burval, D.J., McLain, R.F., Milks, R., Inceoglu, S.: Primary pedicle screw augmentation in osteoporotic lumbar vertebrae: biomechanical analysis of pedicle fixation strength. Spine **32**(10), 1077–1083 (2007)
17. Cook, S.D., Salkeld, S.L., Stanley, T., Faciane, A., Miller, S.D.: Biomechanical study of pedicle screw fixation in severely osteoporotic bone. Spine J. **4**, 402–408 (2004)
18. Lill, C.A., et al.: Mechanical performance of cylindrical and dual core pedicle screws in calf and human vertebrae. Arch. Orthop. Trauma Surg. **126**(10), 686–694 (2006)
19. Chen, H.C., et al.: Effect of different radial hole designs on pullout and structural strength of cannulated pedicle screws. Med. Eng. Phys. **37**(8), 746–751 (2015)

20. Alvine, G.F., Swain, J.M., Asher, M.A.: Treatment of thoracolumbar burst fractures with variable screw placement or Isola instrumentation and arthrodesis: case series and literature review. J. Spinal Disord. Tech. **17**, 251–264 (2004)
21. Gaines, R.W.: The use of pedicle-screw internal fixation for the operative treatment of spinal disorders. J. Bone Joint Surg. Am. **82**, 1458–1476 (2000)

Design and Optical Performance Analysis of a Quasi-stationary Compound Parabolic Concentrator for Photovoltaic Applications

F. Masood[1]([⊠]) [iD], P. Nallagownden[1], I. Elamvazuthi[1], M. A. Alam[2], M. Ali[3], and M. Azeem[2]

[1] Department of Electrical and Electronics Engineering, Universiti Teknologi PETRONAS, 32610 Bandar Seri Iskandar, Malaysia
[2] Department of Mechanical Engineering, Universiti Teknologi PETRONAS, 32610 Bandar Seri Iskandar, Malaysia
[3] Department of Civil and Environmental Engineering, Universiti Teknologi PETRONAS, 32610 Bandar Seri Iskandar, Malaysia

Abstract. This paper presents the design and optical performance analysis of a quasi-stationary low concentration compound parabolic concentrator for potential use in building-integrated and rooftop photovoltaic applications. The compound parabolic concentrator (CPC) belongs to the family of non-imaging concentrators which are used for the concentration of solar radiation over a photovoltaic and/or thermal absorber to obtain a better output with lesser absorber size. The optical performance of a CPC designed for a concentration ration of '2.5×' and a half acceptance angle of 23.6°, with a 50% truncation level, has been evaluated using Monte Carlo ray-tracing simulations, for stationary installations in east-west and north-south directions using Tonatiuh software. The virtual model for the designed collector was developed using CAD modeling software and was later imported to the optical analysis software for performance evaluation. The material and surface properties were specified for different components of concentrating system. The sun shape, position, and height in the sky were defined with clear sky conditions and fixed irradiance. The ray-tracing simulations were performed for different incidence angles to calculate the amount of optical energy collected by the receiver surface. The angular acceptance range of CPC was evaluated. The solar flux distribution on the PV module surface was examined. It was concluded that CPC placed in east-west direction achieved the highest optical efficiency (85%) within its acceptance angle range. The solar flux distribution was found to be uniform in the center of the receiver while flux peaks were observed near its edges for vertically incident solar rays.

1 Introduction

Due to the prevailing situation of sudden growth in energy requirements across the globe together with the substantial pollution caused by carbon emanation from fossil fuels-based plants, the development of clean renewable energy resources is turning out to be very crucial. Solar energy provides infinite potential as an unpolluted energy

© The Author(s), under exclusive license to Springer Nature Singapore Pte Ltd. 2021
M. Awang and S. S. Emamian (Eds.): *Advances in Material Science and Engineering*, LNME, pp. 241–248, 2021.
https://doi.org/10.1007/978-981-16-3641-7_28

source [1]. The key approach for harvesting solar energy is to transform it directly into electrical form. Solar thermal collectors are other available options [2]. The photovoltaic technology enhances photoelectric transformation process which is the prime reason for its universal adoption nowadays. Solar photovoltaic systems have the potential to provide useful energy without employing moving parts. Moreover, they function noiselessly and have low maintenance costs [3]. However, the upfront capital costs of photovoltaic systems limit their large-scale implementation. The techniques for substantially reducing the cost of PV systems therefore need to be identified. The cost reductions can be either achieved by increasing the solar cell's efficiency or by using concentrated photovoltaic technology. The later strategy aims to minimize the cost of the PV cells by reducing the use of semiconductor materials [4]. As the most expensive component in the PV generation system is the semiconductor material, hence reduction in its consumption causes an overall abatement in PV module cost.

Different researchers have investigated the potential of a compound parabolic concentrator (CPC) for reducing the utilization of costly silicon cells in the PV module for lowering the unit price of PV output power. A symmetric 2D CPC can provide concentration ratios ranging from 2–10 with only seasonal tilt angle adjustments [5]. The mathematical model of a CPC based PV system was developed by Tang et al. [6] to estimate optimum values of half acceptance angle and tilt angle for maximizing the annual radiation incident upon its receiver. The numerical model and subsequent experimental validation of a PV module integrated with a 3D cross-compound parabolic concentrator (CCPC) was conducted by Sellami and Mallick to determine optical efficiency and optical flux distribution over the surface of PV receiver for different angles of incidence of sun rays [7]. It was revealed that the CCPC having a concentration ratio of $3.6\times$ resulted in an enhanced geometric profile as compared to 3D CPC for static solar concentrator applications. Guiqiang et al. [8] investigated the performance of a lens walled CPC having an air gap. The authors achieved an improvement of 10% in optical efficiency as compared to premier design. The electrical and thermal outputs of a mirror-symmetrical dielectric totally internally reflecting concentrator (MSDTIRC) based PV system were examined by Sukki et al. [9] for integration with residential buildings. The authors demonstrated that the MSDTIRC with PV cells achieved a concentration gain of $4.2\times$ when compared with an analogous PV cell in the absence of concentrator.

The present research article presents the design and optical performance analysis of a low-profile CPC concentrator designed in-house for low concentration photovoltaic applications. The CPC acts as an optical concentrating element. The ray-tracing simulations were performed for different solar incidence angles to observe the optical power available over the PV receiver surface. The total amount of optical power impinging on the PV receiver was calculated for different solar incidence angles for static installations in east-west and north-south directions. It was estimated that the CPC with 50% truncation, installed in an east-west direction performed better by collecting more optical power and thus possesses larger optical efficiency as compared to the one with the north-south installation.

2 Design of Concentrator for PV Applications

The Compound parabolic concentrator (CPC), conceived by Winston, is a non-imaging concentrator that possesses the largest angle of acceptance for a distinct concentration ratio and aperture width. The terms concentration ratio (C) and half acceptance angle (Θ_a) are the most pivotal design parameters of CPC which govern the performance of the concentrator. The design process starts with selecting the appropriate values of half acceptance angle and the width of the PV receiver for a particular application. The geometric concentration ratio is the ratio of the entry aperture area to the receiver area. The actual concentration ratio is usually less than the geometric concentration ratio due to manufacturing errors in the reflector curves and optical losses. The CPC in its simplest form consists of two parabolic reflectors. The (x,y) coordinates of the right parabolic branch of the CPC profile are determined by using the equation [10]:

$$x_n' = x_n'(\cos\theta_c) - y_n'(\sin\theta_c) + x_0 \tag{1A}$$

$$y_n' = x_n'(\sin\theta_c) + y_n'(\cos\theta_c) + y_0 \tag{1B}$$

Where Θ_c is the half acceptance angle and it is related to the concentration ratio of CPC by the following equation

$$CR = \frac{1}{\sin\theta_c} \tag{2}$$

The left parabolic branch being the mirror image of the right parabolic branch has the following coordinates:

$$y_n(-x_n) = y_n(x_n) \tag{3}$$

A MATLAB code was written to generate the coordinates of the right and left reflectors of CPC and hence plot its profile. The design parameters of CPC are summarized in Table 1.

Table 1. Design parameters of CPC

Parameter	Value	Unit
Acceptance half-angle	23.5	(°)
Concentration Ratio	2.50	-
Width of Absorber	100	mm
Width of Aperture	250.82	mm
Height of CPC	195	mm
Apex	(−22.15,64.34)	mm

3 Optical Analysis of CPC Collector

A comprehensive optical analysis is required to observe the interrelationship among incident solar radiation and the collector components including the receiver surface. It is also used to work out the energy available at the receiver. The ray tracing strategy has proved to be a propitious tool for evaluating the magnitude and disbursement of optical power in solar PV/thermal systems [11]. This research article presents the optical performance evaluation of a CPC collector designed for photovoltaic applications using the Monte Carlo ray-tracing technique. The virtual model of CPC was developed in CAD software using its profile coordinates generated through MATLAB code. The model was then imported into open-source optical software, Tonatiuh for optical analysis and performance evaluation. Material and surface properties were defined; and sun position was specified for the geographical location of the proposed experimental site. The system was configured using optical software environment and relevant atmospheric conditions were specified. The ray-tracing simulations were performed for various angles of incidence, lying inside acceptance angle range and beyond it for collector orientation in both north-south and east-west directions for the geographical coordinates of proposed location. The solar flux distribution on the PV receiver was examined and found to be uniform in the central region while flux peaks were observed near edges for rays falling perpendicular to the aperture. However, as the incidence angle increased, it caused flux distribution to become non-uniform in the center of the receiver.

4 Result and Discussions

4.1 Geometric Profile of CPC

The CPC profile generated through MATLAB code is shown in Fig. 1(a), whereas Fig. 1(b) shows the variation of height, aperture width and concentration ratio with acceptance half-angle of CPC. Figure 1(a) demonstrates a parabolic shape of both reflectors suitable for collecting solar radiations falling within the acceptance range of the concentrator. As shown in Fig. 1(b), CPC height and aperture width decrease with increasing the magnitude of acceptance angle. The concentration ratio also exhibits a similar trend. The maximum value of concentration ratio results for half-acceptance angle of $10°$ but the corresponding value of height is too much for the identical receiver size and at the same time such a lower value of half-acceptance angle is not appropriate for quasi-stationary applications.

4.2 Ray Trace Diagrams

The pathways of both incident and reflected rays were achieved through Monte Carlo ray-tracing simulations. The ray path diagrams of radiations collected by absorber of CPC collector oriented in a north-south direction for incidence angles of $0°$, $15°$ and $30°$ are shown in Fig. 2(a), 2(b) and 2(c) respectively. It is evident that both types of rays i.e. direct and reflected, are collected by the receiver till the incidence angle is less than or equal to the half-acceptance angle of the CPC concentrator. On the other hand, for the incidence

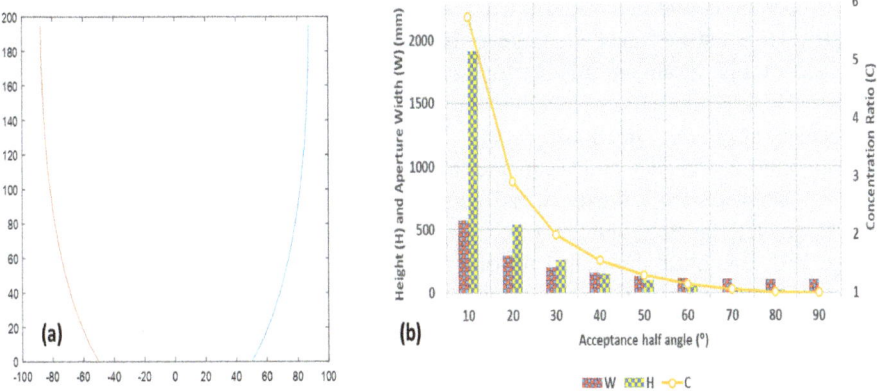

Fig. 1. (a) MATLAB generated profile of designed CPC (b) Variation of height, aperture width and concentration ratio with acceptance half-angle of CPC

angles outside the half-acceptance angle range, only direct rays approach the receiver surface. The optical power collected by the receiver can be substantially increased if the tracker is installed but it increases the cost and complexity of the collector. For the incidence angles higher than the acceptance range, the rays are reflected against the walls of the reflector and are restricted from approaching the receiver surface.

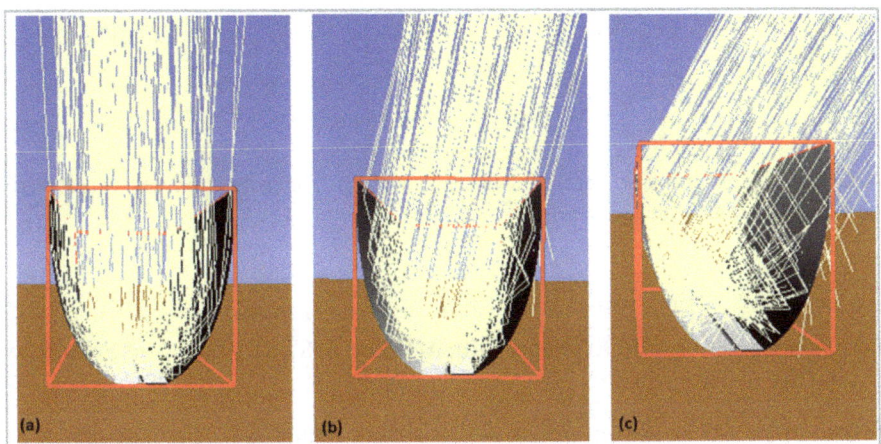

Fig. 2. Ray path diagrams for CPC (a) 0° (b) 15° (c) 30°

4.3 The Acceptance Angle Range, Optical Efficiency and Optical Losses of CPC

In a flat receiver CPC, the place at which the ray is eventually mirrored depends on two parameters namely the angle of incidence of the ray and the geometric profile of the reflector. The effect of the geometry of reflector on ray distribution, for sun rays incident perpendicular to the aperture of CPC, is depicted in Fig. 3(a). The figure shows that all solar radiations reach the target either directly or after one or two reflections. As indicated by Fig. 3, almost 50% of total incident solar flux reach the receiver surface directly and 45% reach after one reflection whereas only 4–5% of total radiation strikes the receiver after two reflections.

The optical efficiency and acceptance angle range in concentrating systems are usually dependent on the reflectivity and geometry of the reflector surfaces. The optical losses, optical efficiency, and acceptance angle range in a symmetric 2D CPC are shown in Fig. 3(b). The figure shows that all solar radiations falling within the acceptance angle range of the concentrator were accepted by CPC. It is also evident that the designed CPC has high optical efficiency since most solar rays approach the PV surface directly or after one or two reflections.

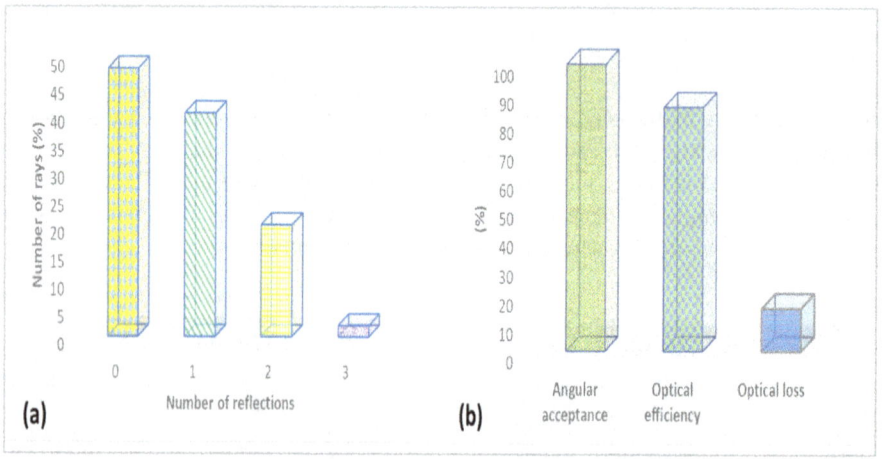

Fig. 3. (a) Number of solar rays reaching the PV module (b) Comparison of angular acceptance, optical efficiency and optical losses for CPC.

4.4 Variation of Optical Efficiency with AOI and DISTRIBUTION of Solar Flux on the Receiver Surface

A stationary CPC aligned in the east-west or north-south directions intercepts solar radiation within its acceptance angle range. The variation of optical efficiency with the angle of incidence (AOI) is illustrated in Fig. 4(a). The optical efficiency attains the highest value of 85% due to low optical losses occurring in the concentrator reflectors within the acceptance angle range. However, beyond the permissible range, the optical efficiency suddenly drops to lower values. The figure also indicates that CPC cannot collect any radiation beyond an acceptance angle range.

The solar flux is concentrated at the corners of the receiver due to crowding of flux lines after reflection from CPC reflectors for lower angles of incidence. The solar flux distribution along the receiver surface is illustrated in Fig. 4(b), which clearly shows that the distribution of solar flux is uniform in the center of the receiver while there is a crowding of flux near the edges. Figure 4(b) also indicates that flux near edges is almost nine times the flux in the least illuminated region. This flux no longer contributes towards electrical power output as the series-connected cells when exposed to non-uniform illumination restrict the current to a value generated by the least illuminated cells. The results depicted by Figs. 4(a) and 4(b) are congruent with those found in existing literature [12], for the same variables.

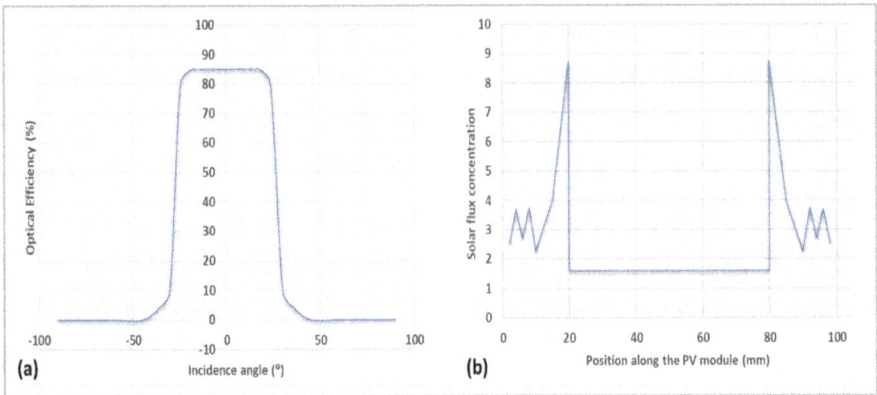

Fig. 4. (a) Change optical efficiency with angle of incidence (b) The distribution of solar flux along the receiver for perpendicular rays

5   Conclusions

Non-imaging CPC collectors are suitable for low concentrating photovoltaic applications as they can collect both direct and diffused radiations and do not require continuous tracking. Such types of collectors are also useful in thermal as well as hybrid PV/thermal applications, whereby they can supply both electricity and process heat for low to medium temperature applications. In this paper, a two-dimensional symmetric CPC collector with a flat absorber was designed particularly meant for PV applications and its optical performance was evaluated using ray-tracing simulations to assess the distribution of solar flux on the receiver. The reflector profiles were generated through a MATLAB program for a half acceptance angle of 23.6° and an absorber width of 100 mm to accommodate a single crystalline solar cell. The resulting optical concentration ratio was found to be '2.5'. The CPC was then truncated to 50% of its full height. Ray tracing simulations were performed for north-south and east-west arrangements of the collector to evaluate its performance for different incidence angles. It was found that more than 50% of total incident rays reach the receiver surface without any reflection from the reflector surfaces. The concentrator exhibited a high value of optical efficiency. The solar flux distribution was found to be uniform in the center region of the receiver whereas crowding of flux lines near the edges caused peak instantaneous flux. This non-uniform flux distribution along the receiver restricts its electrical power output.

Acknowledgment. The authors are highly thankful to Universiti Teknologi PETRONAS (UTP), Malaysia, for providing financial support to conduct this research work under YUTP Scheme (Cost Center: 015LCO-024).

References

1. Kannan, N., Vakeesan, D.: Solar energy for future world: - a review. Renew. Sustain. Energy Rev. **62**, 1092–1105 (2016)
2. Kaloudis, E., Papanicolaou, E., Belessiotis, V.: Numerical simulations of a parabolic trough solar collector with nanofluid using a two-phase model. Renewable Energy **97**, 218–229 (2016)
3. Ullah, A., Imran, H., Maqsood, Z., Butt, N.Z.: Investigation of optimal tilt angles and effects of soiling on PV energy production in Pakistan. Renewable Energy **139**, 830–843 (2019)
4. Baig, H., Sarmah, N., Heasman, K.C., Mallick, T.K.: Numerical modelling and experimental validation of a low concentrating photovoltaic system. Sol. Energy Mater. Sol. Cells **113**, 201–219 (2013)
5. Akhter, J., Gilani, S.I., Al-Kayiem, H.H., Ali, M.: Optical performance analysis of single flow through and concentric tube receiver coupled with a modified CPC collector under different configurations. Energies **12**, 4147 (2019)
6. Tang, R., Wu, M., Yu, Y., Li, M.: Optical performance of fixed east–west aligned CPCs used in China. Renewable Energy **35**, 1837–1841 (2010)
7. Sellami, N., Mallick, T.K.: Optical efficiency study of PV crossed compound parabolic concentrator. Appl. Energy **102**, 868–876 (2013)
8. Guiqiang, L., Gang, P., Yuehong, S., Yunyun, W., Jie, J.: Design and investigation of a novel lens-walled compound parabolic concentrator with air gap. Appl. Energy **125**, 21–27 (2014)
9. Muhammad-Sukki, F., Abu-Bakar, S.H., Ramirez-Iniguez, R., McMeekin, S.G., Stewart, B.G., Munir, A.B., et al.: Performance analysis of a mirror symmetrical dielectric totally internally reflecting concentrator for building integrated photovoltaic systems. Appl. Energy **111**, 288–299 (2013)
10. Paul, D.I.: Optical performance analysis and design optimisation of multisectioned compound parabolic concentrators for photovoltaics application. Int. J. Energy Res. **43**, 358–378 (2018)
11. Hadavinia, H., Singh, H.: Modelling and experimental analysis of low concentrating solar panels for use in building integrated and applied photovoltaic (BIPV/BAPV) systems. Renewable Energy **139**, 815–829 (2019)
12. Paul, D.I.: Theoretical and experimental optical evaluation and comparison of symmetric 2D CPC and V-trough collector for photovoltaic applications. Int. J. Photoenergy **2015**, 1–13 (2015)

Fire Resistance and Mechanical Properties of the Fire-Resistant Board

Kwang Yin Jessica Jong, Ming Chian Yew$^{(\boxtimes)}$, Ming Kun Yew, Chen Hunt Ting, Lip Huat Saw, Tan Ching Ng, Wei Hong Yeo, and Jing Han Beh

Lee Kong Chian Faculty of Engineering and Science, Universiti Tunku Abdul Rahman, Sungai Long Campus, Cheras, 43000 Kajang, Selangor, Malaysia
`yewmc@utar.edu.my`

Abstract. Fire-resistant board is the crucial recent development for fire safety protocol in many buildings. A fire rated board can provide valuable time for the human to evacuate during a fire outbreak. Intumescent fire protection materials provide a wide variety of passive fire protection system with the most efficient utilization. In this research project, the water-based intumescent binder was mixed with vermiculite and perlite to construct the fire-resistant board. Furthermore, the fire-resistant boards were conducted and evaluated by small-scale fire test and three-point flexural test. The best fire protection performance of fire-resistant board was selected and compared with the commercial gypsum board (GB) under the fire test. The experimental result was noticed that this novel fire-rated board (P2) incorporated with the addition of aluminium hydroxide and eggshell flame-retardant fillers in the intumescent binder formulation was consistently proven to be more effective by reducing the temperature up to 78 °C as compared to commercial gypsum board in preventing fires with respect to its sustainability and maintaining the technical integrity throughout the period of 2-h fire test. Lastly, the three-point flexural test observed that P2 reveals the highest flexural stress of 3.542 MPa and the lowest flexural modulus of 208.33 N/mm^2 as compared to other fire-resistant boards by exhibiting the highest ductility characteristics.

1 Introduction

Fire protection systems have become an essence of building nowadays to protect human lives and important assets. It is well known that there are two types of fire protection systems that are used in buildings which are active and passive fire protection systems [1]. Active fire protection (AFP) system includes for example, fire alarms, sprinklers, and fire extinguishers. On the other hand, passive fire protection (PFP) system is for example fire-resistant board, fire doors, fire dampers, and firewalls. During the fire event, AFP system requires some human actions to work efficiently while PFP system does not require any human action but to work and prevent fire automatically if any heat and/or fire contact with the system. In this study, a fire-resistant board can give human enough and valuable times to evacuate when there is any fire emergency as well as prevent fire hazards. It is vital to provide fire safety protection because it contains flame-retardant materials which are known as an intumescent material (IM). According to the results obtained by various

M. Awang and S. S. Emamian (Eds.): *Advances in Material Science and Engineering*, LNME, pp. 249–256, 2021.
https://doi.org/10.1007/978-981-16-3641-7_29

researchers, IM plays an important role in building protection because when it contacts with heat and/or fire, IM will expand to many times of its original thickness and then create an insulation barrier to protect the substrate from being damaged [2–4]. Due to this insulation barrier, it will gain more times for building occupants to evacuate safely during the fire occurrence [5].

Moreover, water-based IM is the latest trend for construction building materials in fire retarding product because of its many advantages such as low-odour, lightweight and environmentally friendly. For this reason, the building industries nowadays have experienced rapid growth in using IM. Besides that, exhaustive investigations from various research reports have discovered that IM has achieved good flammability prevention, physical and chemical performances through many lingers widely on wood and steel structure applications [6–8]. This is now known that IM is designed to retard ignition and slow down the burning rates, and thus to provide the necessary fire insulation barrier. Upon heating, IM swells to form a multi-cellular char layers that provides low thermal conductivity and low density to prevent the rise in temperatures of the substrate from being increased that could cause structural instability and progressive failure.

Following the proven potentials being repeatedly reported by researchers, the community is now focused more on IM aiming for the development of reliable, more efficient, and better flame-retardant formulations. This experimental research is focused on the fire protection performance and the three-points flexural of the fire rated boards which are incorporated with the advanced intumescent binder formulations.

2   Methods

2.1   Samples Preparation

In this experimental research, a dimension of 300 mm (length) × 300 mm (width) × 40 mm (thickness) of fire-resistant boards are designed and fabricated by using the formulations as shown in Table 1. Besides that, the fire protection performance of the commercial GB was compared with the fabricated fire rated boards, P1–P3.

2.2   Fire Test

The small-scale fire test was conducted to determine the fire protective performance of the fire-resistant boards, P1–P3 by using the Bunsen burner flame spray gun blow torch at about 1000 °C for 2 h. The temperature profile (time-temperature curve) of the backside of the fire-rated door prototype is measured and recorded using a digital handheld thermometer. Figure 1 shows a small-scale fire test experimental setup for the fire-rated board.

2.3   Three-Point Flexural Test

Samples were prepared with a dimension of 300 mm × 30 mm × 30 mm respectively in accordance with ASTM D790 [9] is mounted on an Instron Micro Tester with a crosshead speed of 1 mm/min and a 150 mm of support span length is continually drawn apart in

Table 1. Formulations of fire-resistant boards

Ingredients \ Fire-retardant board	Parts by weight for formulations (wt. %)		
	P1	P2	P3
Flame-retardant additives			
APP	20	20	20
PER	10	10	10
MEL	10	10	10
Vinyl acetate copolymer (VAC) emulsion	50	50	50
Flame-retardant filler (pigment)			
TiO_2	4.0	4.0	4.0
Flame-retardant fillers			
$Al(OH)_3$	3.0	3.0	-
$Mg(OH)_2$	-	-	3.0
$CaSiO_3$	3.0	-	3.0
CES	-	3.0	-
Additional Ingredients (flame-retardant materials)			
Vermiculite (ml)	1400	1400	1400
Perlite (ml)	1400	1400	1400

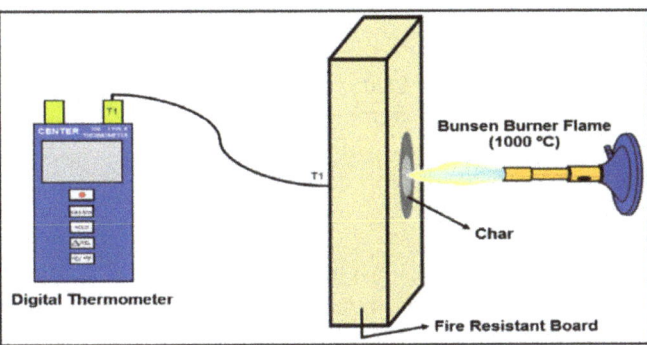

Fig. 1. Schematic of the experimental setup for fire resistant board

compressive mode until the sample is torn apart. Figure 2 indicated the schematic of the three-point flexural test.

The purpose of a flexural test is to determine the flexural strength (i.e., flexural stress) and flexural modulus of the fire-resistant board. Flexural strength is the maximum stress at the outermost fibre on either the compression/tension side of the sample. While flexural modulus is measured from the slope of the stress versus strain deflection curve. Both are used to determine and evaluate the ability of samples to withstand flexure and/or bending forces. The calculation of flexural stress and flexural modulus are shown in Eqs. 1 and 2, respectively.

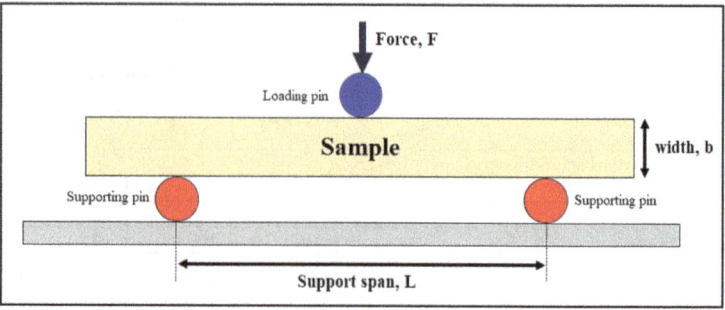

Fig. 2. Schematic diagram of three-point flexural test

Calculation of flexural stress, σ_f (MPa):

$$\sigma_f = \frac{3FL}{2bd^2} \tag{1}$$

Calculation of flexural modulus, E_f (MPa):

$$E_f = \frac{L^3 m}{4bd^3} = \frac{\sigma_f}{\varepsilon_f} \tag{2}$$

Where,
σ_f = stress in outer fibres at midpoint (MPa).
ε_f = strain in outer surface (mm/mm).
E_f = flexural modulus of elasticity (MPa).
F = load at a given point on the load deflection curve (N).
L = support span (mm).
b = width of test sample (mm).
d = depth or thickness of test sample (mm).
D = maximum deflection of the centre of the sample (mm).
m = gradient/slope of the initial straight line of the load deflection curve (N/mm).

3   Results and Discussion

3.1   Fire Test

The main purpose of this test is to characterize the fire protective performance of the fire-rated boards. The results of each fire-resistant boards that are plotted as a function of time are presented in Fig. 3. In this experimental research, the gypsum board acts as the commercial fire-rated door prototype is used to compared with other fire-resistant boards. From the experimental results obtained, all the temperatures of boards increase significantly throughout the test. This phenomenon is because of the good thermal insulation of fire rated boards (P1–P3) with the addition of intumescent binder, vermiculite and perlite, which could dissipate the heat when exposed to fire. The three fabricated fire rated boards have showed a similar rise in temperature at the first 15 min as presented in Fig. 4 After the first 15 min, the gypsum board has started to increase rapidly and then gradually until reaching its maximum temperature of 169 °C at 120 min. For P1, it increased quickly at the initial 15 min and then increased steadily until 100 min. After 100 min of fire test, P1 started to decrease and then rose again at 110 min until 120 min by reaching its maximum temperature of 118 °C. For P3, it revealed a sudden rise at 90 min and then the temperature returned to rise normally at 95 min. This phenomenon might be due to the physical and chemical reactions of the intumescent binder ingredients occurred inside the fire-rated board at 100 min of the fire test until fully decomposed at 120 min with the maximum temperature of 94 °C. In addition, it was noticed that the temperature rises for P2 was considered to have risen slowly until reaching its maximum temperature of 91 °C at 120 min. Therefore, P2 is comparatively considered to be the best fire protective performance in retarding the fire as it has the lowest temperature profile among all fire-resistant boards.

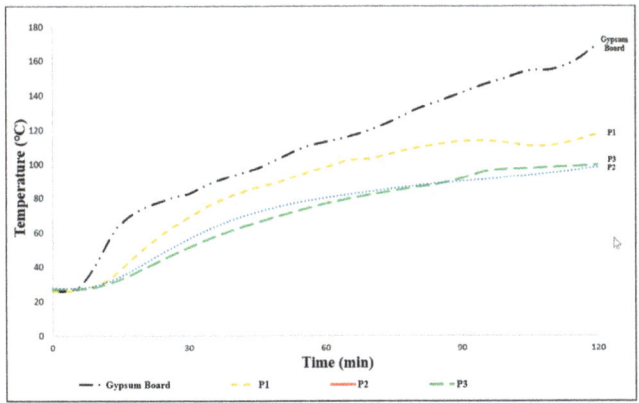

Fig. 3. Evolution of temperature profiles of fire-resistant boards

3.2 Three-Point Flexural Test

This test is used to determine the material behaviour in term of mechanical properties of the fire-rated boards. In this research project, P1–3 have undergone this test to measure the flexural strength (flexural stress) and flexural modulus as shown in Fig. 4. It clearly indicated that the P1 failed in a brittle manner. In contrast, P2 could withstand higher flexural strength after the first crack with higher values of flexural strain, thus exhibiting high ductility characteristics as compared to other fire rated boards. P3 with lower flexural strength failed gradually as it exhibited higher ductility characteristics than P1.

This test is used to determine the material behaviour in term of mechanical properties of the fire-rated boards. In this research project, P1–3 have undergone this test to measure the flexural strength (flexural stress) and flexural modulus as shown in Fig. 4. It clearly indicated that the P1 failed in a brittle manner. In contrast, P2 could withstand higher flexural strength after the first crack with higher values of flexural strain, thus exhibiting high ductility characteristics as compared to other fire rated boards. P3 with lower flexural strength failed gradually as it exhibited higher ductility characteristics than P1.

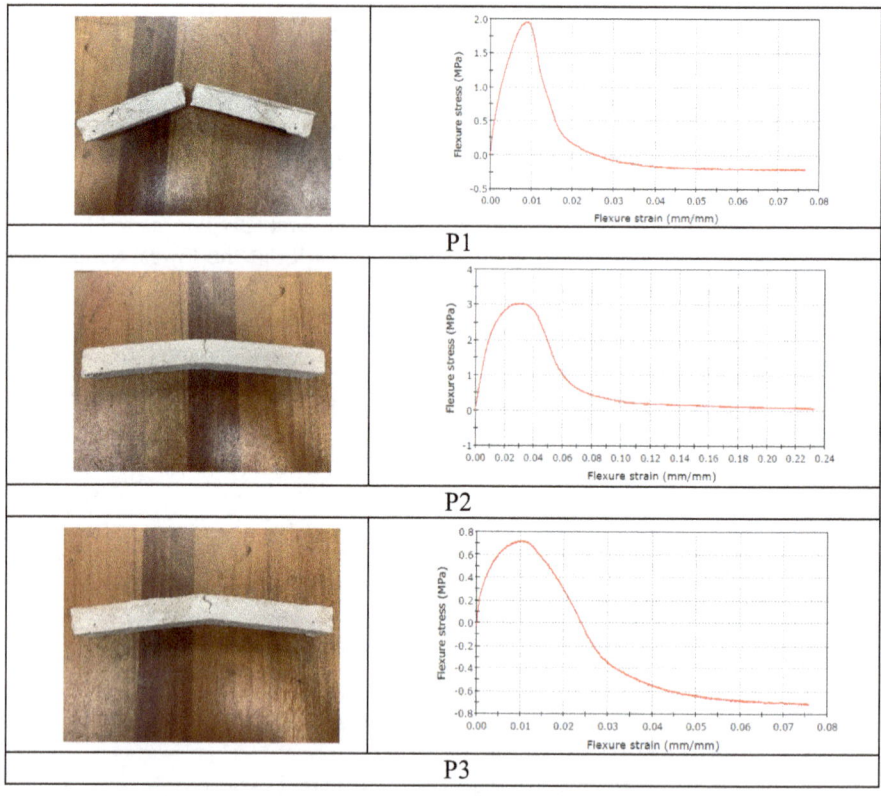

Fig. 4. Flexural stress-strain curve after the three-point flexural test of P1–P3

Table 2 shows the flexural stress and flexural modulus of all the samples that have undergone the flexural test. From the results as tabulated, it is observed that P2 reveals with the highest flexural stress of 3.542 MPa and the lowest flexural modulus of 208.33 N/mm^2 as compared to other fire-rated boards. It has proven that P2 can withstand higher impact of stress by exhibiting the highest ductility characteristics. P1 shows the higher flexural modulus of 416.67 N/mm^2 but with flexural stresses of 2.609 MPa. This flexural test has indicated that P1 is relatively stiff and brittle. However, P3 exhibited the lowest flexural stress of 0.962 MPa as compared to other boards. This indicates that P3 provided weaker flexural properties. As a result, P2 has displayed elastic-plastic behaviour and higher ductility in which demonstrating the good bonding between the VAC binder and flame-retardant materials. It bridged across the cracked section to arrest crack propagation and lead to ductile failure [10].

Table 2. Flexural stress and flexural modulus of P1–P3

Samples	Flexural stress, σ_f (MPa)	Flexural modulus, E_f (N/mm^2)
P1	2.609	416.67
P2	3.542	208.33
P3	0.962	250.00

4 Conclusions

Based on the experimental research findings, it can demonstrate that the tested fire-resistant board of P2 with the addition of 3 wt. % of aluminium hydroxide and 3 wt. % of renewable CES bio-filler has consistently indicated to have a better performance with respect to the fire protection and mechanical properties as compared to the existing commercial GB and other samples P1 and P3. The fire-resistant board, P2 shows the lowest equilibrium temperature of 91 °C by reducing the temperature up to 78 °C as compared to GB throughout the period of the 2-h fire test. On the other hand, the three-point flexural of P2 shows the highest flexural stress of 3.542 MPa, which indicated a significant improvement by up to 280% as compared to P3 (0.962 MPa). As a conclusion, it can be deduced that the selection of suitable combinations of flame-retardant ingredients can directly and effectively influence the technical performances of fire protection and the three-point flexural of fire rated board respect to its sustainability in maintaining the integrity of fire rated board in the event of a fire.

Acknowledgement. The authors would like to thank Universiti Tunku Abdul Rahman for the UTARRF support (Grant number: IPSR/RMC/UTARRF/2019-C2/B01).

References

1. Mróz, K., Hager, I., Korniejenko, K.: Material solutions for passive fire protection of buildings and structures and their performances testing. Procedia Eng. **151**, 284–291 (2016)
2. Yew, M.C., Sulong, N.H.R., Yew, M.K., et al.: Fire propagation performance of intumescent fire protective coatings using eggshells as a novel biofiller. Sci. World J. 2014, 805094 (2014)
3. Beh, J.H., Yew, M.C., Yew, M.K., et al.: Fire protection performance and thermal behavior of thin film intumescent coatings. Coatings **9**, 483 (2019)
4. Yew, M.C., Yew, M.K., Saw, L.H., et al.: Influence of nano bio-filler on the fire-resistive and mechanical properties of water-based intumescent coatings. Prog. Org. Coat **124**, 33–40 (2018)
5. Weil, E.D.: Fire-protective and flame-retardant coatings - a state-of-the-art review. J. Fire Sci. **29**(3), 259–296 (2011)
6. Yew, M.C., Sulong, N.H.R., Yew, M.K., et al.: Investigation on solvent-borne intumescent flame-retardant coatings for steel. Mater. Res. Innov. **18** (sup6):S6–384–S6–388 (2014)
7. Yew, M.K., Yew, M.C., Saw, L.H., et al.: Effects of flame retardant nano bio-based filler on fire behaviors of intumescent coating. Mater. Sci. Forum **947**, 142–147 (2019)

8. Jessica, J.K.Y., Yew, M.C., Yew, M.K.: Preparation of intumescent fire protective coating for fire rated timber door. Coatings **9**, 738 (2019)
9. ASTM: D790-03-Standard Test Method for Flexural Properties of Unreinforced and Reinforced Plastics and Electrical Insulation Materials. ASTM Standards, pp. 1–11 (2015)
10. Aghaee, K., Yazdi, M.A., Yang, J.: Flexural properties of composite gypsum partition panel. Proc. Inst. Civil Eng. Eng. Sustain. **168**(6), 58–263 (2015)

Fire-Resistant Properties of Green Intumescent Coating Incorporated with BioAsh for Steel Protection

Jing Han Beh[1]($\boxtimes$), Ming Chian Yew[2], Lip Huat Saw[2], and Ming Kun Yew[3]

[1] Department of Architecture and Sustainable Design, Lee Kong Chian Faculty of Engineering and Science, Universiti Tunku Abdul Rahman, Cheras, 43000 Kajang, Malaysia
behjh@utar.edu.my
[2] Department of Mechanical and Material Engineering, Lee Kong Chian Faculty of Engineering and Science, Universiti Tunku Abdul Rahman, Cheras, 43000 Kajang, Malaysia
{yewmc,sawlh}@utar.edu.my
[3] Department of Civil Engineering, Lee Kong Chian Faculty of Engineering and Science, Universiti Tunku Abdul Rahman, Cheras, 43000 Kajang, Malaysia
yewmk@utar.edu.my

Abstract. Fire-resistant intumescent coatings applied in steel buildings are important passive fire protection measure to ensure the structural integrity of steel during fire accidents. This study highlighted the use of BioAsh as natural substitute to industrial fillers in the water-based intumescent coating. Fire-resistant properties of water based intumescent coating reinforced with different particle sizes of BioAsh were investigated via fire-resistant test (FRT), carbolite furnace test (CFT), scanning electron microscopy (SEM) and thermogravimetric analysis (TGA). GIC3 sample showed the lowest equilibrium temperature of 115.5 °C and the thickest char expansion to effectively limit the penetration of heat. GIC3 formed the densest and most compacted char layer as shown in the SEM, indicated an excellent char quality and strength to inhibit the heat propagation to the steel substrate. TGA demonstrated the highest residual weight 33.12% of GIC3 at 1000 °C. This research revealed the green intumescent coating incorporated with renewable BioAsh was a promising fire-resistant approach to protect the steel from fire.

1 Introduction

Fire resistant efficacy of building structural system is very crucial to save human lives. Many precedents reported a huge loss of precious human lives and assets in fire accidents [1]. Steel is a major component used in the building structural system and it will become ductile, deform, and collapse eventually when it reaches the critical temperature of 500 °C and more in a fire accident due to the disappearing material strength [2]. Intumescent coating applies onto steel fall under reactive category [3]. Intumescent coating is developed to apply on the steel surface as a fire protective barrier. When the intumescent coating exposed to flame, it will swell and form a carbonaceous char layer that hinder further heat transfer to the steel. This mechanism of intumescent coating is attributed

M. Awang and S. S. Emamian (Eds.): *Advances in Material Science and Engineering*, LNME, pp. 257–264, 2021.
https://doi.org/10.1007/978-981-16-3641-7_30

to the reaction of binder, flame-retardant additives, and mineral fillers upon heating [4]. Numerous studies have been conducted to enhance the formula of current water-based intumescent coating to be more environmentally friendly including the exploration of halogen-free flame-retardant additives and the continuous search of readily available by-products to replace the flame-retardant mineral fillers from industrial sources [5–7]. Flame-retardant mineral fillers such as calcium carbonate, aluminium hydroxide and magnesium hydroxide possess good filling property vitally contribute to a more compact char that better insulate the steel from reaching its critical temperature. Even though mineral fillers share a lower weight proportion in the composition of intumescent coating, the positive impacts to the fire-resistant and mechanical properties of char layer formed are significant. BioAsh is a type of by-product derived from the combustion of rubber-wood biomass in a fuel factory located in Sitiawan Malaysia. Rubber tree, scientifically known as *Hevea brasiliensis*, is a type of hardwood species grow in the tropical climate. Rubberwood biomass is a leftover from rubberwood logging activities after the rubber trees ended the life cycle of latex production. The rubberwood biomass is readily supply to factory to use as biofuel. Combustion of rubberwood biomass generated large amount of rubberwood ash as waste. This far, the knowledge on the utilisation of BioAsh in the water-based intumescent coating is rare. The effects of BioAsh on fire-resistant of intumescent coating is still not well-known. This research seeks to explore the fire-resistant properties of BioAsh incorporated in the water-based intumescent coating.

2   Materials and Methods

Vinyl acetate copolymer (VAC) was supplied by Afza maju trading, Malaysia. VAC was a thick, white milky blend water based polymer used as the core binding agent to other ingredients in this research. Three main flame-retardant additives: ammonium polyphosphate (APP), pentaerythritol (PER) and melamine (MER), and industrial mineral filler: Titanium dioxide (TiO_2) with >99.0% purity were supplied by Synertec Sdn. Bhd, Malaysia. Flame retardant green mineral fillers: BioAsh (agriculture by-product) was obtained from a fuel factory located in Perak, Malaysia. The green intumescent coating samples formula were synthesised from the key ingredients as listed in Table 1. The formula consisted a mixture of vinyl acetate copolymer (VAC) the water based binder and three main flame retardant addtives, flame retardant mineral fillers: titanium dioxide (TiO_2) the white pigment and BioAsh. BioAsh particle size of 600, 425, 300 μm respectively were obtained through sieve analysis. These ingredients were blended homogeneously in a high speed disperse mixer at rotational speed 1500 rpm for one hour.

2.1   Fire-Resistant Test (FRT) and Carbolite Furnace Test (CFT)

Fire resistant efficiency of green intumescent coatings were tested via Bunsen burner experiment setup. GIC sample were hand brushed applied onto a carbon steel plate (100 legnth × 100 width × 10 mm thick) and air dried in room temperature. This procedure was repeated 2 to 4 times until a thin layer ±1.5 mm of GIC dry film was obtained. The steel plate coated with GIC was held vertically by a three fingers clamp attached to

Table 1. Formula of Green intumescent coating samples

Sample	Flame retardant additives	Binder	Flame-retardant mineral fillers:		
			Pigment (%)	Green (%)	
	APP-PER-MEL (%)	VAC (%)	TiO2	BioAsh	Particle size (μm)
GICC	40	50	6.5	–	–
GIC1	40	50	6.5	3.5	600
GIC2	40	50	6.5	3.5	425
GIC3	40	50	6.5	3.5	300

a support stand. A portable EN417 bunsen burner flame gun with 190 g butane gas tank was positioned 7.0 mm distance away from the GIC coated steel plate. The GIC was tested under the exposure of 1000 °C flame from the blowtorch. The time/ temperature of GIC sample was recorded every minute for one hour, recorded by UNI-T UT320D digital thermometer connected to the Type K/J thermocouple of the steel plate. Char formation and thickness of GIC sample was examined via chamber furnace model RHF 15/8. GIC sample was hand brushed onto a steel plate (50 length × 50 width × 10 mm thick) and left to air-dried at room temperature. This step was continued 2–4 times after an end thickness of ±1.5 mm GIC dry film was achieved. A digital Vernier caliper was used to verified the GIC dry film thickness. The GIC sample was heated at an elevating temperature of 50 °C/min until an end temperature of 600 °C was reached. The char thickness and strength of GIC sample was evaluated.

2.2 Scanning Electron Microscopy (SEM)

Char surface morphology and char cell structure of GIC samples were observed under × 5.00 k magnification in a scanning electron microscope. Scanninng electron microscope Hitachi EDAX S3400-N was operated to examine the detail physical properties of the char layer. Thermal conduction and electron emission of the char sample were enhanced with a coat of thin gold film (2–20 nm thick) sputtered onto it. Projection of 1.0 kV low focused beam energy was used to minimise the thermal damage. Efficacy of char formed in the fire-resistant test was evaluated.

2.3 Thermogravimetric Analysis (TGA)

Thermal degradation of the GIC sample was analysed in a Perkin Elmer model STA8000 thermogravimetric analyzer. 5 to 10 mg of GIC dried sample was placed in a crucible supplied with nitrogen gas. The GIC sample was heated from 30 °C to 1000 °C with 20 °C per minute of heating rate. Residual weight and mass change of the GIC sample after completed the heating at 1000 °C was evaluated.

3 Results and Discussion

3.1 Fire-Resistant Performance and Char Expansion and Thickness

FRT were conducted via Bunsen burner setup to investigate the fire protection efficacy
of GIC samples on steel. Temperature profiles of GIC samples was shown in Fig. 2.
Char formation of GIC samples after FRT tests as displayed in Fig. 3. It can be observed
all GIC samples formed a layer of cabonaceous char that was crucial for fire-resistant.
The back temperature of GIC samples after 60 min of FRT were GICC (174 °C), GIC1
(152 °C), GIC2 (140 °C), and GIC3 (115 °C). The incorporation of BioAsh into the GIC
formulas decreased the end temperature from 25 °C to 59 °C in comparison to GICC
(control sample) without any BioAsh. When the BioAsh particle reduced in size, the
fire-resistant performance improved. GIC formula incorporated with 600, 425, 300 µm
of BioAsh induced reduction in temperature by 22, 34, and 59 °C respectively. The fire
protection effect was particular prominent in the lowest end temperature of GIC3 that
showed the best ability to impede fire among all samples.

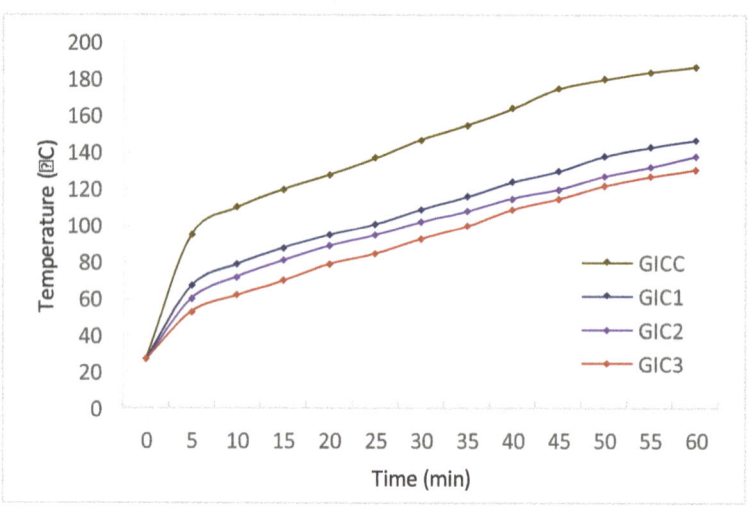

Fig. 2. Temperature profiles of GIC samples

Intumescent char expansion and thickness revealed a remarkable influence on the
flame-retardant efficacy [8]. The reinforcement of BioAsh with different particle sizes
induced various char expansion rate and thickness as shown in Fig. 4.

The char expansion rate of GIC samples were measured when the GIC samples were
heated from 400 °C to 600 °C. The char expansion rate of GICC was 10.5%. GIC1
to GIC 3 respectively indicated an expansion rate of 9.1%, 11.1%, and 12.9%. Char
thickness displayed as GICC (8.0 mm), GIC1 (10.0 mm), GIC2 (12.0 mm) and GIC3
(13.5 mm) at 400 °C. At 600 °C, GICC, GIC1 to GIC3 shown thickness of 9.5 mm,
11.0 mm, 13.5 mm and 15.5 mm. GICC (control sample) had the least char thickness
among all. Eventhough GICC had slightly higher expansion rate than GIC1, the char

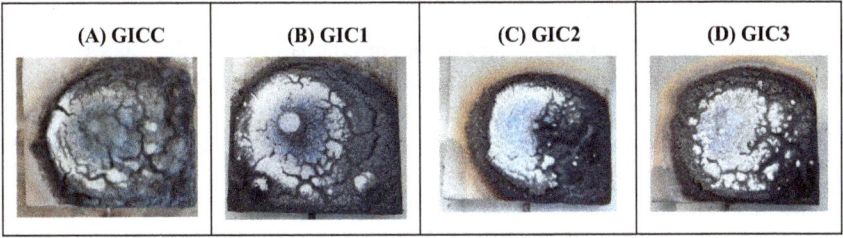

Fig. 3. Char formation of GIC samples after FRT

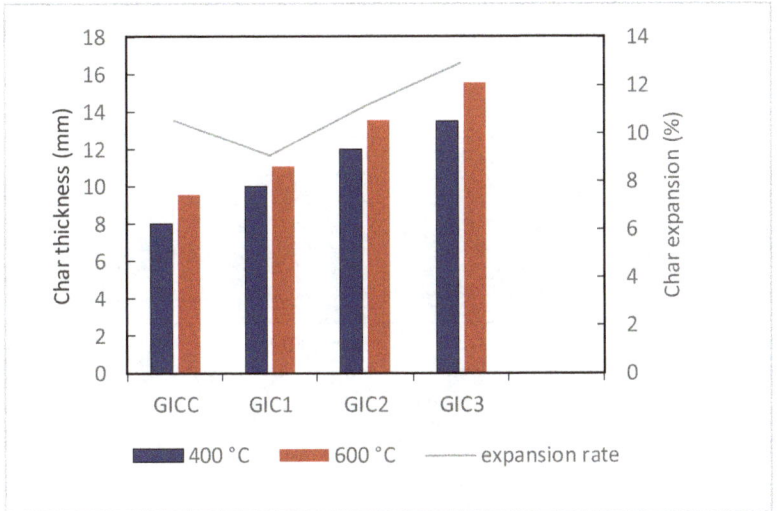

Fig. 4. Char thickness and expansion rate of GIC samples

was weak and less efficient. Continuous increment of char expansion was observed from GIC1 to GIC3. This happened when reinfored BioAsh became smaller in particle size. It can be observed GIC3 generated the highest char thickness which indicated the best fire-resistant performance. Incorporation of BioAsh into the intumescent coating formulations resulted in denser char to efficiently insulate the steel from heat. GIC3 formula reinforcement with 300 µm BioAsh best limited the heat transfer and lowest the heating temperature to steel substrate.

3.2 Char Surface Morphology

Surface morphology of GIC samples before and after the fire-resistant test (FTR) were observed via high magnifications ×500 k in the SEM as shown in Fig. 5. The surface particles distribution and matrix of GIC samples before the FRT were studied. These was followed by the surface analysis of the char structure of GIC samples after the FRT. GIC1 showed the least homogenous particles distribution on the surface dominant by coarse, irregular and obvious coagulum on the surface. GIC2 had similar surface condition as

GIC1, however the coagulum exhibited were smaller in size. GIC3 displayed a smoother and more homogenous particles distributions in comparison to GIC1 and GIC2. The surface matrix of GIC3 was the finest and little coagulum was observed. This distinct charateristic was mainly attributed to the smallest particle size of 300 μm BioAsh mineral fillers reinforced into the GIC3. 300 μm BioAsh particles can better dispersed withitn GIC3 to enhance the surface matrix. The formation of char layer was very curcial to the fire protection efficiency of GIC samples. The char surface of GIC1 displayed intense void content and the char cell was largely occupied by macro cavities. These collapsed voids accelerated the heat transfer to the steel resulted in poor flame-retardant. The char surface of GIC2 and GIC3 were more condense and cavities were declined as compared to GIC1. These char features were noticed when the BioAsh particle sizes were decreased. GIC3 showed the most uniform, densest and finest char structure with the least void and minor cavities to better inhibited the speedy heat transfer.

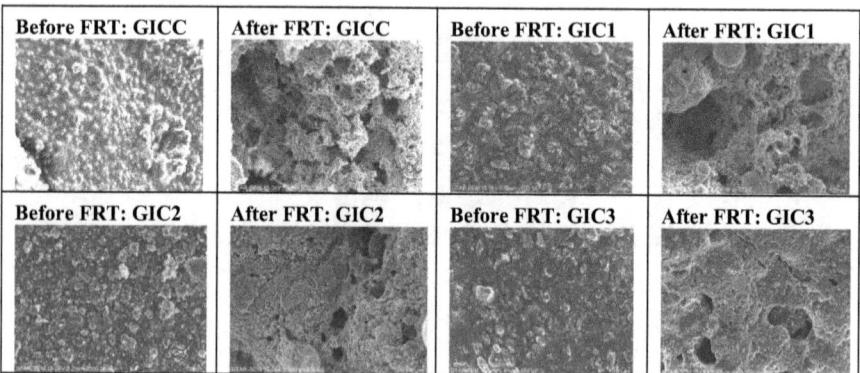

Fig. 5. Char surface micrographs of GIC samples

3.3 Thermal Degradation

Thermal degradation of GIC samples was investigated using TGA. Residual weight of GIC samples after high temperature heating at 1000 °C as presented in Fig. 6.

At temperature between 200 °C to 300 °C, the mass loss of GIC samples were approximate 6 to 12% when the PER decomposed. At 300 °C to 400 °C, APP decomposed to release phosphoric acid, ammonia gas and water to form char. MEL decomposed at about 280 °C to release ammonia gas assisted in the expansion of carbonaceous char. APP encountered second stage degradation at 450 °C. Thus dramatic mass loss can be observed in all GIC samples showed in the peaks. Partial carbonaceous char decomposed at 500 °C and above, left only inorganic phosphate as the major residue. Mass loss of GIC1, GIC2 and GIC3 was 68.94, 67.19, and 66.88% respectively. This scenario reported the addition of BioAsh ameliorate the thermal stability of formula in GIC1 to GIC3. The outcome was particularly prominent when GIC3 reinforced with 300 μm was heated at elevating temperature of 1000 °C, the residual weight remained the highest at

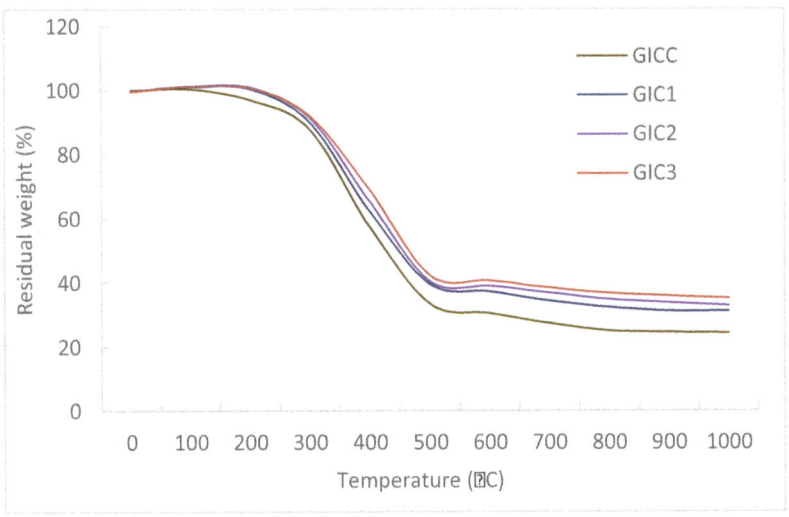

Fig. 6. TGA curves of GIC samples

33.12% compared to the rest. At 1000 °C, residual weights of the rest of GIC samples as GICC (24.26%), GIC1 (31.06%), and GIC2 (32.81%). Thermal degradation of GIC samples reduced when the particle size of reinfored BioAsh decreased. Finer BioAsh particles contributed to a denser and stronger char that lessen the thermal damage at high temperature. GIC3 revealed the best thermal stability when exposed to high heat thus improving the fire-resistant performance.

4   Conclusions

This research investigated the fire-resistant efficacy of green intumescent coating formulations incorporated with different particle size of BioAsh. The FRT demonstrated the lowest equilibrium temperature 115 °C achieved by GIC3 formula incorporated with 300 μm BioAsh particles. This indicated the excellent heat resistant performance of GIC3 with 59 °C difference as compared to the GICC (control sample) without any BioAsh. GIC3 formula activated the thickest char in the CFT evidenced the best flame-retardant performance to limit the heat transfer. Char surface morphology analysis in the SEM revealed the most homogenous distribution of GIC3 formula in the surface matrix, and the formation the densest and most compacted carbonaceous char layer to effectively hinder the speedy heat propagation to the steel substrate. Hence, resulted in the excellent fire-resistant outcome in the FRT. TGA result reported GIC3 had the least damage when exposed to 1000 °C with the lowest mass loss at 64.88%. Residual weight of GIC3 remained the highest at 33.12% after the TGA. Incorporation of BioAsh with 300 μm as formulated in the GIC3 was a promising approach to protect the steel from fire for one hour. It was proven in a series of physical tests that the renewable BioAsh was a realistic natural substitute to replace the exotic industrial mineral fillers in the water-based intumescent coating formulation.

Acknowledgments. The authors would like to thank Universiti Tunku Abdul Rahman for the UTARRF (IPSR/RMC/UTARRF/2019-C2/B01) funding support in this research.

References

1. Sigmann, S.B.: Playing with fire: Chemical safety expertise required. J. Chem. Educ. **95**, 1736–1746 (2018)
2. Farkas, J., Jármai, K.: Fire resistant design. Des. Opt. Met. Struc. Woodhead Publishing, Cambridge, UK (2008)
3. Sutton, I.: Firefighting. Plant Design Operations. 2nd ed. Gulf Professional Publishing, Texas, USA (2017)
4. Anees, S.M., Dasari, A.: A review on the environmental durability of intumescent coatings for steels. J. Mater. Sci. **53**(1), 124–145 (2017). https://doi.org/10.1007/s10853-017-1500-0
5. Wang, Y.Z.: Halogen-Free Flame Retardants. Woodhead Publishing Series in Textiles, Cambridge, UK (2008)
6. Yew, M.C., Yew, M.K., Saw, L.H., Ng, T.C., Durairaj, R., Beh, J.H.: Influences of nano bio-filler on the fire-resistive and mechanical properties of water-based intumescent coatings. Prog. Org. Coat. **124**, 33–40 (2018)
7. Beh, J.H., Yew, M.C., Saw, L.H.: Development of lightweight fire-resistant sandwich panel. In: IOP Conferencs Series on Earth and Environment Science, vol. 476, p. 012031 (2020)
8. Olcese, T., Pagella, C.: Vitreous fillers in intumescent coatings. Prog. Org. Coat. **36**, 231–241 (1999)
9. Delobel, R., Le Bras, M., Ouassou, N., Alistiqsa, F.: Thermal behaviours of ammonium polyphosphate-pentaerythritol and ammomium pyrophosphate-pentaerythritol intumescent additives in polypropylene formulations. J. Fire Sci. **8**, 85–108 (1990)

New Mixing Method of Self-consolidating Concrete Incorporating of Silica Fume

Ming Kun Yew[(⊠)], Ming Chian Yew, Jing Han Beh, Lip Huat Saw, Foo Wei Lee, and Siong Kang Lim

Lee Kong Chian Faculty of Engineering and Science, Universiti Tunku Abdul Rahman, Cheras, 43000 Kajang, Malaysia
{yewmk,yewmc,behjh,sawlh,leefw,sklim}@utar.edu.my

Abstract. Self-consolidating concrete (SCC) is an extensive conventional concrete technology which uses the similar constituent materials but possesses distinctive characteristic compared to conventional concrete. Self-consolidating concrete is highly deformable that can spread into every corner of formwork without any mechanical compaction. The main objective of this research is to study the effect of silica fume on the fresh and hardened properties of self-consolidating concrete in terms of slump diameter, compressive and ultrasonic pulse velocity. Different percentage of silica fume (0, 10, 15 and 20%) as the cement replacement material are incorporated in the self-consolidating concrete to produce self-consolidating concrete with a minimum compressive strength of 50 MPa at 28 days. The significant finding of this study is the implementation of new method in the mixing of concrete constituent materials. This new method resulted in 33% total reduction of time for the mixing process as compared to the conventional mixing method. Hence, this research revealed the incorporating of silica fume with new mixing method was a promising approach to achieve good compressive strength in concrete industry.

1 Introduction

Self-consolidating concrete or self-compacting concrete (SCC) has gained much attention in the concrete world. SCC is an innovative concrete which is able to flow within the entire part of formwork and passing through all the obstacles without any external force provided by the mechanical vibration during the placement of concrete [1]. Through the case studies, a variety of benefits had been identified and certain conditions can be improved using SCC such as reduce labour needed for compacting concrete, increase the speed of construction, improve the surface aesthetics and reduce difficulties in the concrete placing with access constraint [2]. From the point of performance characteristics, SCC can perform the existing concrete's ability to produce the desired engineering properties with a higher constructability compared to conventional concrete. The fresh properties of SCC surpass the performance level of the conventional concrete and this significant result can be showed through the concrete workability [3]. The combination of constituent material plays an important role in the performance of SCC. A deep understanding on the material properties and the chemical reaction within the concrete mixture aids in the designing of SCC mixture proportion. However, generally, a prescribed mix

M. Awang and S. S. Emamian (Eds.): *Advances in Material Science and Engineering*, LNME, pp. 265–272, 2021.
https://doi.org/10.1007/978-981-16-3641-7_31

design of SCC is hard to achieve the desired high-performance properties. Often, pozzolans, containing reactive silica, is always a substitute material for the Portland cement. In conventional concrete, pozzalans is used extensively to enhance the performance and engineering properties of concrete such as heat of hydration reduction, inhabitation of alkali-silica reactivity and cost saving. In the concrete industry, it can be shown that the utilization of pozzolans in the blended cement is with most of the incorporation of fly ash. However, the strength properties with the replacement of silica fume in the SCC are not known commonly. Therefore, SCC incorporating silica fume with new mixing method will be investigated to obtain all the engineering properties of this type of concrete.

Several researches had been conducted to establish SCC mixes in different ways of approach. Three different types of SCC mixes have been identified such as (1) powder-type SCC, (2) viscosity modifying admixture (VMA)-type SCC and (3) combined-type SCC. In this study, the development of combines-type of SCC was used to improve the robustness of powder-type SCC. A small amount of VMA is added into the concrete mix. At the same time, the required cementing materials is lesser compared to the powder-type SCC. Although the water-cement ratio of this SCC is lower than the powder-type SCC, the viscosity of cement paste is enhanced through the provision of VMA. Overall, combined-type SCC has a higher filling ability and stability due to its higher viscosity of the cement paste [4]. This far, the knowledge on the utilisation of new mixing method by incorporating silica fume to achieve SCC is rare. The effects of new mixing method by incorporating silica fume is still not well-known. This research aims to explore the strength properties of silica fume incorporated in the self-consolidating concrete.

2   Materials and Methods

2.1   Cement

Ordinary Portland Cement Type 1 (OPC) was used in this development of SCC mix for the research purpose. This OPC was manufactured by Tasek Corporation Berhad which was available in a bag of 50 kg. The quality of OPC from Tasek Corporation Berhad had been examined through the Malaysian Standard MS 522 which was associated with the British Standard B 12. Table 1 shows the chemical composition of each element presented in OPC.

Table 1. Typical chemical composition of ordinary portland cement

Physical specifications			Chemical composition (%)					
LOI	Specific gravity	Blaine's specific surface area (cm^2/g)	SiO_2	CaO	Fe_2O_3	MgO	Al2O3	SO_3
0.64	3.14	3510	21.28	64.64	3.36	2.06	5.60	2.14

2.2 Mixing Water and Superplasticisers (SP)

Potable water from the municipal tap supplied by the local water supplier was used for mixing and curing. Before placing the water, it is important to ensure the water is free from any contaminants or impurities, which can adversely affect the hydration of cement and the properties of concrete. The specific gravity of the mixing water is taken to be 1 g/cm^3. In this research, the PCE-based superplasticizers used in this production of SCC was known as MasterGlenium SKY 8808 which was manufactured by BASF Company. The recommendation dosage of this superplasticizers to be added was 1.5% by the weight of cementitious material to enhance the workability of fresh concrete.

2.3 Aggregates (Granite and Sand)

The coarse aggregate used to develop SCC in this research was crushed granite which was extracted from a quarry site at Selangor region. The available crushed granites were sieved to a maximum size of 19 mm by using the standard ASTM ¾" sieve and the crushed aggregate which was smaller than 4.75 mm was removed through the sieve No. 4 as well.

The source of fine aggregate used in this research of SCC was from the mining sand which was extracted from the area in Selangor region. All the fine aggregate used was controlled below the maximum size of 4.75 mm and was ensured to be in saturated surface dry by oven-dry the fine aggregate for 24 h. After sieving fine aggregates with the sieve No. 4 according to ASTM standard, fine aggregate was stored in a large plastic container to avoid any ingress of contaminants.

2.4 Silica Fume

Silica fume is one type of mineral admixture which contains cementitious properties that was used in this experiment to produce high strength SCC. A study had shown that addition of about 10% of silica fume to weight of cementitious material in the concrete with low water-powder ratio would decrease the amount of superplasticizers needed for a particular workability [5]. Generally, the reactivity of silica fume was able to alter the rheology of fresh SCC and enhanced the strength and durability of concrete.

2.5 Mix Proportions

Since the optimum water to cement ratio was determined through the preliminary mixture design with four different percentage amount of silica fume which replaced cement content were conducted. Table 2 below outlines the mixture proportions of SCC-SF with 0, 10, 15 and 20% of silica fume as cement replacement at the water to cementitious ratio of 0.32.

Table 2. Mix proportion of silica fume as cement replacement

Specimen	Material (kg/m^3)					
	Cement	Silica Fume	Fine Agg.	Coarse Agg.	Water	Superplasticizers
SCC-0.32	400	0	600	800	128	6.0
SCC^a-SF-10	360	40	600	800	128	6.0
SCC-SF-15	340	60	600	800	128	6.0
SCC-SF-20	320	80	600	800	128	6.0

Note:
^{a}SCC-SF-10 = self-consolidating concrete with 10% of silica fume as partial cement replacement material.

2.6 Testing Methods

The production of SCC was based on the new mixing method instead of the conventional mixing method. The procedure in mixing the concrete constituent materials was different as compared to the conventional method and resulted in the reduction of concrete mixing time. The total time taken for this new SCC mixing method was around 7 to 9 min. In the end, there was a total time reduction up to 33% compared to the conventional mixing method. In parallel with it, a comparison on the mixing steps between these two methods was shown in the Fig. 1 below.

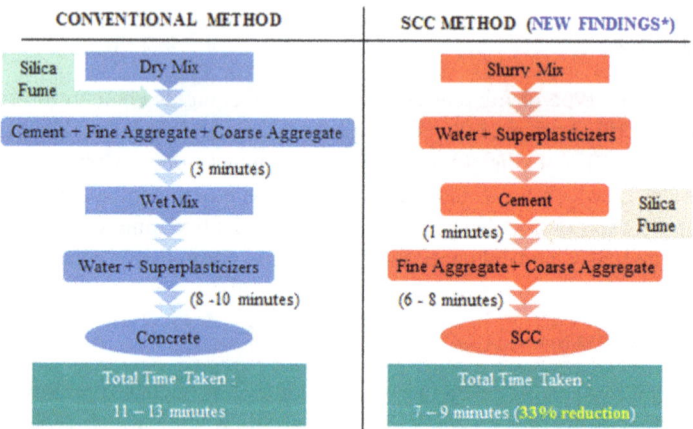

Fig. 1. Comparison on the concrete mixing method between conventional and SCC new method

The specimens were demoulded approximately 24 ± 1 h after casting and were cured in water at 27 ± 2 °C until age of testing. The compression testing machine used was an ELE (Engineering Laboratory Equipment) with a load capacity of 2000 kN running at a pace rate of 2.4 kN/s, in accordance to BS EN 12390–4. To determine the strength properties for each mixture, 18 cubes (100 × 100 × 100 mm) and 18 cylinders (100 × 200 mm) are used to determine the compressive strength and splitting tensile at 1, 7 and 28 days.

3 Results and Discussion

3.1 Properties of Fresh Concrete (Workability)

The results of fresh properties for four different percentage amount of silica fume were obtained and illustrated in Fig. 2. The fresh density of SCC-SF lied in the range of 2200 kg/m^3 to 2400 kg/m^3 which fulfilled the objective of this experimental study. In the comparison between fresh density and hardened density of the trial mixes, all the hardened concrete possessed lighter density. This is mainly due to the free water within the fresh concrete evaporated during the setting of fresh concrete. Reduction of the mass of water through the capillary pores and the formation of C-S-H gel decrease the hardened density of the concrete.

Upon incorporation of silica fume in SCC, the hardened concrete tended to be stable and consistent as all the hardened density of SCC-SF had a standard deviation of below 50 kg/m^3. However, as the percentage amount of silica fume content in the SCC increased, the deviation among the concrete specimens increased as well. This could be explained that the addition of silica fume in the fresh concrete would decrease the free water available for the workability. The reduction of workability resulted in the high possibility in segregation which in the end affects the quality of hardened concrete [6].

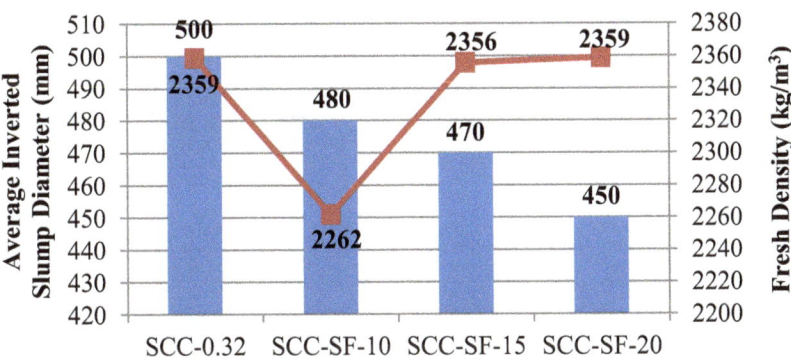

Fig. 2. Average inverted slump diameter and fresh density for trial mixes with different percentage of silica fume as cement replacement material

Figure 2 shows the average inverted slump diameter and fresh density for SCC-0.32 and SCC-SF. Basically, the average inverted slump diameter is an indicator tool which used to interpret the behaviour of flowability as well as the required workability of fresh self-consolidating concrete. For this study, a minimum of 450 mm slump diameter is determined to be the practical limit in the formation of SCC. From the Fig. 2 above, all the mixes attained the minimum workability (450 mm) and the average inverted slump diameter decreased corresponding to the increasing amount of silica fume in the mixes. The reason behind of this relationship is the extreme fineness of silica fume as compared to the particle size of cement. The more the replacement of cement with silica fume, the larger surface area of the powder mixture occupies. Hence, more water is being absorbed by the surface of power which leads to lesser free water available for the workability of fresh concrete. Eventually, internal friction between the constituent materials increases and induces slump loss [7].

3.2 Compressive Strength (Continuous Moist Curing)

Compressive strength is the most essential property of hardened concrete and has its valuable importance in the structural design. In order to provide insight of the compressive strength of concrete, compressive test is the most common evaluation tool to determine the performance and quality of concrete. In this study, the influence of silica fume as the replacement of cement toward the strength of SCC was analysed and discussed. Meanwhile, two different ways of curing methods, for instance, water curing and air dry, were adopted to study their effects toward the development of compressive strength. Basically, all the cubes of specimen were prepared to be tested at 1, 7 and 28 days of water curing. Figure 3 illustrates the growth of compressive strength of SCC-0.32 and SCC-SF at 1, 7 and 28 days of water curing period. Clearly, SCC-SF performed better in their later strength at the age of 28 days. This might be due to the incorporation of silica fume in the SCC which was responsible for the development of late strength. The pozzolanic reaction between the micro silica and calcium hydroxide induces the formation of cementitious products through the secondary hydration process. Hence, strength of concrete can be enhanced at the later stage of curing period [8–10].

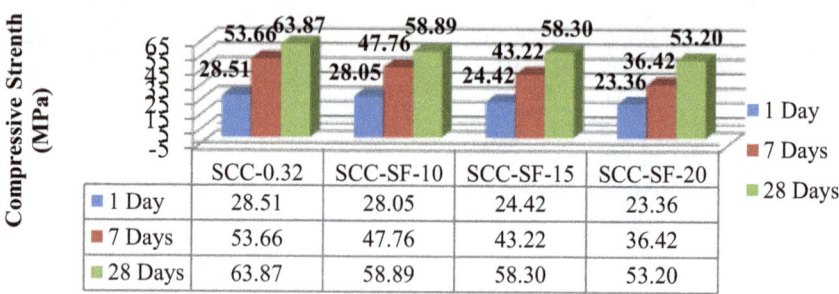

	SCC-0.32	SCC-SF-10	SCC-SF-15	SCC-SF-20
1 Day	28.51	28.05	24.42	23.36
7 Days	53.66	47.76	43.22	36.42
28 Days	63.87	58.89	58.30	53.20

Fig. 3. Compressive strength (water curing) for different percentage of Silica Fume at 1, 7 and 28 days

3.3 Ultrasonic Pulse Velocity

Figure 4 showed that SCC-SF-10 had the highest value of UPV which was 4.798 km/s followed by SCC-0.32 which had 4.640 km/s of UPV. The lowest UPV was recorded by the SCC-SF-20 at 4.215 km/s at the 28 days of ages. Basically, the significance of

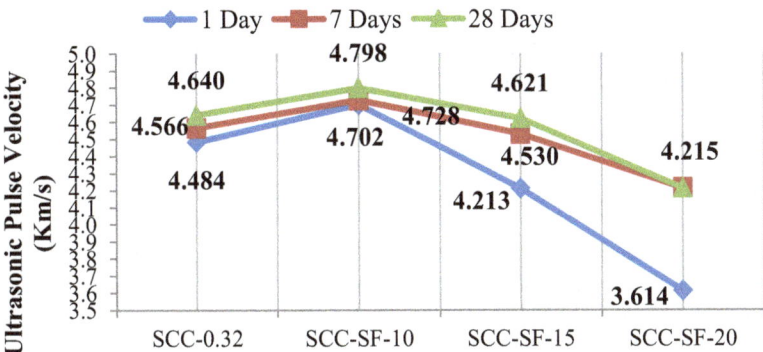

Fig. 4. Ultrasonic pulse velocity development (water curing) for different percentage of Silica Fume at 1, 7 and 28 Days

imperfections of the concrete could be determined by the magnitude of reduction in the UPV. In this case, SCC-SF-10 could be explained in which it had the least amount of voids within the hardened concrete structure and the densest configuration among the constituent materials. The amount of silica fume in SCC-SF-10 were sufficient to develop much C-S-H gel in order to occupy and fill up the voids and pores within the concrete. On the other hand, the lowest UPV in SCC-SF-20 was due to the significant amount of voids within the concrete. The ultrasonic pulse was diffracted around the region of defect and consumed more time to travel across the concrete cube, eventually caused reduction in the pulse velocity. Although SCC-SF-20 had 10% more silica fume as compared to SCC-SF-10, however these excess silica fume did not undergo pozzalanic reaction which induced the secondary hydration process. Hence, the voids in the concrete did not replace with the gel and contributed to a porous structure.

4 Conclusions

Based on the experimental results of this study, the following conclusions can be drawn:

(1) The effect of silica fume on the fresh properties of SCC in terms of workability was investigated. All the SCC control mix and SCC-SF achieved the minimum requirement of inverted slump value (450 mm). However, SCC-SF provided lower value of average inverted slump diameter compared to SCC control mix. Hence, silica fume reduces the workability of SCC.
(2) SCC control mix and SCC-SF had achieved more than 50 MPa of compressive strength at the 28 days of water and air curing period. SCC-0.32 had the highest compressive strength compared to all SCC-SF in which 10% and above of cement replacement level did not improve the compressive strength of SCC.
(3) The other engineering properties of SCC with different amount of incorporation of silica fume in terms of ultrasonic pulse velocity was investigated. SCC-SF-10 had gained the highest UPV at 4.8 km/s after the 28 days of water curing period. This indicated that SCC-SF-10 has the greater durability compared to other mixes.

Acknowledgments. The authors would like to thank Universiti Tunku Abdul Rahman for the UTARRF funding support in this research, Project No. IPSR/RMC/UTARRF/2016-C1/Y03 and IPSR/RMC/UTARRF/2017-C2/Y03.

References

1. Okamura, H., Ouchi, M.: Applications of self-compacting concrete in Japan. In: Wallevik, O., Nielsson, I. (eds.) 3rd International RILEM Symposium on Self-Compacting Concrete. Reykjavik, Iceland. RILEM Publications, pp. 3–5 (2003)
2. Daczko, J.A.: Self-Consolidating Concrete: Applying What We Know. CRC Press, London, United Kingdom (2012)
3. Hackley, V.A., Ferraris, C.F.: Guide to Rheological Nomenclature: Measurements in Ceramic Particulate Systems, NIST SP-946:1-9 (2001)
4. Roziere, E., Granger, S., Turcry, P., Loukili, A.: Influence of paste volume on shrinkage cracking and fracture properties of self-compacting concrete. Cement Concr. Compos. **29**(8), 626–636 (2007)
5. Rougerson, P., Aitcin, P.C.: Optimization of the composition of a high-performance concrete. Cement Concrete Aggregates **16**(2), 115–124 (1994)
6. Gonen, T., Yazicioglu, S.: The influence of mineral admixtures on the short and long-term performance of concrete. Build. Environ. **42**, 3080–3085 (2006)
7. Neville, A.M.: Properties of Concrete. Prentice Hall, Hoboken (1996)
8. Siddique, R., Khan, M.I.: Supplementary cementing materials: silica fume. Engineering Materials. Springer-Verlag, Berlin, pp. 67–119 (2011). https://doi.org/10.1007/978-3-642-17866-5
9. Loh, L.T., Yew, M.K., Yew, M.C.: A new mixing method for lightweight concrete with oil palm shell as coarse aggregate. In: E3S Web of Conferences, vol. 65, 02012 (2018)
10. Yim, H.J., Bae, Y.H., Kim, J.H.: Method for evaluating segregation in self-consolidating concrete using electrical resistivity measurements. Constr. Build. Mater. **232**, 117283 (2020)

Characteristics of Asymmetry Motion in the Axle Vertical Plane on the Dynamic Response of Railway Wagon

Fillemon Nangolo[✉], Mutiu Erinosho, and Ester Angula

Department of Mechanical and Industrial Engineering, University of Namibia, P.O. Box 3624, Ongwediva, Namibia
fnangolo@unam.na

Abstract. The aim of this paper was to analyse the effect of symmetrical and anti-symmetrical excitation modes on the vertical dynamic responses of the railway vehicle. A 9-DOF mathematical model representing the railway wagon was developed using the Lagrange's equation to obtain the system's equations of motion. The equations of motion were uncoupled and solved by numerical integration using piecewise constant as interpolation functions, and thereafter, the model was validated using measured data. The discrepancies between the surveyed paths and the paths predicted by the model were found to be in the range of 6.34% (maximum) to 3.64% (minimum). Tens of reproducible simulation responses were carried out with the model to obtain data for the analysis. The results confirmed that, the effect of symmetrical and anti-symmetrical excitation modes changes the behaviour of the system.

1 Introduction

This paper studies, the effect of symmetrical and anti-symmetrical excitation modes due to bounce, pitch and roll motions of the vertical axle planes in the two-axle bogies have on the vertical response of the railway wagon. The analysis of the vertical dynamic response of the railway wagon depends on the response functions in four reference points (B_{011}, B_{014}, B_{021} and B_{024}) of the car-body (as shown in the vehicle model later), composed by means of these response functions to the symmetrical and anti-symmetrical excitation modes. When the railway wagon travels along on a track with vertical irregularities, the axles make forced movements in the vertical plane, so that the axle plane on a bogie has a translation and rotational motion and due to this combination, the vertical axle plane of each bogie will produce symmetrical and anti-symmetrical motion modes of the vehicle axle plane. And hence, these motions are translated to the suspended masses (bogies and car-body) of the railway wagon, via the suspension elements, thus exciting the symmetrical and anti-symmetrical vertical vibration modes of the railway wagon in the horizontal plane.

From the review of the relevant literatures, it is evident that although different types of railway vehicle models have been developed in order to study the effect of track irregularities on the vertical dynamic responses of the vehicle, very few studies or non-have investigated the effect of symmetrical and anti-symmetrical modes due to bounce,

M. Awang and S. S. Emamian (Eds.): *Advances in Material Science and Engineering*, LNME, pp. 273–280, 2021.
https://doi.org/10.1007/978-981-16-3641-7_32

pitch and roll motion of the vertical planes have on the vertical response of the horizontal plane of the railway wagon. Therefore, it is the intention of this paper to extend preceding analysis by introducing the effect of symmetrical and anti-symmetrical excitation modes due to bounce, pitch and roll motion of the axle vertical planes have on the vertical dynamic response of railway wagon in the horizontal plane. From a theoretical point of view, this section reviews and justifies different representations of linear and non-linear models as well as underlying physical assumptions that will be used throughout this work. Most of the points made have been previously considered by different authors, so that the objective of this section is only to provide a complete and consistent treatment of all the aspects relevant to this research. Over time, there have been several studies [1, 2] which are contributed to by many researchers regarding the dynamic analysis of railway wagons. Some authors [3, 4] investigated the vertical dynamic behaviour of railway bogies moving on rail which is discretely supported by sleepers resting on an elastic foundation. The effects of imperfections on the running surfaces wheel and rail were studied by assigning irregularity functions to these surfaces. Other researchers, [5, 6] studied the dynamic response of a two-bogie vehicle (2D – model) due to symmetrical and anti-symmetrical excitation modes in the vertical plane and concluded, that the excitation modes induced by the track vertical irregularities come, on one hand, from the symmetrical bounce and pitch of the axle planes – symmetrical excitation modes, and on the other hand, from the anti-symmetrical bounce and pitch of the axle planes – anti - symmetrical excitation modes. A vast amount of studies in literature have dealt with 2D – vehicle model that employs the pitch-plane model in order to incorporate the pitch effect of the vehicle, as in [7], on wheel rail impact force. Other studies employed the roll-plane to incorporate the influence of roll dynamics, [8]. However, very few authors or none in literature [9–11] employed 3D – vehicle model to study the vertical dynamic response of the horizontal plane of the railway wagon due to symmetrical and anti-symmetrical excitation modes of the vertical axle planes of the vehicle. Such a model provides all the advantages of roll, pitch-plane models and quite adequate for investigation of the influences of coupled vertical, pitch and lateral motions of the vehicle. Therefore, it is the intention of this paper work to extend preceding analysis by introducing a model which bridges the gap between these exciting classes of models, to obtaining efficient vertical dynamic response of the vehicle model.

2 Vehicle Model

A 9-DOF 3-D mass-spring damper system incorporating bounce, pitch and roll motions of the railway vehicle is developed, as shown in Fig. 1. The model consists of a car body, two bogie frames and four wheel-sets as rigid bodies. The spring and damping elements representing the secondary suspension connecting the car-body to the two bogie frames.

Similarly, the spring and damping elements representing the primary suspension system connect the front and rear bogies to the wheel-sets.

The car body is modeled as a rigid body having a mass m, and mass moment of inertia J_x and J_y about the transverse and longitudinal centroidal horizontal axes, and the static deviation moments of the car-body are given as $D_{xy} = D_{yx}$ to the main axes passing through the center of mass T, respectively. The front and rear bogies are considered as

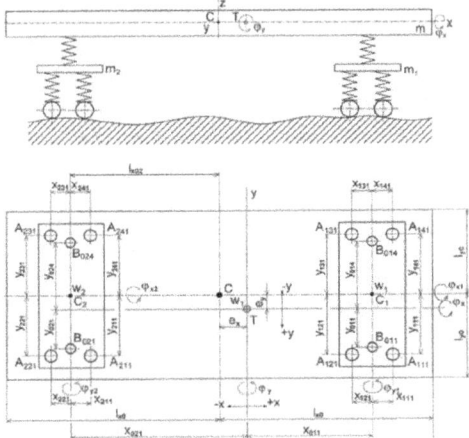

Fig. 1. Analytical model

rigid bodies with masses m_1 and m_2, with mass moment of inertia of the front bogie given as J_{x1} and J_{y1} about the transverse and longitudinal centroidal axes. Similarly, the mass moments of inertia of the rear bogie are given as J_{x2} and J_{y2} about the transverse and longitudinal centroidal axe. The static deviation moments of the front and rear bogies passing through the center of mass T_1, T_2 of the front and rear bogies are given as D_{xy1} = D_{yx1}, $D_{xy2} = D_{yx2}$, respectively. The primary and secondary suspension system are characterized by spring stiffness constant k_{jki} and damping coefficient b_{jki}, where $j = 1, 2$ represent the front and rear bogie, $k = 1, 2, 3, 4$ represent the four quadrant of the railway wagon and $i = 1, ..., n$ represent the reference points of the individual springs on the system. Assuming small vertical and rotational displacement of each body about the x, y, x_1, y_1, x_2, y_2 axes and, that the vehicle car-body and the two (front and rear) bogies be rigid bodies, its motion may be described by the relative vertical displacement of the center of gravity of each body w_T, w_{T1}, w_{T2} and rotational displacements φ_x, φ_y, φ_{x1}, φ_{y1}, φ_{x2}, φ_{y2} of each body about the main longitudinal and transverse axis passing through the center of mass about the x, y, x_1, y_1, x_2, y_2 axes. With these assumptions in mind, nine general coordinates of the system are needed to determine the general displacement of the vehicle as

$$q_j(t) = [w_T, w_{T1}, w_{T2}, \varphi_x, \varphi_y, \varphi_{x1}, \varphi_{y1}, \varphi_{x2}, \varphi_{y2}]^T, \dot{q}_j(t), \ddot{q}_j(t).$$

Therefore, the railway wagon model is thus represented as a 3-D 9-DOF mechanical system.

The car-body vertical motions are described by the equations of rigid vibration modes of the car-body and of the bogies – bounce, roll and pitch, as well as by the symmetrical and anti-symmetrical excitation functions as shown in Fig. 2.

2.1 Formulation of Equations of Motion

To determine the equations of motion, it is necessary to determine the system's kinetic energy E_k, potential energy E_p and Rayleigh dissipation function R_d of the model. The

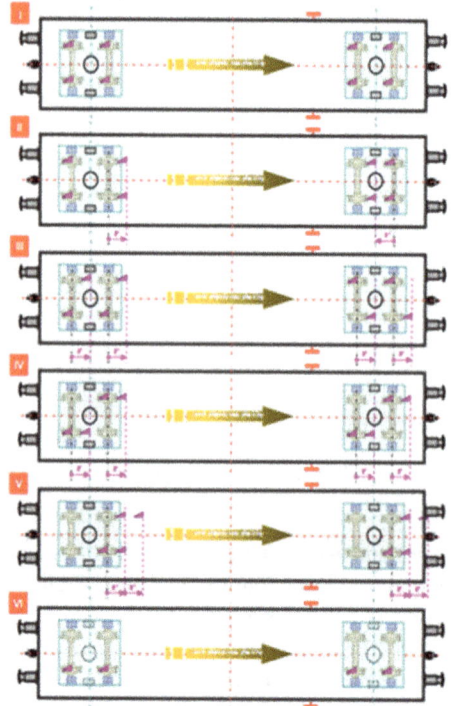

Fig. 2. (I) symmetrical and (II – VI) anti-symmetrical modes

vertical dynamics of arbitrary points $w_j(t)$ of the mechanical system can be obtained by employing the nine components of the generalized coordinates $q_j(t)$. Therefore, the vertical displacement response of the system is the function of the system coordinates and components of the generalized coordinates

$$w_j(t) = w_j\big(w_T, w_{T1}, w_{T2}, \varphi_{x1}, \varphi_{y1}, \varphi_{x21}, \varphi_{y21}, x_i, y_i\big), \dot{w}_j(t), \ddot{w}_j(t) \tag{1}$$

The equations of motion is obtained by using the second order Lagrange equations as follows:

$$\frac{d}{dt}\left(\frac{\partial E_k}{\partial \dot{q}_j}\right) - \frac{\partial E_k}{\partial q_j} + \frac{\partial E_p}{\partial q_j} + \frac{\partial R_d}{\partial \dot{q}_j} = Q_j \tag{2}$$

Where the kinetic energy of the system is given as follows

$$E_k = \frac{1}{2}m\dot{w}^2 + \frac{1}{2}(J_x\dot{\varphi}_x^2 + J_y\dot{\varphi}_y^2 - 2D_{xy}\dot{\varphi}_x\dot{\varphi}_y) + \frac{1}{2}m_1\dot{w}_1^2 + \frac{1}{2}J_{x1}\dot{\varphi}_{x1}^2 + \frac{1}{2}J_{y1}\dot{\varphi}_{y1}^2$$
$$+ \frac{1}{2}m_2\dot{w}_2^2 + \frac{1}{2}J_{x2}\dot{\varphi}_{x2}^2 + \frac{1}{2}J_{y2}\dot{\varphi}_{y2}^2 \tag{3}$$

Hence, the potential energy of the system can be written as follows

$$E_p = \frac{1}{2}\sum_{j=1}^{2}\sum_{k=1}^{4}\sum_{i=1}^{m_k} k_{jki}w_{jki}^2 + \frac{1}{2}\sum_{j=0}\sum_{k=1,2}\sum_{i=1,4} k_{jki}w_{jki}^2 \tag{4}$$

Likewise, the Rayleigh dissipation function is given as

$$D_d = \frac{1}{2}\sum_{j=1}^{2}\sum_{k=1}^{4}\sum_{i=1}^{k_j} b_{jki}\dot{w}_{jki} + \frac{1}{2}\sum_{j=0}\sum_{k=1,2}\sum_{i=1,4} b_{jki}\dot{w}_{jki} \tag{5}$$

Substituting the expression $w_j(t)$ expressing vertical displacement as a function of the generalized coordinate vector $q_j(t)$ as given in Eq. (1) into Eqs. (3), (4) and (5), and after implementation of the relevant derivation and rearrangement of the resulting equations in time domain the equations of motion of the system may be obtained in the general form as

$$\mathbf{M}_h\ddot{\mathbf{q}}_j(t) + \mathbf{M}_b\dot{\mathbf{q}}_j(t) + \mathbf{M}_k\mathbf{q}_j(t) = \mathbf{Q}_j(q_j, \dot{q}_j, t) \tag{6}$$

where; $\mathbf{M}_h$ - is the mass matrix with diagonal elements $\overline{\alpha}_{ii}$ in case of symmetric distribution of the sprung mass, otherwise, the mass matrix is not diagonal – in case of asymmetric distribution of the sprung mass $(\overline{a}_{ij})$, for $i \neq j$. $\mathbf{M}_b$ - damping matrix $(\overline{b}_{ij})$, $\mathbf{M}_k$ - stiffness matrix $(\overline{a}_{ij})$. In this paper, due to generality all the elements of the damping and stiffness matrix are considered not to be zero.

2.2 Solution of Equations of Motion

Transforming both sides of Eq. (6) and after the inverse Laplace transformation, assuming zero initial conditions for the function of the generalized coordinate $q_j(t)$, for $j = 1, 2, \ldots, n/2$, the form of the sum of convolution integral is obtained as follows,

$$q_j(t) = \sum_{l=1}^{n/2}(-1)^{j+i}\sum_{k=1}^{n/2}\left[K_{ji,k}\int_0^t F_i(\tau)e^{-\beta_k(t-\tau)}\cos\Omega_k(t-\tau)d\tau \right.$$

$$\left. +\frac{L_{ji,k} - \beta_k K_{ji,k}}{\Omega_k}\int_0^t F_i(\tau)e^{-\beta_k(t-\tau)}\sin\Omega_k(t-\tau)d\tau \right] \tag{7}$$

Where; $\Omega_k^2 = \Omega_{0k}^2 - b_k^2$ is the natural damped frequency, β_k- denotes the damping ratios, and Ω_{0k}^2 is the undamped natural frequency. K_{jik} and L_{jik} are unknown coefficients of amplitude, depending on the mechanical properties of the system under consideration. The solutions to Eq. (7) are obtained by formulating a numerical integration of the convolution integral. The method uses an interpolation function of the excitation function $f_j(t)$ from which an approximation solution to the convolution integral is obtained.

3 Model Validation

To validate the mathematical model, an experiment was set-up to measure vertical dynamic response of a real railway wagon. A Shmmps – No. 8154 4722 2200 four-axle railway vehicle was selected for this investigation. The vehicle was modified to be in accordance with the given requirements in the theoretical model. The original bogie frames were removed and replaced by another bogie frame type Y25 from a passenger freight wagon. Each test was repeated 2 – 3 times and a total of 89 tests were measured. Twenty quantities were measured, these are; – front and rear bogie frames relative vertical displacement response with to the wheelsets within each bogie (9 sensors were used).

4   Results and Discussion

A MATLAB code of the railway vehicle model in Fig. 4 was developed and calibrated against experimental results conducted for different excitation models. Model A-I was taken as the baseline in this paper. Figures 3 show the vertical dynamic and frequency responses function of the car-body driving over symmetrical excitation.

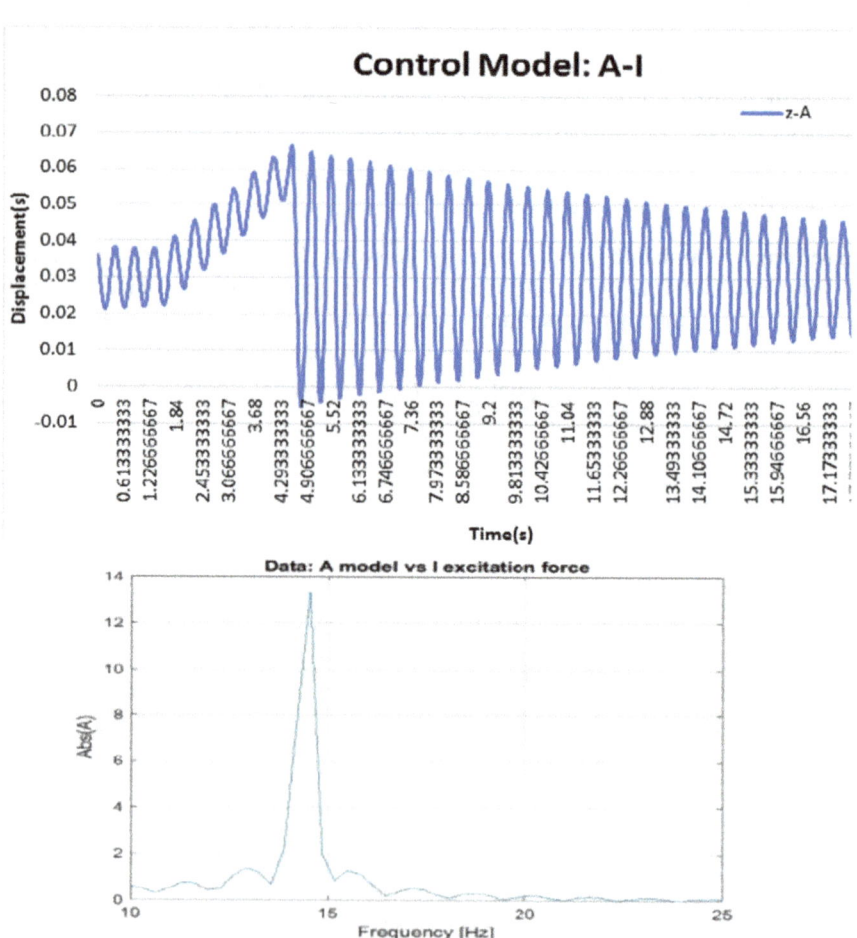

Fig. 3. Vertical dynamic and frequency responses function: car-body

A comparison of the control model A-I (blue) time response to the other cases (B-II and C-III) is shown in Fig. 4. The frequency responses of the vehicle under symmetrical and anti-symmetrical excitations (I–VI) are showed in Fig. 5. From the results obtained, it is clear that the introduction of anti-symmetrical excitation changes the behaviour of the system. A second peak in model B (red) appears at low frequency and with model E (green) a third peak appears, all due to the introduction of anti-symmetrical excitations or due to frequency/amplitude modulation.

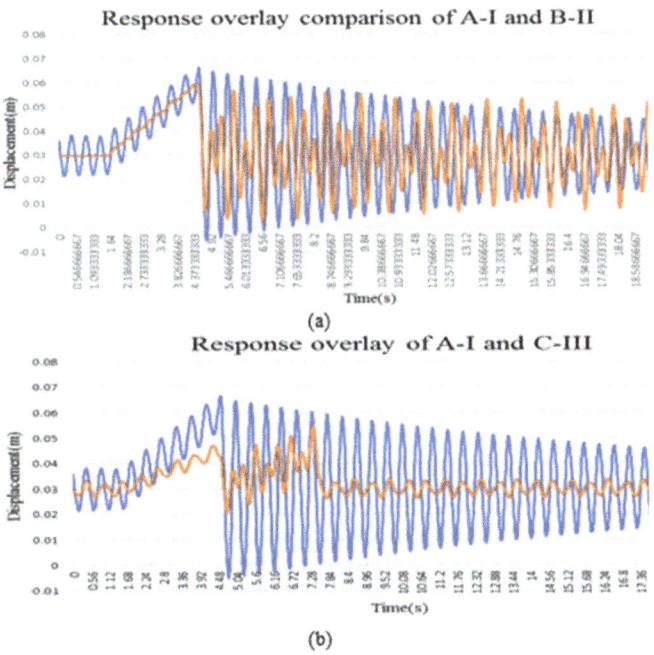

Fig. 4. Time response of model vehicle due to anti-symmetrical excitation (II–VI)

From the results obtained, model A is only performing bounce motion and filtering out the roll and pitch motion about x and y–axes, hence the response of the system is dominated by the vertical dynamic response (bounce) of the horizontal plane of the car-body. Same is true for model B, C and D, the anti-symmetrical excitation introduce the pitch motion of the car-body, filtering out the roll-motion. The response (two peaks) clearly shows the pitching about y-axis and the bounce in the z-direction with a reduced in amplitude and phase shift due to damping. A third peak is recorded with model E (green), indicating the introduction of roll motion about the x-axis. A combination of roll, pitch and bounce motion – has a great effect on damping as shown in Fig. 9(b) as compared to Fig. 9(a). It is also evident from the results obtained that, for excitation modes that are symmetrical about the x-axis, the axle plane can only have pitch motion, since the bounce motion is not conveyed to the bogies due to the axle filtering out the bounce motion. Same is true, that the axle plane can only perform bounce motion by not

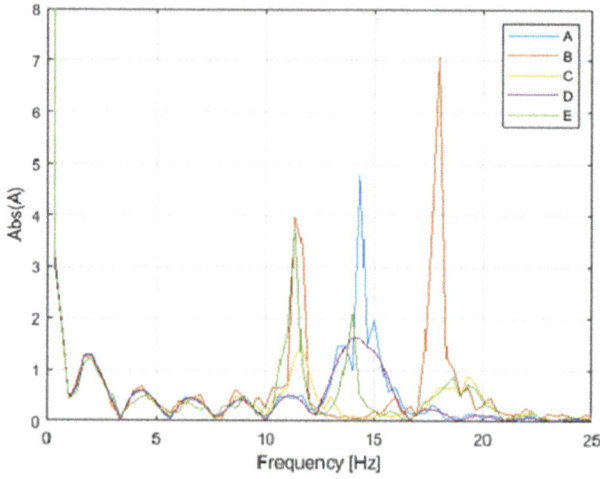

Fig. 5. Frequency response of model vehicle

transmitting pitch motion to the bogie. This effect is the result of movements between vertical dynamic response of the axles and the velocity of the vehicle.

5 Conclusion

In this work, a method to evaluate the effect of symmetrical and anti-symmetrical excitation mode of the vertical axle plane on the vertical dynamic response of the railway wagon has been studied. The model generally confirmed that symmetrical and anti-symmetrical excitation mode has an influence on the behaviour of the railway wagon under different loading. It is important to note that the level of agreement achieved between measured data and mathematical model very strongly depends on the quality of test data and on factors that are related to the model, such as the level of detail included in the model and integrity of the input data, such as dimensions, inertia properties and system's characteristics.

References

1. Soukup, J., Volek, J.: Dynamic of Rigid and Deformable Bodies, Usti nad Labem, Czech Republic
2. Nangolo, F., Soukup, J.: Journal of Machine Modeling and Simulation
3. Woodhouse, J.: Journal of Sound and Vibration **215**, 547–569
4. Scanlan, A.H.: Journal of Sound and Vibration **13**, 499–503
5. Sun, Y.Q., Dhanasekar, M., Roach, D.: Proceedings of institution of Mechanical Engineers
6. Dikmen, F., Bayraktar, M., Guclu, R.: Indian Academy of Sciences, pp. 265–280
7. Ellis, J.R. (ed.): Road Vehicle Dynamics, Ahron, USA
8. Pust, L., Lada, M. (ed.): Pruzne ukladani stroju, SNTL, Prague, Czech Republic
9. Daniel, J.I. (ed.): Engineering Vibration II, New Persey, USA
10. Mitschke, M. (ed.) Influence of Road and Vehicle Dimensions on the Amplitude of Body Motions and Dynamic Wheel Loads. SAE, Preprint 310C
11. Jimn, H., Zhi-Fang, F. (ed.): Modal Analysis

Design of a Walkway Using Renewable Energy Harvesters Within Pavement

Elham Maghsoudi Nia[1]($\boxtimes$), Noor Amila Wan Abdullah Zawawi[1],
and Balbir Singh Mahinder Singh[2]

[1] Department of Civil and Environmental Engineering, Universiti Teknologi PETRONAS,
Kuala Lumpur, Malaysia
[2] Department of Fundamental and Applied Sciences, Universiti Teknologi PETRONAS,
Kuala Lumpur, Malaysia

Abstract. Generally, pedestrian walkway are designed to provide an accessible and safe path for people who prefer to walk. The walking activity can benefit both people and society in term of health and environment conservation through reducing usage of vehicle transpiration. Hence, the pedestrian walkway can serve more than providing a safe path and are potential to be a pedestrian and environmentally friendly walking space. This study aims to design a pedestrian and environmentally friendly walkway. The results of experiment results showed that 48.88 kWh energy is produced by pedestrian footsteps and synthetic rubber can provide an ergonomic pavement for pedestrian. Hence, the study is contributed in developing environmentally friendly walkway using renewable energy system as the contributor factor in reducing fossil fuel consumption of urban spaces sector. Moreover, merging pedestrian friendly walkway with hybrid renewable energy technologies (for generating electricity of lighting and providing users' comfort as well) would develop green concepts for pedestrian paths in cities.

Keywords: Walking energy harvesting · Piezoelectric mechanism · Hybrid system · Footsteps · User comfort

1 Introduction

Fossil fuels are the main source of energy which may deplete in the future. Moreover, they cause climate change issues, greenhouse gas emission and consequently, global warming. Increasing use of renewable energies is suggested as a solution in response to sustainability concerns. Lighting system is also positively impacted by green technology [1]. However, currently, solar, wind or other renewable energy equipment of lighting is in the initial stages [2]. This study explores the issues and problems in the design of a pedestrian and environment friendly walkway. The study addressed the question of; "How to design and construct a pedestrian friendly walkway using renewable energy technologies for lighting system?" Therefore, kinetic energy of footstep as a source of renewable energy were selected. Harvesting walking energy is a sustainable method that converts wasted energy of body into electric power. There are different types of mechanism for walking energy conversion that this study focused on piezoelectric transduction

M. Awang and S. S. Emamian (Eds.): *Advances in Material Science and Engineering*, LNME, pp. 281–288, 2021.
https://doi.org/10.1007/978-981-16-3641-7_33

as a desirable mechanism [3]. The study selected a walkway with high foot traffic as a case study to investigate the amount of generated energy by the pedestrian. Then, the generated energy will be used for public consumption, which in this study will supply lighting system of the walkway. In addition to harvesting renewable energy, the study intends to provide a pedestrian friendly walkway and user comfort. Therefore, in order to design a walkway that provides accessibility, thermal comfort, safety and security, the path will be paved with flexible and anti-fatigue materials.

2   Methods

As aforementioned, a case study technique was selected as the general approach. The number of footsteps along the walkway was counted during a day. The study proposed a pedestrian walkway, which equipped with a pavement consists of piezoelectric sensors. The top layer of the pavement also selected as a flexible and green material. Then, the scaled prototype fabricated according to the mechanism of walking. Finally, the scaled prototype tested in term of power generation by the piezoelectric system within the pavement when people step on it. After finalizing the system design and improving prototype according to the results, the prototype is tested for one week for power generation in an open space with the most footstep traffic. The place that is selected to test the prototype is the crowded way at Universiti Teknologi PETRONAS. According to the variables gained from the literature, the prototype of a pedestrian walkway using equipped pavement with harvesters is designed. The paving of the prototype consists of sixteen tiles 335 mm × 335 mm. each tile includes a Plywood as a base that 36 pieces of piezoelectric materials is placed on it. In order to fix the piezoelectric sensors and make them immovable, an acrylic layer with 36 holes as size as sensors is used. A layer of flexible materials such as recyclable rubber covers the piezoelectric sensors at the top. The prototype includes the equipped pavement with the piezoelectric materials, which, the generated voltage is tested by multi meter. The size of the tile is according to the spatial structure of the walking cycle and normal range of motions assuming a velocity of 1.3 m/s [5–7]. The scaled prototype of the pedestrian walkway pavement is shown in Fig. 1.

3   Results and Discussion

The results are described in two sections of flexibility of the prototype and power generation of the system as follows:

3.1   Results of the Flexibility of the Prototype Pavement

In order to test the flexibility of the material, ten men and women are weighed and step on three different materials including PVC mat, rubber mat and wooden board. The output power generated by them is shown in Table 1. The results show that the PVC mat, rubber and wooden board generates 0.5, 1.96, 0.99 mV in average, respectively. As a result, the rubber mat and wooden board generates 75%, 50% more power than PVC mat, respectively. As a conclusion, although rubber mat generates more power, PVC mat, wooden board are more flexible and user friendly than the rubber. Therefore, PVC mat is used for the experimental test due to the availability of the material and flexibility of it [4].

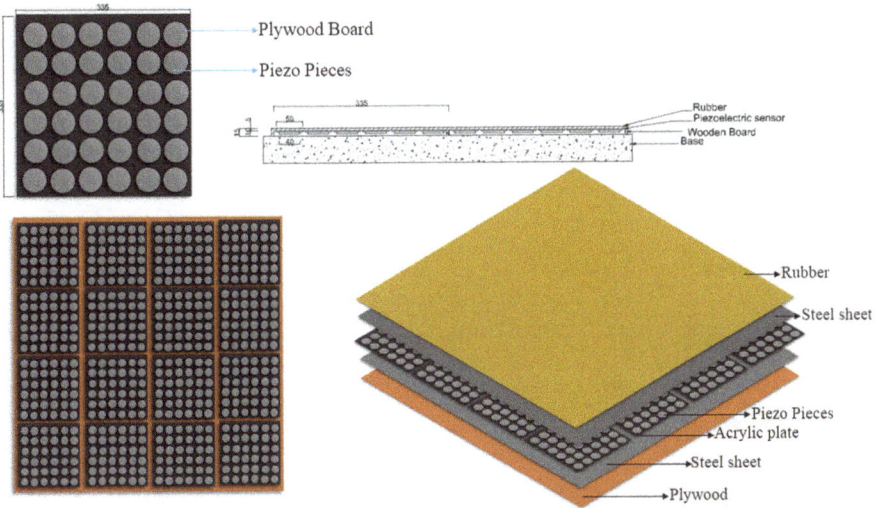

Fig. 1. The scaled prototype of the pavement of the pedestrian walkway

Table 1. The output power generated by ten men and women on different materials

Female					Male				
No.	Weight (kg)	Voltage (V)			Weight (kg)	Voltage (V)			
		PVC	Rubber	Wood		PVC	Rubber	Wood
1	73.4	0.5	1.2	0.7	69.3	0.5	3.9	1.5
2	55.6	0.4	3.2	1.5	122.5	0.6	2.3	1.5
3	45.4	0.9	1.2	1.3	75.8	0.4	3.2	1.02
4	64	0.6	2.3	1.2	67.9	0.9	3.5	0.7
5	63.1	0.5	1.4	0.6	77.1	0.6	2.9	0.9
6	59.2	0.3	0.4	0.1	61.9	0.7	1.5	1.2
7	57	0.3	0.4	0.2	68.4	0.2	1.3	0.9
8	59	0.3	1.2	0.8	61.5	0.2	2.7	1.9
9	62.1	0.6	1.6	0.6	68.7	0.2	2.3	1.6
10	49.5	0.6	0.8	0.6	87.3	0.7	1.9	1.01

3.2 Results of Power Generation by Piezoelectric Tile

In order to measure the out power generated by footstep on piezoelectric sensors, data logger set to record the data each 50 ms. The Figs. 2 and 3 shows the voltage generated by the pedestrian in each day. As the results shows the generate voltage differs according to the weight of people and their velocity. The current also measured for one day by data logger. The shunt resistor 1 Ω is used for this purpose. The results show the constant

Table 2. The output power generated by piezoelectric material in each day

Results of testing the prototype for six days			
No.	Number of pedestrian per hour in average	Voltage (V)	Output power (W) Current = 0.0002 A
1st day, 1st week			
	100	376.95	0.08
2nd day, 1st week			
	176	1731.17	0.35
1st day, 2nd week			
	109	2073.67	0.42
2nd day, 2nd week			
	99	845.84	0.17
1st day, 3rd week			
	141	1396.78	0.28
2nd day, 3rd week			
	140	988.14	0.20

Table 3. The power output calculation for the case study by piezoelectric pavement

Output power (Watt per day)					
1st day	**2nd day**	**3rd day**	**4th day**	**5th day**	**6th day**
0.08	0.35	0.42	0.17	0.28	0.20
Output power per hour					
0.02	0.05	0.06	0.03	0.04	0.03
Output power per person (W)					
0.0002	0.0003	0.0005	0.0002	0.0003	0.0002
Output power per hour for 2706 person (W)					
0.51	0.76	1.47	0.66	0.77	0.55
For 220 m of the case (721 feet) (W)					
92.02	137.23	265.41	119.20	138.20	98.47
For 17 h (W)					
1564.51	2332.83	4512.01	2026.33	2349.45	1673.97

current, which is in average 0.23 mA. Therefore, the current is assumed as constant factor with the value of 0.0002 A.

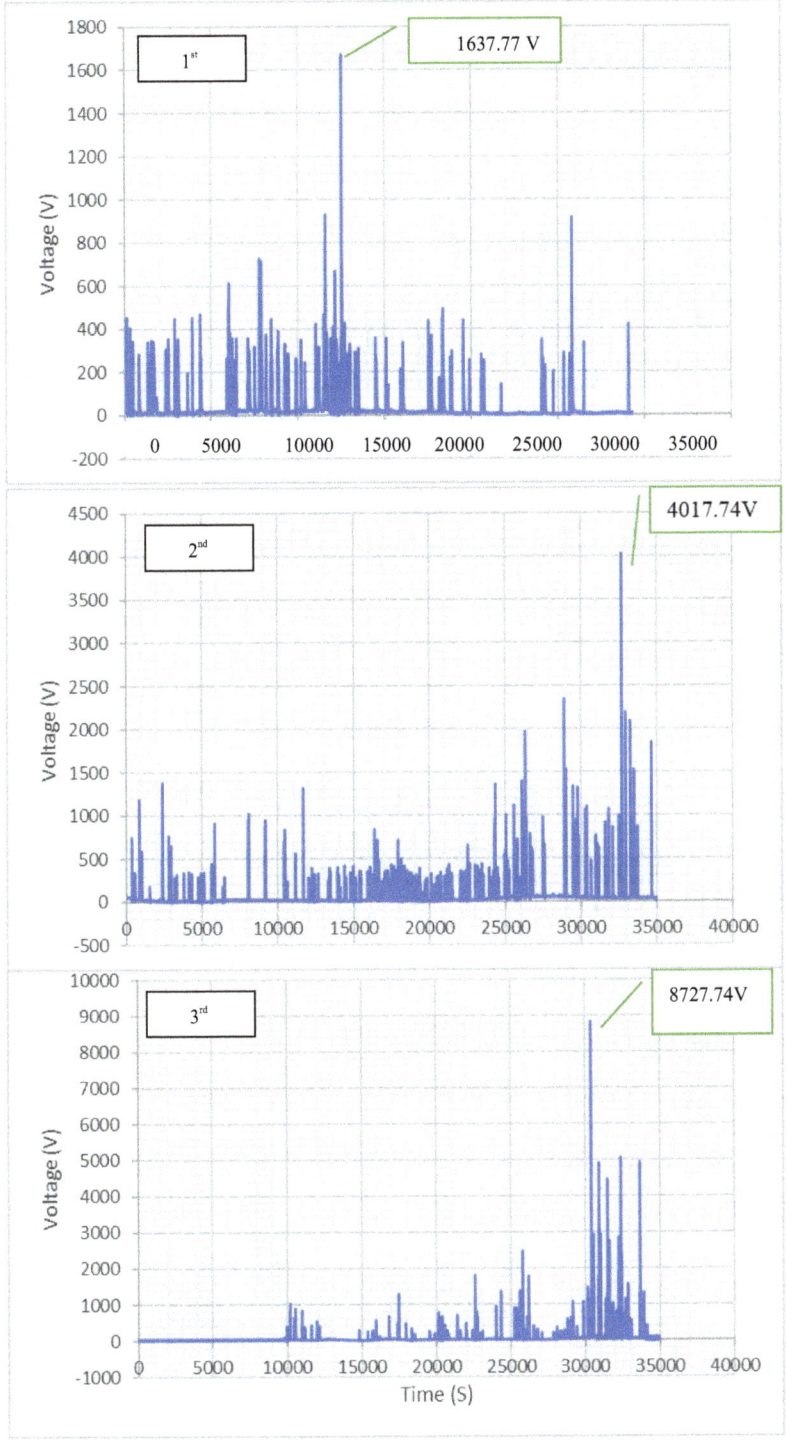

Fig. 2. The voltage generated for first, second and third days by the pedestrian

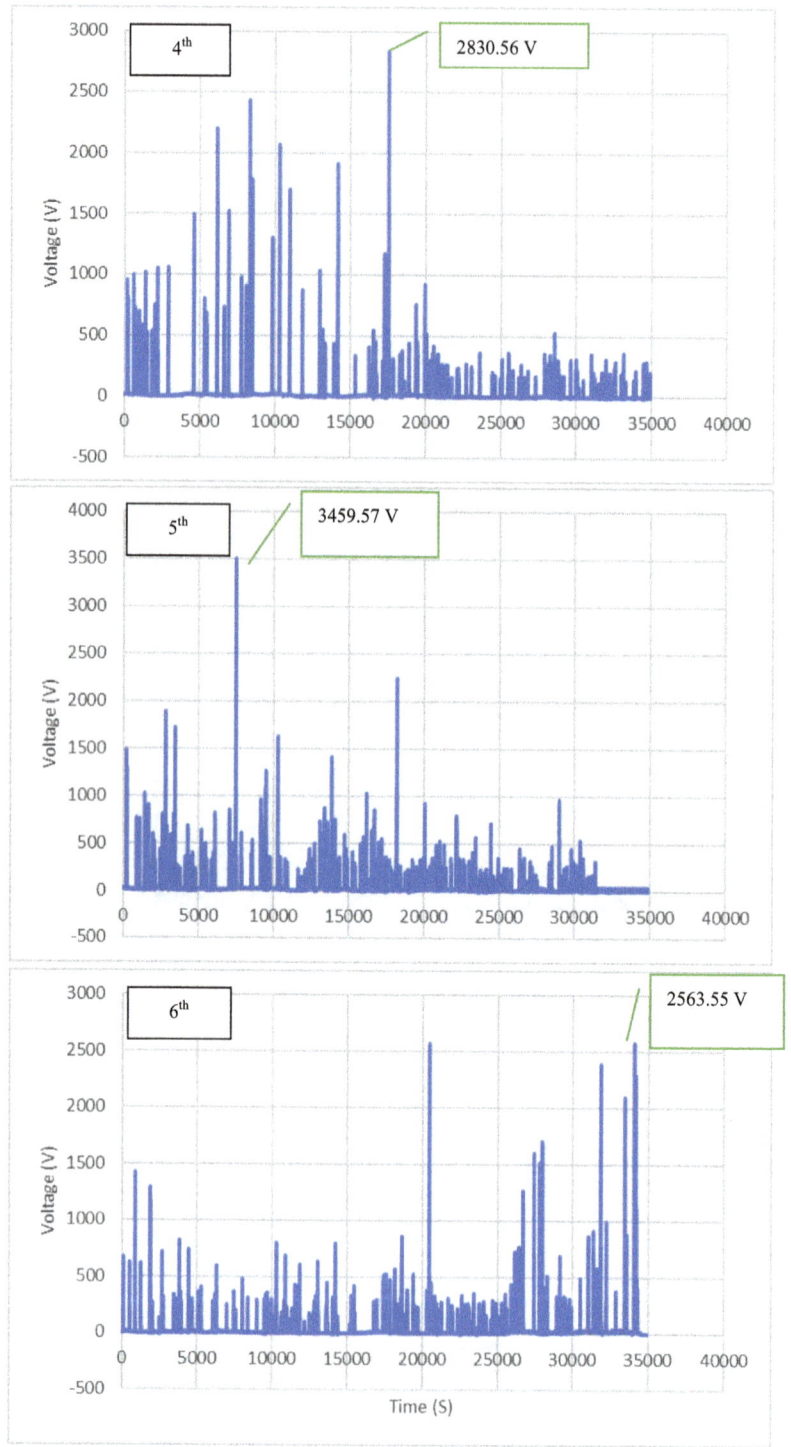

Fig. 3. The generated voltage by pedestrian in fourth, fifth and sixth days

The number of pedestrian and voltage generated in each day is summarized in Table 2 and output power is calculated for each day. Accordingly, Table 3 shows the power generated by the pedestrian is calculated for the length and number of the pedestrian in the case study. Therefore, in average each person generates 0.0003 W power. If we assume to use this pavement along 220 m length of KLCC-Pavilion pedestrian bridge, with 2706 number of people in each hour and for 17 opening hours [8], 2409.85 W energy would be generated. This amount of energy could light 542 LED lamps (18 W) along the walkway during night time.

4 Conclusions

The study designed a walkway that can provide comfortable surface for pedestrians. The results show that covering the pavement with rigid material leads to generating more voltage. In comparison, rubber, wooden board and PVC mat can generated more voltage, which means PVC mat is more flexible and can provide user friendly pavement. It also designed a walkway that can harvest kinetic energy of pedestrian. The proposed prototype consists of sixteen pavement tiles equipping with piezoelectric sensors and covered with a steel sheet and a rubber layer at the top. The study examined different connections and arrangements of the materials and layers to obtain the optimum condition in term of sufficient power generation. The results show that the by arranging positive pole of piezoelectric materials at the top and covering it with a big size of steel sheet can distribute the loads better and subsequently, generate higher voltage. Assuming the case study of pedestrian walkway between Pavilion and KLCC for implementing the prototype, hybrid system can generate 48.88 kWh per day, using kinetic energy of 2607 pedestrian in average per hour (according to the observation). Totally, in the short-term benefits, this power can light 542 LED (18 W) along the walkway for five hours at night for equipping only 220 m of the case study walkway. It can be concluded that by increasing efficiency of piezoelectric materials, length of walkway and number of pedestrians more power would be generated by piezoelectric tiles.

Acknowledgments. The author would like to thank Univesiti Teknologi PETRONAS for providing graduate assistantship to do PhD studies with Civil and Environment Engineering Department.

References

1. Patterson, T., Gillespie, R., O'Donnell, E.J.: Pedestrian-lighting options and roles of responsibility within unincorporated delaware communities. Institute for Public Administration, University of Delaware (2011)
2. Higgins, J., I.o.L. Engineers: Street lighting policy document, H.C. Council, Editor. The Director of Environment (2009)
3. Nia, E.M., Zawawi, N.A.W.A., Singh, B.S.M.: A review of walking energy harvesting using piezoelectric materials. In: IOP Conference Series: Materials Science and Engineering. IOP Publishing (2017)

4. Maghsoudi Nia, E., Wan Abdullah Zawawi, N.A., Singh, B.M.: Design of a pavement using piezoelectric materials. Materialwissenschaft und Werkstofftechnik **50**(3), 320–328 (2019)
5. Vaughan, C.L., Davis, B.L., O'connor, J.C.: Dynamics of human gait. Human Kinetics Publishers, Champaign (1992)
6. Boulic, R., Thalmann, N.M., Thalmann, D.: A global human walking model with real-time kinematic personification. Vis. Comput. **6**(6), 344–358 (1990)
7. Gate, C.: Assessment of Gait (2015). http://clinicalgate.com/assessmentofgait/
8. Elham, M.N., Noor Amila, W.A.Z., Balbir Singh, M.S.: Energy efficient lighting system design for pedestrian walkway using footsteps energy harvesters. Int. J. Ecol. Environ. Conserv. **23**, 89–92 (2017)

Short Review on Silicone Rubber Based Composites as High Voltage Insulation Material

Nornazurah Nazir Ali[1], Hidayat Zainuddin[1], and Jeefferie Abd Razak[2(✉)]

[1] Fakulti Kejuruteraan Elektrik, Universiti Teknikal Malaysia Melaka, Hang Tuah Jaya, 76100 Durian Tunggal, Melaka, Malaysia
[2] Fakulti Kejuruteraan Pembuatan, Universiti Teknikal Malaysia Melaka, Hang Tuah Jaya, 76100 Durian Tunggal, Melaka, Malaysia
jeefferie@utem.edu.my

Abstract. Silicone rubber (SiR) was established as one of the most favorable high voltage (HV) insulator for outdoor application. It possesses unlimited advantages such as undeniable heat resistance, lightweight and versatility in outside environment with outstanding electrical insulation behavior. The progress of SiR utilization as HV insulator was continuously advanced since its first establishment, seven decades ago. The manipulation of this functional macromolecules is diversified due to its customizable, manufacturability, easy processing, low-cost and modifiable. Extensive research has been performed all around the globe to discover the potential of SiR for various other applications. In this review, special attention to SiR types, modification using filler addition, SiR based nanocomposites, electrical insulation performance, and related issues on SiR as high voltage insulation materials were emphasized. This is significance to establish a standard guideline to optimize SiR potential for HV application.

Keywords: Silicone rubber · High voltage insulation · Composite · Sustainable materials · Outdoor insulator

1 Introduction

Insulation perform a vital role in electrical transmission and distribution system. The insulation part in high voltage (HV) transmission systems are slightly intricate due to strict requirement to withstand electrical and environmental stresses. Historically, the high-voltage insulators were first made from porcelain and vitreous materials. However, the porcelain and glass-based insulator for HV possessed limitations such as brittleness, difficult to manufacture and tedious assembly, heavy-weight and somewhat expensive. Hence, the silicone rubber (SiR) was invented as replacement materials for HV application, considering its robustness and other advantages.

SiR macromolecules was first established in United States of America (USA) as early as 1948 as new composite insulator concept [1]. In 1970, Switzerland had first introduced SiR based weather shed hollow core insulator, which led into the discovery of SiR robustness against impact loads and pollution [2]. Later, mass production of SiR

M. Awang and S. S. Emamian (Eds.): *Advances in Material Science and Engineering*, LNME, pp. 289–295, 2021.
https://doi.org/10.1007/978-981-16-3641-7_34

based insulators by Germany and USA was taken in 1976 and 1983, respectively [2]. SiR was eventually becoming popular than ceramic insulator because of their higher strength to weight ratio, superior performance against contamination and lighter weight. SiR is not only used as a polymeric insulator in high-voltage equipment but also in other fields such as biomedical sciences [3], aeronautics [4], synthesis of nanomaterials [5], and robotics [6].

What makes SiR as promising leading HV insulator is its capability to exhibit hydrophobicity that hinders the formation of conductive path in insulator, that ensuring lower leakage current [7]. SiR is not only able to withstand at wide range of temperatures, but able to improve the surface properties that resulting in lower adhesion [8]. HV insulator coated with nano powder filled SiR, possessed lower ice adhesion which slightly delayed the accumulation of ice on an insulator surface. Janssen et al. [9] had proposed the high temperature vulcanizing (HTV) SiR, as preferable insulator core sheath and shed material for high polluted areas due to its superior hydrophobicity.

Despite of the favorable properties of SiR, this material also got limitation where it tends to age due to weathering. Hence, several research studies have been carried out to tackle this issue. Modification of SiR elastomer through filler addition in composite preparation was favorable and easy to execute. Thus, this short review provides special attention to this approach, by emphasizing the SiR types, SiR modification, nanofiller, vulcanization and hydrophobicity issue of SiR based composites for HV insulator.

2 Common SiR Types for HV Insulation

The main component of SiR are the polymerized siloxanes with different functional group, such as methyl group. SiR is easily identified based on their functional group. SiR without a functional group was named as polysiloxane, due to the Q class elastomeric type. This SiR is heterochain with alternating silicone (Si) and oxygen (O) chain atoms. Polydimethylsiloxanes (MQ) are SiR elastomer with two methyl functional group. While VMQ are SiR elastomer with vinyl and methyl functional groups, whereas PVMQ are SiR elastomer with phenyl, vinyl and methyl functional groups [10]. FVMQ are SiR elastomer with vinyl, methyl and 3,3,3-trifluoropropyl side groups (refer Fig. 1). The difference in SiR elastomeric composition may result in different vulcanization rates. The presence of other functional group had influenced the chemical composition and reaction rate of SiR toward certain chemicals.

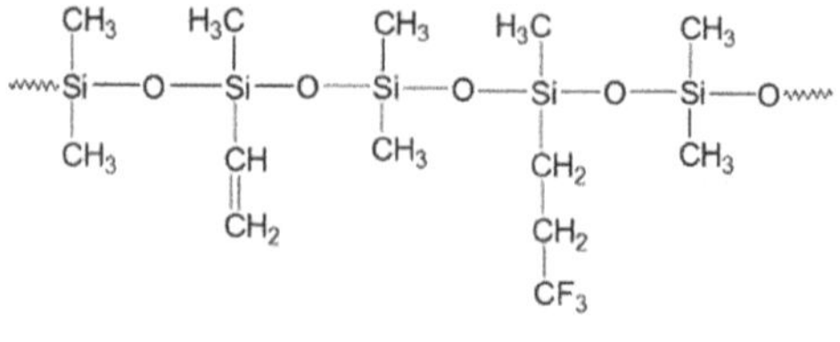

Fig. 1. FVMQ with functional groups substituted by the phenyl, vinyl, or 3,3,3-trifluoropropyl functional group [11]

The quality and performance of SiR are basically depending to the composition of SiR [12]. Hence, it is crucially important to identify SiR composition, prior of insulator development using SiR as main raw material. Modification of SiR is required to customize this material for favorable insulation application. One of the simplest modification routes of SiR is via filler addition to produce SiR based composite insulator.

3   SiR Modification Trough Filler Addition

SiR based composite preparation and development as insulator for HV application are mainly influenced by the filler addition factor. Fillers are added to SiR rubber to enhance the resultant strength, durability, and processing characteristics. In general, the filler behavior in composite is mainly influenced by the filler type, filler concentration, size, shape, particle size distribution, the surface chemistry, and others. Filler added has normally occupying the polymer matrix volume and most probably will establish the chemical bond between the filler-matrix [13]. Silica and alumina trihydrate (ATH) are two examples of mineral fillers that commonly added to SiR [14]. Selection of silica and ATH are due to their flame retardancy advantage which able to suppress the tracking and erosion. The hydration water from this inorganic filler will be hydrated at higher temperature which providing resistance for tracking and erosion [15].

Filler concentration factor in SiR based composite has proven to influence insulation performance [16]. Filler addition should not too excessive as this condition will lead into agglomeration due to poor dispersion. Reduction in material tracking and erosion which resulting to higher leakage current are due to filler agglomeration. Hence, optimization of filler concentration for SiR based composites need to be performed. Addition of 20 – 25 wt.% of silica and surface-modified ATH has improved the inclined plane test performance for SiR [17]. For both ATH and silica fillers, at 1.50 μm of particle size, it was found that the silica filled composites are performed better than ATH filled SiR composites [18]. Different types of filler added has resulted different performance to the SiR based composites.

However, filler sizes also need to be considered in the production of composite insulators. Semi-reinforcing filler with size of 10–100 nm has significantly increased the strength, tearing and wear resistance of produced composites [13]. Filler size of more than 10 μm were pronounced as flaws which initiated crack during the bending, while filler size at range of 1 – 10 μm were describe as diluent with minor effects. This was attributed to the reduction of total surface area which limits the interaction between the filler and matrix. Hence, it can be deduced that the filler particle size could provide a significance influence to the performance of SiR based composite insulator for HV application. Due to the advanced in nanotechnology, filler size effect at down to nanometer scale range in SiR based nanocomposites was reviewed at the following section.

4   SiR Based Nanocomposites for HV Insulation Application

Polymer nanocomposites have gained much interest due to their potential of improving the resulted mechanical, thermal and electrical properties of pure polymer matrix [19].

Owing to the higher surface area of nanofiller, only smaller amount of nanofiller concentration has been added to SiR based nanocomposites. The added nanofiller prone to promote better interaction and bonding with polymer matrix at lower filler addition [20]. Hag et al., has found that the SiR performance with addition of only 10 wt.% of nanofiller was comparable to the SiR performance at 50 wt.% of micron size filler addition [21].

It has been reported that silica nanofiller helps to extend the inception time of SiR nanocomposites, where the inception time increases with an increase of nanofiller loadings. Nanofiller slows down the broadening of the tree channel by restraining the electrical field distribution around the needles, which increases the tree inception time. In addition, treeing forms easily in neat polymer compared with that in nanocomposites [22]. Nanofiller has shown to obstruct the morphological and organic structural damage compared with micro-filled SiR and pure SiR. Therefore, nano silica filler is believed to be a reasonable approach to enhance the corona resistance of micro silica-filled SiR [23]. In another study, the use of 1 vol% of nano alumina has significantly improved the dielectric strength of the silicone rubber/ethylene propylene diene monomer (SiR/EPDM) composite [24].

In another study, has proven that the nano silica filled SiR composite had higher resistance to erosion due to the inclined plane tracking test, as compared than micro silica filled SiR composites. In addition, the hydrophobicity recovery of nano filled SiR based composites was found to be lower than that of micro-filled composites, owing to the shielding effect of the nanofiller. The contact angle of pure SiR was higher below the area of needle tip after 48 h of exposure to corona [24]. In general, nanofiller can help to improve the partial discharge resistance, but it may reduce the hydrophobicity recovery of the composites. Hence, utilization of nanomaterial for SiR based nanocomposites are very much important for improving the micro-scale filler filled SiR based composites performances.

For SiR based nanocomposites production, the most important challenge is on the selection of correct processing method which allowing good particle dispersion to improve the resulted properties of SiR matrix. For instance, electrospinning of SiR based nanocomposites, has promoted uniform dispersion of nanofiller which at the end, increases the erosion resistance [25]. Mixing of hybrid micro filler and nanofiller is another efficient method to reduce cost and agglomeration problems of nanofiller. The hybrid mixture was found improving the eroded mass of sample by 85%, which marked significance improvement in erosion resistance as compared than using micro filler or nanofiller alone [25].

All in all, choosing correct filler for SiR based composites or nanocomposites for insulation application, can be very complicated. Filler cost must be affordable for cost efficient manufacturing. It is important to choose suitable filler with an appropriate filler particle size for optimum performance of resulted composite. The filler loading must be controlled as the effects of filler may be severe if the filler concentration is too much.

5 SiR Based Composites for HV Insulator

SiR for high voltage application is installed as an insulator along the transmission line in the form of long rod such as the hollow core insulator. This is preferably important due to its simplicity than cap and pins type insulator. Design modification made by increasing the insulator length and adding the shed part could endure higher voltage level. The large, shed design for SiR may eschew the water cascading [26]. For tropical Malaysia weather which having high average annual rainfall, it is important for insulator to resist with water absorption for resistance to physical degradation. SiR based insulator is suitable for outdoor application, but it is still lacking with aging and weathering degradation, which require the SiR to be produced as SiR based composite or nanocomposites. Hence, the hydrophobicity of SiR for outdoor application is becoming main issues when manufacture this advanced material for HV application.

Cyclic low molecular weight (LMW) SiR are advantageous in terms of hydrophobicity recovery due to their faster diffusion rate [27]. The hydrophobicity recovery rate of SiR has basically influenced by the filler concentration. SiR based composites with higher filler loadings had lower increase in voltage flashover as compared than low filled samples. SiR composite sample with higher filler loading shows maximum increase in voltage flashover of 23%, while maximum increase up to 9% flashover for lower filled SiR [28]. Thus, it can be deduced that the hydrophobicity recovery rate are correlates well with the SiR based composite formulations.

SiR also has another unique property of hydrophobicity transfer, which improves its ability as an insulator. The hydrophobicity of SiR may transfer to the pollutants attached on the SiR surface and causes the pollutant to inherit the same hydrophobicity. In other words, the pollutant becomes a part of the insulation. However, hydrophobicity transfer is a very slow process, which will not play a significant role if the pollutant layer is too thick and if the build-up of pollutants is far more blistering. At present, the hydrophobicity transfer of SiR has not been studied extensively [29]. The ability to restore hydrophobicity after a pollutant layer has built up on the surface allows the SiR to suppress the development of leakage current, dry-band arcing, and flashover.

Latest development on SiR based composites are on super-hydrophobicity which resulted by the wet or dry anisotropic etching of insulator surface [30]. Super-hydrophobic produces self-cleaning surface due to excellent water repellant effects that eliminate surface contamination, reduce ice accumulation, promotes corrosion resistance, and extend the lifetime of insulator products.

6 Conclusions

In conclusion, SiR based composite or nanocomposites are promising alternative of unfilled SiR for HV application. However, in producing good performance of filled SiR, various factors need to be taken into consideration, such as the filler type, filler size, filler loading, other processing variables etc. Extensive research in developing the best SiR filled composites and nanocomposites has been overwhelmingly conducted throughout the world due to encouraging performance of filled SiR for outdoor environment for HV application. Hence, several matters related with SiR performance, such as hydrophobicity, physical degradation, HV voltage performance, environmental pollution and related others are needed to deal with extra consideration.

Acknowledgments. Authors would like to acknowledge Universiti Teknikal Malaysia Melaka (UTeM) for sponsoring this research. Thanks to COSSID, CERIA, FKE and FKP for technical support in completing this study.

References

1. Papailiou, K.O: Composite insulators are gaining ground-25 years of Swiss experience. In: 1999 IEEE Transmission and Distribution Conference, vol. 2, pp. 827–833 (1999)
2. Hall, J.F.: History and bibliography of polymeric insulators for outdoor applications. IEEE Trans. Power Delivery. **8**(1), 376–385 (1993)
3. Janeiro-Arocas, J., et al.: Creep analysis of silicone for podiatry applications. J. Mech. Behav. Biomed. Mater. **63**, 456–469 (2016)
4. Xu, R., et al.: Novel bilayer wound dressing composed of silicone rubber with particular micropores enhanced wound re-epithelialization and contraction. Biomaterials **40**, 1–11 (2015)
5. Liu, W., He, G.: Storage life of silicone rubber sealing ring used in solid rocket motor. Chinese J. Aeronaut. **27**(6), 1469–1476 (2014)
6. Berahman, R., Raiati, M., Mehrabi Mazidi, M., Paran, S.M.R.: Preparation and characterization of vulcanized silicone rubber/halloysite nanotube nanocomposites: effect of matrix hardness and HNT content. Mater. Des. **104**, 333–345 (2016)
7. Chang, W., Gorur, R.S.: Hydrophobicity of silicone rubber used for outdoor insulation. In: Proceedings of 1994 4th International Conference on Properties and Applications of Dielectric Materials (ICPADM), vol. 1, no. 1–2, pp. 266–269 (2007)
8. Arianpour, F., Farzaneh, M., Kulinich, M.A.: Hydrophobic and ice-retarding properties of doped silicone rubber coatings. Appl. Surf. Sci. **265**, 546–552 (2013)
9. Janssen, H., Seifert, J.M., Karner, H.C.: Interfacial phenomena in composite high voltage insulation. IEEE Trans. Dielectr. Electr. Insul. **6**(5), 651–659 (2003)
10. Mowrer, N.R.: Polysiloxanes coatings innovations. In: Ameron International, pp. 1–16 (2003)
11. Harper, C.A.: Handbook of Plastics, Elastomers, and Composites, 4th edn. McGraw Hill Companies, Inc., New York (2002)
12. Papailiou, K.O., Schmuck, F.: Silicone Composite Insulators Materials, Design, Applications. Power Systems. Springer, Heidelberg (2013)
13. Harper, C.A.: Handbook of Plastics Technologies, 2nd edn. McGraw Hill Companies, Inc., New York (2006)
14. Koshino, Y., Umeda, I., Ishiwari, M.: Deterioration of silicone rubber for polymer insulators by corona discharge and effect of fillers. In: Materials Science - 1998 Annual Report Conference on Electrical Insulation and Dielectric Phenomena, vol. 1, pp. 72–79 (1998)
15. Gorur, R.S., Cherney, E.A., Hackam, R.: Performance of polymeric insulating materials in salt-fog. IEEE Power Eng. Rev. **PER-7**(4), 58–59 (1987). https://doi.org/10.1109/MPER. 1987.5527194
16. Ghunem, R., Jayaram, S., Cherney, E.: Suppression of silicone rubber erosion by alumina trihydrate and silica fillers from dry-band arcing under DC. IEEE Trans. Dielectr. Electr. Insul. **22**(1), 14–20 (2015)
17. Ansorge, S., Schmuck, F., Papailiou, K.O.: Improved silicone rubbers for the use as housing material in composite insulators. IEEE Trans. Dielectr. Electr. Insul. **19**(1), 209–217 (2012)
18. Meyer, L., Jayaram, S., Cherney, E.A.: Thermal conductivity of filled silicone rubber and its relationship to erosion resistance in the inclined plane test. IEEE Trans. Dielectr. Electr. Insul. **11**(4), 620–630 (2004)

19. Tanaka, T.: Aging of polymeric and composite insulating materials aspects of interfacial performance in aging. IEEE Trans. Dielectr. Electr. Insul. **9**(5), 704–716 (2002)
20. Vas, J.V., Venkatesulu, B., Thomas, M.J.: Tracking and erosion of silicone rubber nanocomposites under DC voltages of both polarities. IEEE Trans. Dielectr. Electr. Insul. **19**(1), 91–98 (2012)
21. El-Hag, A.H., Simon, L.C., Jayaram, S.H., Cherney, E.A.: Erosion resistance of nano-filled silicone rubber. IEEE Trans. Dielectr. Electr. Insul. **13**(1), 122–128 (2006)
22. Venkatesulu, B., Thomas, M.J.: Erosion resistance of alumina-filled silicone rubber nanocomposites. IEEE Trans. Dielectr. Electr. Insul. **17**(2), 615–624 (2010)
23. Du, B.X., Ma, Z.L., Gao, Y., Han, T., Xia, Y.S.: Effects of nano filler on treeing phenomena of silicone rubber nanocomposites. In: 2011 Annual Report Conference on Electrical Insulation and Dielectric Phenomena 2011, pp. 788–791 (2011)
24. Nazir, M.T., Phung, B.T., Hoffman, M.: Performance of silicone rubber composites with SiO_2 micro/nano-filler under AC corona discharge. IEEE Trans. Dielectr. Electr. Insul. **23**(5), 2804–2815 (2016)
25. Bian, S., Jayaram, S., Cherney, E.: Erosion resistance of electrospun silicone rubber nanocomposites. IEEE Trans. Dielectr. Electr. Insul. **20**(1), 185–193 (2013)
26. Krystian, L.C.: Influence of profile on the pollution performance of ceramic longrod insulators. Ukr. J. Tech. Elektrodynamika **1**(2), 113–116 (2008)
27. Krivda, A., Hunt, S.M., Cash, G.A., George, G.A.: MALDI-TOF/MS characterisation of LMW PDMS in high voltage HTV silicone rubber insulators. In: 2000 Annual Report Conference on Electrical Insulation and Dielectric Phenomena, vol. 2, pp. 703–708 (2000)
28. Gutman, I., DernfalkA. : Pollution tests for polymeric insulators made of hydrophobicity transfer materials. IEEE Trans. Dielectr. Electr. Insul. **17**(2), 384–393 (2010)
29. Swift, D.A., Spellman, C., Haddad, A.: Hydrophobicity transfer from silicone rubber to adhering pollutants and its effect on insulator performance. IEEE Trans. Dielectr. Electr. Insul. **13**(4), 820–829 (2006)
30. Vazirinasab, E., Jafari, R., Momen, G.: Evaluation of atmospheric-pressure plasma parameters to achieve superhydrophobic and self-cleaning HTV silicone rubber surfaces via a single-step, eco-friendly approach. Surf. Coatings Technol. **375**(July), 100–111 (2019)

Co-pyrolysis of Empty Fruit Bunches with Palm Kernel Shell, Palm Leaves and Sawdust to Produce Fine Chemicals

Nurul Asyikin Binti Badir Noon Zaman, Noridah Binti Osman[✉], and Aqsha Aqsha

Centre of Biofuel and Biochemical Research (CBBR), Chemical Engineering Department, Universiti Teknologi PETRONAS, Seri Iskandar, Malaysia
noridah.osman@utp.edu.my

Abstract. Agriculture materials of oil palm tree and forestry such as rubber tree are commonly utilised and this can be contributed to the wastes. Therefore, this research is looking for value-added to these types of materials from the production of pyrolysis oil. This paper was intended to determine the best ratio of feedstock to produce high quantity of bio-oil yield and to determine properties of chemicals produced in bio-oil. The samples of empty fruit bunches (EFB) with palm kernel shell (PKS), palm leaves (PL) and sawdust (SDT) were co-pyrolyzed with different ratios (25:75, 50:50 and 75:25) to produce liquid oil, and the yield was analyzed and characterized by using GC-MS. The results show the mixture of EFB:PKS with ratio of 25:75 produced the highest yield of 40.53%, meanwhile, the co-pyrolysis of EFB:SDT produced the lowest yield which was 19.25%. The chemicals for each bio-oil were determined and the major organic chemicals such as phenols, aldehydes group, methyl alcohol and acetic acids were found for the purpose products of our fine chemicals.

1 Introduction

Malaysia is one of the well-known countries that produce a huge amount of palm oil every year. As has been reported before, the production of palm oil increased tremendously from about 4 million tonnes in 1990 to 18.8 million tonnes in 2012. However, this huge production would produce a huge amount of wastes too. To solve this issue, many researchers have proposed various methods of substituting these biomass wastes to the alternative energy sources with high efficiency and more environmental-friendly features. The use of renewable energy sources such as solar, wind and hydro energy, not only has a low maintenance cost but the emission of greenhouse gases such as carbon dioxide could be reduced by 70% by 2050, as reported by International Renewable Energy Agency (IRENA).

In recent years, research on biomass energy has become very popular. Biomass is defined as an organic material which is derived from plants or animal sources. Biomass is fractionated into lignin, cellulose and hemicellulose [1]. It is very abundant and can be easily found in diverse forms such as agriculture residues, wood residues, dedicated energy crops and municipal solid wastes. Palm oil tree is one of the biggest plantations in Malaysia. It has been reported that biomass in palm oil tree has produced 10% of oil, meanwhile, 90% belongs to wastes biomass such as empty fruit bunches (EFB), palm oil mill effluent (POME), palm mesocarp fibre, palm oil fronds, palm oil trunks and palm kernel shell (PKS). Empty fruit bunches (EFB), palm kernel shell (PKS), palm leaves (PL) and sawdust are biomass feedstocks used in this research. EFB is a bulky and voluminous brown bunch left over at palm oil mills after the removal of sterilized fruit by a rotary thresher drum [2]. The same situation goes with rubber trees, in which the estimated yield obtained by the process of converting the raw logs into the sawn timber was only 20% and the remainder is available as a residual biomass [3]. Previously, production of fuel from pyrolysis is extensively studied as a potential solution. Pyrolysis is a thermo-degradation of biomass or waste in the absence of oxygen which is performed under temperature of 400 to 700 °C and produces three final products which are bio-oil, bio-char and non-condensable gases such as CO, CO_2, H_2 and light hydrocarbons [4]. Bio-oil has various applications such as in fuels or precursor in petroleum refinery feedstock. Moreover, it was reported to contain valuable chemicals in significant concentrations and this may contribute to another benefit as a chemical feedstock [5, 6]. Even though the oil produced by pyrolysis of biomass can be used as a substitute for fossil fuels, but previous research has shown that the oil produced needs to be upgraded since it contains high levels of oxygen. Previous research indicated that biomass contains 47–51 wt.% carbon and 42–46 wt.% oxygen [7]. The oxygenated pyrolysis oil produced will have low calorific value which could cause corrosion problems and instability. Bio-oil can be upgraded to become a transportation fuel and a source of value-added chemicals and there are few ways to upgrade the pyrolysis oil such as hydrodeoxygenation, catalytic cracking and steam reforming. However, these methods are costly.

Thus, to overcome this problem, many researchers have shown that the use of co-pyrolysis technique is able to improve the characteristics of pyrolysis oil such as increasing the oil yield, reducing the water content and increasing the calorific value of the oil. Co-pyrolysis is a process which involves two or more materials as feedstocks. The liquid oil produced from co-pyrolysis, then, will undergo the distillation process to produce a final product which is fine chemical. Fine chemical is a chemical compound made in relatively small amounts that can be fully characterized and specified. It has been used in many applications such as in production of resins, in pharmaceuticals, fragrances, lubricant and also food flavoring agents.

This paper basically has two main parts which are determination of the best ratio to produce the higher quantity of bio-oil and the determination of chemicals in bio-oil produced. Based on the approach presented, the major aim of this research is to discover an optimum liquid oil for chemicals production.

2 Methods

2.1 Raw Materials

EFB, PKS and leaves from oil palm were collected from the nearest farm while sawdust was collected from a local company at Seri Iskandar, Perak, Malaysia. These samples were heated in an oven for 24 h at 105 °C and were grinded and sieved into smaller size, which was 2 mm.

2.2 Co-pyrolysis Process

The samples were co-pyrolyzed with different ratios which for EFB with PKS, SDT and PL, the ratios were 25:75, 50:50 and 75:25, respectively, by using a fixed-bed reactor (drop-type pyrolyzer), as shown in Fig. 1. The temperature and reaction time were fixed at 450 °C and 20 min, respectively. A 20 g of feedstock was inserted into the biomass holder zone of drop-type pyrolyzer and reaction temperature was set-up. After the pyrolysis temperature was achieved, the feedstock was dropped down to the reactor to start the reaction for 20 min. After the reactor was cooled for 30 min, the liquid was collected and was kept in freezer.

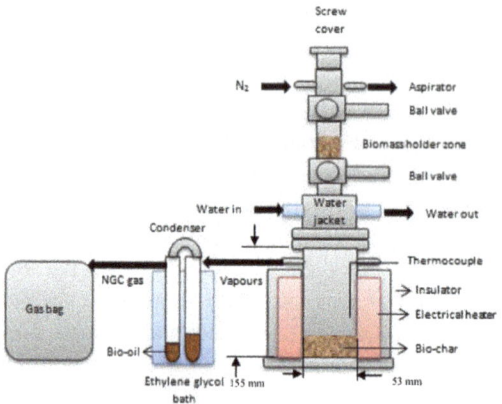

Fig. 1. The schematic diagram of fixed-bed reactor drop-type pyrolyzer [20]

2.3 *Bio-oil* Characterization

2.3.1 Gas Chromatography – Mass Spectrometer (GC-MS)

The characterization of chemicals compound in bio-oil was performed by the gas-chromatography-mass spectrometry (GC-MS) technique. The separation was made on a HP-5MS capillary column (30 m × 0.25 mm × 0.25 μm). The He gas was used as a carrier with flow rate of 1 mL/min. The GC temperature was held at 40 °C for 2 min, before raising it up to 260 °C for 10 min with ramping of 5 °C/min. The injector temperature of 240 °C with an injection volume of 1 μL were adopted and the split ratio was set at 50:1. The mass spectrometer was operated in full scan mode and its mass range was set at 20–400 Da.

3 Results and Discussions

The proximate and ultimate analysis of raw feedstock have been done to determine the percentage of moisture, volatile matter, ash and fixed carbon, respectively.

Table 1. The proximate and ultimate analysis of feedstock

Feedstock	EFB	SDT	PKS	PL
Proximate analysis				
Moisture (%)	9.589	10.245	6.451	8.056
VM (%)	73.255	70.111	71.101	65.033
Ash (%)	8.633	5.989	5.002	9.885
FC (%)	8.523	13.655	17.446	17.026
Ultimate analysis				
C	24.31	47.55	44.38	42.37
H	7.15	6.13	2.34	3.91
N	0.11	0.12	0.58	2.11
S	0.04	0.09	n.d	n.d
O	68.39	46.11	52.70	51.61

From the results presented in Table 1, the moisture of feedstock has reached a value below 10 wt.%, except for sawdust, which is an optimum condition for doing pyrolysis. PL and EFB have shown the highest values of ash content which are 9.885% and 8.633%, respectively, as compared to the PKS and sawdust which both are 5.002 and 5.989%, respectively. The percentage of the ash value in raw feedstock should be noteworthy since the high ash value will lead to a reduction of liquid yields in pyrolysis [6]. All of the feedstocks have shown a higher percentage of volatile matter, between 65.033–73.255% which is an optimum condition for high production of bio-oil [7].

After the characterization of the raw feedstock, the feedstocks were co-pyrolyzed to produce liquid yield. The products were collected after the cooling process and weighted. The percentage of yields were determined as shown in Fig. 2.

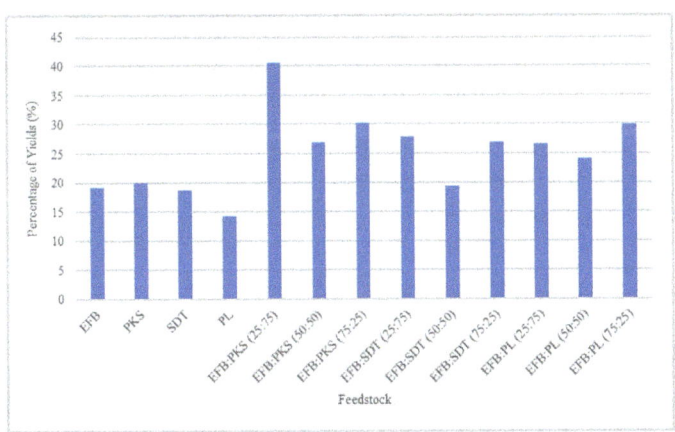

Fig. 2. The percentage of bio-oil produced by single feedstock, EFB:PKS, EFB:SDT and EFB:PL

Figure 2 shows that the co-pyrolysis of feedstock with different ratios have produced slightly different amounts of yield as compared to the single feedstock. Single feedstock for each material had produced oil with amounts lower than 20% compared to when it has been mixed with another feedstock. For co-pyrolysis of EFB:PKS with ratio of 25:75, 50:50 and 75:25, the percentage of bio-oil produced were 40.53%, 26.91% and 30.28%, respectively. Meanwhile, the amount of bio-oil produced by EFB:SDT and EFB:PL resulted in lower percentages as compared to the EFB:PKS for each ratio which were 27.78%, 19.25%, 26.90% and 26.61%, 24.02% and 29.90%, respectively. These differences have been studied from numerous researchers that found that the quantity of bio-oil obtained from co-pyrolysis are higher than those from pyrolysis of single feedstock [6, 13]. As mentioned in previous research, these differences happened due to the synergetic or the interactive effects such as type and contact of feedstock, temperature and heating rates and removal or equilibrium of volatiles formed among the feedstocks [14].

There are about hundreds of chemicals that have been found in bio-oil, but according to the GC-MS analysis summarized in Fig. 3, the dominant chemicals found are phenols, acetic acid, methyl alcohol, ketones aldehydes and hydrocarbon groups. The presence of many aldehydes, ketones, alcohols and carboxylates is due to the cracking of hemicellulose and cellulose, meanwhile, the formation of phenols is resulted from the cracking of lignin in biomass [15]. The composition of cellulose, hemicellulose and lignin in biomass have played an important role for producing the needed chemical because the chemicals composition produced in bio-oil depends on the type of biomass used. As mentioned in previous paper, the chemicals produced came from decomposition and depolymerization of cellulose, hemicellulose and lignin [16].

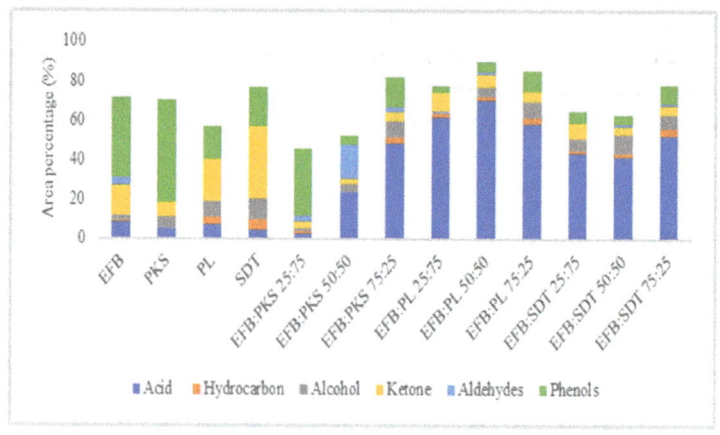

Fig. 3. Chemicals formed in bio-oil for each feedstock

Co-pyrolysis of 25:75 EFB:PKS shows the highest percentage of phenol with percentage area of about 33.92% compared to the other mixtures with phenol percentage ranged from 3.48 to 15.08%. The higher percentage of phenols might be due to the high composition values of lignin in both feedstock of PKS and EFB which are in between

46.3–50.7% and 17.99–22.10%, respectively, compared to the percentage of lignin in PL and sawdust, as mentioned by previous researchers [16–18]. The mixture of EFB:PL shows the highest percentage area of acetic acid for each ratio of 25:75, 50:50 and 75:25 which are 61.38%, 70.9% and 57.89%, respectively. The higher amount of acid in leaves might be due to the presence of glucuronic acid that formed many in cellulose of the leaves [19]. Many alcohol group chemicals have been found in bio-oil, and methyl alcohol is one of the highest chemicals that is formed in almost co-pyrolysis oil except for the mixture of 25:75 EFB:PL. The co-pyrolysis of EFB:SDT shows the highest percentage of alcohol which is 10.8%. Besides the aldehydes and ketone, the percentage of hydrocarbon group found in every bio-oil is also higher.

4 Conclusions

The results have proven that the ratio of feedstock is an important parameter to produce high quantities of oil. The co-pyrolysis of EFB:PKS with ratio of 25:75 and 75:25 shows the highest percentages of bio-oil which are 40.53% and 30.28%, respectively. Meanwhile, the co-pyrolysis of 50:50 EFB:SDT produced the lowest amount of yield which was 19.25%. This is due to several factors such as low percentage of ashes, high volatile matter value and also the composition of the cellulose, hemicellulose and lignin. The components for each bio-oil were determined by using GC-MS and the organic components such as phenols, aldehydes group, methyl alcohol, acetic acids and a lot of alkane groups have been found in bio-oil. The mixture of 25:75 EFB:PKS shows the highest percentage of phenol with percentage area of about 33.92% .

Acknowledgments. This study was funded by Yayasan Universiti Teknologi PETRONAS (YUTP – NO. 0153 AA-H45).

References

1. De Wild, P.: Biomass pyrolysis for chemicals. Dissertation, University of Groningen (2011)
2. Chang, S.H.: An overview of empty fruit bunch from oil palm as feedstock for bio-oil production. Biomass Bioenerg. **62**, 174–181 (2014). https://doi.org/10.1016/j.biombioe.2014.01.002
3. Ghani, W.A.W.A.K., da Silva, G., Alias, A.B.: Physico-chemical characterizations of sawdust-derived biochar as potential solid fuels. Malays. J. Anal. Sci. **18**, 724–729 (2014)
4. Balat, M., Mehmet Balat, E., Kirtay, H.B.: Main routes for the thermo-conversion of biomass into fuels and chemicals. Part 1: pyrolysis systems. Energy Convers. Manage. **50**, 3147–3157 (2009). https://doi.org/10.1016/j.enconman.2009.08.014
5. Rutkowski, P., Kubacki, A.: Influence of polystyrene addition to cellulose on chemical structure and properties of bio-oil obtained during pyrolysis. Energy Convers. Manage. **47**, 716–731 (2006). https://doi.org/10.1016/j.enconman.2005.05.017
6. Abnisa, F., Daud, W.M.A.W., Ramalingam, S., Naqiuddin, M., Azemi, B.M., Sahu, J.N.: Co-pyrolysis of palm shell and polystyrene waste mixtures to synthesis liquid fuel. Fuel **108**, 311–318 (2013). https://doi.org/10.1016/j.fuel.2013.02.013
7. Brebu, M., Ucar, S., Vasile, C., Yanik, J.: Co-pyrolysis of pine cone with synthetic polymers. Fuel **89**, 1911–1918 (2010). https://doi.org/10.1016/j.fuel.2010.01.029

8. Abdullah, N., Sulaiman, F., Gerhauser, H.: Characterisation of oil palm empty fruit bunches for fuel application. J. Phys. Sci. **22**, 1–24 (2011)

9. Abnisa, F., Wan Daud, W.M.A.: A review on co-pyrolysis of biomass: an optional technique to obtain a high-grade pyrolysis oil. Energy Convers. Manage. **87**, 71–85 (2014). https://doi.org/10.1016/j.enconman.2014.07.007

10. Aqsha, A., Mahinpey, N., Mani, T., Salak, F., Murugan, P.: Study of sawdust pyrolysis and its devolatilisation kinetics. Can. J. Chem. Eng. **89**, 1451–1457 (2011). https://doi.org/10.1002/cjce.20584

11. Cepeliogullar, O., Putun, A.E.: Thermal and kinetic behaviors of biomass and plastic wastes in co-pyrolysis. Energy Convers. Manage. **75**, 263–270 (2013). https://doi.org/10.1016/j.enconman.2013.06.036

12. Oyedun, A.O., Tee, C.Z., Hanson, S., Hui, C.W.: Thermogravimetric analysis of the pyrolysis characteristics and kinetics of plastics and biomass blends. Fuel Process. Technol. **128**, 471–481 (2014). https://doi.org/10.1016/j.fuproc.2014.08.010

13. Hua, D., Wu, Y., Chen, Y., Li, J., Yang, M., Lu, X.: Co-pyrolysis behaviors of the cotton straw/pp mixtures and catalysis hydrodeoxygenation of co-pyrolysis products over Ni-Mo/Al2O3 catalyst. Catalysts **5**, 2085–2097 (2015). http://www.mdpi.com/2073-4344/5/4/2085

14. Hassan, H., Lim, J.K., Hameed, B.H.: Recent progress on biomass co-pyrolysis conversion into high-quality. Bioresour. Technol. **221**, 645–655 (2016). https://doi.org/10.1016/j.biortech.2016.09.026

15. Upramono, D.S., Usrini, E.K., Uana, H.Y.: Yield and composition of bio-oil from co-pyrolysis of corn cobs and plastic waste of HDPE in a fixed bed reactor. J. Jpn. Inst. Energy **95**, 621–628 (2016)

16. Kabir, G., Hameed, B.H.: Recent progress on catalytic pyrolysis of lignocellulosic biomass to high- grade bio-oil and bio-chemicals. Renew. Sustain. Energy Rev. **70**, 945–967 (2017). https://doi.org/10.1016/j.rser.2016.12.001

17. Ninduangdee, P., Kuprianov, V.I., Young, E.: Thermogravimetric studies of oil palm empty fruit bunch and palm kernel shell: TG/DTG analysis and modeling. Energy Procedia **79**, 453–458 (2015). https://doi.org/10.1016/j.egypro.2015.11.518

18. Solikhah, M.D., et al.: Characterization of bio-oil from fast pyrolysis of palm frond and empty fruit bunch. In: IOP Conference Series: Materials Science and Engineering, vol. 349, p. 012035 (2018). https://doi.org/10.1088/1757-899X/349/1/012035

19. Sukiran, M.A., Kartini, N.O.R., Bakar, A.B.U., Chin, C.M.E.E.: Optimization of pyrolysis of oil palm empty fruit bunches optimization of pyrolysis of oil palm empty fruit bunches. Am. J. Appl. Sci. **21**, 653–658 (2009). https://doi.org/10.3844/ajas.2009.869.875

20. Izzatie, N.I., et al.: Co-pyrolysis of rice straw and polypropylene using fixed-bed pyrolyzer. In: IOP Conference Series: Materials Science and Engineering, vol. 160, p. 012033 (2016).https://doi.org/10.1088/1757-899X/160/1/012033

Material Characterization of Linear Low Density Polyethylene Blended with Synthetic Fibers Using FTIR for Rotational Molding Process

Nikita Gupta and PL Ramkumar[✉]

Mechanical and Aerospace Engineering Department, Institute of Infrastructure Technology Research and Management, Ahmedabad, Gujarat, India

Abstract. Rotomoulding is an evolving industry for the manufacture of hollow plastic items of various sizes. Linear Low Density Polyethylene (LLDPE) is commonly used for this process, but when the end properties of the product require an increase in mechanical strength, additives play a vital role. The present investigation opts for glass fibre (GF) and confibre (polypropylene-based fibre) from the list of synthetic additives. The present study characterizes the various prepared LLDPE/GF blends (20%, 25%, 30% and 35%) and LLDPE/Confiber blends (30%, 35%, 40% and 45%). Fourier Transform Infrared Spectroscopy analysis was experimented to analyse the characteristics of the blend to verify proper miscibility of the distinct mix. The data revealed that 25% LLDPE/Glass fiber and 35% LLDPE/Confiber have proved to fulfil the criteria of appropriate mixing.

1 Introduction

An abundant community of materials available to date for users has been taken over by Polymers. This is due to their excellent properties such as light weight, less reactivity, efficient colour dye pigmentation, and so on. Polymeric product can be manufactured from distinct plastic processing techniques, from which rotational molding has gained popularity in last few decades as it can manufacture a comparatively stress free plastic product [1, 2]. The use of this tool encompasses a wide variety of uses, from a toy to immense vessels. Over the last few decades, the rotational moulding process has achieved ubiquity, and so a great deal of research is related to this process.

Around 80% of polyethylene from the thermoplastics group is possibly favoured for roto molding the products. Linear low-density polyethylene (LLDPE) is more ideally used for this process, due to its requisite melt flow properties and less shear affectability [3, 4]. LLDPE coordinates all requirements but lacks the essential mechanical consistency (strength being the key criteria) needed in some applications such as armors, combat tanks, and so on and in such cases, added substances or fillers will function subsequently. In order to maximise the strength of any end product, fibres have exceptional consistency. As an additive, natural and synthetic fibres may be used. But natural fibre has a comparatively lower propensity to help the study of processability as needed

M. Awang and S. S. Emamian (Eds.): *Advances in Material Science and Engineering*, LNME, pp. 303–308, 2021.
https://doi.org/10.1007/978-981-16-3641-7_36

for the process of rotational moulding [5, 6]. Synthetic fibres provide a viable alternative in terms of strength [7]. Distinct literatures has provided the benefits of adding synthetic fibers with the base resin for enhancing the strength. Haque et. al, studied the effect of glass fiber on the Polyethylene (PE) composites, for which they observed that on increasing the glass fiber content, an increase in mechanical properties could be observed [8]. Confiber hardly have any applicable literature available regarding the rotational moulding process, contrary to the literature available for glass fibre.

In mixing with LLDPE for verifying roto moldability, an attempt is made to potentially use Confiber and Glass fiber. Tests based on FTIR was analysed in order to confirm proper miscibility of the blends. LLDPE and glass fiber were mixed from 20 to 35%, and LLDPE/confiber were blended from 30 to 45%. The idea is to observe the dominance of both the peaks through FTIR, which confirms the miscibility of the blends as a part of primary investigation, which is analysed in the present study.

2 Methods

LLDPE was used as a base resin in this analysis, while Glass Fiber and Confiber were used as two separate additives for the preparation of two distinct blends. Greenage Industries, Ahmedabad, provided pure LLDPE of rotomolded grade Ge3645 with MFI of 4.5 g/10 min and 0.936 g/cm^3 density. The 2.55 g/cm^3 density of E-glass fibre was acquired from HS Enterprise, Ahmedabad. Confiber was supplied from Mass India Inc., Ahmedabad, with a density of 0.916 g/cm^3 and MFI 20 g/10 min.

Fourier Transform Infrared Spectroscopy has potentially been utilized for verifying the compounds present in the blend, which then confirms the miscibility of the mix [9]. Basically, by transmitting infrared light at the atomic level of the material, a Perkin Elmer FTIR Spectrum Two Universal ATR was used to obtain the peaks. Transmission data was collected in the 450 cm^{-1} to 4000 cm^{-1} range. The data interval was kept as 1 cm^{-1}. For the specific experiment, a prior resolution set beforehand was 4 cm^{-1}. With every 5% rise, separate blends (LLDPE/GF) were prepared from a 20% to 35% weight ratio. Similarly, with an increasing phase of 5%, different blends (LLDPE/Confiber) from 30% to 45% were prepared.

3 Results and Discussions

Using FTIR, characterization is accomplished by transmitting light through the atoms of the sub-stances. With this phenomenon, which gives information on the current bond in the material, several characteristic peaks are obtained. The Perkin Elmer library, for example, recognises major peaks for pure LLDPE at 2915, 2845, 1467, 1462, 1377, 730, 717 cm^{-1}. These peaks suggest some peculiar bonds in pure LLDPE, such as H = O, C = O, CH2 and other related bonds [10]. So for pure LLDPE, one may describe the characteristic peak curves. Glass fibre, similarly, has distinctive peaks ranging from 3200–3600 cm^{-1}. When bonded to C and Si, the N-H / O-H compound shows significance at 3200–3600 cm^{-1} peaks. Compared to the expansion of C = O in esters/carbonyl, the band at 1735 cm^{-1} yields C-O and Si-O bonds at 1100 cm^{-1}. For Confiber, peak values at 2917 cm^{-1}, 2838 cm^{-1}, 1455 cm^{-1} resembles the stretching bond of CH2.

Characterization peaks of both LLDPE/Glass Fiber and LLDPE/Confiber are described from the information available from the literature. The graphs are plotted in terms of the wavenumber obtained for each percentage of rays transmitted (%T) into the atoms of the substance as shown in the graphs obtained below.

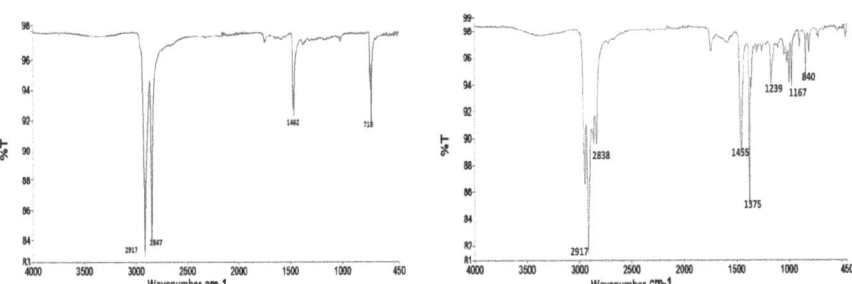

Fig. 1. FTIR Peaks of 30% LLDPE/Confiber   **Fig. 2.** FTIR Peaks of 45% LLDPE/Confiber

The FTIR peaks of 30% LLDPE/Confiber as shown in Fig. 1, resembles the peaks as that of pure LLDPE. This depicts that the confiber peaks are rarely evident on 30% confiber blended with pure LLDPE. Figure 2 resembles that 45% and above confiber when mixed with base resin, shows the peak dominance of pure confiber only, thus proving less significant in terms of miscibility as LLDPE peaks could hardly be observed. On the other hand, as shown in Fig. 3, for 35% confiber concentrated with LLDPE, provides a significant remark of the peak values of both pure LLDPE and Confiber, thus making it more suitable in terms of appropriate mixing. Again for 40% as shown in Fig. 4, a similar trend as of confiber was evinced which makes it less appropriate to be considered for further experimentation.

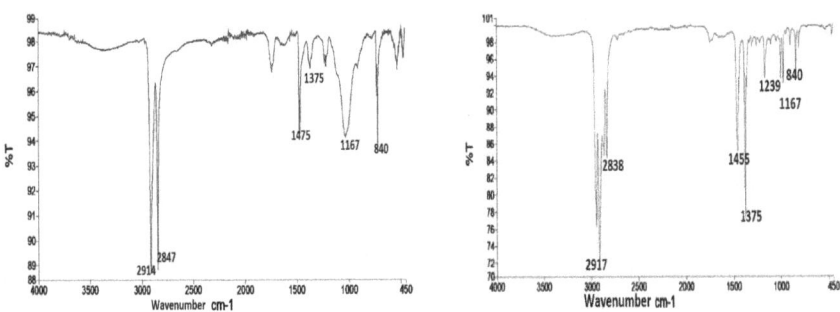

Fig. 3. FTIR Peaks of 35% LLDPE/Confiber   **Fig. 4.** FTIR Peaks of 40% LLDPE/Confiber

Figure 5 and 6 shows the mixing of 20% and 35% Glass fiber blended with LLDPE respectively. The mixture of 20% to 35% glass fibre combined with LLDPE by weight is now verified by FTIR. In LLDPE, where evidence of peaks of both LLDPE and glass fibre is substantially found, the precise concentration of glass fibre is considered to be optimum.

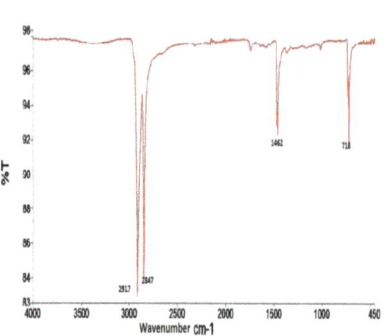

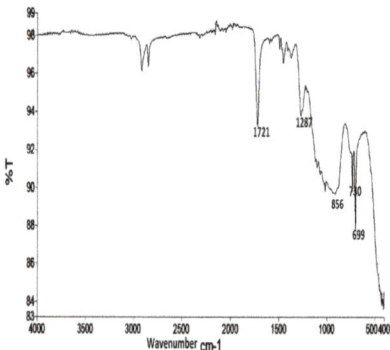

Fig. 5. FTIR Peaks of 20% LLDPE/Glass Fiber

Fig. 6. FTIR Peaks of 35% LLDPE/Glass Fiber

Figure 5 demonstrates the peaks of LLDPE only, thus making the blend incompatible in terms of proper mixing as glass fiber peaks are rarely visible. As shown in Fig. 6, for 35% LLDPE/GF blend, LLDPE compounds cannot be traced for which the blend makes it unsuitable for further experimentation.

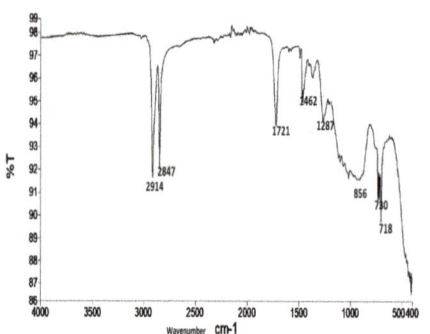

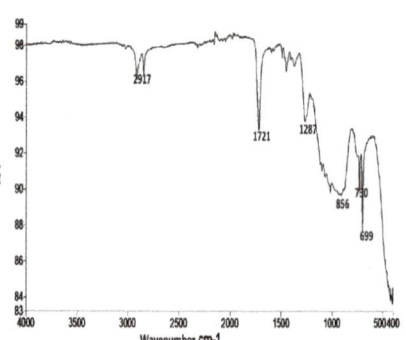

Fig. 7. FTIR Peaks of 25% LLDPE/Glass Fiber

Fig. 8. FTIR Peaks of 30% LLDPE/Glass Fiber

As shown in Fig. 7, the significance of both LLDPE and Glass fiber peaks are dominant for 25% LLDPE/GF, thus making it a requisite blend as needed for verifying a proper mixing. On the other hand 30% Glass fiber when blended with LLDPE, made it unsuitable as peaks of LLDPE was not significantly available as evident from Fig. 8. Similar sort of work was reported by Meireless et.al. where Polystyrene (PS) and cellulose acetate (CA) were mixed and tested with FTIR, where it was observed that below 30% Polystyrene/CA, makes the blend suitable in terms of evincing the peaks of both PS and CA [11].

4 Conclusions

Hollow seamless products can be easily manufactured using rotational molding process. Various materials are associated to be used as base resin for this technique, but LLDPE is greatly utilized due to its unique flow properties. Few applications where strength is on high demand, therein mixing additives with LLDPE for rotational molding process can potentially be useful. FTIR serves as a preliminary investigation so as to verify an appropriate mixing of the blends by evincing the peak values of the materials. LLDPE and Glass fiber were mixed from 20% to 35%, whereas LLDPE/Confiber were blended from 30% to 45% in the present study. The FTIR analysis confirmed that for 25% LLDPE/GF and 35% LLDPE/Confiber, the peak values of both the materials were significantly observed. Thus 25% LLDPE/Glass fiber and 35% LLDPE/Confiber can be considered for further experimentation as they sustain an appropriate mixing based on FTIR analysis.

References

1. Ramkumar, P.L., Waigaonkar, S.D., Kulkarni, D.M.: Effect of oven residence time on mechanical properties in rotomoulding of LLDPE. Sādhanā **41**(5), 571–582 (2016). https://doi.org/10.1007/s12046-016-0489-4
2. Gupta, N., Ramkumar, P., Sangani, V.: An approach toward augmenting materials, additives, processability and parameterization in rotational molding: a review. Mater. Manuf. Process. **35**, 1539–1556 (2020). https://doi.org/10.1080/10426914.2020.1779934
3. Nikita Gupta, P.L., Ramkumar, : Analysis of synthetic fiber-reinforced LLDPE based on melt flow index for rotational molding. In: Praveen Kumar, A., Tatacipta Dirgantara, P., Krishna, Vamsi (eds.) Advances in Lightweight Materials and Structures: Select Proceedings of ICALMS 2020, pp. 599–606. Springer Singapore, Singapore (2020). https://doi.org/10.1007/978-981-15-7827-4_61
4. Ramkumar, P.L., Kulkarni, D.M., Chaudhari, V.V.: Fracture toughness of LLDPE parts using rotational moulding technology. Int. J. Mater. Prod. Technol. **58**, 305 (2019). https://doi.org/10.1504/IJMPT.2019.100003
5. Abhilash, S.S., Singaravelu, D.L.: Effect of fiber content on mechanical and morphological properties of bamboo fiber-reinforced linear low-density polyethylene processed by rotational molding. Trans. Indian Inst. Met. **73**(6), 1549–1554 (2020). https://doi.org/10.1007/s12666-020-01922-y
6. Cisneros-López, E.O., Pérez-Fonseca, A.A., González-García, Y., et al.: Polylactic acid–agave fiber biocomposites produced by rotational molding: A comparative study with compression molding. .Adv Polym. Technol. **37**, 2528–2540 (2018). https://doi.org/10.1002/adv.21928
7. Kirsanov, A.I., Stolyarov, O.N.: Mechanical properties of synthetic fibers applied to concrete reinforcement Механические свойства синтетических волокон для армирования бетона **4**, 15–23 (2018). https://doi.org/10.18720/MCE.80.2
8. Haque, M.M., Hasan, M.: Influence of fiber surface treatment on physico-mechanical properties of betel nut and glass fiber reinforced hybrid polyethylene composites. Adv. Mater. Process. Technol. **4**, 511–525 (2018). https://doi.org/10.1080/2374068X.2018.1465322
9. Gupta, N., Ramkumar, P.: Experimental investigation of linear low density polyethylene composites based on coir for rotational molding process. Polym. Polym. Compos. **096739112095324** (2020). https://doi.org/10.1177/0967391120953246

10. Jung, M.R., Horgen, F.D., Orski, S.V., et al.: Validation of ATR FT-IR to identify polymers of plastic marine debris, including those ingested by marine organisms. Mar. Pollut. Bull. **127**, 704–716 (2018). https://doi.org/10.1016/j.marpolbul.2017.12.061
11. Meireles, S., Filho, G.R., De Assunc, R.M.N., et al.: Blend compatibility of waste materials—cellulose acetate (from sugarcane bagasse) with polystyrene (from plastic cups): diffusion of water, FTIR, DSC, TGA, and SEM study. J. Appl. Polym. Sci. **104**(2), 909–914 (2007). https://doi.org/10.1002/app.25801

Analysis of Spread Moored FPSO-Tanker Side-by-Side System with Offloading Operability

Ruly Irawan[✉], N. L. Azizan, M. S. Liew, A. M. Al-Yacouby, and Kamaluddeen Usman Danyaro

Universiti Teknologi PETRONAS, 32610 Seri Iskandar, Perak, Malaysia
ruly_17005131@utp.edu.my

Abstract. Dynamic load acting on the side-by-side mooring system is one of the important factors to estimate the offloading operability of the FPSO. In this paper, a numerical mooring analysis for the side-by-side moored FPSO-tanker system is investigated. Floating Structure analysis is divided into two parts. The first is to calculate the hydrodynamic coefficient, and the second part is calculating the system responses in the time-domain analysis. The prediction of FPSO-Tanker responses is necessary during offloading. This research presents the possibility of the use of Artificial Neural Networks model for the prediction of the responses of FPSO-Tanker system with spread-mooring configuration. The ANN model considering various metocean data such as azimuth angles, significant wave height, wind load, swell, hawser and the mooring configurations. The results of this study indicated that the ANN model shows an accurate prediction of the FPSO-tanker response that are required for the design of side-by-side vessels during offloading.

1 Introduction

The deep-sea oil exploration is increasing as the demand for the fuel is increasing rapidly. Floating Production Storage and Offloading (FPSO) facilities are found to be very effective for the production, storage, and offloading of oil and gas in much higher depth oil fields [1]. There is a significant demand for solving the sea keeping problem of the FPSO mooring system during production operations [2]. The hawsers are polyester cables connected between vessels. It helps limit the relative distance of the vessels during offloading operation. During the offloading procedure design, a determination of a system that is stable and does not experience large motions assures the safe operation [3, 4]. Relative vessel headings, hawsers, and mooring azimuth are the important variables that affect the relative displacement between vessels which cause collision during the period of transient weather [5, 6]. Therefore, it is important for the engineer to have an application to design the side-by-side vessels efficiently for all design ranges parameters [7–9].

The analysis of the side by side FPSO-Tanker system includes the calculation of hydrodynamic coefficients, and simulation of dynamic responses of FPSO-Shuttle tanker system in a time domain, and the validation of numerical and software simulation [10–12]. Rini [13], investigate the dynamic response and hydrodynamic interactions of an

© The Author(s), under exclusive license to Springer Nature Singapore Pte Ltd. 2021
M. Awang and S. S. Emamian (Eds.): *Advances in Material Science and Engineering*, LNME, pp. 309–317, 2021.
https://doi.org/10.1007/978-981-16-3641-7_37

FPSO for different location of Malaysian waters. The effect of mooring stiffness, damping, viscous damping, roll damping, sloshing, and other hydrodynamic factors affect the dynamic response of the FSRU-LNGC system is also investigated [14].

The design of side-by-side FPSO-Tanker during offloading, involved various environmental loads, such as various azimuth angles, met-ocean data, and characteristics of the platform hull [15–17]. The designs also have to analyze a large number of the hawser and the mooring configurations, increase the design space where the number of the design variables is significantly higher. With this large design space, we want to use of the Neural Network model as a surrogate model to replace the expensive direct simulation is important to find the dynamic responses during offloading operability for side by side FPSO-Tanker system which requires a large number of simulations [18–20]. The result of this research is shown the ability of the Neural Network model for practical use by the engineer to solve the offshore problems with respect to lower computational times, especially in conceptual and preliminary design stages.

2 Mathematical Modelling and Analysis

An Artificial Neural Network is an information processing model that is inspired by the way biological nervous systems to process information. A basic and simplest neural model of a neuron is shown in Fig. 1, known as McCulloch–Pitts neuron [16]. The most common sigmoid function is the logistic function, which will be illustrated in Fig. 2. Upon receiving a given number of inputs x_i, $i = 1, N$, it calculates a linear combination of these inputs using weights w_i to generate the weighted input z:

$$z = \sum_{i=1}^{N} w_i x_i(t) \tag{1}$$

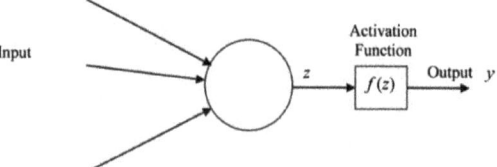

Fig. 1. McCulloch–Pitts neuron.

Next, it provides an output y via an activation function $f(z)$. Several functions can be used [17], the linear function being the simplest one, regarding a decision making (1 or 0). This function is defined as:

$$y(t) = f(z) = \begin{cases} 0, & z < \beta \\ 1, & z \geq \beta \end{cases} \tag{2}$$

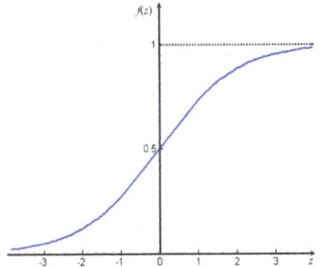

Fig. 2. Logistic activation function

Where the parameter β determines the transition point of the neuron. The sigmoid functions are continuous activation functions widely used in ANN application is expressed by:

$$y(t) = f(z) = \frac{1}{1 + e^{-\alpha z}} \tag{3}$$

The assumption of the vessel motion responses is linear. Therefore, the coupled motion equation of vessels can be expressed using this following equation.

$$\sum_{j=1}^{12} \left[-\omega^2 (M_{ij} + A_{ij}) - i\omega B_{ij} + C_{ij} \right] = F_i \tag{4}$$

Where: M_{ij} is the Mass matrix, A_{ij} is the Added mass coefficient matrix, B_{ij} is the Damping coefficient matrix, C_{ij} is the Restoring matrix, ξ_{ij} is the Complex amplitude of the motion response, and F_i is the Complex amplitude of the wave force and moment. The time-domain motion responses for side-by-side FPSO-Tanker system can be written using the following equation.

$$\sum_{j=1}^{12} \left[(M_{ij} + A_{ij})\ddot{x} + \int_{-\infty}^{t} R_{ij}(t - \tau)x(\dot{\tau})d\tau + C_{ij}x_j \right] = F_i(t) \tag{5}$$

Where R_{ij} is the Retardation function matrix and F_i includes the following forces; First and second-order wave forces, Wind forces, Current forces, Restoring forces of mooring lines, and fenders.

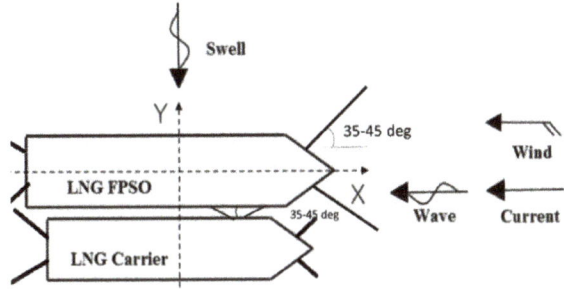

Fig. 3. The direction of the environment for operability analysis

Analysis of two side-by-side mooring configurations is considered for the operability analysis as shown in Fig. 3. The ranges of azimuth angles used in this study are 35 and 45°. The horizontal angle of the hawser and mooring lines against the x-axis is 35 and 45°, while the position of the fenders is remaining the same. the principal characteristics of the FPSO and Tanker are presented in Table 1.

Table 1. Calculation conditions for two-body motion analysis

Main Particulars	FPSO	Tanker
Length (m)	213.9	171.2
Width (m)	44.8	35.8
Depth (m)	25.5	20.4
Draft (m)	10.8	9
Displacement (m)	98923.1	52821.3
COG above base	17.14	12.53
Roll gravity radius (m)	16	10.2
Pitch gravity radius (m)	60	50
Yaw gravity radius (m)	60	50
Vertical gravity center (m)	13.8	12

Table 2. Environmental conditions for operability analysis

No. of condition		1	2	3	4	5	6
Wind-wave	Hs [m]	1.0	2.0	3.0	4.0	5.0	6.0
	Tp [sec]	7.5					
	Heading [deg]	180					
Swell	Hs [m]	1.00					
	Tp [sec]	12.0					
	Heading [deg]	270					
Wind	Vw [m/s]	10.0					
	Heading [deg]	180					
Current	Vc [m/s]	1.0					
	Heading [deg]	180					

Table 2 presents the range of values for the design variables with six ranges of the input variables (Hs) defined, and four fixed variables (wave, wind, current, swell). These six input variables and four defined parameters are not enough to predict the output value between the input variables. This is because the valid output is only available for the specific case, and the interpolation between these values may be not accurate since it involved many parameters. Therefore, the ANN model has been proposed in this study to be used as a surrogate model to predict the infeasible solution.

3   Results and Discussions

This study using a dataset with a total of 404 cases. This research has 70%, 15%, and 15% division of cases between the training, validation and testing respectively. Table 3 shows the statistical results for the ANN model both training and testing set. Table 3 and Table 4 shown the database has similar ranges of variation, mean and standard deviation. Thus, it indicates the consistency of the database in the design space.

Analysis of the offloading operability is shown in Figs. 4, 5, 6 and 7. Based on the time-domain simulation, each value of graph denotes the maximum, fender mooring forces, hawser forces and relative displacement for six cases (Hs 1 m to 6 m) as shown in Fig. 4 and Fig. 5. According to the results, it is found that the operability of side-by-side with spread mooring lines is enhanced linearly with the increasing of mooring angle. Detail results for the environmental conditions 1 to 6 are shown in Figs. 4, 5 and 6. As the angle of mooring lines is decreased, dynamic forces acting on the mooring lines are increased.

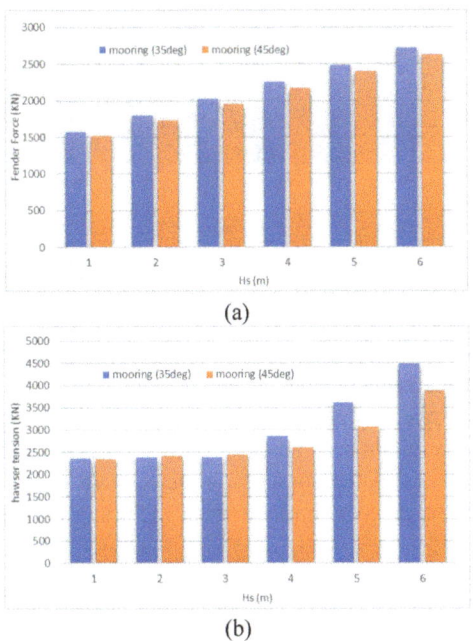

(a)

(b)

Fig. 4. (a) Fender forces and (b) Hawser forces for different significant wave height (m)

The analysis was performed in the same parametric study for all cases using different Hs values. The maximum and minimum mooring and hawser forces are shown in Fig. 6, in which the minimum fender forces and relative displacement between vessels are found when the mooring angle is around 45°. In Fig. 4 and Fig. 5, show the find the best configuration of moorings angle which can be identified directly the minimum value.

Recalling that the training/validation of the ANN model using random initialization of the weights. The error measured (RMSE) using the ANN standard fitting model for the 20 independent runs of each ANN is relatively small as shown in Table 4. Considering these errors, the correlation $(1 - r)$ and CPUs time for the ANN model are shown in Table 3 and Table 4 respectively.

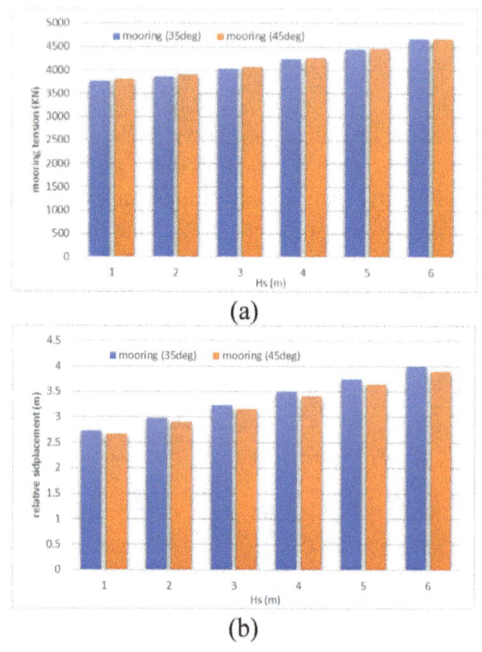

(a)

(b)

Fig. 5. (a) Fender forces and (b) Hawser forces for different significant wave height (m)

Finally, the performance of the ANN model to predict the mooring, hawser, fender forces, and relative displacement between vessels in the overall design space are shown in Fig. 6 and Fig. 7. The relative differences between the ANN model prediction with the actual simulation results within the range $[-5\%, +5\%]$. All errors measured in Fig. 6 and Fig. 7 are small, indicating the good performance of the ANN model. As expected, the graph of the prediction curve is similar to those in the simulation curve, all the results are in an equal range.

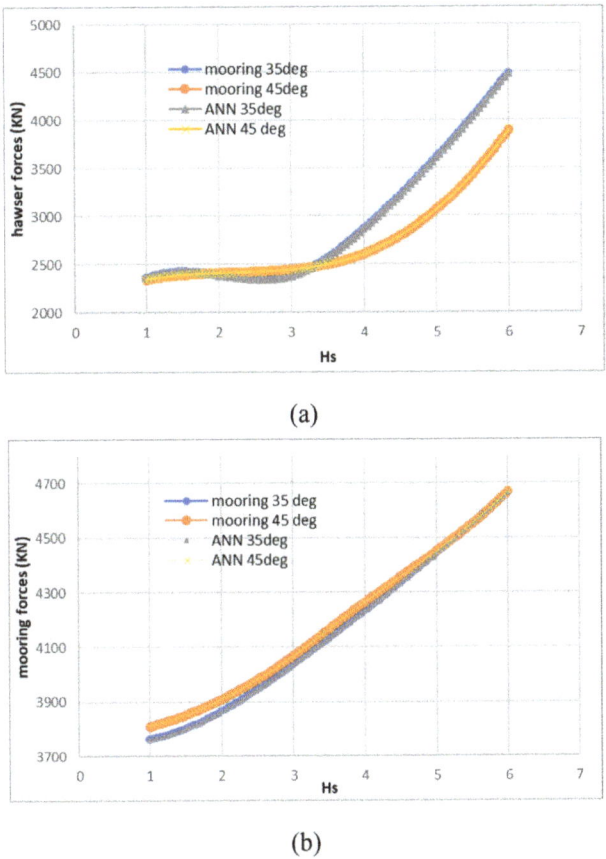

(a)

(b)

Fig. 6. (a) hawser forces and (b) Mooring forces (Simulation-ANN model comparison)

The small standard deviation values presented in Table 3 and Table 4 also indicate the robustness of the ANN model. More importantly, the performance of the ANN model in this study for overall design space is acceptable for practical applications, especially in conceptual and preliminary design stages. Finally, the last column of Table 4 corresponds to the CPU calculation times required for the training of each ANN model. Clearly, those values are more than 200 times faster compared to the times required for a full dynamic coupled analysis.

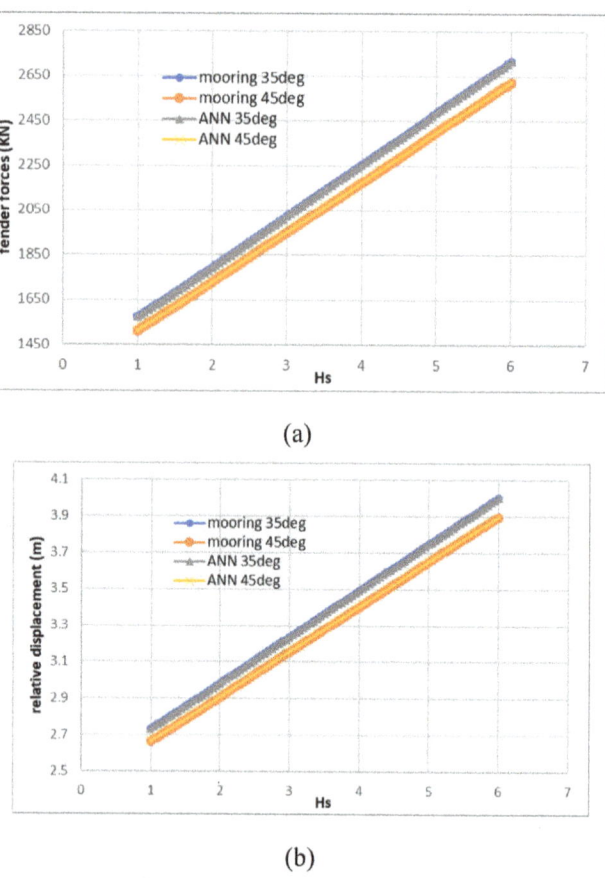

Fig. 7. (a) hawser forces and (b) Mooring forces (Simulation-ANN model comparison)

4 Conclusions

In this paper, the mooring analysis for the side-by-side moored FPSO and Tanker carrier was carried out to estimate the side by side offloading operability. Through the presented study, the following points can be concluded that the side-by-side hawser and mooring configuration is an important factor for the side-by-side offloading operability. The operability is highly affected by the arrangement of hawsers and moorings. To ensure safe loading during the offloading operation, a motion prediction methodology, such as ANN is required. The results of this study indicated that the ANN model shows an accurate prediction of the FPSO-tanker response that are required for the design of side-by-side vessels during offloading.

Acknowledgments. The authors would like to thank Yayasan Universiti Teknologi Petronas (YUTP) and Universiti Teknologi PETRONAS for funding this project.

References

1. Sorensen, L.T., Jorgensen, T., Kirkeby, L.T., et al.: Smoking and alcohol abuse are major risk factors for anastomotic leakage in colorectal surgery. Br. J. Surg. **86**, 927–931 (1999)
2. Harrison, D.K., Hawthorn, I.E.: Amputation level viability in critical limb ischaemia: setting new standards. Adv. Exp. Med. Biol. **566**, 325–332 (2005)
3. Hong, S.Y., Kim, H., Cho, S.K., et al.: Numerical and experimental study on hydrodynamic interaction of side-by-side moored multiple vessels. Ocean Eng. **32**, 783–801 (2005)
4. Lima, B., Jacob, B.P., Ebecken, N.F.F.: A hybrid fuzzy/genetic algorithm for the design of offshore oil production risers. Int. J. Numer. Methods Eng. **64**, 1459–1482 (2005)
5. Farid, M., Halim, A., Koto, J.: Hydrodynamic interaction of three floating structures. J. Ocean. Mech. Aerosp. **8**, 1–11 (2014)
6. Zhao, W.H., Yang, J.M., Hu, Z.Q.: Hydrodynamic interaction between FLNG vessel and LNG carrier in side by side configuration. J. Hydrodyn. **24**, 648–657 (2012)
7. Oil Companies International Marine Forum. Ship to Ship Transfers – Considerations Applicable to Reverse Lightering Operations (2009)
8. Buchner, B., Van, D.A., Wilde, J.D.: Numerical multiple-body simulations of side-by-side mooring to an FPSO. Proc. Int. Offshore Polar Eng. Conf. **1**, 343–353 (2001)
9. Nishanth, R.: School of Civil and Mechanical Engineering. Dynamic Response and Life-Cycle Analysis of Floating Production Storage and Offloading Systems (2018)
10. Zhao, W., Yang, J., Hu, X.Z.: Hydrodynamics of an FLNG system in tandem offloading operation. Ocean Eng. **57**, 150–162 (2013)
11. Van, D.V.T.N., Hove, D.T.: FPSO offloading concepts, their nautical feasibility and safety assessment. In: Proceedings of Annual Offshore Technology Conference, pp. 2623–2630 (2002)
12. Ormberg, H., Larsen, K.: Coupled analysis of floater motion and mooring dynamics for a turret-moored ship. Appl. Ocean Res. **20**, 55–67 (1998)
13. Jacob, B.P., Ebecken, N.F.F.: Adaptive reduced integration method for nonlinear structural dynamic analysis. Comput. Struct. **45**, 333–347 (1992)
14. Jacob, B.P., Bahiense, R.D.A., Correa, F.N.: Parallel implementations of coupled formulations for the analysis of floating production systems, part I: Coupling formulations. Ocean Eng. **55**, 206–218 (2012)
15. Jacob, B.P., Franco, L.D., Rodrigues, M.V.: Parallel implementations of coupled formulations for the analysis of floating production systems, part II: Domain decomposition strategies. Ocean Eng. **55**, 219–234 (2012)
16. Yue, J., Kang, W., Mao, W.: Prediction of dynamic responses of FSRU-LNGC side-by-side mooring system. Ocean Eng. **195**, 106731 (2020)
17. Kim, M.S., Ha, M.K., Kim, B.W.: Relative motions between LNG-FPSO and side-by-side positioned LNG carrier in waves. Proc. Int. Offshore Polar Eng. Conf **2003**, 210–217 (2003)
18. Kim, Y.: Finite memory quadratic Volterra model for the response prediction of a slender marine structure under a Morison load. J. Fluids Struct. **56**, 75–88 (2015). https://doi.org/10.1016/j.jfluidstructs.2015.05.003
19. Aggarwal, N., Manikandan, R., Saha, N.: Predicting short term extreme response of spar offshore floating wind turbine. Procedia Eng. **116**, 47–55 (2015)
20. Adeli, H.: Neural networks in civil engineering. Comput. Civ. Infrastruct. Eng **16**(126–142), 2001 (2001)
21. Waszczyszyn, Z., Ziemiański, L.: Neural networks in mechanics of structures and materials - New results and prospects of applications. Comput. Struct **79**, 2261–2276 (2001)
22. Quéau, L.M., Kimiaei, M., Randolph, M.F.: Artificial neural network development for stress analysis of steel catenary risers: sensitivity study and approximation of static stress range. Appl. Ocean Res. **48**, 148–161 (2014)

Stability and Thermal Conductivity Evaluation of Less Concentration Surfactant Wrapped Functionalized Graphene Dispersed in Ethylene Glycol

Balaji Bakthavatchalam[1]($\boxtimes$), Khairul Habib[1], Pugazhandhi Bakthavatchalam[2],
B. Keerthana[3], Sundarajoo Thulasiraman[4], and R. K. Pongiannan[5]

[1] Department of Mechanical Engineering, Universiti Teknologi PETRONAS,
32610 Bandar Seri Iskandar, Perak Darul Ridzuan, Malaysia
[2] Manipal Academy of Higher Education, Manipal, Karnataka, India
[3] Manipal School of Information Sciences, MAHE, Manipal, Karnataka, India
[4] Department of Electrical and Electronics Engineering, Universiti Teknologi PETRONAS,
32610 Bandar Seri Iskandar, Perak Darul Ridzuan, Malaysia
[5] Department of Electrical and Electronics Engineering, SRM Institute
of Science and Technology, Kattankulathur Campus, Chennai 603203, India

Abstract. Graphene, a single-layer hexagonal carbon atom lattice, has recently emerged as an intriguing framework for basic experiments of condensed matter physics, as well as an exciting candidate for potential usage of heat transfer systems. Moreover, graphene nanofluid has excellent anisotropic behavior regarding thermal conductivity. The most critical concerns for usage in diverse graphene nanofluid applications are inadequate dispersibility and insufficient stability in solvents. This study explores the stability and thermal conductivity of functionalized graphene wrapped with octadecylamine (ODA) surfactant and sonicated in ethylene glycol base fluid. In this study, graphene nanoparticles are functionalized with chemicals like nitric acid and sulphuric acid, and it is wrapped with ODA, and its thermal conductivity and dispersion behavior is evaluated. The results show that the maximum thermal conductivity was 0.43 W/m.K at 55 °C while the formulated nanofluid is stable for six days.

1 Introduction

Nanofluid is the current trend of heat transfer fluids where nano-sized particles are sonicated with base fluids to attain superior thermophysical properties, especially thermal conductivity [1–4]. Due to nanofluids' significance, they are incorporated with common fluids for different heat transfer applications [5, 6]. Consequently, graphene's remarkable properties such as high heat conductivity, tunable bandgap, room temperature hall effect, high charge carrier conduction, and elasticity made researchers focus its application in formulating heat transfer fluids [7]. The significant features of graphene nanoparticles are presented in Fig. 1. Single-layer graphene, bilayer graphene, and few-layer graphene

M. Awang and S. S. Emamian (Eds.): *Advances in Material Science and Engineering*, LNME, pp. 318–326, 2021.
https://doi.org/10.1007/978-981-16-3641-7_38

are the different types of graphene used in the present researches. Graphene is a two-dimensional (2D) monolayer of sp2 carbon atoms arranged in a honeycomb lattice. Due to the high thermal conductivity (>5000 W/m. K) and 2D structure, graphene nanoparticle is combined with conventional base fluids which are called as Graphene nanofluids. Graphene nanofluid has become one of the most desirable working fluids for heat transfer applications in the past decade due to its fascinating thermal, mechanical, electrical, and optical properties. These characteristics drawn considerable interest in research and industry from its invention in 1994 by Boehm, Setton and Stumpp. The in-plane metallic properties make them extremely conductive in parallel direction to graphene layers, nevertheless it has low conductivity in the perpendicular layers owing to weak interactions of vander waals forces. However, the application of graphene nanofluids to thermal systems, electronics and photovoltaics have been limited due to its less stability and volatile structure.

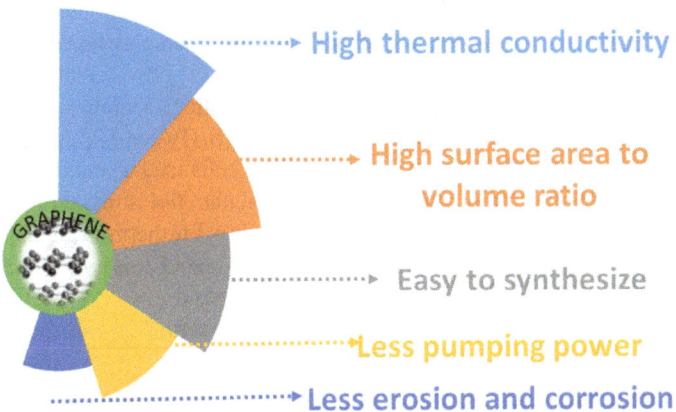

Fig. 1. Important features of graphene

Many studies reported on improving the dispersion stability and thermophysical properties of graphene nanofluids with a functionalization approach. Functionalization of graphene with chemicals is a primary method for improving the dispersion stability and structure modification. For instance, Sadeghinezhad et al. [7] improved the dispersion stability and thermal conductivity of graphene/water nanofluids through covalent functionalization. The authors modified the graphene nanoparticles' surface that led to a maximum thermal conductivity improvement of 17%, and there was not any sedimentation till 840 h. Karami et al. [8] obtained a maximum zeta potential value of 20.9 mV via covalent functionalization of the graphene nanoparticles. Unfortunately, the viscosity and shear stress of the formulated nanofluid increased, leading to high pumping power. Ghozatloo et al. [9] conducted an experimental study to investigate the stability and thermal conductivity of functionalized graphene synthesized by an alkaline approach. The result shows clearly that graphene nanofluids' thermal conductivity strongly depends on functional groups, and it is also recommended to use a low concentration of graphene nanoparticles to achieve excellent dispersion. In a different study, Agromayor et al. [10]

analyzed the thermal conductivity of functionalized graphene from 0.25 to 1 wt% concentration in water base fluid. They reported a 12% thermal conductivity enhancement at 0.5 wt% with a minimum pressure loss. Ma et al.[11] performed oxidation-reduction process to functionalize graphene (FG) and synthesized graphene/silicone oil nanofluid. Adding FG to silicone oil increased thermal and electrical conductivity. They also found that temperature, concentration, and loading of nanoparticles attributed to the enhancement of thermal conductivity. On the other hand, researchers focused on adding different surfactants to the nanofluids to attain good stability and thermophysical properties [12]. Some of the studies are as follows. Sarsam et al. [13] used different surfactants such as SDBS, GA, CTAB, and SDS, where all the surfactants resulted in enhanced dispersion stability. Unexpectedly, the addition of all these surfactants resulted in a large amount of foam formation that affected the heat transfer efficiency. To overcome foam formation, Bakthavatchalam et al. [14] used the ionic liquid as a surfactant and compared its performance with conventional surfactants, where ionic liquid-based nanofluids were found to be effective. Cakmak et al. [15] studied the effect of anionic, non-ionic, and cationic surfactants in deionized water. Nanofluid with anionic surfactant proved better stability than the other two surfactants. Wlazlak et al. [15] proved that SDS surfactant does not affect the geyser boiling as they have the better capability to attach to the graphene surface. Hussein et al. [16] used SDS, CTAB, Tween-80, and Triton X-100 surfactants in Graphene/distilled water nanofluids to enhance the stability and thermophysical properties. The zeta potential of the studied nanofluids ranged from -18.4 to -34.7 mV that can be considered as moderate stability. Furthermore, Wang et al. [17] proved that repulsive forces could be developed by ionic and non-ionic surfactants that help maintain the dispersion stability of graphene nanofluids.

The above literature confirms that both functionalization and surfactant addition influenced the colloidal stability and thermophysical properties of graphene nanofluids. Simultaneously, each method has its limitations like shape change, foam formation, etc. In this research, graphene nanoparticles are functionalized, and the Octadecylamine surfactant is wrapped on the FG surface to improve the formulated fluids' stability and thermal conductivity. Graphene-containing nanofluids are synthesized and dispersed in ethylene glycol to prepare homogeneous samples of 0.04 wt%. Inherently, the visual inspection method is used to evaluate the dispersion stability. Besides, thermal conductivity is measured at different temperatures using a thermal analyzer.

2   Materials and Methods

2.1   Materials

Graphene particles with a lateral thickness of 1.5 nm and Octadecylamine surfactant are purchased from Sigma Aldrich, USA. The characteristics of the studied nanoparticle are presented in Table 1. Ethylene glycol, sulphuric acid, and nitric acids are acquired from the Merck group.

2.2   Functionalization of Graphene Nanoparticles

One of the methods to enhance the colloidal stability of nanoparticles in the base fluid is chemical functionalization. In this research, 200 ml of sulphuric acid and 100 ml of nitric

Table 1. Specifications of graphene nanoparticles

Parameters	Specifications
Purity	> 98%
Colour	Black
Form	Powder
Particle size	< 2 μm
Thickness	1.5 nm
Surface area	500 m^2/g
Density	0.2–0.4 g/cm^3
Relative gravity	2–2.25 g/cm^3

acid is blended with 10 g of graphene nanoparticles. A small proportion of potassium chlorate is added to the above mixture under stirring, which is refluxed at a temperature of 80 °C and 300 RPM speed for 24 h. This mixture is then subjected to magnetic stirring at 300 RPM without heating for 12 h in a hot plate magnetic stirrer. Besides, the obtained solvent is now inflicted to ultrahigh centrifuge (Sorvall LYNX 6000, Thermoscientifc) three times with the configuration of 7000 RPM, 10 min, and 24 °C. At this stage, the graphene nanoparticles are washed with distilled water, and the deposits are collected. Finally, these deposits are heated in an oven at 50 °C and 100 mbar for 14 h. The obtained functionalized graphene is presented in Fig. 2.

Fig. 2. Raw image of the synthesized functionalized graphene

2.3 Preparation of SWFG Nanofluid

The stability enhancement of nanofluids is mainly dependent on the preparation method. The two-step method is a simple approach to prepare nanofluids, followed in the present study with the nanoparticle concentration of 0.04 wt%. The synthesized functionalized graphene (0.016 g) is combined with ethylene glycol solution (40 ml) along with Octadecylamine surfactant (0.008 g) and homogenized in a magnetic stirrer for 3 h, 1000 rpm, and 25 °C. Using a probe sonicator (Sonics, VCX 750 sonicator), the resulted colloid is sonicated for 1.5 h at room temperature. The probe sonicator was set to operate at every 50 s (ON) and 10 s (OFF) of pulses with an amplitude of 60% that allows the nanoparticles to break into tiny pieces.

2.4 Characterisation

High-resolution Transmission Electron Microscope (Zeiss Libra 200 FE) is used for analyzing the morphological characteristics of the prepared nanoparticles. KD2-Pro thermal analyzer (Decagon devices, USA) is utilized for measuring the thermal conductivity of the studied sample. The experimental setup for thermal conductivity measurement is depicted in Fig. 3. Finally, dispersion stability is evaluated by the sedimentation analysis of the Visual Inspection approach, where the sedimentation of nanoparticles is monitored at different intervals of time.

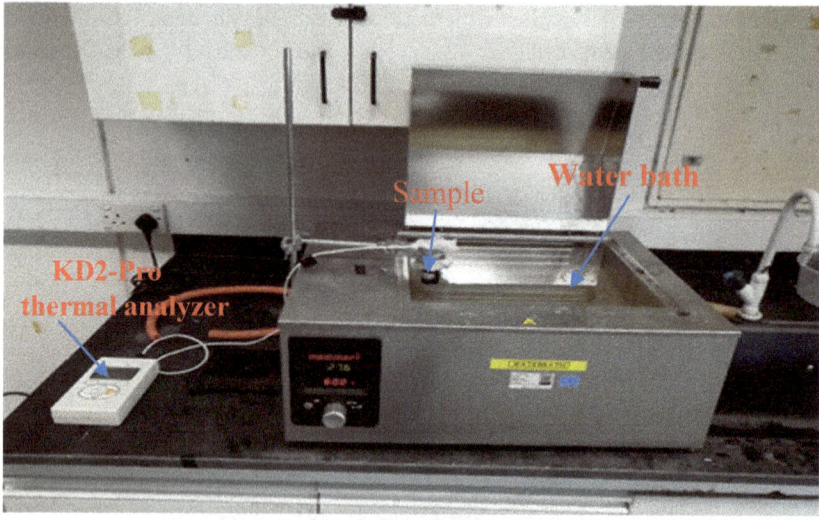

Fig. 3. Experimental setup for thermal conductivity measurement

3 Results and Discussion

3.1 Morphological Characteristics

Figure 4 shows the TEM image of the synthesized surfactant wrapped functionalized graphene at different magnifications. The presence of grains and crystallite structure is

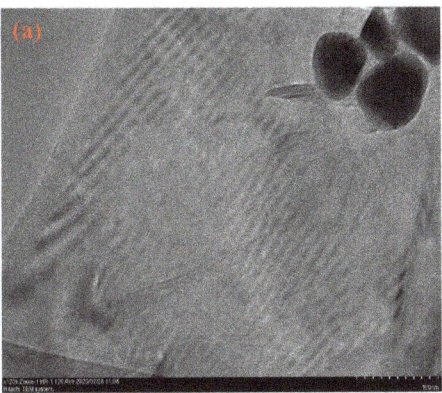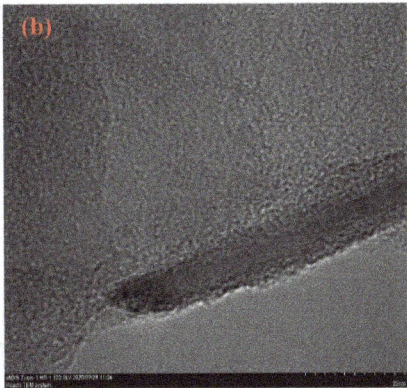

Fig. 4. TEM image of the surfactant wrapped functionalized graphene a) 120 kx b) 800 kx zoom

seen in the images. Figure 3 (a) also shows a plane that is oriented in a single direction. Thus, this shows that the surfactant-wrapped functionalized graphene is the single crystalline particle. When the nanoparticles are zoomed at x800k (Fig. 3(b)), a highly porous and hollow structure is visible. Certain morphological modifications (waviness) and degradation of the surface may be viewed as proof of functionalization.

3.2 Thermal Conductivity

Thermal conductivity is the ability of a material to conduct heat from one source to another source. The obtained thermal conductivity with respect to temperature is shown in Fig. 5. The figure shows that the thermal conductivity of SWFG + EG nanofluid tends to increase when the temperature increases, which is consistent with previous literature [18, 19]. In simple, the thermal conductivity of the SWFG + EG sample increases when the temperature was set approximately at 25°C to 50°C, but the thermal conductivity dropped when the temperature was greater than 55°C. Thermal conductivity values ranged from 0.2 W/m. K to 0.43 W/m.K.

3.3 Stability Analysis

The thermophysical properties of nanofluids are mainly dependent on the dispersion stability of nanoparticles in base fluids. The visual inspection method is one of the easy techniques to examine the stability of nanofluids. The formulated sample is poured in a vial, and its dispersion stability is determined for six successive days, shown in Fig. 6. From the image, it can be seen that the prepared sample is exceptionally stable for the first three days. However, the sample's color degrades to become transparent black for the last three days, indicating the sample's sedimentation rate. This study recommends increasing the ultrasonication time and surfactant type to enhance the dispersion stability further.

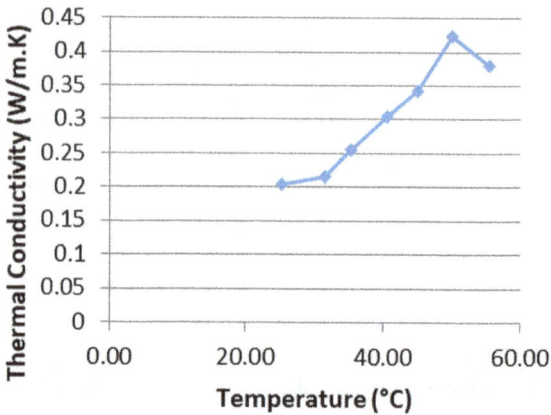

Fig. 5. Graph of thermal conductivity versus temperature of SWFG dispersed in EG

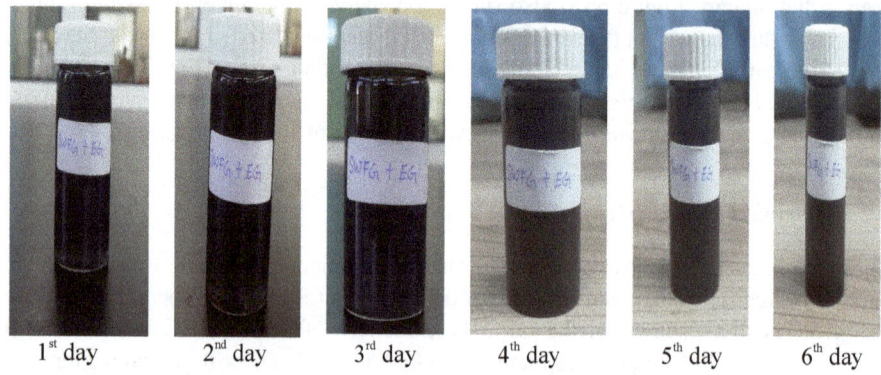

| 1ˢᵗ day | 2ⁿᵈ day | 3ʳᵈ day | 4ᵗʰ day | 5ᵗʰ day | 6ᵗʰ day |

Fig. 6. Sedimentation analysis of the formulated sample at different time intervals

4 Conclusions

This research examined the dispersion stability and thermal conductivity performance of nanofluids containing surfactant wrapped functionalized graphene in ethylene glycol solution. The maximum thermal conductivity obtained is 0.43 W/m.K at 55 °C. The moderate thermal conductivity may be due to the low concentration of graphene nanoparticles in the nanofluid. Eventually, less concentrated nanofluids are more essential for thermal systems to avoid coagulation and high pumping power. Based on the sedimentation analysis, the nanoparticles start aggregating to form the third day. On the other side, even on the sixth day, there was no complete sedimentation of graphene nanoparticles in the base fluid, enabling it to be a potential candidate for low-temperature applications.

Acknowledgments. The author would like to acknowledge Universiti Teknologi PETRONAS for their research support through YUTP grant (015LC0–118).

Conflicts of Interest. Authors have no conflicts of interest to declare.

References

1. Bahiraei, M., Heshmatian, S.: Graphene family nanofluids: A critical review and future research directions. Energy Convers. Manage. **196**, 1222–1256 (2019). https://doi.org/10.1016/j.enconman.2019.06.076
2. Bakthavatchalam, B., Habib, K., Saidur, R., Saha, B.B., Irshad, K.: Comprehensive study on nanofluid and ionanofluid for heat transfer enhancement: a review on current and future perspective. J. Mol. Liq. **305**, 112787 (2020). https://doi.org/10.1016/j.molliq.2020.112787
3. Arshad, A., Jabbal, M., Yan, Y., Reay, D.: A review on graphene based nanofluids: preparation, characterization and applications. J. Mol. Liq. **279**, 444–484 (2019). https://doi.org/10.1016/j.molliq.2019.01.153
4. Sadeghinezhad, E., et al: A comprehensive review on graphene nanofluids: recent research, development and applications. Energy Convers. Manag. **111**, 466–487 (2016). https://doi.org/10.1016/j.enconman.2016.01.004.
5. Hussein, O., Habib, K., Nasif, M., Muhsan, A., Bakthavatchalam, B.: Turbulence combined convective heat transfer and nanofluids flows over double forward facing steps. MATEC Web Conf. **225**, 01007 (2018). https://doi.org/10.1051/matecconf/201822501007
6. Dawooda, H.K., Hussein, O.A., Abdullah, A.Q., Bakthavatchalam, B.: Thermal and rheological properties of metallic oxides based nanorefrigerants: effect of nanoparticles type and shape. Solid State Technol. **63**, 4405–4423 (2020) http://solidstatetechnology.us/index.php/JSST/article/view/5350
7. Arora, N., Gupta, M.: An updated review on application of nanofluids in flat tubes radiators for improving cooling performance. Renew. Sustain. Energy Rev. **134**, 110242 (2020). https://doi.org/10.1016/j.rser.2020.110242
8. Karami, H., et al.: The thermophysical properties and the stability of nanofluids containing carboxyl-functionalized graphene nano-platelets and multi-walled carbon nanotubes. Int. Commun. Heat Mass Transf. **108**, 104302 (2019). https://doi.org/10.1016/j.icheatmasstransfer.2019.104302.
9. Ghozatloo, A., Shariaty-Niasar, M., Rashidi, A.M.: Preparation of nanofluids from functionalized graphene by new alkaline method and study on the thermal conductivity and stability. Int. Commun. Heat Mass Transf. **42**, 89–94 (2013). https://doi.org/10.1016/j.icheatmasstransfer.2012.12.007.
10. Agromayor, R., Cabaleiro, D., Pardinas, A., Vallejo, J., Fernandez-Seara, J., Lugo, L.: Heat transfer performance of functionalized graphene nanoplatelet aqueous nanofluids. Materials (Basel). **9**, 455 (2016). https://doi.org/10.3390/ma9060455.
11. Ma, W., Yang, F., Shi, J., Wang, F., Zhang, Z., Wang, S.: Silicone based nanofluids containing functionalized graphene nanosheets. Colloids Surf. A Physicochem. Eng. Asp. **431**, 120–126 (2013). https://doi.org/10.1016/j.colsurfa.2013.04.031
12. Bakthavatchalam, B., Habib, K., Saidur, R., Shahabuddin, S., Saha, B.B.: Influence of solvents on the enhancement of thermophysical properties and stability of MWCNT nanofluid. Nanotechnology **31**, 235402 (2020). https://doi.org/10.1088/1361-6528/ab79ab
13. Sarsam, W.S., Amiri, A., Kazi, S.N., Badarudin, A.: Stability and thermophysical properties of non-covalently functionalized graphene nanoplatelets nanofluids. Energy Convers. Manag. **116**, 101–111 (2016). https://doi.org/10.1016/j.enconman.2016.02.082.
14. Bakthavatchalam, B., Habib, K., Wilfred, C.D., Saidur, R., Saha, B.B.: Comparative evaluation on the thermal properties and stability of MWCNT nanofluid with conventional surfactants and ionic liquid. J. Therm. Anal. Calorim. (2020). https://doi.org/10.1007/s10973-020-10374-x

15. Keklikcioglu Cakmak, N.: The impact of surfactants on the stability and thermal conductivity of graphene oxide de-ionized water nanofluids. J. Therm. Anal. Calorim. **139**(3), 1895–1902 (2019). https://doi.org/10.1007/s10973-019-09096-6
16. Hussein, O.A., Habib, K., Saidur, R., Muhsan, A.S., Shahabuddin, S.: Alawi, OA: The influence of covalent and non-covalent functionalization of GNP based nanofluids on its thermophysical, rheological and suspension stability properties. RSC Adv. **9**, 38576–38589 (2019). https://doi.org/10.1039/C9RA07811H
17. Wang, S., Yi, M., Shen, Z.: The effect of surfactants and their concentration on the liquid exfoliation of graphene. RSC Adv. **6**, 56705–56710 (2016). https://doi.org/10.1039/C6RA10 933K
18. Sarbolookzadeh Harandi, S., Karimipour, A., Afrand, M., Akbari, M., D'Orazio, A.: An experimental study on thermal conductivity of F-MWCNTs–Fe3O4/EG hybrid nanofluid: effects of temperature and concentration. Int. Commun. Heat Mass Transf. **76**, 171–177 (2016). https://doi.org/10.1016/J.ICHEATMASSTRANSFER.2016.05.029
19. Vajjha, R.S., Das, D.K.: Experimental determination of thermal conductivity of three nanofluids and development of new correlations. Int. J. Heat Mass Transf. **52**(21–22), 4675–4682 (2009). https://doi.org/10.1016/j.ijheatmasstransfer.2009.06.027

The Investigation of Hot Extrusion Process Parameters on Microstructure and Mechanical Performance of Aluminium Alloy in Biomedical Industry

Nor Fadhilah Mohamad Khalili$^{(\boxtimes)}$, Hoo Jian Jun, Azlan Ahmad, Nabihah Sallih, Mohd Amri Lajis, and Shazarel Shamsudin

Universiti Teknologi PETRONAS, Seri Iskandar, Malaysia
fadhilah_19000247@utp.edu.my

Abstract. Aluminium alloy 7075 is chosen as the material to be investigated to replace stainless steel as the material of hyperthermic needle. Hot Direct Extrusion process is famous for reducing the size of the grain to enhance the mechanical properties of the alloy. The process parameters are investigated through several extrusion process related researches. Speed of extrusion was operated within preheat temperature of 350 to 450 °C and pre-heat time of 1 to 2 h. The effect of extrusion parameters were analyzed by using design of experiment method including Taguchi method and Full Factorial method. Furthermore, density test, Vickers hardness test, compression test and microscopic visual inspection such as optical microscopic and scanning electron microscopic (SEM) were conducted. Several alternatives such as twist extrusion, heat treatment and aging must be done to improve the mechanical properties of extrudates.

1 Introduction

In biomedical industries, hypothermic needle is the one-use product which creates a lot of wastage despite having high money investment on the product by using stainless steel as the material of hypothermic needle. Aluminium alloy is proposed to replace stainless steel in producing hypothermic needles to reduce the cost of production and ease the recycling process of hypothermic need.

Aluminium alloy is the third most abundant element in earth's crust [1]. Aluminium alloys are suitable to be applied in the automobiles and aircraft industries due to its good mechanical performance and weight saving because they have high strength to weight ratios. Aluminium alloys have wide range of chemical compositions and product forms which can be manufactured by all available metal working techniques and cutting processes [2]. There are several aluminium alloys' series that are available in the market such as 1xxx, 2xxx, 3xxx, 4xxx, 5xxx, 6xxx, and 7xxx series. Different series are having different alloying compositions and different mechanical properties which are suitable for different applications [3, 5].

Alloying elements are playing a critical role in modifying the mechanical properties of the aluminium alloys. To increase the strength to weight ratio, aluminium alloys will

M. Awang and S. S. Emamian (Eds.): *Advances in Material Science and Engineering*, LNME, pp. 327–332, 2021.
https://doi.org/10.1007/978-981-16-3641-7_39

be a promising choice among all metals [5]. The most commonly used alloy in this series for the aircraft industry is AA7075 [6]. AA7075 has been chosen to replace stainless steel in producing hypothermic needles [7]. The common disadvantage of 7xxx series alloy is it reduces formability in the room temperature. Hence, the alloy must be formed an elevated temperature. This alloy can perform age hardening method to transform into a better mechanical performance alloy [8].

Extrusion process has been chosen as the manufacturing process to produce the aluminium alloy for the grain refinement to improve the mechanical and corrosion resistance properties of the aluminium alloy [4]. Different parameters will be investigated by Design of Experiment method such as Taguchi method and the best combination of parameters to extrude the aluminium alloy with the results of the best mechanical properties.

This research is focusing on identifying the appropriate extrusion process parameters in producing aluminium alloy, analyzing the mechanical performance, the microstructure of the material, and determining the relationship between parameters by using Design of Experiment (DOE) method.

2 Methods

2.1 Process Parameters

Extrusion parameters play a crucial part to ensure the quality consistency of the extrudates. Before constructing the design of experiment, parameters must be set up first. Table 1 shows the process parameters of the hot extrusion.

Table 1. Process parameter of hot extrusion

Parameters	Value
Pre-heated temperature (°C)	350~450
Pre-heat time (h)	1~2
Speed of ram (mm/s)	1
Pressure of ram (ksi)	100
Extrusion ratio	2:1

2.2 Design of Experiment (DOE)

Method used in DOE is factorial experiment which studies the effect of each factor on response and the intersection between on response. The controlled parameters are the pre-heat temperature and the pre-heat time. This project involves two factors and two levels of experiments. This experiment will have four samples produced as listed below in Table 2.

Table 2. Four possible runs of the experiments

Run	Factors			
	Pre-Heat Temperature (°C)		Pre-Heat Time (hr)	
1	1	350	1	1
2	1	350	2	2
3	2	450	1	1
4	2	450	2	2

2.3 Experiment Procedure

The extrusion process started by pre-heat the cast AA7075 with 350 °C for one and two hours and follow up with temperature of 450 °C for one to two hours.

All extruded sample then were cut for metallurgy sample preparation. Samples were mounted for grinding and polishing process. The samples were grind using 400, 600 and 800 SiC grit paper and were polished using metaDi fluid of 6 and 3 microns. After polishing sample were etched with Keller's etchant (190 ml H_2O, 5 ml HNO_3, 3 ml HCl, 2 ml HF) for 20 s. then the samples are ready for OM and SEM to obtain microstructure images.

For Vickers hardness test, force applied is 200gs with dwell time, 15s. The extrudates and cast AA7075 were cut into dimension of (15 mm × 30mm) for compression test referring by ASTM E9 and speed of compression test is 0.005m/min.

3 Result

3.1 Density Test

The extrudates have reductions in their densities compared to the cast alloy. The cast AA7075 has density of 2.8 g/cm^3 meanwhile density of Run 1, 2, 3 and 4 are 2.78 g/cm^3, 2.796 g/cm^3, 2.798 g/cm^3, 2.794 g/cm^3. The reduction of density indicates porosities or impurities in between grains (Fig. 1).

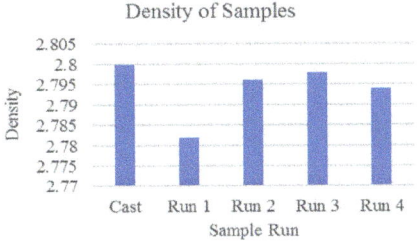

Fig. 1. Bar chart of sample's density

3.2 Vickers Hardness

From the Vickers hardness test in Fig. 2, the average hardness of extrudates is lower tat casted with 185.56 HV and standard deviation of 9.5, 50.7% in average. The highest values among extrudates is Run 3 while the least value is Run 2.

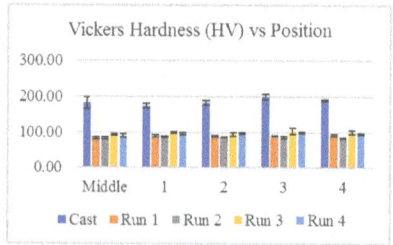

Fig. 2. Bar chart of Vickers hardness test vs position

Throughout the Vickers hardness testing, the middle position had lower hardness value compare to the position 1, 2, 3, and 4. The porosity and impurities in between the grains cause severe reduction in hardness of the extrudates.

3.3 Compression Test

Yield strength of the extrudates was reduced by 65.67% compared to the cast sample. Run 3 has the highest yield strength while run 2 has the lowest yield strength (Fig. 3).

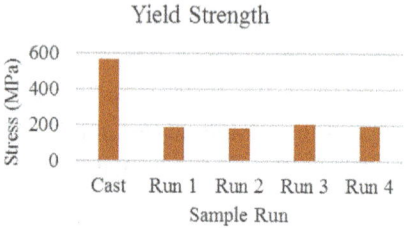

Fig. 3. Bar chart of sample's yield strength

3.4 Scanning Electron Microscopic (SEM)

The sequence of microstructure images as shown in Fig. 4 (a), (b), (c), (d) and (e) are cast AA7075, run 1, 2, 3, and 4. The microstructure in run 1 and 2 have more porosity while in run 3 and 4, there are grains orientation and less porosity. To conclude run 3 and 4 have better mechanical performance compare to run 1 and 2.

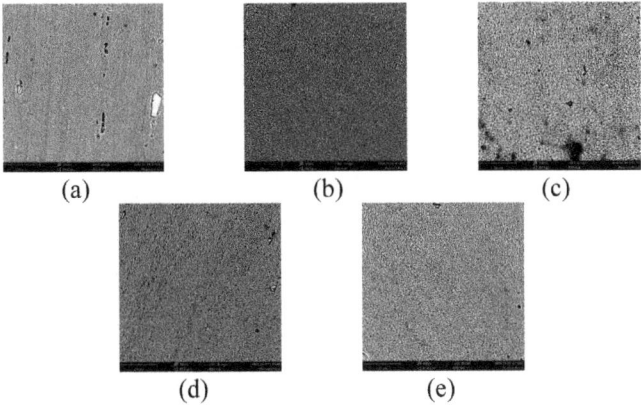

Fig. 4. SEM images for 3000x magnification

4 Conclusions

The best parameter to produce the extrudates with better mechanical performance are pre-heat temperature of 450 °C and one hour of heating time. Pre-heat temperature has higher effect than pre-heat time on the mechanical performance [9]. The more factors and levels of DOE will increase accuracy in the effect of parameter. The reduction in hardness and yield strength is caused by presence of porosity in grain boundaries. There are several methods to reduce porosity in extrudates such as twist extrusion and forging process. As conclusion, AA7075 extrudates are not achieving grains refinement which could lead to better performance. Hence, AA7075 extrudates is not capable of replacing the stainless steel in producing hypothermic needle unless the secondary process such as twist extrusion heat treatment is done.

References

1. Chang, Y., Meng, Z.-H., Ying, L., Li, X.-D., Ma, N., Hu, P.: Influence of hot press forming techniques on properties of vehicle high strength steels. J. Iron. Steel Res. Int. **18**(5), 59–63 (2011). https://doi.org/10.1016/S1006-706X(11)60066-6
2. Chahare, A., Inamdar, K.H.: A review on process parameters affecting aluminium extrusion process. Int. J. Innov. Res. Sci. Eng. **2**(12), 193–198 (2016)
3. Gharde, A.N.: Review on the microstructure and microhardness of extruded aluminum alloys. JETIR **2**(4), 1097–1102 (2015)
4. Sheppard, T., Tunnicliffe, P.J., Patterson, S.J.: Direct and indirect extrusion of a high strength aerospace alloy (AA 7075). J. Mech. Work. Technol. **6**(4), 313–331 (1982)
5. Ng, C.H., et al.: Reviews on aluminum alloy series and its applications. Acad. J. Sci. Res. **5**(12), 708-716 (2017)
6. Carle, D., Blount, G.: The suitability of aluminium as an alternative material for car bodies. Mater. Des. **20**(5), 267–272 (1999)
7. Jawalkar, C.S., Kant, S.: A review on use of aluminium alloys in aircraft components. i-manager's J. Mater. Sci. **3**(3), 33–38 (2015)

8. Shriwas, A.K., Kale, V.C.: Impact of aluminum alloys and microstructures on engineering properties-review. IOSR J. Mech. Civ. Eng. **13**(3), 2320–2334 (2016)
9. Shamsudin, S., Zhong, Z.W., Rahim, S.N.A., Lajis, M.A.: The influence of temperature and preheating time in extrudate quality of solid-state recycled aluminum. Int. J. Adv. Manuf. Technol. **90**(9–12), 2631–2643 (2016)

Influence of Graphene Nanoplatelets Loading into Physical and Morphological Characteristics of Polyaniline Nanocomposites

Jeefferie Abd Razak[1(✉)], Hazman Hasib[1], Nor Aisah Khalid[1], Noraiham Mohamad[1], Mazlin Aida Mahamood[1], Mohd Shahadan Mohd Suan[1], and Mohd Muzafar Ismail[2]

[1] Fakulti Kejuruteraan Pembuatan, Universiti Teknikal Malaysia Melaka, Hang Tuah Jaya, Durian Tunggal, 76100 Melaka, Malaysia
jeefferie@utem.edu.my

[2] Fakulti Teknologi Kejuruteraan Elektrik and Elektronik, Universiti Teknikal Malaysia Melaka, Hang Tuah Jaya, Durian Tunggal, 76100 Melaka, Malaysia

Abstract. This study has evaluated the effects of graphene nanoplatelets loading to the physical and morphological characteristic of polyaniline-based nanocomposites. The synthesis of PANI was conducted through an oxidative aniline polymerization method in an acidic medium. GNPs loadings are varied into 0, 0.25, 0.50, 0.75 and 1.00 wt.% addition. It was found that the presence of GNPs within PANI matrix has three-time enhanced the DC conductivity. Platelet like GNPs nano disc-flake and establishment of conductive network among them had successfully assisted the PANI polymer conductivity which mainly caused by the polar and bipolar movement along the conjugated polyaniline backbone. Addition of GNPs as an alternative charge carrier helps to reduce the dependency of dopant utilization in increasing the electro-conductivity of conductive PANI polymers.

Keywords: GNPs · PANI · Filler loadings · Physical · Morphological · Electrical conductivity

1 Introduction

In this work, modification to the polyaniline (PANI) conductive polymer by adding the graphene nanoplatelets (GNPs) at various filler loading was performed. The effects of GNPs addition into the physical and morphological properties of PANI/GNPs nanocomposites has been evaluated. Improved properties of PANI filled composites could be potentially manipulated for many demanding applications which require electro-conductive polymeric material as main substrate. Modified PANI or PANI based composite could potentially be viable for various multitude applications such as for gas sensing, sensor and actuators, microelectronics, electro-chromic devices, smart fabric, rechargeable battery, antenna, electromagnetic protecting gadget and many more [1, 2].

M. Awang and S. S. Emamian (Eds.): *Advances in Material Science and Engineering*, LNME, pp. 333–340, 2021.
https://doi.org/10.1007/978-981-16-3641-7_40

PANI is a common green protonated emeraldine that has conductivity on a semi-conductor level of the order of 10^0 Scm^{-1}, higher than the other common polymers ($<10^{-9}$ Scm^{-1}) but lower than any of typical metals ($>10^4$ Scm^{-1}). PANI polymer are becoming attractive due to ease of synthesis and possibility of controlling their electrical conductivity by changing the state of protonation or oxidation state [3, 4]. Modification of PANI through composite strategy had previously researched, such as on the development of micro strip antenna segment [5, 6], while development of electron storage through in-situ polymerization of TiO_2-TVC/PANI composite film had increased the electron storage [7].

GNPs has 2D structure where the carbon atoms were arranged in honeycomb network, which at the end resulting less resistance of electrical current flow and greater stiffness up to 1 TPa [8]. The GNPs can conduct electricity better than copper with improved surface barrier attributes due to their platelets like morphology. The electrical conductivity of GNPs at parallel to surface is 10^7 S/m while for perpendicular to surface is 10^2 S/m [9]. In addition, the planar structure of GNPs, provide 2D pathway for phonon transport for efficient thermal flow [10].

In this study, in-situ incorporation of GNPs with PANI during polymerization in an acidic medium was performed. Physical properties of produced PANI/GNPs nanocomposites were accessed through the XRD, Raman spectroscopy and Four Point Probe DC conductivity tests, while the morphological behavior of them was evaluated by the SEM microscope observation.

2   Materials and Methods

In this study, pristine PANI and PANI/GNPs nanocomposites were synthesized by aniline oxidative in-situ polymerization in an acidic medium. For pristine PANI preparation, about 22.84 g of ammonium peroxydisulfate (APS) and 10.36 g of aniline hydroxide were dissolved in 200 mL of distilled water, separately. Later, the pre-cooled step was followed for both solution at 9 °C for 12 h. Then, the precooled solutions were mixed in a beaker and stirred for one hour by using a magnetic stirrer. The stirred mix solution was left in a chiller at 9 °C for 24 h. The precipitated was then filtered by using a vacuum filter at 50 kPa. While filtering process was on-going, the mixture has been washed with 200 mL of 0.2 M of hydrochloric acid and 200 mL of acetone. Finally, the drying of filtered mixture was performed in a vacuum oven for 24 h at 60 °C.

For synthesis of PANI/GNPs nanocomposites at different loadings of GNPs (0.25, 0.50, 0.75 and 1.00 wt.%), the GNPs content was based on 100% of APS and aniline hydrochloride total weight. The respective GNPs weight amount was added after the first pre-cooled, during the solution mixing at one hour, before it being left to cool again at 9 °C for 24 h. The subsequent following steps are similar like the preparation of pristine PANI, until the nanocomposite powder of PANI/GNPs has been completely prepared.

The effect of GNPs loading towards the crystallinity and morphological structure of PANI/GNPs nanocomposites was evaluated by the XRD analysis. For this analysis, it was carried out by using PanAnalytical diffractometer which operated at 40 kV and 30 mA, using Cu Kα radiation with λ = 0.154 nm of wavelength. The scans were taken at between 10 - 80° ranges at a scan rate of 0.05° per second with a continuous scan step size of 0.0170°.

Raman spectrometer analysis was performed to analyze the structural properties of PANI and PANI/GNPs nanocomposites. The analysis was conducted by using UniRAM-3500 set-up with argon-ion laser at <0.02 mW and 514.5 nm excitation within the experimental ranges at between of 900–1900 cm^{-1}.

DC-conductivity measurement by using a four-point probe method was applied to measure the electrical conductivity of pristine PANI and PANI/GNPs nanocomposites samples. The test was conducted in accordance to ASTM D999–89 standard. A four-point probe machine model Jandel RM3-AR Sample of PANI and PANI/GNPs nanocomposites were pressed into pallet with dimension of 2.80 mm thick and 9.00 mm diameter by using a hydraulic press machine, at room temperature setting. The electrical conductivity was measured based on the following Eq. 1 and Eq. 2, where ρ is the resistivity in ohm.cm, while S is the distance between two inner probes at four-point probe set-up, and σ is the conductivity in Scm^{-1}.

$$\rho \, = \, 2\pi.S \, (V/I) \tag{1}$$

$$\sigma \, = \, 1/\rho \tag{2}$$

For morphological observation of produced pristine PANI and PANI/GNPs nanocomposites powder, it was performed by using the scanning electron microscope (SEM) model Zeiss EVO50. The observation on the selected samples (pristine PANI as controlled sample, PANI/0.20 wt.% GNPs, PANI/0.75 wt.% GNPs and PANI/1.00 wt.% GNPs nanocomposites was performed at the magnification of 1000x under the accelerating voltage of 15 kV. Both samples were coated beforehand with gold-palladium coating using sputter coater model Polaron to prevent the electrostatic charging phenomena and poor image resolution.

3 Results and Discussion

Figure 1 shows the overlaid XRD curves for pristine PANI and PANI/GNPs nanocomposites at various GNPs loadings. Pristine PANI shows one broad and diffused peak at $2\Theta = 21°$ which indicate the presence of crystallinity behavior of produced PANI. Several important peaks for pristine PANI at $2\Theta = 15°, 21°$ and $26°$ are corresponded to 011, 020 and 200 diffraction planes, respectively. By adding the GNPs nanofiller at 0.25, 0.50, 0.75 and 1.00 wt.% has increased the intensity at $2\Theta = 26°$ that correspond to the crystallinity of graphene. The intensity was found becoming intense with the increase of GNPs loadings. This situation has clearly suggested the strong interaction between the PANI backbone with the GNPs nanofiller, during the aniline oxidation polymerization. In addition, this also suggest the homogeneous dispersion of GNPs within or in-between of PANI polymer macromolecules.

Raman spectroscopy analysis was performed to evaluate the information on PANI/GNPs physico-chemical structure. Raman spectra of pristine PANI and PANI/GNPs nanocomposites with different GNPs loadings are depicted in the following Fig. 2. Basically, for PANI based samples, there are three important vibrational range, which are C-H bending mode at between 1100 – 1210 cm^{-1}, several C-N stretching mode

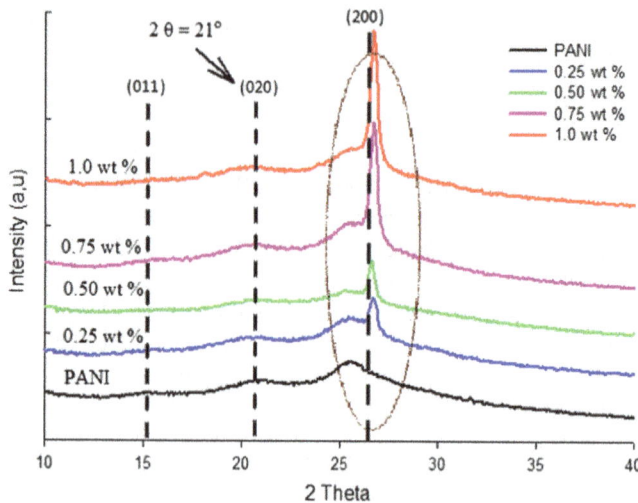

Fig. 1. X-Ray Diffraction (XRD) pattern of pristine PANI and PANI/GNPs nanocomposites

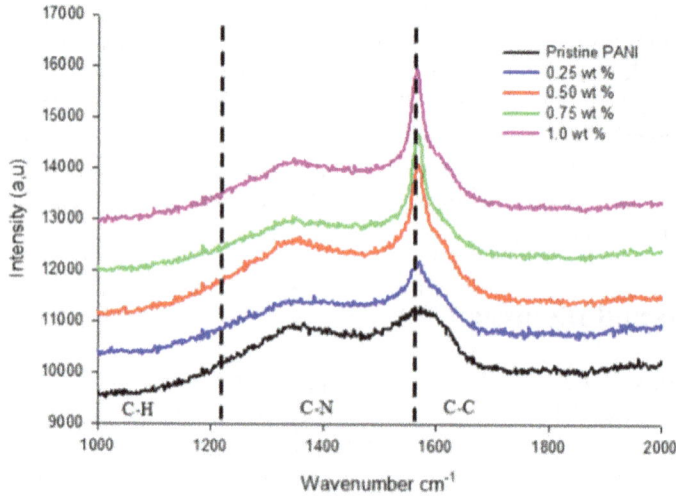

Fig. 2. Raman spectra of pristine PANI and PANI/GNPs nanocomposite

(polaron, imines and amines) at between of $1210-1520\,\mathrm{cm}^{-1}$ and C-C stretching modes at between $1520\,\mathrm{cm}^{-1}$ that attribute to benzoid C-C stretching mode, into $1650\,\mathrm{cm}^{-1}$ which attributed to quinoid C-C stretching mode. In this Raman spectra also represent two strong bands corresponds to D and G band at $1300\,\mathrm{cm}^{-1}$ and $1600\,\mathrm{cm}^{-1}$, respectively. A strong π-π conjugation between the PANI and GNPs phases has occurred that resulting from an increase in I_d/I_g ratio of PANI/GNPs nanocomposites with increase of GNPs loadings. This improved interaction was enhanced the electrical properties of

PANI/GNPs nanocomposites as it comprised larger amount of charge carrier transport at higher GNPs addition.

In this study, the electrical conductivity performance of pristine PANI and PANI/GNPs nanocomposites with various loading of GNPs was explained based on sheet resistance measurement. PANI based powder were first pressed using a hydraulic press to prepare a disc-shaped pellet with thickness of less than 3.00 mm. The following Fig. 3 depicts the DC conductivity plots of pristine PANI and PANI/GNPs nanocomposites, measured by the four-point probe method. Based on the plots, it was found that, at lower content of GNPs (0.25 wt.%) has generated lower conductivity value as compared than the conductivity of pristine PANI sample. This condition was resulted due to lower concentration of GNPs which disrupted the PANI molecular backbone. Therefore, the lower GNPs loadings has basically prevented the formation of complete micelle when the aniline polymerization was taking place. However, the conductivity values have proportionally increased with the increase of GNPs content, especially at higher loading addition. The presence of GNPs at higher loading helps to promotes the polymerization by stabilizing and increasing the PANI micelles size which induced uninterrupted polymerized condition that increase the conductivity. In addition, good contact between GNPs flakes also promotes the current transport due to less resistance between platelet which interlinked between each other to establish the alternative current network path for better electrical conductivity [13]. PANI electric conductivity is caused by conjugation in the polyaniline backbone, where PANI has both benzoid and quinoid unit. The polar and bi-polar movement along the polymer backbone has responsibly caused the polymer electro-conductivity mechanism [14].

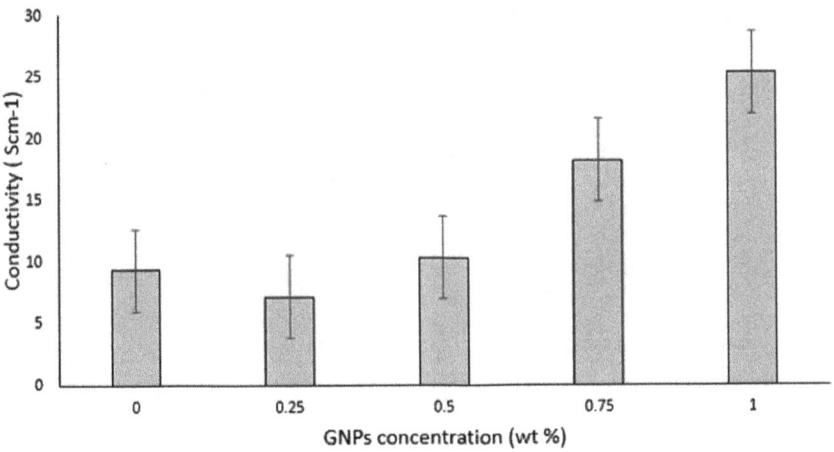

Fig. 3. DC-Conductivity plots of pristine PANI and PANI/GNPs nanocomposite

SEM observation was performed to observe the morphological characteristic of pristine PANI and PANI/GNPs nanocomposites at various GNPs loading. Pristine PANI as controlled sample, while PANI/GNPs nanocomposites at 0.25 and 0.75 wt.% to represent filled PANI at lower GNPs addition and PANI/1.00 wt.% GNPs for nanocomposites

at higher GNPs loading. Pristine PANI as depicted in Fig. 4(a) possessed agglomerated globular shape. However, the morphology of PANI globular has been significantly
changed after the presence of GNPs in PANI/GNPs nanocomposites for entire loadings. At 0.25 wt.% of GNPs addition, presence of graphene flakes has destructed the
agglomerates globular of PANI polymer. At lower loading, better dispersion of GNPs
within PANI has separated the macromolecules and prevent the agglomerates formation.
This condition explained the significant reduction of DC conductivity value in comparison to pristine PANI sample. However, by increasing further the GNPs loading more

Fig. 4. SEM images of pristine PANI and PANI/GNPs nanocomposite (1000x of magnification);
(a) pristine PANI; (b) PANI/0.25 wt.% GNPs; (c) PANI/0.75 wt.% GNPs; and (d) PANI/1.00 wt.%
GNPs

than 0.25 wt.%, it was found that very well integration between GNPs nanofiller and PANI macromolecules cluster, due to successful in-situ oxidative aniline polymerization reaction. Flaky morphology with several flat and edges surfaces can be clearly detected throughout the PANI cluster, which indicating better dispersion of GNPs. It proved that the aniline monomers were diffused and polymerized into graphene nano-sheets layer by layer during in situ polymerization [15]. At higher loadings (1.00 wt.%) of GNPs, the SEM micrograph of PANI/GNPs nanocomposite (Fig. 4d) also revealed that the GNPs were homogeneously distributed within the PANI matrix. The presence of GNPs within the PANI matrix has led to enhance the electro-conductivity characteristic of PANI/GNPs nanocomposites as proven in this study. Presence of GNPs has successfully added the alternative pathway for current movement, due to network inter-linkage between the graphene platelets together with PANI macromolecules clusters.

4 Conclusions

In this work, pristine PANI and PANI/GNPs nanocomposites with various loadings of GNPs nanofiller addition has been successfully synthesized through an oxidative aniline in situ polymerization method. This has been confirmed by the output findings from Raman spectroscopy analysis. The XRD and Raman findings again has confirmed the dispersion nature and well-integration of GNPs within the PANI macromolecules. This has further proven by the SEM micrograph which also explained by the positive DC conductivity results pattern, that was proportionally increased with increase of GNPs loadings. In overall, addition of GNPs into conductive PANI, are potentially able to further improve the electrical conductivity properties of pristine PANI and PANI/GNPs nanocomposites. This will certainly be benefited for various impending high-impact applications or future utilization.

Acknowledgments. Authors would like to acknowledge Universiti Teknikal Malaysia Melaka (UTeM) for sponsoring this research under the short-term research grant: *PJP/2019/FKP(4B)/S01662*. Thanks to COSSID and FKP for facilities and technical support in completing this study.

References

1. Tsolis, C., Whittow, W., Alexandridis, A., Vardaxoglou, J.: Embroidery and related manufacturing tecniques for wearable antennas: Challenges and opportunities. Electronics **3**(2), 314–338 (2014)
2. Hassan, W.M., Attiya, A.: Textile antenna integrated with a cap for L1 GPS application. In: 35th National Radio Science Conference (NRSC), 38–46 (2018)
3. Ding, X., Han, D., Wang, Z., Xu, X., Niu, L., Zhang, Q.: Micelle-assisted synthesis of polyaniline magnetite nanorods by in situ self – assembly process. J. Coll. Interf. Sci. **320**, 341–345 (2008)
4. Hasoon, S.A., Abdullah, I.A.: Optical and electrical properties of thin films of polyaniline and polypyrrole. Int. J. Electrochem. Sci. **7**, 10666–11067 (2012)
5. Frederick, A.: Smart Nanotextiles: Inherently Conducting Polymers in Healthcare. Da Vinci's Notebook, vol. 3 (2011)

6. Ashok, Y., Vinod, K.S., Manu, C., Himanshu, M.: A review on wearable textile antenna. J. Telecommun. Switching Syst. Netw. **2**(3), 37–41 (2015)
7. Wancai, B., Fongsamut, C., Ngaotrakanwiwat, P.: Development of TiO_2/TiO_2-V_2O_5 compound with polyaniline for electron storage. Energy Procedia **79**, 903–909 (2015)
8. Kausar, A., Ur Rahman, A.: Effect of graphene nanoplatelets addition on properties of thermoresponsive shape memory polyurethane-based nanocomposite. Fullerenes, Nanotubes, Carbon Nanostruct. **24**(4), 235–242 (2016)
9. Mehrali, M., et al.: Investigation of thermal conductivity and rheological properties of nanofluids containing graphene nanoplatelets. Nanoscale Res. Lett. **9**(1), 1–12 (2014). https://doi.org/10.1186/1556-276X-9-15
10. Wang, R.-X., Huang, L.-F., Tian, X.-Y.: Understanding the protonation of polyaniline and polyaniline-graphene interaction. J. Phy. Chem. C **116**(24), 13120–13126 (2012)
11. Ben-Valid, S., et al.: Polyaniline-coated single-walled carbon nanotubes: synthesis, characterization, and impact on primary immune cells. J. Mat. Chem. **20**(12), 2408–2417 (2010)
12. Badi, N., Khasim, S., Roy, A.S.: Micro raman spectroscopy and effective conductivity studies of graphene nanoplatelate/polyaniline composites. J. Mater. Sci. Mater. Electron **27**, 6249–6257 (2016)
13. Jun, Y.S., Um, J.G., Jiang, G., Yu, A.: A study on the effects of graphene nano platelets (GnPs) sheet sizes from a few to hundred microns on the thermal, mechanical, and electrical properties of polypropylene (PP)/GnPs composites. eXPRESS Polymer Lett. **12**(10), 885–897 (2018)
14. Molapo, K.M., et al.: Electronics of conjugated polymers (I): polyaniline. Int. J. Electrochem. Sci. **7**(12), 11859–11875 (2012)
15. Chin, S.Y., Abdullah, T.K., Mariatti, M.: One-step synthesis of conductive graphene/polyaniline nanocomposites using sodium dodecyl benzene sulfonate: preparation and properties. J. Mater. Sci. Mater. Elect. **28**, 18418–18422 (2017)

Evaluating the Impact Toughness of Mild Steel Welded Joint Using Taguchi Method in Automated Shielded Metal Arc Welding

Shamini Janasekaran[1]([✉]), Walisijiang Tayier[1], Hong Seng Guan[2], and Teo Hiu Hong[2]

[1] Centre for Advanced Materials and Intelligent Manufacturing, Faculty of Engineering, SEGi University Sdn Bhd, Built Environment & IT, 47810 Petaling Jaya, Selangor, Malaysia
shaminijanasekaran@segi.edu.my
[2] Centre for Advanced Engineering Design, Faculty of Engineering, SEGi University Sdn Bhd, Built Environment & IT, 47810 Petaling Jaya, Selangor, Malaysia

Abstract. The impact toughness is commonly used in industrial and manufacturing fields for testing of materials properties. The impact toughness is also known as the Charpy V-notch testing which emphasis systematized of strain rate test and its influences over energy observed from the specimen during fracture. In this research, the mild steel (AISI 1080) used to weld for impact toughness (Charpy V-notch testing). The mild steel has a good mechanical property and low cost so it's commonly used in the manufacturing and welding processing fields. Due to the critical issue of welding quality and welding speed accuracy, the electrode E6013 was used to filler metal on the specimens by automated based shielded metal arc welding (ABSMAW) which also can improve welding quality and reduce welding errors. In this study parameters set up with speed (range from 3.19 to 4.6 mm/sec) using the different range of angle and electrode diameter (range from 45° to 60°, and 2.6 mm, 3.2 mm, 4.0 mm) also configuration of the welding specimen was clamped in the butt joint of double grooving which allows automatic based shielded metal arc welded to took place. The welding specimens were cut according to Charpy V-notch testing requirement with the ASTM E23 standard, the V-notch point was considered in the welded pool and heat-affected zone (HAZ), the specimens were ground properly while the Taguchi method used to analyze responses data from the experimental testing. The experiment found a direct relationship between welding angle, impact strength, and absorbed energy that the welding angle increases, the impact strength and absorbed energy increases; the welding speed is also the most effective factor in this testing. In this study also found the toughness of the welded pool is much lower than the toughness of HAZ. The optimum values are produced through the welding speed of 3.19 mm/sec, welding angle of 50° and electrode diameter of 2.6 mm.

1 Introduction

The mild steels are commonly used in industrial and manufacturing fields relating to its good mechanical properties while it has good weldability and thermal conductivity [1]. The shielded metal arc welding (SMAW) is a process of joining method for similar

M. Awang and S. S. Emamian (Eds.): *Advances in Material Science and Engineering*, LNME, pp. 341–348, 2021.
https://doi.org/10.1007/978-981-16-3641-7_41

and dissimilar materials [2]. Recently, different researchers conducted on the impact strength using different parameters and achieved different results. The research done by the Ambuj Saxena et al. also on the impact strength in the fusion zone (FZ) is significantly lower than the heat-affected zone (HAZ) and the fracture surface of FZ and HAZ are different according to the using different parameters such as electrode diameter (range from 3.2 mm to 4.0 mm), current (range from 78A to 110A), and welding speed (range from 100 mm/min to 130 mm/min) [3]. The research done by the RajaKumar et.al also investigated impact strength using two different welding techniques such as SMAW and gas tungsten arc welding (GTAW) and they used to weld in the metal plates of 304B4 (300 mm * 70 mm * 10 mm, ASTM 23 (55 mm × 10 mm × 10 mm)). As a result, the 304B4 metal plate produced good toughness at 37J using SMAW [4]. Furthermore, some studies conducted on impact toughness of HSLA-80 steel that they reported the toughness of the welded pool is significantly higher than the toughness of the HAZ [5]. However, the issue of welding speed accuracy and welding quality in manual welding processing has been investigating until the present. Thus, the aim of this study is to investigate the effects of electrode parameters on the impact toughness of mild steel welded joint in the automatic based shielded metal arc welding (ABSMAW) using Taguchi technique. The ABSMAW can improve welding speed accuracy and welding quality while it can enhance to good joining between the two specimens.

2   Experimental Procedure

2.1   Material and Sample Preparation

Figure 1 reveals sheets of AISI1080 mild steel (34 mm × 30 mm × 4.5 mm) were well welded via the ABSMAW on double grooving open square of butt joint configuration. The electrode E6013 used filler metal on the base metal. The welded specimens were ground as Silicon Carbide (SiC) abrasive paper, such as 220 grit. It removes burrs at the joining interface.

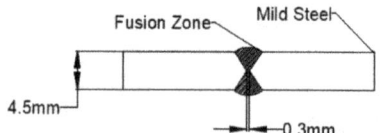

Fig. 1. Configuration joint (double grooving in butt jointing)

2.2   Automated Based Shielded Metal Arc Welding (ABSMAW)

The ABSMAW is performed by the Arduino board which connects with PC and the motherboard can interpret data systematically and then transfers to data information from PC to stepper motor for machine processing, as shown in Fig. 2. This article executed to the Arc200 SMAW machine for welding purposes. The machine current ranges from 0 to 160 A at a constant voltage are 26.4 V.

In this study, the 45°, 55° and 60° of welding angle were conducted at a constant current of 90 A and voltage 26.4 V while the effects of electrode diameters (2.6 mm to 4.0 mm) and welding speed (3.19 mm/sec to 4.6 mm/sec) investigated at a constant machine parameter (voltage of 26.4 V and welding current of 90 A). According to the current setting, this paper referred to the previous research that it reported the medium welding current produces to expectable impact strength between the 80 A and 90 A [6]. The all of parameters are obtained via the many of preliminary testing that it can avoid some unexpectable results. The welding parameters were listed in Table 1.

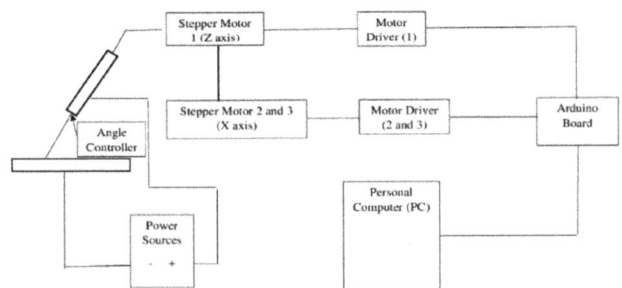

Fig. 2. Schematic drawing for experimental testing

Table 1. Shielded metal arc welding process parameters

Welding speed (mm/sec)	Electrode diameter (mm)	Electrode angle (°)	Trial no
4.6	2.6	45	1
4.6	3.2	50	2
4.6	4	60	3
3.9	2.6	50	4
3.9	3.2	60	5
3.9	4	45	6
3.19	2.6	60	7
3.19	3.2	45	8
3.19	4	50	9

2.3 Metallographic Characterizations

The specimens were well welded with double grooving of the butt joint. The cross-section of welded metal is shown in Fig. 3. The two specimens (34 mm × 30 mm × 4.5 mm) were welded by the ABSMAW and then the welded metals machined with (55 mm × 10 mm × 4.5 mm) according to the ASTM E23 (Charpy test) standard, as shown in Fig. 3. The specimens were cut into two parts for testing of welded pool and

HAZ and then the welded specimens were ground in the grinding machine as the 200grit of SiC paper. After that, these specimens were etched by Ferric Chloride ($FeCl_3$). The preparation of welded specimens was angling notched in the welded surface as the 2 mm deep, 0.25 mm radius along the base, and 45° angle (ASTM E23 standard); the welded specimens were tested separately from the welded pool and Heat Affected Zone (HAZ).

2.4 Impact Toughness Testing Design

This research investigated the impact of toughness for welded metal using different electrode parameters. The Charpy (impact) testing performed in the JBW-300 pendulum Charpy v-notch machine. The impact testing executed under the environmental conditions that included impact velocity of 5.2 m/s, 22 C room temperature and 54% of humidity [7]. The testing parameters are set by the Taguchi orthogonal arrays (OAs). Each parameter tested two times for the accuracy of testing. In this study, the impact strength was calculated according to the following Eq. (1) [8].

$$\text{Impact strength} = \frac{E}{A} \tag{1}$$

Where E-observed energy (J) that directly obtained from pendulum dial, A- Area of the V-notch specimen (area of the notch in the specimen is 0.4 cm^2).

2.5 Taguchi Orthogonal Arrays (OAs)

The Taguchi method is an essential technique to improve process performance and productivity. The Taguchi method supports to use of an orthogonal array for the design of the different parameters in the experiment [9]. This article used the orthogonal array of L9 (3^3) that the parameters automatically designed with the factors and levels from the preliminary testing.

3 Result and Discussion

3.1 Results of Testing for Impact Toughness

In Table 2, it shows results of impact toughness from the experimental testing that the different parameters produced a variety of impact strength. Meanwhile, the result shows the trial (9) to produce the highest impact strength as the low welding speed.

3.2 Analysis of Impact Toughness

This study investigated the impact toughness of mild steel using the different parameters. Figure 3 illustrated the brittle fracture surface to produce in the welded metal using the different parameters while the experiment also found the different welding parameters to produce the different rate of observed energy so the impact strength and observed energy are possible to relate the carbon particles and phase of grain structure. This result is also supported by various researchers and authors in the fields of welding [10–12].

Table 2. The results of impact toughness

Trial	Observed Energy in HAZ (J)			Impact Strength of HAZ (J/cm^2)	Observed Energy in Welded Pool (J)			Impact Strength of Welded Pool (J/cm^2)
	Test 1	Test 2	Average		Test 1	Test 2	Average	
1	119	111	115	287.5	78	62	70	175
2	159	154	157	392.5	81	68	75	187.5
3	63	57	60	150	93	78	86	215
4	154	141	148	370	95	84	90	225
5	52	40	46	115	94	87	91	227.5
6	138	131	135	337.5	48	34	41	102.5
7	68	62	65	162.5	103	91	97	242.5
8	107	94	101	252.5	60	49	55	137.5
9	77	72	75	187.5	148	133	141	352.5
					Base Metal		200	500

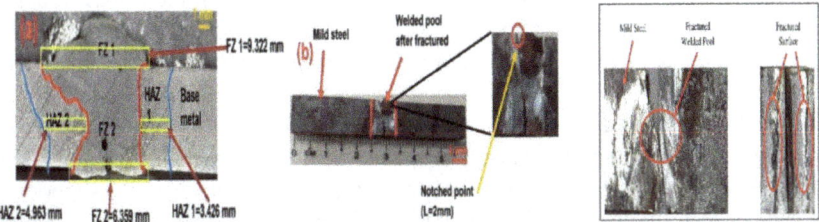

Fig. 3. a) Cross-sectioned configuration (FZ – Fusion Zone, HAZ – Heat Affected Zone), b) Welded specimens and fracture surface for weld bead.

Furthermore, when the observed energy is higher, the impact strength will be increased at the same proportion. The AISI 1080 mild steel has 0.14–0.2% carbon content which have high impact strength. However, the electrode E6013 is consist of 0.07% carbon content is much lower than the base metal. Thus, the impact strength of the welded pool is lower than the impact strength of the base metal. Figure 4 shows the Signal – Noise ratio and Means of impact toughness of welded pool that the different parameters significantly influenced the impact strength of weldment at the room temperature (21C). The diagram depicts the impact toughness is significantly influenced by the welding angle at 50°. The experiment found the direct relationship between the welding angle and impact strength, that means the welding angle increases, then the impact strength increases from 41 J to 141 J. This study found better result than Randy Chiong et al. (2019) study that they obtained impact strength is from 44 J to 86 J [13]. The welding speed also is an essential factor in the impact strength at 3.19 mm/sec. However, the welding speed increases, then the impact of toughness decreases. This study found the electrode diameter is not a significant factor in the testing.

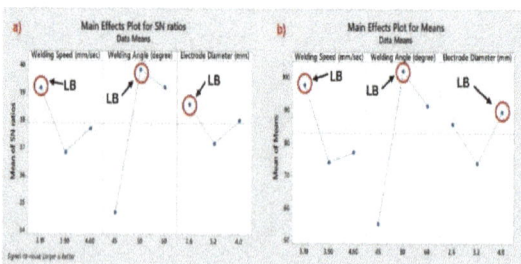

Fig. 4. a) Signal /Noise ratio and b) Means of impact toughness for the welded pool (LB – larger is better).

In the previous research, the different researchers reported the different parameters to influences the grain size of FZ and HAZ. That the size of grains increases, then the impact strength decreases [14]. For the part of ductility, the ductility of the HAZ is higher than ductility of FZ [15]. The graph shows the welding angle is a significant factor at 50°. The experiment also found the inverse relationship between the welding angle and impact toughness, means the welding angle increases, then the impact toughness decreases from 157 J to 46 J. In contrast, the electrode diameter and welding speed less influenced to impact strength at 4.0 mm and 4.6 mm/sec, respectively. Moreover, the welding angle is essential role-playing on the impact toughness of the welded pool and HAZ that the welding angle can control the crater of the welded pool to increase deep penetration. However, the figure also shows that the toughness in HAZ is much higher than the welded pool, the relationship between grain size and HAZ is also supported by the Leandro de Jesus Jorge et al. research [16]. They reported the grain size generally increases in the HAZ with the low cooling rate; the carbon content and fractured energy are much different in the welded pool section as the high heat input.

According to the Taguchi technique, it recommended to the optimum values of parameters that the best value of welding speed at 3.19 mm/sec. Meanwhile, it also gave significant levels of welding angle and electrode diameter at 50° and 2.6 mm, respectively.

4 Conclusion

The research testing is illustrated in the following summary:

1. The sheets of mild steel AISI1080 are successfully welded by ABSMAW.
2. The results explained the different electrode parameters influenced the impact toughness of welded metal and fracture energy.
3. When the welding angle increased, the impact toughness increases in the welded pool from 41 J to 141 J. The 50° of welding angle is the best-recommended value in this research.

4. The welding angle increases, then the impact of toughness and fracture energy decrease in HAZ from 157 J to 46 J.
5. The welding speed increase, then the impact strength decreases in the impact strength of welded pool.

Acknowledgement. This research was supported in part by the SEGi University Sdn Bhd and the University of Malaya.

Conflicts of Interest. The authors declare that they have no conflicts of interest to report regarding the present study.

References

1. Nathan, R., Balasubramanian, V., Malarvizhi, S., Rao, A.G.: Effect of welding processes on mechanical and microstructural characteristics of high strength low alloy naval grade steel joints. Defence Technol. **11**, 308–317 (2015)
2. Vashishtha, H., Taiwade, R., Sharma, S., Patil, A.: Effect of welding processes on microstructural and mechanical properties of dissimilar weldments between conventional austenitic and high nitrogen austenitic stainless steels. J. Manuf. Process. **25**, 49–59 (2017)
3. Saxena, A., Kumaraswamy, M.R., Madhu, V.: Influence of welding consumables on tensile and impact properties of multi-pass SMAW mild steel joints vis-a-vis base metal. Defence Technol. **14**, 188–195 (2018)
4. RajaKumar, R., Rao, : Microstructure and Mechanical properties of Borated Stainless Steel (304B) GTA and SMA welds. La Metallurgia Italiana. **5**, 47–52 (2018)
5. de Jesus Jorge, L., et al.: Microstructure and mechanical properties of borated stainless steel (304B) GTA and SMA welds. La Metallurgia Italiana. **5**, 47–52 (2018)
6. Pradana, A., Suryanto, L.: Hardness distribution and impact toughness of carburized steel welded by SMAW. In: International Conference on Mechanical Engineering Research and Application, 2019, pp. 1-7 (2019)
7. Shin, H.-S., Park, K.-T., Lee, C.-H., Chang, K.-H., Do, V.N.V.: Low temperature impact toughness of structural steel welds with different welding processes. KSCE J. Civ. Eng. **19**, 1431–1437 (2015)
8. Yang, Z., Ha, S., Jang, B.-S., Lee, Y.: Effect of welding residual stress redistribution on the Charpy absorbed energy. J. Mech. Sci. Technol. **32**(9), 4345–4356 (2018). https://doi.org/10.1007/s12206-018-0832-2
9. Zerti, O., Yallese, M.A., Khettabi, R., Chaoui, K., Mabrouki, T.: Design optimization for minimum technological parameters when dry turning of AISI D3 steel using Taguchi method. Int. J. Adv. Manuf. Technol. **89**(5–8), 1915–1934 (2016). https://doi.org/10.1007/s00170-016-9162-7
10. Zhang, L., Pittner, A., Michael, T., Rhode, M., Kannengiesser, T.: Effect of cooling rate on microstructure and properties of micro alloyed HSLA steel weld metals. Sci. Technol. Weld Join. **20**, 371–377 (2015)
11. Pirinen; Marticanena, Layusa, Karkhinb, Ivanob, : Effect of heat input on the mechanical properties of welded joints in high-strength steels. Weld Int. **2**, 14–17 (2016)
12. Jorge, J.C.F., et al.: Influence of welding procedure and PWHT on HSLA steel weld metals. J. Mater. Res. Technol. **8**, 561–571 (2019)

13. Chiong, R., Khandoker, N., Islam, S., Tchan, E.: Effect of SMAW parameters on microstructure and mechanical properties of AISI 1018 low carbon steel joints: an experimental approach. IOP Conf. Series: Mater. Sci. Eng. **495**, 1–9 (2019)
14. Bodude, M.A., Momohjimoh, I.: Studies on effects of welding parameters on the mechanical properties of welded low-carbon steel. J. Miner. Mater. Characterization Eng. **3**, 142–153 (2015)
15. Mohammed, R.A., Abdulwahab, M., Dauda, E.T.: Properties evaluation of shielded metal arc welded medium carbon steel material. Int. J. Innov. Res. Sci. Eng. Technol. **2**, 1–8 (2013)
16. de Jesus, L., et al.: Mechanical properties and microstructure of SMAW welded and thermically treated HSLA-80 steel. J. Mater. Res. Technol. **7**, 598–605 (2018)

The Influence of Electrode Parameters on the Size of Heat Affected Zone (HAZ) of Mild Steel Welded Joint Using Taguchi Technique in Automated Shielded Metal Arc Welding

Walisijiang Tayier[1], Shamini Janasekaran[1(✉)], and Teo Hiu Hong[2]

[1] Centre for Advanced Materials and Intelligent Manufacturing, Faculty of Engineering, Built Environment & IT, SEGi University Sdn Bhd, 47810 Petaling Jaya, Selangor, Malaysia
shaminijanasekaran@segi.edu.my

[2] Centre for Advanced Engineering Design, Faculty of Engineering, Built Environment & IT, SEGi University Sdn Bhd, 47810 Petaling Jaya, Selangor, Malaysia

Abstract. The mild steel AISI1080 has good characteristics of mechanical properties and weldability and commonly used in different types of industrial applications. In this research, the high heat temperature transfers to the welding area and formed the heat-affected zone (HAZ). The reducing size of HAZ is a key issue due to the welding because the properties of the materials such as grains structures were changed by the high heat input energy in HAZ so this zone properties weaken than base metal also the mild steel AISI1080 is used to weld by the automated based shielded metal arc welding for increasing welding accuracy and efficiency. The size of HAZ is observed in this research. The welding parameters are set by the Taguchi orthogonal arrays (OAs), that included welding speed (range from 3.19 to 4.6 mm/s), welding angle (range from 45° to 60°) and electrode diameter (range from 2.6 to 4.0 mm). The welding samples were clamped in the butt joint of closed square configuration to weld by ABSMAW and these samples were ground with Silicon Carbide paper (SiC) and etched with Ferric Chloride ($FeCl_3$) solution. The Heat Affected Zones (HAZ) were measured and calculated by the digital microscopy (DM). According to the Taguchi analysis, the experiment found the inverse relationship between the welding speed and size of HAZ, means the welding speed increases, then the size of HAZ decreases as the low heat input and slow cooling rate. In contrast, the research found the direct relationship between the welding angle and the size of HAZ that means when size of HAZ increased, the welding angle increases.

1 Introduction

Mild steel commonly used for industrial and manufacturing fields due to its high tensile and impact strength. The mild steel is a low carbon content steel with good features for the welding process also the low carbon content steel has the capability of good weldability [1–3]. the mild steel AISI1080 is characterized by good mechanical properties in the heat-affected zone (HAZ) [4]. The strength of the welded joint is determined by the

© The Author(s), under exclusive license to Springer Nature Singapore Pte Ltd. 2021
M. Awang and S. S. Emamian (Eds.): *Advances in Material Science and Engineering*, LNME, pp. 349–355, 2021.
https://doi.org/10.1007/978-981-16-3641-7_42

HAZ property that is important to know phase transition of materials [5] while size of HAZ is determined by the heat input energy [6], which affects to the mechanical properties as the changing the microstructure of weldment [7]. In terms of mechanical failures, it tends to occur in HAZ, where the phase of microstructures totally different from the base metal. Some researchers reported that mechanical failures occur in HAZ. Ren, Lu, Yang and Liu reported the behavior of grains boundary of cracking and found that the heat input increases, then the size of cracking decreases [8]. Xu, Liu, Lu, Wang and Ding reported that equiaxed grains formed near the HAZ that high-temperature essential role plays in the transformation of microstructure [9]. At the same time, the heat input energy during the welding significantly influences to size and microstructure change in HAZ, which determines the strength of weldment [10]. Moreover, Honggang Dong, Xiaohu Hao, Dewei and Deng revealed that the martensite and ferrite formed in the microstructure of HAZ as the high heat transfers from the fusion zone (FZ) to HAZ [11]. However, the size (properties) of HAZ is significantly influenced by the welding parameters and high heat input energy. Thus, in this study, investigated the effects of welding parameters on the size of HAZ of mild steel AISI1080 welded joint using automated based shielded metal arc welding (ABSMAW). The responses data are analyzed by the Taguchi method as the optimum values.

2 Experimental Procedure

2.1 Material and Sample Preparation

Figure 1 shows sheets of mild steel AISI 1018 (34 mm × 30 mm × 6 mm) were welded successfully by ABSMAW on a single closed square of butt joint configuration. The samples are ground through three different types of Silicon Carbide (SiC) abrasive paper, namely, 220 grit, 1000 grit, and 2000 grit. It removes burrs on the joining interface.

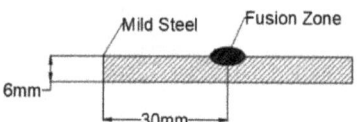

Fig. 1. Configuration joint (closed square in butt jointing)

2.2 Automatic Based Shielded Metal Arc Welding (ABSMAW)

The Automatic Based Shielded Metal Arc Welding is controlled automatically by Arduino control panel (Uno and Mega) that it transfers to data information from a personal computer (PC) with C++ computer language to a stepper motor for the machining process of moving. Thus, this article chooses the Automated process with Arc 200 Shielded Metal Arc Welding Machine (26.4 V).

The effect of welding parameters on the size of HAZ investigated at constant current 90 A and constant voltage 26.4 V. The welding speeds are set at 3.19 mm/sec, 3.9 mm/s, and 4.6 mm/s. The welding angles are conducted on testing at 45°, 50°and 60°. The electrode diameters are set by the parameter design that includes 2.6 mm, 3.2 mm, and 4.0 mm. The all of parameters are obtained via many of preliminary testing that it can avoid some unexpectable results. Table 1 below shows the given parameters design.

Table 1. Shielded metal arc welding process parameters

Welding speed (mm/sec)	Electrode diameter (mm)	Electrode angle (°)	Trial no.
4.6	2.6	45	1
4.6	3.2	50	2
4.6	4	60	3
3.9	2.6	50	4
3.9	3.2	60	5
3.9	4	45	6
3.19	2.6	60	7
3.19	3.2	45	8
3.19	4	50	9

2.3 Testing Design

This study conducted on HAZ over welded metal. The heat-affected zone (HAZ) that observed and calculated from the digital microscopy (DM) and measured the thickness of value in HAZ. This research testing data obtained from measured HAZ thickness by this automatic calculator system in Digital Microscopy. Each parameter tested three times for the accuracy of testing.

2.4 Metallographic Characterizations

Visual inspections were conducted to estimate the welded appearance of the closed square in the butt joint. In the analysis of Metallographic, the welded metals were cut perpendicular to the welding direction. First, the specimens were ground with three types of abrasive paper, such as 220 grit, 1000 grit, and 2000 grit. Second, it was polished with a polishing machine and then etched with Ferric Chloride ($FeCl_3$). The weld pool cross-sections were identified using digital microscopy from Celestron Micro Direct 1080 HDMI. The digital microscopy can measure and calculate accurately with the dimension of HAZ.

2.5 Taguchi Method and Confirmation Testing

The Taguchi method is an essential tool for Parameter design, which offers a systematic design technique for optimizing the design for a variety of characters in the experimental testing. The Taguchi method included identifying parameter characteristic, determining factors and levels, identify the experimental procedure for data, executing of the experiment, data analysis with signal/noise ratio (S/N), identify optimum values and confirmation testing. The confirmation testing can prove the Taguchi method accuracy for the prediction of optimum parameters [12]. The confirmation testing for the size of HAZ is calculated by the following Eq. 1 [13]:

$$\text{Predicted Size of HAZ (SH)} = WS + WA + ED - 2S \tag{1}$$

Where WS is average welding speed from selected optimum welding speed; WA is average welding angle from selected optimum welding angle; ED is average electrode diameter form selected optimum electrode diameter; S is the average size of HAZ from the list.

3 Result and Discussion

3.1 Analysis of HAZ

Table 2 shows the results of HAZ that the dimensions of HAZ is lower at 4.6 mm/sec of welding speed because the high welding speed produces low heat input energy, which can reduce the size of HAZ with the slow cooling rate. The experiment found the inverse relationship between the welding speed and heat input means the welding speed increases, then the heat input energy decreases.

Table 2. Results for HAZ

Trial	Size of HAZ (mm)				Heat input (kJ/mm)
	Test 1	Test 2	Test 3	Average	
1	1.338	1.311	1.334	1.328	0.5165
2	1.342	1.327	1.321	1.33	0.5165
3	1.314	1.331	1.311	1.319	0.5165
4	1.393	1.303	1.358	1.351	0.6092
5	1.629	1.647	1.612	1.629	0.6092
6	1.351	1.319	1.334	1.334	0.6092
7	1.569	1.501	1.593	1.554	0.7448
8	1.352	1.307	1.292	1.339	0.7448
9	1.633	1.556	1.569	1.586	0.7448

Figure 2 shows that when the size of HAZ mild steel AISI1080 decreases then the welding speed increased, and heat input decreased. The microstructure of HAZ in mild steel AISI 1080 was typically coarse grains which consisted of martensite (black) and ferrite (bright) and are related to heat input energy. The size and microstructure of HAZ under different range of welding speed from 3.19 to 4.6 mm/sec and heat input from 0.5165 to 0.7448 kJ/mm are also shown in Fig. 2. It also found that when the thickness of HAZ decreases from 1.339 mm to 1.328 mm and increase the welding speed (because the increase in welding speed produces low heat input energy) can change size of HAZ smaller. It was also found that the microstructure of HAZ in mild steel AISI1080 mainly consisted of coarse grains of ferrite leads to an increase in material strength. The number of martensite appeared in HAZ when the welding speed at 4.6 mm/sec also can increase material hardness. With increasing the heat input also observed the growth of fine grains in HAZ. The mechanical failures occur in the HAZ with increasing the heat input. High heat input increases the fine grains and increases the hardness in the weldment. During the research observed that the low welding speed produces high heat input energy which increasing cooling rate faster and produces large size of HAZ. The size of HAZ increases as the welding speed decreases. Typically, the microstructure and property of HAZ totally different from base metal. Thus, the changed of microstructure and property in HAZ leads to residual stress reduces materials strength increases materials brittle and decreases materials resistance to corrosion which causes of mechanical failures, so it means mechanical failures easy to occur in HAZ. This phenomenon can be concluded that the reducing size of HAZ can lead to increase materials strength and reducing mechanical failures. For furthermore, Fig. 2 shows the relationship between the size of HAZ and welding parameters. Found that the direct relationship between the welding angle and size of HAZ, means the welding angle increases, then the size of HAZ increases In the research found the inverse relationship between the welding speed and size of HAZ means when the size of HAZ decreased, the welding speed increases also the Taguchi diagram gave the optimum values for parameters that included 4.6 mm/sec of welding speed, 45° of welding angle and 2.6 mm of electrode diameter.

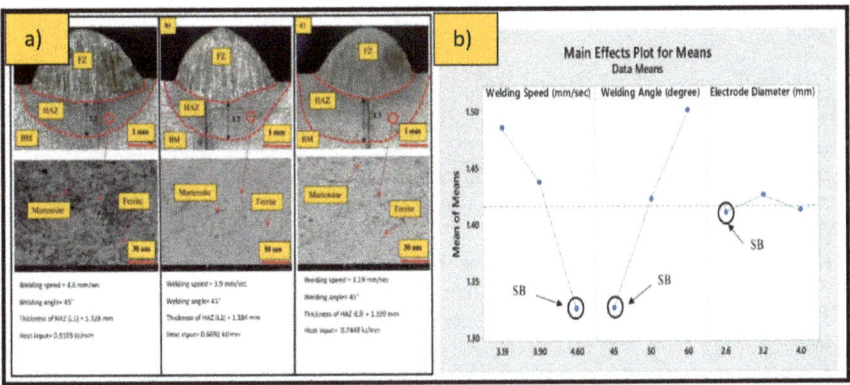

Fig. 2. a) SEM and cross-sectioned images for HAZ and b) Signal/ noise ratio diagram for HAZ.

3.2   Confirmation Testing

The confirmation testing was executed through the optimum values of size of HAZ, which evaluated the errors between the predicted value and experimental value. The optimum values are 4.6 mm/s of welding speed, 45° of welding angle and 2.6 mm of electrode diameter. In confirmation testing, the research found a 1.227 mm thickness of HAZ. The error of between predicted value and experimental value not existed which is a good agreement for the experimental testing. Thus, the results indicated that the Taguchi method has the ability of good predictable for experimental testing and it can minimize experimental cost and procedure, so this is reasons for using Taguchi method in this study.

4   Conclusion

The experimental testing and the result are the following summary:

1. The sheets of mild steel were successfully welded by the ABSMAW.
2. The research found the thickness of HAZ decreases from 1.339 mm to 1.328 mm with increasing the welding speed.
3. The amount of martensite appeared in HAZ when the welding speed at 4.6 mm/sec, it can increase material hardness. With increasing the heat input, the growth of fine grains in HAZ was observed.
4. The experiment found the direct relationship between the welding angle and size of HAZ

Acknowledgement. This research was supported in part by the SEGi University Sdn Bhd, grant number: SEGiIRF/2018–10/FoEBE-17/80, and in-kind contribution of the University of Malaya.

Conflict and Interest. The authors declare that they have no conflicts of interest to report regarding the present study.

References

1. Kokabi, A.K.: Effects of heat inputinfriction stir weldingon microstructure and mechanical properties of AA3003H18 plates.Trans. Nonferrous Met. Soc. China. **25,** 2147–2155 (2015)
2. Khodamorad, A., Hassan, S., Shahram, K.: Effect of bainite morphology on mechanical properties of the mixed bainite– martensite microstructure in D6AC steel. J. Mater. Sci. Technol. **28,** 336–342 (2012). https://doi.org/10.1016/S1005-0302(12)60065-6
3. Srivastava, S., Garg, R.K.: Process parameter optimization of gas metal arc welding on IS: 2062 mild steel using response surface methodology. J. Manuf. Process. **25,** 296–305 (2017)
4. Li, S., Li, K., Mengjia, H., Yao, W., Cai, Z., Pan, J.: The mechanism for HAZ liquation of nickel-based alloy 617B during gas tungsten arc welding. Metals. **94,** 1–13 (2019)

5. Yin, L., et al.: Microstructures and their distribution within HAZ of X80 pipeline steel welded using hybrid laser-MIG welding. Weld. World **62**(4), 721–727 (2018). https://doi.org/10.1007/s40194-018-0582-x

6. Shi, Y., Cui, S., Zhu, T., Gu, S., Shen, X.: Microstructure and intergranular corrosion behavior of HAZ in DP-TIG welded DSS joints. J. Mater. Process. Technol. **256**, 254–261 (2018)

7. Li, Z.: The influenceofniobiumontheplasticdeformationbehaviorsof310saustenitic stainless steel weld metals at different temperatures. Mater. Sci. Eng. **743**, 648–655 (2019)

8. Lu, R., Liu, Y.: Liquation cracking in fiber laser welded joints of inconel 617. J. Mater. Process. Technol. **226**, 214–220 (2015)

9. Xu; Liu, Lu, Wang and Ding, : Evolution of carbides and its characterization in HAZ during NG-TIG welding of Alloy 617B. Mater. Charact. **130**, 270–277 (2017)

10. Moravec, J., Novakova, I., Sobotka, J., Neumann, H.: Determination of grain growth kinetics and assessment of welding effect on properties of S700MC steel in the HAZ of welded joint. Metals. **707**, 1–20 (2019)

11. Dong, H., Hao, X., Deng, D.: Effect of welding heat input on microstructure and mechanical properties of HSLA steel joint. Metallogr. Microstruct. Anal. **3**(2), 138–146 (2014). https://doi.org/10.1007/s13632-014-0130-z

12. Asadi, P., Akbari, M., Besharati, M.K., Panahi, M.S.: Optimization of AZ91 friction stir welding parameters using Taguchi method. J. Mater. Des. Appl. **24**, 1–13 (2015)

13. Kumar, S., Sing, D.P., Singh, V.: Optimization of process parameters for friction stir welding of cast alloy AA7075 by Taguchi method. Int. J. Mech. Mater. Eng. **13**, 543–549 (2019)

The Optimization of Automated Based Shielded Metal Arc Welding Parameters on Welded Pool Geometry of Mild Steel Welded Joint Using Taguchi Technique

Walisijiang Tayier[1], Shamini Janasekaran[1(✉)], Liew Kah Fook[2], and Teo Hiu Hong[2]

[1] Centre for Advanced Materials and Intelligent Manufacturing, Faculty of Engineering, Built Environment and IT, SEGi University, Sdn Bhd, 47810 Petaling Jaya, Selangor, Malaysia
shaminijanasekaran@segi.edu.my

[2] Centre for Advanced Engineering Design, Faculty of Engineering, Built Environment and IT, SEGi University, Sdn Bhd, 47810 Petaling Jaya, Selangor, Malaysia

Abstract. Shielded metal arc welding (SMAW) primarily used for metal processing fields with advantages of low costing. However, the low welding quality and welding speed accuracy are main issues on the conventional SMAW. These issues can significantly influence to geometry shapes and appearance such as bead width and penetration, which determines degree of strength in the mechanical properties of weldment. Thus, in this study, the automated based shielded metal arc welding (ABSMAW) used to weld sheets of mild steel AISI1080 to evaluate bead width and penetration using different welding parameters. During the research, the parameters focused on welding speed range from 3.19 to 4.6 mm/s with manipulating the welding angle and electrode diameter (45° to 60° and 2.6 to 4.0 mm). The configuration of workpiece was clamped in the butt joint of closed square to weld. The samples were ground and cross-sectioned to view the welded pool geometry with etching material. The Taguchi method used to analyze the bead width and penetration using the signal/noise (S/N) ratio diagram. This experiment found the inverse relationship among the welding speed, bead width and penetration that means the welding speed increases, then the bead width and penetration decrease. In contrast, the welding angle increases, then the bead width and penetration increase. The optimum value of bead width and penetration are produced by welding speed of 3.19 mm/s, welding angle of 60° and electrode diameter of 4.0 mm.

1 Introduction

Automated based welding technique has received great attention as a high efficiency and accuracy to weld similar and dissimilar metal in the industrial and manufacturing fields. Some experiments have been performed to study the welded geometry (geometry shapes) such as bead width and penetration during the arc welding process [1–3]. The deeper penetration and wider bead with were the important phenomena during the welding process. Which could determine welding quality and appearance [4, 5]. The flux covered filler wire (electrode) used to weld during the shield metal arc welding (SMAW), it

M. Awang and S. S. Emamian (Eds.): *Advances in Material Science and Engineering*, LNME, pp. 356–363, 2021.
https://doi.org/10.1007/978-981-16-3641-7_43

totally avoids some oxidation on the welded pool and enhances the penetration level in the welded pool. Hence, single pass weld with lower welding speed can be achieved [6]. Ravinder Pal Singh et al. found the welding parameters which obtained optimum value for bead width and penetration during the arc welding. They reported Taguchi method used to predict optimal setting for each welding parameters and it can analyze behavior of welding parameters [7]. Senthilkumar et.al reported the bead width and penetration were influenced by different parameters such arc length, welding speed and welding current [8]. Rajeev Kumar et al. found the better bead width and penetration were found in welded joint at 100 A of welding current [9]. Improved welding parameters to achieve deeper penetration and observed that it can reduces welding defects on the welded surface [10, 11]. Jyotirmaya Kar also found the welding voltage influences the penetration depth, welding speed determines the shape of welded pool [12]. However, conventional arc welding such as SMAW has some of issues. For example, low welding quality and welding speed accuracy because it is performed by manually. Thus, in this study, the automated based shielded metal arc welding (ABSMAW) exploited on mild steel AISI1080 and geometry shapes of welded pool was studied experimentally. The effects of welding parameters on geometry shapes was discussed. And the relationship between the welding parameters and geometry shapes were discussed using Taguchi method.

2 Experimental Procedure

2.1 Material and Sample Preparation

The sheets of AISI 1018 mild steel (34 mm × 30 mm × 6 mm) used in the base metal. Figure 1 indicates that these metal sheets were welded successfully by ABSMAW on a single closed square of butt joint configuration. The samples were ground with three types of Silicon Carbide (SiC) abrasive paper, namely, 220 grit, 1000 grit, and 2000 grit for removing burrs at the joining interface.

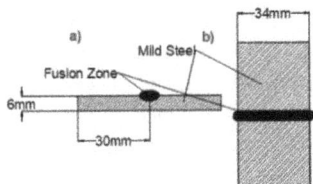

Fig. 1. Configuration joint (closed square in butt jointing). a) Font view b) Top view

2.2 Automated Based Shielded Metal Arc Welding (ABSMAW)

The SMAW machine fixed with open source structure, which is controlled by the Arduino board. The Arduino board transfers and compiles data from a personal computer (PC) to machine using C++ code. The Arc200 welding machine (0–160 A, 26.4 V) was used to weld in the experimental testing.

The influence of welding speed of 3.19 mm/s, 3.9 mm/s and 4.6 mm/s studied at constant welding current of 90 A and constant welding voltage of 26.4 v while the electrode angle (45° to 60°) and electrode diameter (2.6 mm to 3.2 mm) also investigated in this research at the constant machine parameters. The all of parameters are obtained via many of preliminary testing that it can avoid some unexpectable results. For the accuracy of the research data, each set of parameters were welded or tested three times as shown in Table 1.

Table 1. Shielded metal arc welding process parameters

Welding speed (mm/s)	Electrode diameter (mm)	Electrode angle (°)	Trial no.
4.6	2.6	45	1
4.6	3.2	50	2
4.6	4	60	3
3.9	2.6	50	4
3.9	3.2	60	5
3.9	4	45	6
3.19	2.6	60	7
3.19	3.2	45	8
3.19	4	50	9

2.3 Metallographic Characterizations

Visual inspections were carried out to evaluate the weld appearance of a butt joint. In the analysis of metallographic, the welded metals were cut perpendicular to the welding direction. Firstly, the specimens were ground into three types of paper, namely, 220 grit, 1000 grit, and 2000 grit. Secondly, the ground specimens polished in the polishing machine. Finally, these specimens were etched with Ferric Chloride ($FeCl_3$) in 10 s and brushed to surface two times. The weld pool cross-sections were observed using a digital microscope from Celestron Micro Direct 1080 HDMI.

2.4 Taguchi Method

The Taguchi method typically uses two main steps in the optimization process. In first step, the Taguchi method uses signal/noise ratio (S/N) to determine control factors and variables. Secondly, identifies target for control factors such as "larger is better", "normal is better" and "small is better" which determines response varies [13]. Present work was used to target of "larger is better" for bead width and penetration. The confirmation testing is an important step for Taguchi method that it determines confidential of Taguchi technique. The confirmation testing divided into two main parts that includes predicted value and experimental value. The experimental value is obtained by the retesting from the optimum values of parameters. The predicted value is calculated by the following Eq. 1 and 2 [14]:

$$\text{Predicted Bead width (BW)} = \text{WS} + \text{WA} + \text{ED} - 2\text{B} \qquad (1)$$

$$\text{Predicted Penetration (Pe)} = \text{WS} + \text{WA} + \text{ED} - 2\,\text{P} \qquad (2)$$

Where, WS is average welding speed from selected optimum welding speed; WA is average welding angle from selected optimum welding angle; ED is average electrode diameter form selected optimum electrode diameter; B and P are average size of bead width and penetration from the list.

2.5 Testing Design

This study investigated the geometry shapes of the welded pool, which is observed by the Digital Microscope (DM). The geometry shapes such as bead width and penetration were automatic measured and calculated by the DM that the configuration is given in Fig. 2. This method can be guaranteed to the accuracy of testing value and achieved good results. For the accuracy of testing, each parameter tested three times.

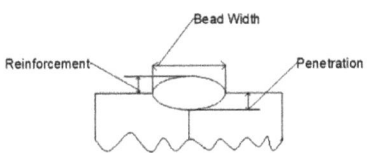

Fig. 2. Geometry shape of the welded pool

3   Result and Discussion

3.1 Analysis of Bead Width

Table 2 shows results for geometry shapes that indicates trial 7 produces largest bead width and deeper penetration at 3.19 mm/s of welding speed and 60° of welding angle. The experiment found the inverse relationship among the welding speed, bead width and penetration, means the welding speed increases, then the bead width and penetration decrease.

In consideration of effect of welding parameters of ABSMAW to the bead width, the detailed experiment was implemented and the metal droplets behavior at the bead width was observed, as shown in Fig. 3. The big bead width of cross section occurred when the speed was slower (3.19 mm/s). In contrast, when the welding speed was higher (4.6 mm/s), the bead width decreases. And, when the welding speed became slower, the bead width increases. The observed images illustrate that the welded pool was extremely

Table 2. Results for geometry shapes

Trial	Bead width (mm)				Penetration (mm)			
	Test 1	Test 2	Test 3	Average	Test 1	Test 2	Test 3	Average
1	4.904	4.839	4.845	4.863	1.136	1.107	1.114	1.119
2	5.589	5.537	5.548	5.558	0.796	0.772	0.731	0.766
3	5.853	5.832	5.818	5.834	0.735	0.679	0.694	0.703
4	5.795	5.763	5.748	5.769	1.594	1.581	1.585	1.587
5	6.351	6.307	6.318	6.325	1.347	1.316	1.323	1.329
6	5.723	5.689	5.695	5.702	0.809	0.768	0.747	0.775
7	7.136	7.122	7.096	7.118	1.693	1.617	1.637	1.649
8	5.806	5.739	5.756	5.767	1.023	0.983	0.966	0.991
9	6.541	6.483	6.498	6.507	0.853	0.782	0.804	0.813

instable when the welding speed was higher (<4.6 mm/s) and the amount of metal droplets flowing out from the welding surface to generated column. When the welding speed slower (>4.6 mm/s), the generated column was flowing back to the welded pool. The following back metal droplets accumulated and high temperature of metal droplets not only on the welding surface but also the inside of the welded pool, which caused a wider bead width. There were significant factors for the bead width. The welding speed and electrode diameter were observed. When the welding speed and electrode diameter decreased, the bead with become wider. Furthermore, the experiment found the flow of metal droplets takes place from center to the edges when the welding speed became lower because the surface tensions is higher at the edges of welded pool that caused bead width wider. Figure 3 shows the signal/noise ratio that as the welding speed went higher, the bead width decreased. This confirmed that the heat input energy on the welded pool was decreased. On the other hand, the bead width occurred wider when the welding angle increased.

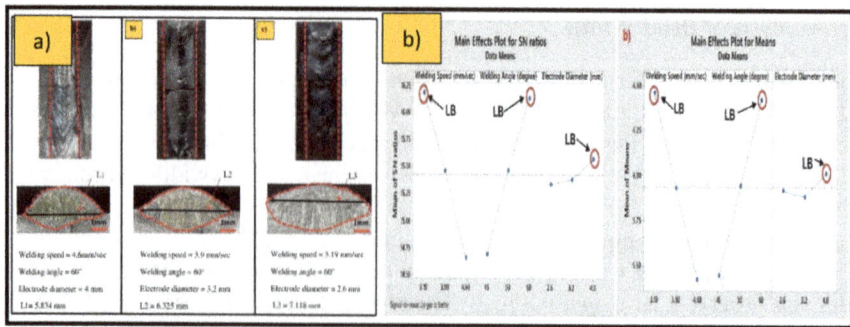

Fig. 3. a) Surface appearance and cross section of bead width and b) Three a) S/N ratio value in bead width (LB – larger is better)

3.2 Analysis of Penetration

Figure 4 shows the geometry shapes cross sections for penetration during the ABSMAW process. The increase of penetration at constant welding angle of 60° was indicated clearly in the images. The increased penetration was located at lower welding speed of 3.19 mm/s. With the welding speed decreased from 4.6 mm/s to 3.19 mm/s, the penetration moved deeper into the metal. And this increased penetration occurred during ABSMAW process using different welding speed and electrode diameter. According to the observed results, the metal droplets quick accumulated under the lower welding speed (>4.6 mm/s) at the certain region, as indicated Fig. 3. Such accumulation could bring the high temperature metal droplets and enhance the heat input energy on the welded pool. Thus, the metal droplets could transfer more heat to melt the welding specimen to get a deeper weld. In Fig. 4, the increased penetration was observed. This was because the welding speed was lower, so the temperature of metal droplets was higher which could cause a high heat input of metal droplets. Therefore, with the welding speed went slower from 4.6 mm/s to 3.19 mm/s, the penetration became deeper.

3.3 Confirmation Testing

The confirmation testing was performed via the optimum value of bead width, and penetration which referred to signal/noise ratio diagram and compares with the predicted value as shown in Table 3. The optimum values are 3.19 mm/s of welding speed, 60° of welding angle and 4.0 mm of electrode diameter. In the confirmation testing, the research found 8.946 mm of bead width and 1.704 mm of penetration. The error of between predicted value and experimental value not existed which is good agreement for the experimental testing. Thus, the results indicated that the Taguchi method has ability of good predictable for experimental testing and it can minimize of experimental cost and procedure, so this is reasons for using Taguchi method in this study.

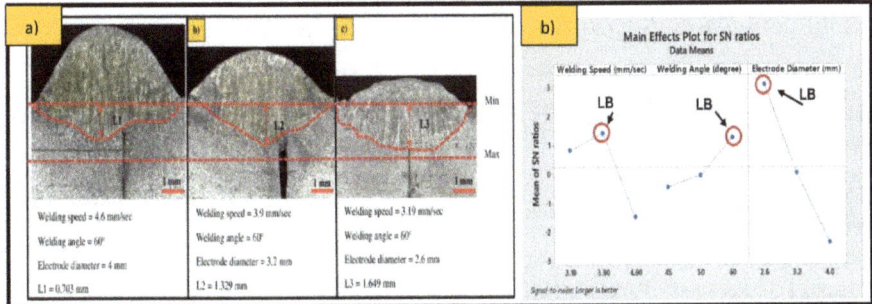

Fig. 4. a) Cross section of penetration and S/N ratio of parameters value in penetration (LB – larger is better)

Table 3. Confirmation testing for bead width and penetration

	Optimum parameters		
	Predicted value	Experimental value	Errors
Levels	3.19 mm/s (WS), 60° (WA), 4.0 mm (ED)	3.19 mm/s (WS), 60° (WA), 4.0 mm (ED)	
Bead width (mm)	7.028	8.946	No
Penetration (mm)	0.982	1.704	No

4   Conclusion

The experimental test and Taguchi factorial design are drawn in the following summary:

1. The mild steel sheets were successfully welded by the ABSMAW.
2. The experiment found the different welding parameters influenced to welded pool geometry, such as bead width and penetration.
3. The research found the inverse relationship among the welding speed, bead width and penetration, means the welding speed increases, then the bead width and penetration decrease dramatically.
4. The experiment found the flow of metal droplets takes place from center to the edges when the welding speed became lower because the surface tensions is higher at the edges of welded pool that caused bead width wider.
5. According to the confirmation testing, it confirmed that the Taguchi method has ability of good predictable for experimental testing and it can minimize of experimental cost and procedure.

Acknowledgement. This research was supported in part by the SEGi University, grant number: SEGiIRF/2018-10/FoEBE-17/80, and in kind contribution of the University of Malaya.

Conflicts and Interest. The authors declare that they have no conflicts of interest to report regarding the present study.

References

1. Son, J.S., Lee, J.-P., Park, M.-H., Jin, B.-J., Yun, T.-J., Kim, S.: A study on on-line mathematical model to control of bead width for arc welding process. Proc. Eng. **174**, 68–73 (2017)
2. Tseng, K.-H., Wang, N.-S.: Research on bead width and penetration depth of multicomponent flux-aided arc welding of grade 316 L stainless steel. Powder Technol. **311**, 514–521 (2017)
3. Srivastava, S., Garg, R.K.: Process parameter optimization of gas metal arc welding on IS:2062 mild steel using response surface methodology. J. Manuf. Process. **25**, 296–305 (2017)

4. Chandrasekhar, N., Vasudevan, M., Bhaduri, A.K., Jayakumar, T.: Intelligent modeling for estimating weld bead width and depth of penetration from infra-red thermal images of the weld pool. J. Intell. Manuf. **26**(1), 59–71 (2013). https://doi.org/10.1007/s10845-013-0762-x
5. Dhandha, K.H., Badheka, V.J.: Effect of activating fluxes on weld bead morphology of P91 steel bead-on-plate welds by flux assisted tungsten inert gas welding process. J. Manuf. Process. **17**, 48–57 (2015)
6. Nagaraju, S., Chandrasekhar, N., Jayakumar, T., Vasantharaja, P., Vasudevan, M.: Optimization of A-TIG welding process parameters for 9Cr–1Mo steel using response surface methodology and genetic algorithm. In: International Welding Conference (2014)
7. Singh, R.P., Garg, R.K., Shukla, D.K.: Mathematical modeling of effect of polarity on weld bead geometry in submerged arc welding. J. Manuf. Process **21**, 14–22 (2016)
8. Senthilkumar, B., Kannan, T.: Effect of flux cored arc welding process parameters on bead geometry in super duplex stainless steel claddings. Measurement **62**, 127–136 (2015). https://doi.org/10.1016/j.measurement.2014.11.007
9. Kumar, R., Chattopadhyaya, S., Kumar, S.: Influence of welding current on bead shape, mechanical and structural property of tungsten inert gas welded stainless steel plate. Mater. Today **2**, 3342–3349 (2015)
10. Luo, Y., Tang, X., Fenggui, L.: Experimental study on deep penetrated laser welding under local subatmospheric pressure. Int. J. Adv. Manuf. Technol. **73**(5–8), 699–706 (2014). https://doi.org/10.1007/s00170-014-5870-z
11. Siddaiah, A., Singh, B.K., Mastanaiah, P.: Prediction and optimization of weld bead geometry for electron beam welding of AISI 304 stainless steel. Int. J. Adv. Manuf. Technol. **89**, 27–43 (2016)
12. Kar, J., Mahanty, S., Roy, S.K., Roy, G.G.: Estimation of average spot diameter and bead penetration using process model during electron beam welding of AISI 304 stainless steel. Trans. Indian Inst. Met. **68**(5), 935–941 (2015). https://doi.org/10.1007/s12666-015-0529-5
13. Eduardo Izeda, A., Pascoal, A., Simonato, G., Mineiro, N., Gonçalves, J., Ribeiro, J.E.: Optimization of robotized welding in aluminum alloys with pulsed transfer mode using the Taguchi method. Proceedings **426**, 1–6 (2018)
14. Ibrahim, M., Maqsood, S., Khan, R.: Optimization of gas tungsten arc welding parameters on penetration depth and bead width using Taguchi method. J. Engg. Appl. Sci. **35**, 51–59 (2016)

Optimization of Tensile Strength in Automated Shielded Metal Arc Welding Using Taguchi Technique

Shamini Janasekaran[1]([✉]), Walisijiang Tayier[1], Gan Jin Hoe[1], and Teo Hiu Hong[2]

[1] Centre for Advanced Materials and Intelligent Manufacturing, Faculty of Engineering, Built Environment and IT, SEGi University Sdn Bhd, 47810 Petaling Jaya, Selangor, Malaysia
shaminijanasekaran@segi.edu.my

[2] Centre for Advanced Engineering Design, Faculty of Engineering, Built Environment and IT, SEGi University Sdn Bhd, 47810 Petaling Jaya, Selangor, Malaysia

Abstract. The shielded metal arc welding (SMAW) is commonly used to weld for similar and dissimilar metals in the manufacturing fields. However, SMAW is a manual metal joining processing technique, the low welding quality and welding speed accuracy are both the main problems when using SMAW. So for a better result, the automatic based shielded metal welding (ABSMAW) was used to weld mild steel plates AISI1018 as the different electrode parameters in this study. The parameters focus on welding speed range from 3.19 mm/s to 4.6 mm/s and welding angle range from 45° to 60° by manipulating the electrode diameter from 2.6 mm to 4.0 mm for the welding process. The workpieces were clamped at the flat position in a customized jig for welding and all samples were cut with ASTM E8 standard before starting the tensile test. The tensile testing performed with the Huang Ta H500 universal testing machine which has the capability of 500 kN. The ABSMAW machine not only supports welding speed accuracy and smooth welding but also allow the better quality. The test result was analyzed with the Taguchi method while the result shows the welding angle is the most significant value for the tensile testing. The experiment found that there is an inverse relationship between welding angle and tensile strength means the welding angle increases then the tensile strength decreases dramatically also found out that there is the direct variation found in tensile strength welding speed and electrode diameter mean when the welding speed and electrode diameter increase, the tensile strength increases. The optimum values for tensile strength were produced by welding speed of 4.6 mm/s, welding angle of 45° and electrode the diameter of 3.2 mm.

1 Introduction

The mild steel is widely used in the manufacturing and industrial fields that can be categorized for its high ductile and tensile strength, and it is most suitable for mechanical testing [1]. Previously, the number of researchers conducted at mild steel welded joint. Rohit Jha et al. investigated the influence of different parameters such as currents (from 100 A to 140 A), on tensile strength of mild steel using conventional SMAW. They reported

M. Awang and S. S. Emamian (Eds.): *Advances in Material Science and Engineering*, LNME, pp. 364–369, 2021.
https://doi.org/10.1007/978-981-16-3641-7_44

the welding current increases, then the ultimate tensile strength (UTS) increases [2], and the effects of different electrode types on mechanical properties is studied by the Sudhin et al. that they found the electrode E6011 produced maximum UTS and ultimate yield strength (UYS) of 358 MPa and 421 MPa, respectively. The electrode E6013 generated to maximum UTS and UYS of 383.2 MPa, and 319.7 MPa, respectively. That means the electrode E6013 is producing high tensile strength than the electrode E6011 when the welding process [3]. Jafarlou et al. investigate the of low carbon steel sheets (100 mm × 37.5 mm × 5.5 mm) welded joint using 80 A of welding current and 150 min/mm of welding speed. They reported the tensile strength is determined by the carbon content in the grains structure. Therefore, the carbon content in weldment increases, then the ductile and tensile strength decrease [4]. Moreover, Bodude and Momohjimoh studied at the effect of welding current (100 A to 150 A) on mechanical properties that they demonstrated the welding current is an essential factor for the increasing tensile strength [5]. In contrast, some researchers reported that the welding current increases, then the tensile strength decreases [6]. Oluwasegun Biodun Owolabi et al. investigated on tensile strength of the various type of steel that the tensile strength of low carbon steel is higher than the middle carbon steel [7]. However, the low welding quality and welding speed accuracy are the main problems in manual welding. It influences mechanical properties. Thus, the aims of this study are to investigate the effects of electrode parameters on tensile strength of mild steel in the ABSMAW using Taguchi technique in which used to analyze output data systematically from the experimental testing.

2 Experimental Procedure

2.1 Material and Sample Preparation

Figure 1 shows the plates of AISI 1018 mild steel (225 mm × 34 mm × 4.5 mm) were successfully welded by ABSMAW on the double groove of butt joint configuration.

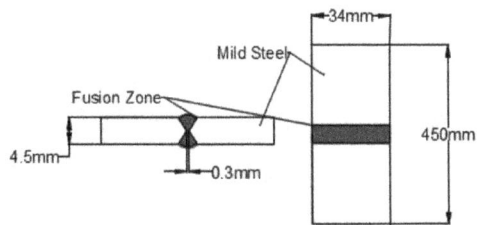

Fig. 1. Configuration joint (double grooving in butt joint)

2.2 Automated Based Shielded Metal Arc Welding (ABSMAW)

The ABSMAW is controlled by the Arduino control panel which interprets to data systematically from the specific code such as C++, and it can control the stepper motors and parameters for machine processing, as shown in Fig. 2. This study used the Arc200

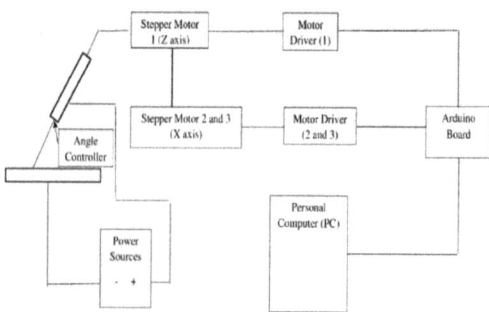

Fig. 2. Schematic drawing for experimental testing

SMAW machine for welding purposes. The welding machine current ranges from 0 to 160 A and a constant voltage are at 26.4 V.

The effects of electrode diameters (2.6 mm to 4.0 mm), welding speed (3.19 mm/s to 4.6 mm/s) and electrode angle (45° to 60°) on tensile strength of mild steel welded joint was conducted at a 90 A of constant welding current and 26.4 V of welding voltage. The all of parameters are obtained via many of preliminary testing that it can avoid some unexpectable results. The welding parameters are listed in Table 1.

Table 1. Shielded metal arc welding process parameters

Parameters	Level 1	Level 2	Level 3
Welding speed (mm/sec)	3.19	3.9	4.6
Welding angle (°)	45	50	60
Electrode diameter (mm)	2.6	3.2	4.0

2.3 Metallographic Characterizations

In this study, the plates of mild steel AISI1080 were successfully welded as the double groove of the butt joint. The cross-sectioned is shown in Fig. 3. The two specimens (450 mm * 34 mm * 4.5 mm) were welded by the ABSMAW that the machine recommended straight welding method; the welded specimens were cooled by freshwater or air for 10 min and then the welded samples fabricated with grinding machine according to the ASTM E8-04 standard that is given in Fig. 4 [1].

2.4 Taguchi Method

The Taguchi method offers the engineers about the efficient approach and systematic data for determining the optimum value and reducing cost. The Taguchi method uses Orthogonal Arrays (OAs) from Designs of Experiment (DOE) with factors and parameters that it can significantly decrease the experimental configurations and production cost [8].

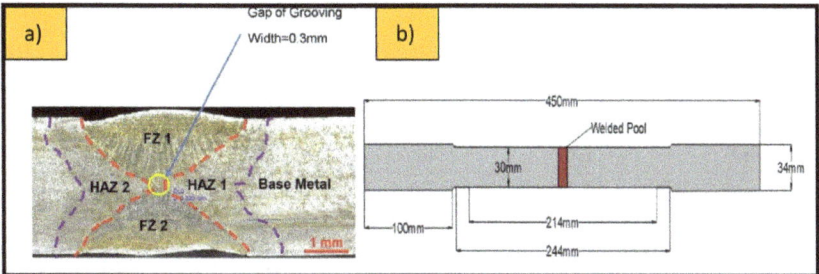

Fig. 3. a) Cross-sectioned configuration (double grooving of the butt joint; FZ-Fusion Zone, HAZ- Heat Affected Zone) and b) ASTM E8-04 standard for tensile specimen

This article conducted on Taguchi orthogonal arrays L9(3^3), and this testing tested nine times for study. The experimental testing performed by the three different parameters in the Taguchi orthogonal arrays (OAs) that it can evaluate the welding characterizations and analyze output values with Signal to Noise (S/N) ratio.

2.5 Testing Design

This research investigated on tensile strength for welded specimens. The welded specimens are tested by the Universal Tensile Machine (Hung Ta H500, capable of 500 kN force). The specimen fitted in two jaws of testing machine to run tensile stress. During the tensile test, the stress-strain diagram drawn per specimen where the tensile load is determined.

3 Result and Discussion

3.1 Results of Tensile Testing

Table 2 depicts the results of tensile strength from experimental testing. The table shows the high welding speeds (4.6 mm/s) produce high tensile strength. Meanwhile, the tensile strength is the highest point at 45°.

3.2 Analysis of Tensile Strength

From the previous studies, the number of researchers reported that the tensile failures consistently occurred at the heat-affected zone (HAZ) using a multi-pass welding method with high tensile strength electrode (E7018). Since the carbon content more produces in fusion zone (FZ), but the ferrite (Fe) mostly produces in HAZ. Thus, the ductility in HAZ is higher than FZ [9]. However, this research used to single-pass welding method to weld with the lower tensile strength electrode (E6013) that the failure of the tensile region is between the FZ and HAZ, and the obtained average of tensile strength are 342 MPa. As compared with the results from the previous studies, the result of testing in this paper can more economical and more efficiency. Figure 4 shows the tensile strength diagram for the different welding angles that the welding angle increases, then the tensile strength

Table 2. Testing response data for tensile strength and loading force

Trial no	Welding speed (mm/sec)	Electrode diameter (mm)	Welding angle (°)	Loading force (kN)	Tensile strength (MPa)
1	4.6	2.6	45	52.7	380
2	4.6	3.2	50	50.9	370
3	4.6	4	60	47.9	340
4	3.9	2.6	50	37.1	270
5	3.9	3.2	60	43.3	310
6	3.9	4	45	52.2	370
7	3.19	2.6	60	44.8	320
8	3.19	3.2	45	54.7	390
9	3.19	4	50	46	330

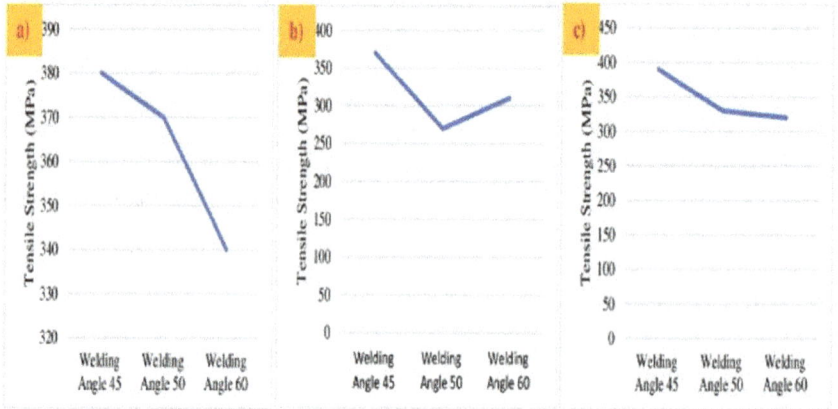

Fig. 4. Tensile strength diagram for different welding speed

decreases at the different welding speeds. Trial 8 produces the highest tensile strength (390 MPa) at 3.19 mm/s of welding speed, 3.2 mm of electrode diameter, and 45° of welding angle. The result of tensile strength is better than Ogbunnaoffor C. K, Odo J.U, and Nnuka, E. E research that they obtained maximum tensile strength is 383 MPa [10].

According to the Signal to Noise (S/N) ratio for tensile strength, it recommended to the optimum values for this study that includes 4.6 mm/s of welding speed, 45° of welding angle and 3.2 mm of electrode diameter.

4 Conclusion

The summary can be illustrated from the experimental results of this study:

1. The AISI 1018 mild steel plates successfully welded with the double groove of butt joining by the ABSMAW
2. During the welding process, the tensile strength influenced by selected parameters.
3. The welding angle is the most significant factor for tensile strength and loading the force that it produced the maximum tensile strength and loading force of 390 MPa and 54.7 kN, respectively, at 45°.
4. The welding speed is the second most significant factor that the welding speed increases, then the tensile strength increases.
5. The electrode diameter influenced to tensile strength and loading force at 3.2 mm.

Acknowledgement. This research was supported in part by the SEGi University Sdn Bhd, grant number: SEGiIRF/2018-10/FoEBE-17/80 and the University of Malaya.

Conflicts and Interest. The authors declare that they have no conflicts of interest to report regarding the present study.

References

1. Nassar, A.A., Lefta, R.M., Abdulsada, M.J.: Experimental study of the effect of welding electrode types on tensile properties of low carbon steel AISI1010. Kufa J. Eng. **9**, 163–173 (2018)
2. Jha, R.: Investigating the effect of welding current on the tensile properties of SMAW welded mild steel joints. Int. J. Eng. Res. Technol. (IJERT) **3**, 1304–1307 (2014)
3. Sudhin, C., Nagarajan, N.M.: Comparative studies on joining of structural steels using MIG and ARC welding processes. Int. J. Recent Sci. Res. **7**, 12405–12410 (2016)
4. Jafarlou, H., Hassannezhad, K., Asgharzadeh, H., Marami, G.R.: Enhancement of mechanical properties of low carbon steel joints via graphene addition. Mater. Sci. Technol. **6**, 1–13 (2017)
5. Bodude, M.A., Momohjimoh, I.: Studies on effects of welding parameters on the mechanical properties of welded low-carbon steel. J. Minerals Mater. Charact. Eng. **3**, 142–153 (2015)
6. Tahir, A.M., Lair, N.A.M., Wei, F.J.: Investigation on mechanical properties of welded material under different types of welding filler (shielded metal arc welding). In: AIP Conference Proceeding, Sabah (2018)
7. Owolabi, O.B., Aduloju, S.C., Metu, C.S., Chukwunyelu, C.E., Okwuego, E.C.: Evaluation of the effects of welding current on mechanical properties of welded joints between mild steel and low carbon steel. Am. J. Metall. Mater. Eng. **1**, 1–4 (2016)
8. Chen, W., Allen, J.: A procedure for robust design: minimizing variations caused by noise factors and control factors. ASME J. Mech. Des. **118**, 478–484 (1996)
9. Chiong, R., Khandoker, N., Islama, S., Tchan, E.: Effect of SMAW parameters on microstructure and mechanical properties of AISI 1018 low carbon steel joints: an experimental approach. IOP Conf. Ser. Mater. Sci. Eng. **495**, 1–9 (2019)
10. Ogbunnaoffor, C.K., Odo, J.U., Nnuka, E.E.: The effect of welding current and electrode types on tensile properties of mild steel. Int. J. Sci. Eng. Res. **7**, 1–6 (2016)
11. Rizvi, S.A., Tewari, S.P.: Effect of different welding parameters on the mechanical and microstructural properties of stainless steel 304H welded joints. Int. J. Eng. **30**, 1–8 (2017)
12. Okonji, P.O., Nnuka, E.E., Odo, J.U.: Effect of welding current and filler metal types on macrostructure and tensile strength of gtaw welded stainless steel joints. Int. J. Sci. Res. Eng. Trends **1**, 9–12 (2015)

Study of Round and Square Mechanical Clinching Joint Strength for SPCC Steel and Aluminium Alloy 5052 Sheet Metal

Chanchin Wang[1(✉)], Hengkeong Kam[2], Kentwee Tan[2], and Xin Wang[3]

[1] Department of Mechanical Engineering, SEGi University, 47810 Petaling Jaya, Malaysia
wangchan@segi.edu.my
[2] Department of Mechanical and Material Engineering, Universiti Tunku Abdul Rahman, 43000 Kajang, Malaysia
kamhk@utar.edu.my, kentwee96@1utar.my
[3] School of Mechanical Engineering, Shanghai Dianji University, No. 300 Shuihua Road, Pudong, Shanghai, China
wangx31915@sdju.edu.cn

Abstract. Round mechanical clinching is commonly used in joining sheet metals. In this research, round and square clinching joints strength were carried to study the strength of both clinching joints. SPCC steel and aluminium alloy 5052 were used for evaluation in the experiments. These materials were clinched with different punch loads by using round and square clinching dies of same geometry parameters. Shearing test and peeling test were performed to study the round and square clinched joint strength. Failure mode of the joints were examined. Optimal forming load for the clinched joint were investigated and higher forming load will lead to defects. The experimental results show that the square clinched joint is 5%–70% stronger than conventional round clinched joint but the square joint strength in peeling test for the SPCC steel is 49% weaker.

1 Introduction

The weight of automobile is preferred to be reduced in order to reduce the cost, consumption of fuel and emission of carbon dioxide. Besides welding, riveting and hemming, clinching process is commonly used in joining parts in automobile [1] and electrical devices [2]. The main tools used in the process are a punch, a die, a holder and a press. The principle of this process is by applying force to plastically deform the sheet metals and create an interlock to joint these sheet metals together. This process can join two steel of material together without adding a joining element like rivet and is able to join dissimilar materials such as joining aluminium and steel [3]. Other advantages include damage of this process to surface coating of the workpiece is relatively small and less energy required in the process when compared to spot welding and self-pierce riveting. Common clinching process are round and rectangular clinching [4]. The main parameters which will affect the joint strength are neck thickness, interlock thickness and material properties [5]. This joining method was normally applied on low density

M. Awang and S. S. Emamian (Eds.): *Advances in Material Science and Engineering*, LNME, pp. 370–375, 2021.
https://doi.org/10.1007/978-981-16-3641-7_45

metal or alloy with high ductility. However, mechanical clinching can also be used on joining ultra-high strength and low ductility steel by modifying the clinching punch size and the die width and depth to control the metal flow to avoid defects during the clinching process [6]. Joint strength of two ultra-high strength steel sheets can be further increased by using a two-stage clinching process [7]. Aluminium alloy sheet of square and rectangular mechanical clinching joint strength were studied and it was found that the rectangular clinched joint is stronger than square clinched joint [8]. Even though the static joint strength of a resistance spot welding joint is higher than mechanical clinching, but mechanical clinching has a higher fatigue limit than a resistance spot welding joint [9]. In this research, the round and square clinched joints strength were investigated by experiment to serve as a reference for the design in selecting suitable clinching tools to optimize joint strength.

2   Experiment Setup and Joint Strength Evaluation

Sheet metal of aluminium alloy A5052 with 1.25 mm thickness and SPCC cold rolled steel with 1.00 mm thickness were used in this research. The mechanical properties of the materials are shown in Table 1 for comparison purpose. Figure 1 shows the shape and dimensions of upper punch and lower die used for the mechanical clinching experiment, both round and square clinching tools are having the same cross section dimensions. Forming load of 15 kN–55 kN was applied to clinch the sheet metals.

Table 1. Material properties and blank thickness

Materials	Thickness (mm)	Tensile strength (MPa)	Flow stress (MPa)	Elongation (%)
SPCC steel	1.00	309	$\bar{\sigma} = 390\,\bar{\varepsilon}^{0.17}$	25
A5052	1.25	241	$\bar{\sigma} = 354\,\bar{\varepsilon}^{0.16}$	8

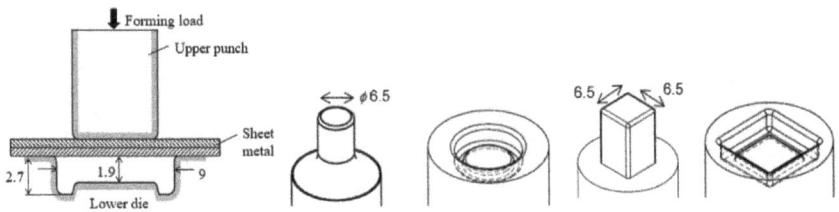

Fig. 1. Round and square clinching tools with same cross-section geometry

Tensile tests were tested on a universal testing machine at 2 mm/min and at room temperature to evaluate the round and square clinched joints strength. Based on Japanese Industrial Standard Committee, two type of loading mode (see Fig. 2), i.e., peeling test and tension-shearing test are considered for the evaluation. The maximum force in peeling test, F_o and tension-shearing test, F_s were measured until the clinched joint failed.

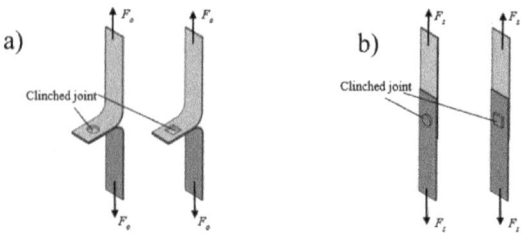

Fig. 2. Joint strength tests a) peeling test, b) shearing test

3 Results and Discussion

Figure 3 (a) and (b) show the graph of forming load vs punch stroke for round and square clinching process respectively. It was found that square clinching required higher forming load to produce the same measure of bottom thickness for both materials. This is because the cross sectioned area of a square punch has larger contact surface than a round punch. Figure. 4 (a) and (b) show the punched shape from the top and bottom view for round and a square clinched joint.

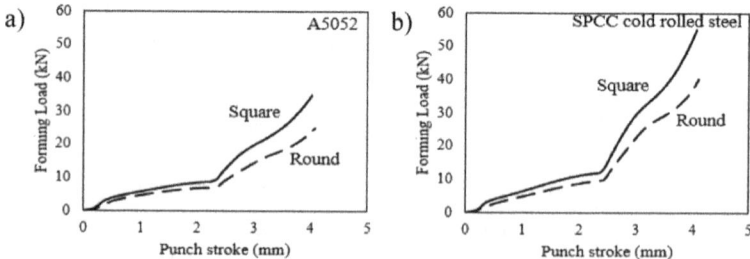

Fig. 3. Forming load vs punch stroke for round and square clinching a) A5052, b) SPCC steel

Fig. 4. Clinched joints from experiment a) round, b) square clinched joint

Figure 5 (a) and (b) show the strength for shearing and peeling test for round and square clinched joints produced by different forming load in clinching process for A5052 sheet metal respectively. Clinched joints are formed after 18 kN for round clinched joint and 25 kN for square clinched joint. When forming load increased to 30 kN on the round clinching specimen, crack can be observed on the neck of the round clinched joint because too much of plastic deformation take place at this location (see Fig. 7(a)).

As there is defect on the clinched joint, therefore the joint strength at 30 kN forming load drops drastically. However, both peeling and shearing tests show that the square clinched joint exhibits stronger than round clinched joint. This is because of the neck and interlock formed along the perimeter of a square clinched had larger areas than of a round clinched joint, therefore it can resist larger external load. In shearing test (see Fig. 5(a)), the forming load to produce maximum joint strength for square and round clinched joint is 25 kN. At this forming load, square clinched joint is stronger than round clinched joint by about 70%. Hence, the square clinch joint shows stronger joint strength to resist shear force than round clinched joint for A5052. Figure 5(b) also shows that the optimal peeling joint strength is achieved at 25 N of forming load. Square clinched joint is stronger than round clinched joint by about 42%.

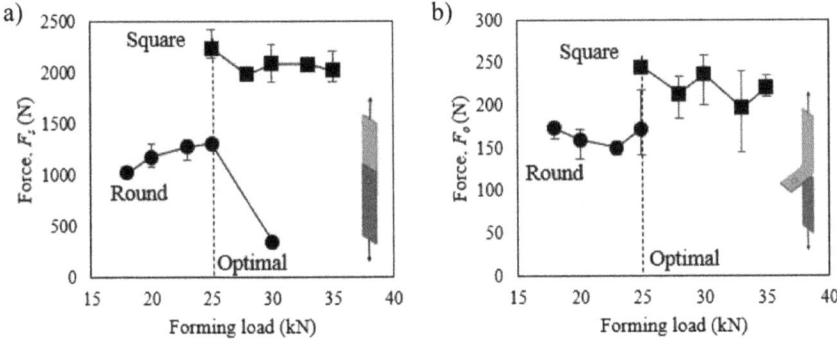

Fig. 5. Joint strength of A5052 a) shearing test, b) peeling test

Figure 6 (a) and (b) show the joint strength in shearing and peeling test for round and square clinched joint produced by different forming load in clinching process for SPCC cold rolled steel sheet metal respectively. In Fig. 6(a), round clinched joint exhibits stronger joint strength than square clinched joint at lower forming. The optimal forming load to obtain maximum shearing joint strength for round and square clinched joint is 50 kN and 55 kN respectively. Meanwhile, maximum shearing joint strength for round and square clinched joint is 1881 N and 1984 N. Square clinched joint exhibits stronger joint strength than round clinched joint by about 5% in shearing mode. From Fig. 6(b), the optimal forming load to produce maximum shearing joint strength for round and square clinched joint is 50 kN and 53 kN respectively. Square clinched joint has peeling joint strength of 378 N and round clinched joint has peeling joint strength of 565 N when the optimal forming load is applied to form the clinched joint. Hence, the round clinched joint had relatively stronger joint strength than square clinched joint for SPCC steel by about 49% in peeling mode.

From Fig. 7 (b) shows the SPCC cold rolled steel round clinched joint specimen that formed after 55 kN forming load. It was observed that material flow backward causing thinning at the bottom. This phenomenon leads to the joint strength drop after 55 kN. Therefore, further increase the forming load will not strengthen the joint but weaken it.

Figure 8 shows the types of failure mode on the peeling and shearing test for both materials were observed from the joint strength test. It was found that all the A5052

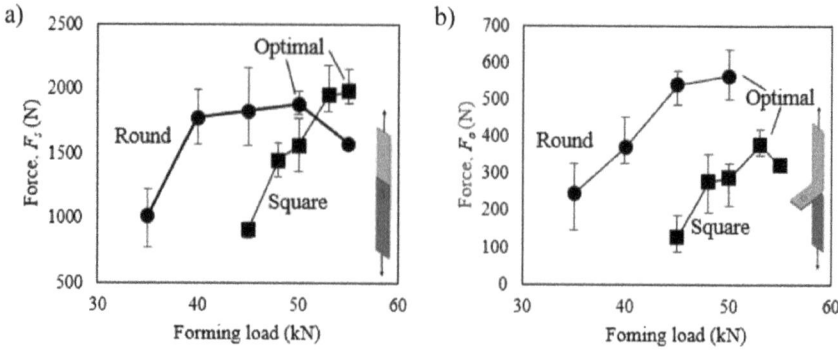

Fig. 6. Joint strength of SPCC steel a) shearing test, b) peeling test

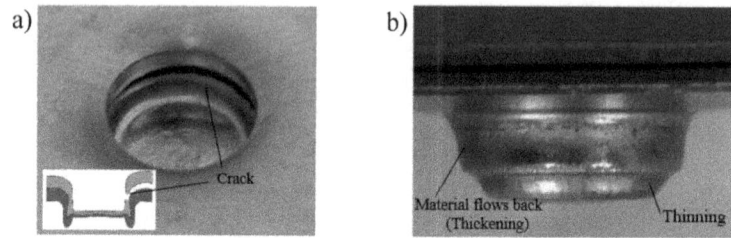

Fig. 7. Round clinched joint defects at high forming load a) A5052, b) SPCC steel

clinched joint specimens failed at the neck region due to insufficient of neck thickness and low ductility. On the other hand, button separation failure mode was found for all the SPCC cold rolled steel clinched joint specimens due to insufficient of interlock to hold the upper and lower sheet metals. Higher forming load should be applied on SPCC cold rolled steel for larger interlock formation but it was limited to 55 kN to avoid the punch from failure.

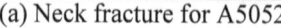

(a) Neck fracture for A5052   (b) Button separation for SPCC steel

Fig. 8. Failure mode

4   Conclusion

The present research is intended to provide a reference on the behaviour of joint strength by considering the tool shape for mechanical clinching. From the results of shearing and peeling tests, square clinched joint is 70% and 42% respectively stronger than conventional round clinched joint for A5052. For SPCC cold rolled steel, the square clinched joint is 10% stronger than round clinched joint in shearing but 49% weaker in peeling mode.

References

1. Barnes, T.A., Pashby, I.R.: Joining techniques for aluminium spaceframes used in automobiles: Part II adhesive bonding and mechanical fasteners. J. Mat Proc. Tech. **212**, 884–889 (2000)
2. Krause, A.R., Chernenkoff, R.A.: A comparative study of the fatigue behaviour of spot welded and mechanically fastened aluminium joints. SAE Technical Paper Series, 950710 (1995)
3. Mori, K., Abe, Y.: A review on mechanical joining of aluminium and high strength steel sheets by plastic deformation. Int. J. Lightweight Mater. Manuf. **1**, 1–11 (2018)
4. Mucha, J.: The analysis of rectangular clinching joint in shearing test. Maint. Reliab. **51**, 45–50 (2011)
5. Varis, J.P., Lepisto, J.: A simple testing-based procedure and simulation of the clincing process using FEA for establishing clinching parameters. Thin-Walled Struct. **41**, 691–709 (2003)
6. Abe, Y., Saito, T., Nakagawa, K., et al.: Rectangular shear clinching for joining of ultra-high strength steel sheets. Procedia Manuf. **15**, 1354–1359 (2018)
7. Abe, Y., Ishihata, S., Maeda, T., et al.: mechanical clinching process using preforming of lower sheet for improvement of joinability. Procedia Manuf. **15**, 1360–1367 (2018)
8. Abe, Y., Kato, T., Mori, K.: Joining of aluminium alloy sheets by rectangular mechanical clinching. In: 14th International ESAFORM Conference on Material Forming, pp. 1253–1258 (2011)
9. Abe, Y., Kato, T., Mori, K., et al.: Mechanical clinching of ultra-high strength steel sheets and strength of joints. J. Mat Proc. Tech. **214**, 2112–2118 (2014)

Feasibility of Forming Aluminium-Polymer Clinched Joint with Round Grooved Clinch Set

Hengkeong Kam[1(✉)], Chanchin Wang[2], Yiying Tan[1], and Xin Wang[3]

[1] Department of Mechanical and Material Engineering, Universiti Tunku Abdul Rahman,
43000 Kajang, Malaysia
kamhk@utar.edu.my, tonnitan96@1utar.my
[2] Department of Mechanical Engineering, SEGi University, 47810 Petaling Jaya, Malaysia
wangchan@segi.edu.my
[3] School of Mechanical Engineering, Shanghai Dianji University, No. 300 Shuihua Road,
Pudong, Shanghai, China
wangx31915@sdju.edu.cn

Abstract. Application of lightweight materials such as metals and polymers has become more important in aerospace and automotive industries. Joining of dissimilar materials is relatively challenging as compared to conventional joining of similar materials due to the differences of mechanical properties. Thus, the present research is carried out experimentally to investigate the feasibility of joining aluminium alloy A5052 with 3 different types of polymer, i.e., polycarbonate (PC), polyvinyl chloride (PVC) and high impact polystyrene (HIPS) by using round-grooved clinch die. The clinched joints were produced by applying various punch loads on the combinations of the materials at room temperature. The joint strength was examined by performing single lap shear test to investigate the relationship between clinching forming load and joint strength. It was found that the forming load exhibits great influence on the joinability, joint strength and failure modes of the hybrid structure. Optimal forming load for each hybrid joint was investigated, further increase in forming load does not increase the joint strength. The maximum load of Al-PC clinched joint is the highest among the 3 types of polymers, follow by Al-PVC and Al-HIPS clinched joint.

1 Introduction

The awareness of sustainable development and green environment have brought great impact on the automotive industry nowadays. The car manufacturers started to design for lighter products in order to reduce fuel consumption as to directly reduce the weight of the vehicles [1]. Thus, aluminium and polymer are the two common lightweight materials which have been widely used in automotive industry to reduce weight while enhancing performance and safety [2]. Polymers are available in variety of colours, this can beautify the appearance of a product. Assembly of aluminium and polymer parts represents a key method of mechanical design in enhancing product's performance while minimizing the overall weight of the products [3]. However, joining dissimilar materials such as metal and polymer is very challenging because of the large difference in material behaviours

M. Awang and S. S. Emamian (Eds.): *Advances in Material Science and Engineering*, LNME, pp. 376–381, 2021.
https://doi.org/10.1007/978-981-16-3641-7_46

[4]. To overcome the challenges, mechanical fastening, adhesive bonding and clinching can be the solution to this problem [5]. However, clinching has not gained much attention from the industry in its application such as metal-polymer hybrid joint [6] due to lack of studies on the feasibility of joining polymer with metal using mechanical clinching method. Clinching punch and die geometry contribute greatly in controlling the material flow to make a joining becomes possible [7, 8]. Various clinching tools such as round split, round flat and rectangular shear were applied to join aluminium and polymer. However, it was reported that the round grooved clinch dies are not suitable to be used for joining aluminium alloy A6082-T6 with polycarbonate as the bulge was torn out from the joint after joining and thus, die with movable wall is preferred in joining aluminium alloy with polymer to avoid the tearing [9]. Pre-heating on the thermoplastic polymers has been studied for clinch joining with aluminium alloy [10]. The outcome of this study will identify suitability of using round grooved clinch die at room temperature to form hybrid joint, the effect of materials selection and joining force on joint formation, joint strength and failure mode.

2   Mechanical Clinching Setup and Joint Strength Evaluation

Sheet metal of aluminium alloy A5052 with 1.25 mm thickness and polycarbonate (PC), polyvinyl chloride (PVC), high impact polystyrene (HIPS) with thickness of 1.8 mm, 1.85 mm and 1.6 mm respectively were prepared in this research. Figure 1 shows the stress-strain curve of the materials obtained from tensile test at room temperature. Figure 2 shows the tooling layout used for the mechanical clinching experiment. The clinching die in this research is a modified round grooved die compare to the round grooved die used by other researcher [9] as it was concluded that deeper die depth and bigger die diameter will promote the formation of interlock [11]. Aluminium alloy is used as the upper sheet and the polymers are used as the lower sheet. Forming loads from 10kN to 50kN were applied at 5 mm/s to form the hybrid joints at room temperature by using a universal testing machine. 3 samples were collected for each combination of hybrid materials (Al-PC, Al-PVC and Al-HIPS) at each punching load applied.

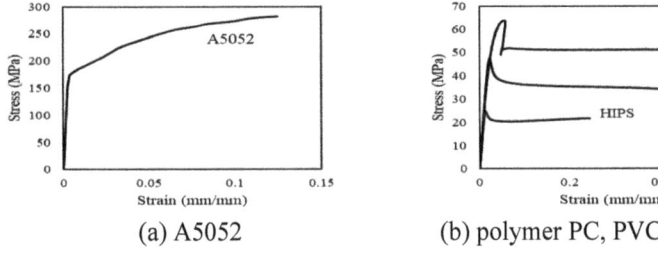

(a) A5052   (b) polymer PC, PVC and HIPS

Fig. 1. Engineering stress-strain curve at room temperature

Figure 3 shows the schematic diagram of single lap shear test. Shear strength of the hybrid joint was assessed by single lap shear test using universal testing machine with load capacity of 100kN at the control shear rate of 1mm/min. The shear strength and failure modes for each hybrid joint were evaluated.

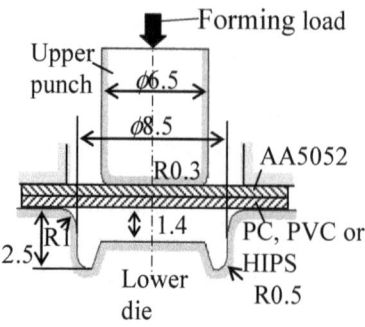
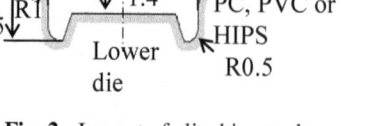

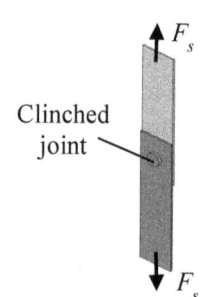

Fig. 2. Layout of clinching tools

Fig. 3. Shearing test specimen

3   Results and Discussion

The forming load (up to 20 kN) versus punch stroke data of clinching aluminium alloy A5052 with PC, PVC and HIPS are presented in Fig. 4. The forming history curves are similar to each other as A5052 is much stronger than the polymers, so most of the forming load were used to deform the A5052. Figure 5 shows the cross-sectional view and geometry parameters, i.e., neck thickness, interlock thickness and bottom thickness of each hybrid material clinched structure produced by 20 kN of punching load. By examining the cross-section view, the results suggest polymer with higher strength and ductility such as polycarbonate will have relatively larger interlock and bottom thickness, on the other hand, hybrid joint formed with lower strength and ductility polymer such as high impact polystyrene will have smaller interlock and minimum bottom thickness clinched connections. The findings show that hybrid clinched joint is possible to be formed with the round grooved die. Die geometry is playing an important role in material flow control.

Figure 6 shows the hybrid clinched joint strength produced by different forming load. Maximum joint strength for Al-PC, Al-PVC and Al-HIPS is 1.19 kN, 1.14 kN and 0.75 kN respectively. It was found that the forming load affects the joinability of the hybrid structure and the corresponding failure mode. The clinched joints are formed after applying minimum load of 20 kN, 18 kN and 15 kN for Al-PC, Al-PVC and Al-HIPS respectively. It shows that the minimum forming load to form the hybrid joint is proportional to the material strength. When the forming load is relatively low, the shear strength of the clinched joint increases proportionally. However, further increase of forming load does not increase the joint strength significantly. Optimal forming load to produce maximum joint strength for Al-PC, Al-PVC and Al-HIPS are 45 kN, 35 kN and 45 kN respectively.

During the single lap shear test, four different types of failure mechanisms were observed, namely button separation (BS), pull-out (PO), polymer fracture (PF) and neck fracture (NF) as shown in Fig. 7. Button separation is a typical failure mode occurs on clinched connections which happens when low forming load is applied to clinch the joint. The aluminium button slipped out from the polymer button due insufficient interlock to hold the upper and lower sheet together. As the forming force increases, pull-out failure

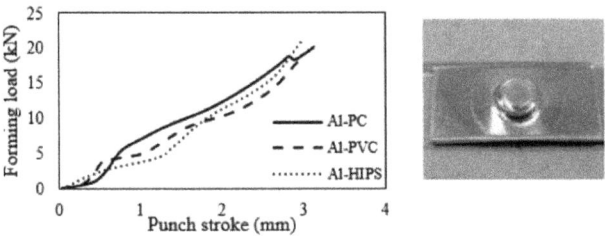

Fig. 4. Clinching forming load vs stroke

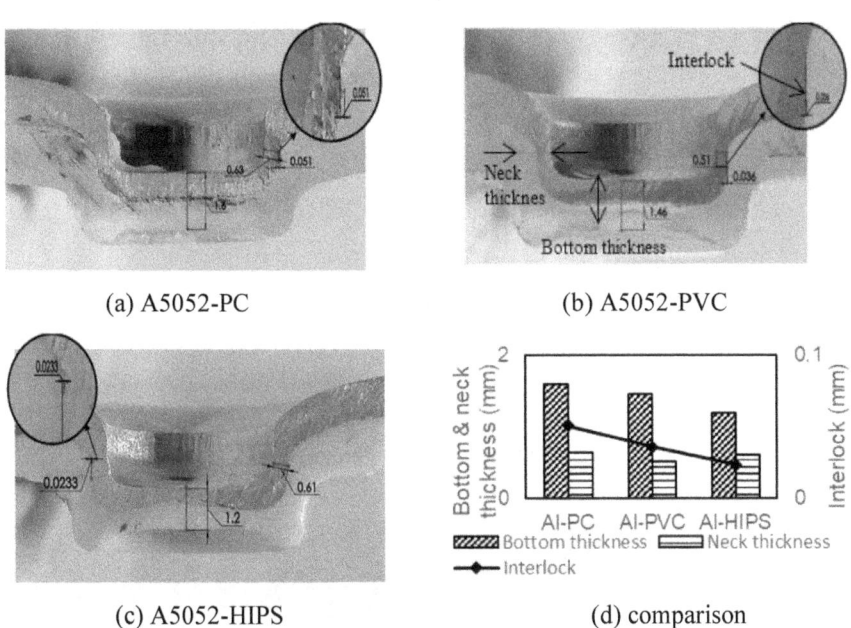

(a) A5052-PC   (b) A5052-PVC

(c) A5052-HIPS   (d) comparison

Fig. 5. Cross sectional views of hybrid structures (Forming load: 20 kN)

mode takes place when shearing the clinched joint. This failure mode happens when the polymer button deforms plastically. The polymer button diameter becomes larger and allows the aluminium button to slip off from it. Crazing (whitening) is observed on the polymer sheet during the pull-out as an indicator of the plastic deformation on the polymer sheet. Polymer fracture takes place when the elongation on the polymer sheet comes to limit of the ductility. This failure mode takes place on Al-HIPS joint as HIPS has a relatively lower ductility and strength compared to PC and PVC. Neck fracture failure mode is found on shearing Al-PVC clinched joint that is produced by high forming load. Higher forming load can produce a deeper interlock on the clinched joint, however, it will also reduce the neck thickness of the upper sheet.

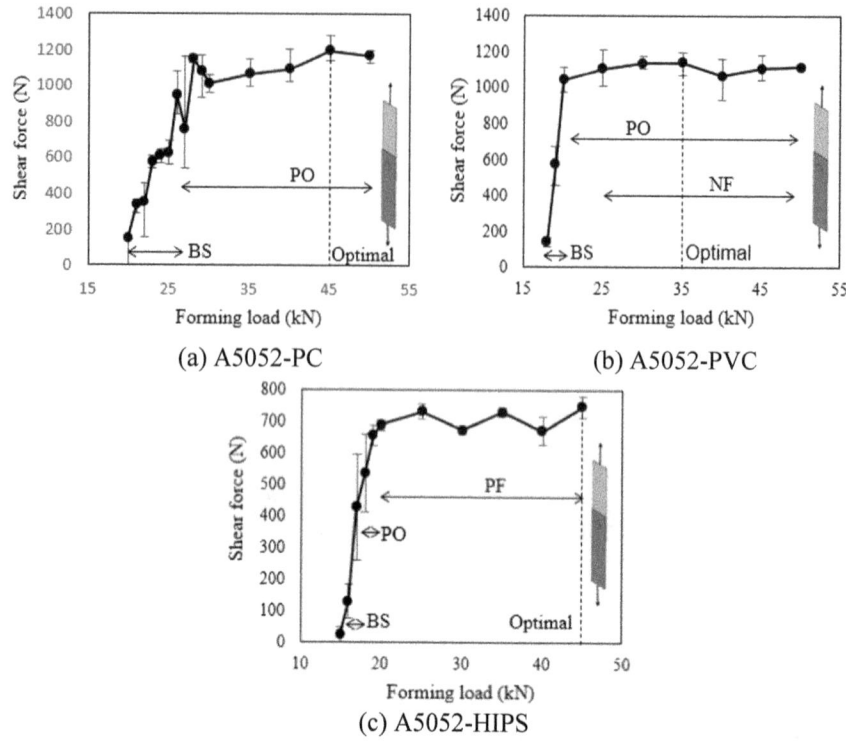

Fig. 6. Shear strength vs forming load

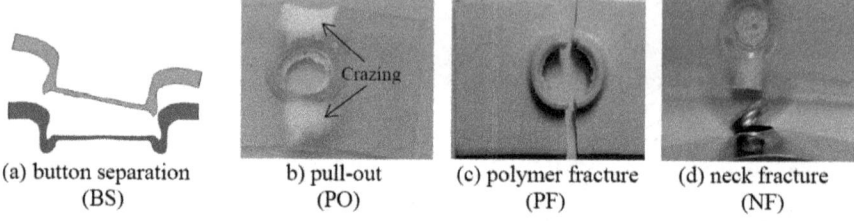

Fig. 7. Failure modes

4 Conclusion

The deforming behaviour of aluminium-polymer hybrid clinched joint has been investigated. Clinching forming load has great influence on the joinability and failure modes of the hybrid structure, insufficient forming load will cause less interlock to be formed which leads to button separation, while too much forming load does not increase the joint strength significantly. Optimal forming load to produce maximum joint strength for Al-PC, Al-PVC and Al-HIPS are 45 kN, 35 kN and 45 kN respectively. Maximum shearing joint strength for Al-PC, Al-PVC and Al-HIPS are 1.19 kN, 1.14 kN and 0.75 kN respectively. Al-PC is the strongest hybrid clinched joint among the 3 hybrid clinched joints, follow by Al-PVC and Al-HIPS. Round grooved clinch die can be applied to produce aluminium-polymer hybrid clinched joint at room temperature.

References

1. Del Pero, F., Delogu, M., Pierini, M.: The effect of lightweighting in automotive LCA perspective: estimation of mass-induced fuel consumption reduction for gasoline turbocharged vehicles. J. Clean. Prod. **154**, 566–577 (2017)
2. Hirch, J.: Recent development in aluminium for automotive applications. Trans. Nonferrous Metals Soc. China **24**, 1995–2002 (2017)
3. Sakundarini, N., Taha, Z., Abdul-rashid, S.H., et al.: Optimal multi-material selection for lightweight design of automotive body assembly incorporating recyclability. Mater. Des. **50**, 1461–1476 (2013)
4. Martinsen, K., Hu, S.J., Carlson, B.E.: Manufacturing technology joining of dissimilar materials. CIRP Ann. Manuf. Technol. **64**, 1461–1476 (2015)
5. Amancio-Filho, S.T., Santos, J.F.: Joining of polymers and polymer-metal hybrid structures: recent development and trends. Polym. Eng. Sci. **49**, 1461–1476 (2009)
6. Modi, S., Stevens, M., Chess, M.: Mixed material joining advancements and challenges. Center for Automotive Research, MI (2017)
7. Mucha, J.: The analysis of lock forming mechanism in the clinching joint. Mater. Des. **32**, 4943–4954 (2011)
8. Abe, Y., Mori, K., Kato, T.: Joining of high strength steel and aluminium alloy sheets by mechanical clinhing with dies for control of metal flow. J. Mater. Process. Technol. **212**, 884–889 (2012)
9. Lambiase, F.: Mechanical behaviour of polymer-metal hybrid joints produced by clinching using different tools. Mater. Des. **87**, 606–618 (2015)
10. Lambiase, F.: Joinability of different thermoplastic polymers with aluminium AA6082 sheets by mechanical clinching. Int. J. Adv. Manuf. Technol. **80**(9–12), 1995–2006 (2015). https://doi.org/10.1007/s00170-015-7192-1
11. Abe, Y., Kishimoto, M., Kato, T., et al.: Joining of hot-dip coated high-strength steel sheets by mechanical clinching. Int. J. Mater. Form. **2**, 291–294 (2010)

Design for Multiple Life Cycles Support Tool

Tze Fong Go[1(✉)], Dzuraidah Abd Wahab[2], Hawa Hishamuddin[3],
and Lip Kean Moey[4]

[1] Centre for Advanced Materials and Intelligent Manufacturing, Faculty of Engineering,
Built Environment and IT, SEGi University, Jalan Teknologi, Kota Damansara,
47810 Petaling Jaya, Selangor, Malaysia
gotzefong@segi.edu.my
[2] Centre for Automotive Research, Faculty of Engineering and Built Environment,
Universiti Kebangsaan Malaysia, UKM, 43600 Bangi, Malaysia
[3] Department of Mechanical and Manufacturing Engineering, Faculty of Engineering and Built
Environment, Universiti Kebangsaan Malaysia, UKM, 43600 Bangi, Malaysia
[4] Centre for Modelling and Simulation, Faculty of Engineering and the Built Environment,
SEGi University, Jalan Teknologi, Kota Damansara, 47810 Petaling Jaya, Selangor, Malaysia

Abstract. Malaysia Automotive Institute launched the National Automotive Policy 2014 to enhance the awareness of remanufacturing. However, the lack of awareness to the practice of Design for Remanufacturing and designers are not supported by specific tools to evaluate the important design elements of remanufacturing lead to limitation of remanufacturing activities in Malaysia. Therefore, Design for Multiple Life-Cycles support tool (DFMLCST) is proposed in this paper to evaluate the design parameters of a product at the early stage of design using a modified Analytical Hierarchy Process (AHP). The design tool was validated based on expert opinion via user acceptance test on the support tool. Based on the case study on threshing machine, disassemblability and serviceability (>20%) was found to be the most two important design elements followed by reassembly (12%). As a conclusion, DFMLCST has successfully developed and performed as a platform to implementing DFMLC into a product.

1 Introduction

The awareness of develop environmentally friendly via remanufacturing started in Malaysia due to the National Automotive Policy (NAP) 2014 launched by Malaysia Automotive Institute (MAI) [1], however, the lack of awareness to the practice of Design for Remanufacturing leads to limitation of remanufacturing activities in Malaysia [2]. Therefore, life-cycle thinking in the development of new products has led design engineers to rethink the way they design sustainable products [3]. One of the most viable approaches is designing products for multiple life-cycles as mentioned in [4]. A future trend of sustainable product might emphases remanufacturing or multiple life cycles of product (MLCP) [5]. However, product designers are rarely supported by specific methodologies or tools to evaluate the important design elements of remanufacturing. Furthermore, traditional design tools most are focused on functionality and cost at the expense of environmental issues [6], and designers lack knowledge and awareness of

M. Awang and S. S. Emamian (Eds.): *Advances in Material Science and Engineering*, LNME, pp. 382–387, 2021.
https://doi.org/10.1007/978-981-16-3641-7_47

multiple life-cycle considerations in their work [2]. Therefore, it is imperative for a design engineer to acquire the capability to build components or modules that will support MLCP. In this paper, Design for Multiple Life-Cycles support tool (DFMLCST) was developed to provide design guidelines and platform to analyse design strategies by a modified Analytical Hierarchy Process (AHP) method. The aim of the development of DFMLCST was to examine the extent of product remanufacturability adoption in Malaysia industry.

DFXs (Design for X) have been studied for purposes of supporting MLCP are Design for Environment (DFE) [7], Design for Material Recovery (DFMR) [8], Design for Component Recovery (DFCR) [9], Design for Assembly (DFA) [10], Design for Disassembly (DFD) [11], Design for Upgrade [12], Design for Durability (DFR) [13], Design for Modularity (DFMo) [14] and Design for Maintainability (DFMa) [15]. Design for Multiple Life-Cycles (DFMLC) is a relatively new DFX because it combines the eco-design strategies. But, DFMLC can optimise choices when it comes to environmentally friendly EOL strategies such as remanufacturing, reuse and recycling. Therefore, a MLCP should be expected to be designed for longevity, component recoverable, modular and serviceable in order to enhance the sustainability of the product.

The factors that lacks of implementation of DFMLC concept in the local industry are poor consumer perceptions of remanufactured components, lack of knowledge and technology in the remanufacturing process, lack of policies and laws for the remanufacturing process and lack of research and development programs for the remanufacturing process. Moreover, the conflicts of design properties, lack of existing standards and assisting tool in the DFMLC model are the challenges in the field of sustainable component design. Therefore, emphasis on development and research on MLCP aspects are needed to address these issues. Therefore, a design analysis tool named as DFMLC support tool is presented in this paper.

2 Methods

A design support tool named as "Design for Multiple Life-Cycles Support Tool" (DFMLCST) was developed using Guide Matlab R2013a in this paper. Therefore, this section presents two parts. Guide Matlab® provides a graphical user interface (GUI) design platform, additional features can be modified with coding in the Matlab editor. This section consists of two section: description the development of DFMLCST using Matlab and description the validation of DFMLCST.

2.1 Development of DFMLCST Using Matlab

DFMLCST consists of three main parts with composed by 5 user interfaces (UI). The main UIs that developed for DFMLCST with their respective explanation are shown in Table 1. The main interface for this support tool are shown in Fig. 1 and there are three parts in this UI with the respectively push buttons to activate the functions accordingly.

The quality of a product design can be improved by proper DFX guidelines then increase the reliability of the product. Typically, "X" represents the design guidelines are grouped according to the requirement of product, such as ease of disassembly, ease of manufacturing, ease of assembly and so on [16]. Therefore, DMLCPST provides the design guidelines in the form of either DFX or product feature requirements. By pressing the design guidelines button, an UI is opened to call guidelines based on user preference as shown in Fig. 2.

Figure 2(a) shows the pressing the button design evaluation in Part 3 to activate a new UI. The main function of this UI is to ensure the design concern on design criteria and elements of MLCP during product design stage. In DFMLCST, the AHP method was deployed to establish the weightage for each of the design elements. Press the "Enter Design Strategies/Idea" button to fill in the design strategies. Then, complete the mapping between design strategies with appreciate design criteria, DFX and design element. By using "1" to indicate the design strategy does effect on appreciate design criteria, DFX and design element, whereby "0" indicates that the design strategy does not effect on appreciate design criteria, DFX and design element.

Table 1. Main graphical user interface of DFMLCST

No	GUI	Explanation
1	DFMLCST	• Main User Interface of the design support tool • Prepare the push buttons to activate the following three parts: Part 1 provides two push buttons to get the recommended DFX for either manufacture driven or remanufacture driven; Part 2 provides the appreciate design guidelines and Part 3 provides a push button which link to the UI for design evaluation
2	Part_2	• One of the interfaces for Part 2 • Provides two pop-up manuals to select DFX or product features respectively to get the recommended design guidelines
3	Part_2_2	• Second interfaces for Part 2 • Provides two push buttons, one to call all the design guidelines that assist design for multiple life-cycles and another one to call the diagram of AHP hierarchy that analysis design strategies
4	Part_3	• User interface for Part 3 • Provides *"Enter Design Strategies/Idea"* button for filling in the relevant design strategies/idea and complete the pairwise comparison matrices for each AHP level • Provides *"Display Weightage"* button to display the result of design element weightage • Provides *"Display AHP Hierarchy"* button to display diagram of AHP hierarchy for specific analysis • Provides *"Tutorial"* button to get the guidance to use this software
5	Tutorial	• User interface for "Tutorial" • Provides two push button *"Tutorial 1"* and *"Tutorial 2"* • Provides eight steps to apply this design support tool

If consistency ratio (CR) for each matrix is ensured less than 20%, press the "Display Weightage" button to display the weightage of all element hierarchy AHP as shown in Fig. 2(b). If diagrammatic result is preferable, press the "Display AHP Hierarchy" button to get the diagram of AHP hierarchy with respectively weightage. DFMLCST does provide two tutorials with guidance of step-by-step to use this design support tool by pressed the "Examples" button.

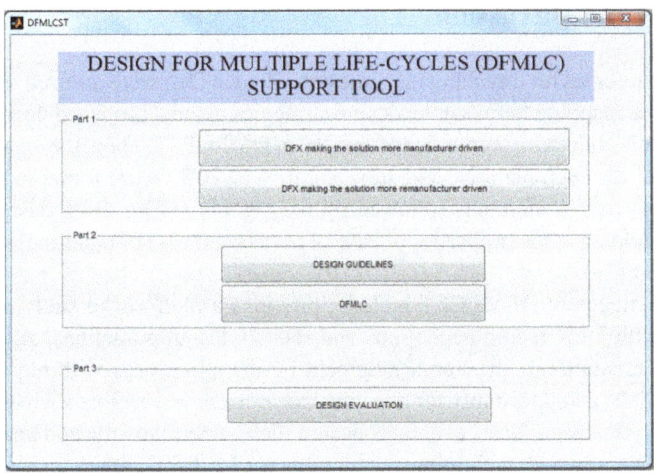

Fig. 1. Main user interface of the design support tool

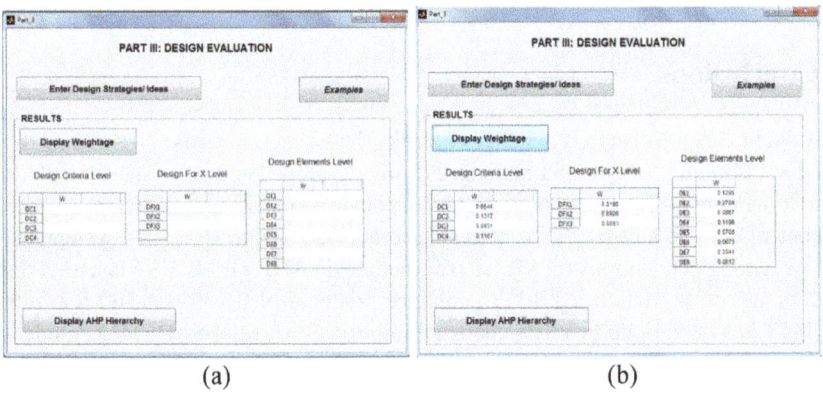

(a)   (b)

Fig. 2. User interface for part 3: design evaluation (a) before design evaluation and (b) after design evaluation

2.2 Tool Validation via User Acceptance Testing

In this study, the reliability of the modified AHP model determined via the validation techniques structural walk through and face validity. Then, usability of DFMLCST validated via User Acceptance Testing (UAT). In the process of software development, UAT involves validating the software in a real setting by the intended audience to check the requirements and ensure that the software meets user's needs [17]. Four respondents were invited to the UAT session, including an Original Equipment Manufacturer (OEM) engineer, an automotive component design consultant and two academic researchers.

3   Results and Discussion

The design strategies for threshing machine described in [18] were referred as case study. After done the mapping between the design strategies against proposed design elements to determine the pair-wise comparison for AHP in DMLCPST. Then, the obtained results indicated that disassemblability and serviceability (>20%) were found to be the most two important design elements followed by reassembly (12%). Both design elements are strongly related to the properties of ease of access and ease of handling, as suggested by research in [19].

The advantages of DFMLCST include: provides a platform to study design ideas for MLCP with DFX recommendations and related design guidelines, allows product designers to create more structured judgment by mapping ideas with required MLCP criteria, provides an opportunity for designers to consider key factors such as design criteria to generate design ideas, evaluates design ideas quantitatively, and analyse design ideas systematically with weighting calculations for MLCP parameters. The case study result indicated that a successful MLCP does needs the eight design elements: assemblability, disassemblability, cleanability, durability, accessibility, modularity, serviceability and parts commonability. Disassemblability and serviceability are the main factors that been considered during design the threshing machine to make it be a MLCP.

4   Conclusions

As a conclusion, to solve the challenges of lack of existing standards and assisting tool in the DFMLC model, DFMLCST has successfully developed and performed as a platform to implementing DFMLC into a product by considering the design guidelines and design element of MLCP. DFMLCST is also able to analyse design strategies systematically with weighting calculations for MLCP parameters via AHP. DFMLCST has been developed using Guide Matlab R2013a to enhance practical of developed DFMCL model. The UAT has been done to verify the feasibility of the developed DFMCL model and DFMLCST. Lastly, disassemblability and serviceability was found to be the most two important design elements for a MLCP.

References

1. Information on http://muvata.org.my/uploads/3/1/7/5/31759905/a9rb701.pdf
2. Information on https://www.ncapec.org/docs/USAID%20Study%20on%20Malaysian%20R emanufacturing.pdf
3. Information on https://www.gcsm.eu/Papers/28/0.3%20Jawahir.pdf

4. Go, T.F., Wahab, D.A., Hishamuddin, H.: Multiple generation life-cycles for product sustainability: the way forward. J. Clean. Prod. **95**, 16–29 (2016)
5. Paterson, D.A., Ijomah, W.L., Windmill, J.F.: End-of-life decision tool with emphasis on remanufacturing. J. Clean. Prod. **148**, 653–664 (2017)
6. Ijomah, W.L., McMahon, C.A., Hammond, G.P., et al.: Development of design for remanufacturing guidelines to support sustainable manufacturing. Robot. CIM-INT Manuf. **23**, 712–719 (2007)
7. Fiksel, J.: Design for Environment: A Guide to Sustainable Product Development. McGraw Hill, New York (2012)
8. Ilgin, M.A., Gupta, S.M.: Environmentally conscious manufacturing and product recovery (ECMPRO): a review of the state of the art. J. Environ. Manag. **91**, 563–591 (2010)
9. Go, T.F., Wahab, D.A., Rahman, M.A., et al.: Disassemblability of end-of-life vehicle: a critical review of evaluation methods. J. Clean. Prod. **19**, 1536–1546 (2011)
10. Bras, B.: Design for remanufacturing processes. In: Kurt, M. (ed.) Mechanical Engineers' Handbook, pp. 301–328. Wiley (2014)
11. Soh, S.L., Ong, S.K., Nee, A.Y.C.: Design for disassembly for remanufacturing: methodology and technology. Procedia CIRP **15**, 407–412 (2004)
12. Aziz, N., Wahab, D., Ramli, R.: Establishment of engineering metrics for upgradable design of brake caliper. In: Campana, G., Howlett, R.J., Setchi, R., Cimatti, B. (eds.) SDM 2017. SIST, vol. 68, pp. 87–97. Springer, Cham (2017). https://doi.org/10.1007/978-3-319-57078-5_9
13. Rios, F.C.: Beyond recycling: design for disassembly, reuse, and circular economy in the built environment. Doctoral dissertation, Arizona State University (2018)
14. Yang, S.S., Ong, S.K., Nee, A.Y.C.: A decision support tool for product design for remanufacturing. Procedia CIRP **40**, 144–149 (2016)
15. Sassanelli, C., Pezzotta, G., Pirola, F., et al.: Design for product service supportability (DfPSS) approach: a state of the art to foster product service system (PSS) design. Procedia CIRP **47**, 192–197 (2016)
16. Matsumoto, M., Yang, S., Martinsen, K., Kainuma, Y.: Trends and research challenges in remanufacturing. Int. J. Precis. Eng. Manuf.-Green Technol. **3**(1), 129–142 (2016). https://doi.org/10.1007/s40684-016-0016-4
17. Otaduy, I., Díaz, O.: User acceptance testing for Agile-developed web-based applications: empowering customers through wikis and mind maps. J. Syst. Softw. **133**, 212–229 (2017)
18. Dunmade, I.: Design for multi-lifecycle: a sustainability design concept. Int. J. Eng. Res. Appl. **3**, 1413–1418 (2013)
19. Sundin, E.: Product and Process Design for Successful Remanufacturing. Linköping University Electronic Press (2004)

Photovoltaic Array Interconnection Optimization Based on Cloud Cover

Fei Lu Siaw[✉] and Yong Zhen Ooi

Centre for Advanced Electrical and Electronic Systems (CAEES), Faculty of Engineering, Built Environment and Information Technology, SEGi University, Kota Damansara, 47810 Petaling Jaya, Malaysia
siawfeilu@segi.edu.my

Abstract. The impact of mismatch losses can be minimized by optimizing the interconnection configuration of a photovoltaic (PV) array. This paper proposes an interconnection using splitting technique to mitigate mismatch losses in an effective manner. This design consists of twenty-four PV modules in the arrangement of 4 × 6. The array is then split into three sub-groups (eight cells in each group) that has even number of rows and an optimum number of parallel branching. Series-parallel (SP), total-cross-tied (TCT) and the proposed interconnection are studied through modelling and simulations in MATLAB Simulink. The proposed interconnection generated 3594 W when not subjected to any shading. Next, cloud cover shading scenarios with different irradiation values are presented. In the first case study, the proposed interconnection delivered the highest output power, P_{mp} at 1700 W. This maximum output power is 44.1% and 38.9% higher than SP and TCT configurations. In the second case study, the proposed interconnection generated the highest output power at 2663 W which is 35.6% and 28.9% higher than both SP and TCT configurations. The results show that the proposed interconnection is the most favourable out of the three as it mitigated mismatch losses better than its counterparts.

1 Introduction

Solar energy is an uncertain and variable energy source as compared to conventional energy sources such as fossil fuel. This is especially true on cloudy days as solar energy received is fluctuating and substantially lower due to mismatch losses caused by non-uniform solar irradiation [1, 2].

The term interconnection refers to the configuration or topology of solar PV system. The study of an optimised interconnection design is important to generate maximum power. An optimized interconnection can minimize mismatch losses. Mismatch losses cause solar arrays to generate electricity at a lower capacity. Typically, partial shading is the major contributor of mismatch losses [3, 4]. Partial shading can be contributed by factors such as trees, nearby building, dust, soil and clouds [5]. Trees, nearby infrastructure, dust and soil are classified as secondary cause of mismatch losses where the impact is less severe than clouds. Clouds are considered as the primary cause of mismatch losses because of large solar irradiation variation due to random cloud sizes and thickness.

M. Awang and S. S. Emamian (Eds.): *Advances in Material Science and Engineering*, LNME, pp. 388–394, 2021.
https://doi.org/10.1007/978-981-16-3641-7_48

It is very common to integrate a bypass diode in parallel to solar cells in a module [6, 7]. The function of the bypass diode is to help mitigate mismatch losses by becoming forward bias when the corresponding solar cells are shaded and so providing an alternative escape pathway for electric current to be channelled to the neighbouring cells [8–10]. In addition, enhancing PV array interconnections from basic configurations such as series and parallel into more sophisticated and effective topology further improves system output performance [11, 12].

2 Methods

The methodology of this study focuses on the arrangement of solar modules. There are eight possible arrangements, i.e. $1 \times 24, 2 \times 12, 3 \times 8, 4 \times 6, 6 \times 4, 8 \times 3, 12 \times 2$, and 24×1. The basic concept of the matrix format is denoted as $p \times q$ where p is the solar cells connected in a series string (rows) and q is the number of parallel branches (columns). For example, a matrix format of 6×4 can be interpreted as 4 parallel branches with each branch consisting of 6 solar modules that are connected in series.

Table 1. Summary of arrangements for twenty-four PV modules

Arrangements	Characteristics
1×24 2×12 3×8	• High branching capability • Generates high current • Generates low voltage
4×6 6×4	• Satisfactory branching capability • Generates moderately high current • Generates moderately high voltage
8×3 12×2 24×1	• Low branching capability • Generates low current • Generates high voltage

An arrangement with higher parallel branches will generate higher current at a lower voltage. On the other hand, an arrangement that consists of fewer parallel branches will generate lower current at a higher voltage while achieving similar output power. Hence, the number of possible arrangements can be further reduced as shown in Table 1. From the table, 4×6 and 6×4 arrangements are both suitable for the final design of the solar system. This study focuses on the 4×6 arrangement because a general perception of clouds being shaped horizontally is considered.

Partial shading effect is dispersed by parallel branching out the PV modules into different rows or columns and splitting the array into different portions. The number of PV modules that can be grouped together is examined. There are many ways to split the twenty-four PV modules into sub-groups. However, not all groupings will generate the same output power. To achieve maximum output power, the splitting technique ensures that each subgroup must have the same number of rows as the neighbouring group.

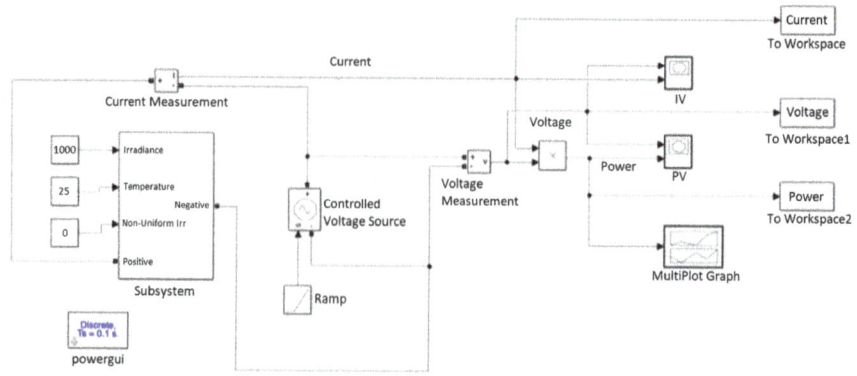

Fig. 1. Main system design in MATLAB Simulink

MATLAB Simulink is used to model interconnections such as SP, TCT, as well as a newly proposed interconnection. Simulation results are compared to identify the best interconnection that can mitigate mismatch losses under non-uniform irradiation conditions. The simulation work and can be separated into two sections: (1) Main System (2) Sub-system. The subsystem houses twenty-four PV modules that are interconnected with each other to form a solar array. The main system on the other hand is responsible of receiving the temperature and irradiance input as well as the analytical components to determine the performance of the array (Fig. 1). The analytical components include controlled voltage source, voltage measurement, current measurement, IV graph plotter and PV graph plotter.

A subsystem block contains twenty-four PV modules, connected according to the proposed interconnection to form an array. The uniform irradiation is 1000 W/m^2 as default and the value for nonuniform irradiance varies from 0–999 W/m^2 according to partial shading conditions. This study focuses on the effect of partial shading and not the effect of temperature, therefore all cells' temperature is controlled at 25 °C.

The proposed interconnection is an integration of all important considerations mentioned above to ensure that the impact of partial shading is minimized (Fig. 2). It consists of twenty-four PV modules with each module rated at 150 W. The modules are connected in the arrangement of 4×6. The array, consisting twenty-four PV modules, is split into three different sections. Within each sub-group, there are eight PV modules connected in the TCT configuration. Each sub-group has identical number of rows, to ensure that the output power is maximized.

3   Results and Discussion

A well-designed interconnection is able to mitigate partial shading effect when the number of parallel branching is high. The proposed design presented in Fig. 2 consists of six parallel branches, i.e. L1 to L6. The simulated maximum output power is 3594 W, maximum output current, I_{mp} is 48.56 A, and the maximum output voltage, V_{mp} is 74 V when not subjected to any shading.

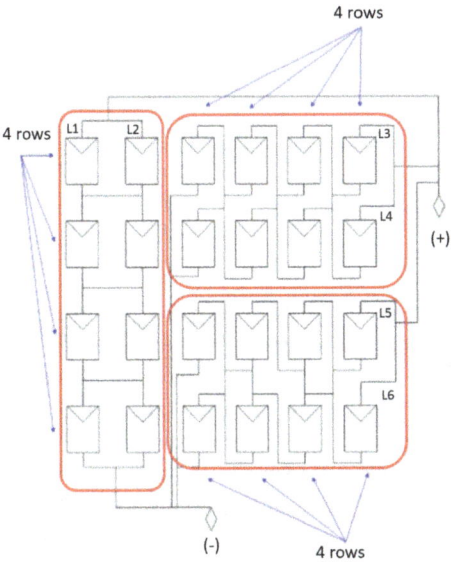

Fig. 2. Proposed 4 × 6 interconnection

Next, random shading pattern is taken into consideration. Multiple shading patterns from different classes are overlaid on top of one another, resulting a random overall shading pattern. Modules that were heavily affected by thicker clouds due to several overlapping clouds receive lower irradiation value at 100 W/m², hence the darker colour coded PV modules 11 to 15. On the other hand, light grey colour is used to represent PV modules 21 to 25 with the input irradiation of 500 W/m² to indicate that more sunlight irradiance is received from these modules as a result of lesser overlapping clouds. In addition, a lighter shade of grey is illustrated for PV modules 31, 32 and 41 with input irradiation set to 700 W/m². Simplified simulation models for all three competing interconnections and further analysis are presented in Fig. 3 and their corresponding results are summarized in Table 2.

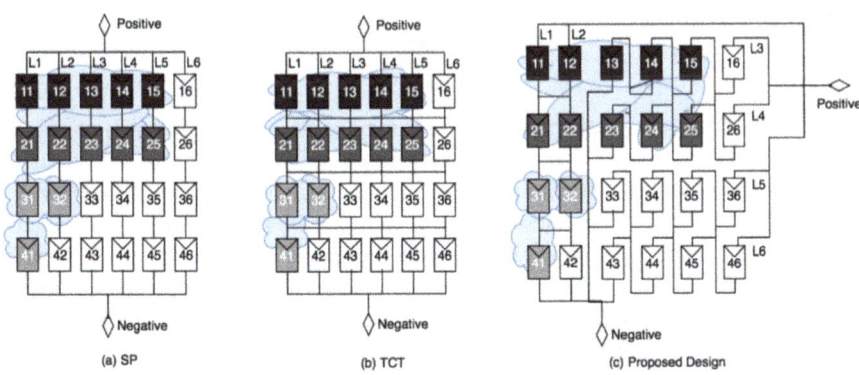

Fig. 3. Case study 1 - First shading pattern

Table 2. Results for random shading pattern from Fig. 3

Connection type	I_{mp} [A]	V_{mp} [V]	P_{mp} [W]
SP	12.84	74.00	950
TCT	12.75	81.40	1038
Proposed design	22.78	74.60	1700

The proposed interconnection design delivered the highest output power, P_{mp} at 1700 W. This maximum output power was 44.1% and 38.9% higher than SP and TCT configurations accordingly. The availability of two available parallel branch L5 and L6 with PV modules 33 to 36 and 43 to 46 in the proposed design, enables that system to generate higher output power. Due to the shading area presented in Fig. 3, SP interconnection was limited to only one available current pathway which is parallel branch L6. The first row of SP interconnection (PV modules 11 to 15) were highly shaded and produced very low current. This current became the limitation of the whole circuit. In the TCT configuration, the partial shading effect was mitigated slightly more effectively than SP and this account for the slight increment in overall output power in TCT over SP.

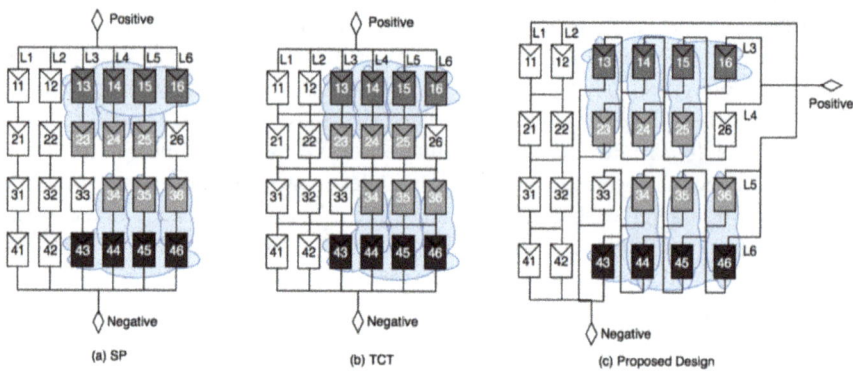

(a) SP   (b) TCT   (c) Proposed Design

Fig. 4. Case study 2 - Second shading pattern

The second random shading pattern has three distinct shading intensities with the shading region leaning towards the right side of the circuit (Fig. 4). The top shaded region has intermediate shading intensity therefore the input irradiation was set to 600 W/m^2 (PV modules 13 to 16). The centre shaded region received 800 W/m^2 to indicate the highest sunlight collection due to less cloud overlapping. The bottom shaded region received 200 W/m^2 because that region is saturated with clouds overlapping one another. Figure 4 presents simplified simulation models for all three competing interconnections mentioned and the results are tabulated in Table 3 below.

The proposed interconnection generated the highest power at 2663 W which is 35.6% and 28.9% higher than both SP and TCT (Table 4). The interconnection in TCT and SP

Table 3. Results for random shading pattern from Fig. 4

Connection type	I_{mp} [A]	V_{mp} [V]	P_{mp} [W]
SP	22.80	75.20	1715
TCT	23.73	79.80	1894
Proposed design	35.88	74.20	2663

configuration generated the least amount of output power because the bottom shaded area limited current flow. On the other hand, the bottom shaded region has only affected parallel branch L6 in the proposed design. In addition, the barrier and current limitation in the centre region of the solar array did not restrict the flow of current between terminals at branch L4 and L5 due to the soft shading condition (800 W/m^2).

4 Conclusions

One of the most common challenge associated with solar PV system is partial shading, in which power generation is significantly reduced when subjected to mismatch losses. In this paper, an optimised interconnection is proposed where splitting technique is incorporated to disperse partial shading in a more effective manner. The interconnection consists of twenty-four PV modules in the arrangement of 4 × 6. The array is then split into three sub-groups (8 in each group) with even number of rows and an optimum number of parallel branches. The proposed interconnection minimized mismatch losses and hence improved the overall power output. Three configurations (SP, TCT and proposed interconnection) are studied using simulation models. The proposed design generated P_{mp} of approximately 3594 W when it is not subjected to any shading. Next, random cloud cover scenarios with different irradiance values were simulated and the proposed interconnection achieved the highest output power. These results proved that the proposed design is the most favourable out of the three as it mitigated mismatch losses better than its counterparts, SP and TCT topologies.

Acknowledgments. This research was fully funded by SEGi University.

References

1. Shams El-Dein, M., Kazerani, M., Salama, M.: An optimal total cross tied interconnection for reducing mismatch losses in photovoltaic arrays. IEEE Trans. Sustain. Energy **4**, 99–107 (2013)
2. Vijayalekshmy, S., Bindu, G., Rama Iyer, S.: A novel zig-zag scheme for power enhancement of partially shaded solar arrays. Sol. Energy **135**, 92–102 (2016)
3. Villa, L., Picault, D., Raison, B., Bacha, S., Labonne, A.: Maximizing the power output of partially shaded photovoltaic plants through optimization of the interconnections among its modules. IEEE J. Photovolt. **2**, 154–163 (2012)

4. Satpathy, P.R., Jena, S., Sharma, R.: Power enhancement from partially shaded modules of solar PV arrays through various interconnections among modules. Energy **144**, 839–850 (2018)
5. Belhachat, F., Larbes, C.: Modeling, analysis and comparison of solar photovoltaic array configurations under partial shading conditions. Sol. Energy **120**, 399–418 (2015)
6. Nasiruddin, I., Khatoon, S., Jalil, M.F., et al.: Shade diffusion of partial shaded PV array by using odd-even structure. Sol. Energy **181**, 519–529 (2019)
7. Carannante, G., Fraddanno, C., Pagano, M., et al.: Experimental performance of MPPT algorithm for photovoltaic sources subject to inhomogeneous insolation. IEEE Trans. Industr. Electron. **56**, 4374–4380 (2009)
8. Jazayeri, M., Uysal, S., Jazayeri, K., et al.: Experimental analysis of effects of connection type on PV system performance. In: Proceedings of 2013 International Conference on Renewable Energy Research and Applications (ICRERA), Madrid, Spain (2013)
9. Quaschning, V., Hanitsch, R.: Numerical simulation of current-voltage characteristics of photovoltaic systems with shaded solar cells. Sol. Energy **56**, 513–520 (1996)
10. Woyte, A., Nijs, J., Belmans, R.: Partial shadowing of photovoltaic arrays with different system configurations: literature review and field test results. Sol. Energy **74**, 217–233 (2003)
11. Chong, K.K., Siaw, F.L.: Electrical characterization of dense array concentrator photovoltaic system. In: Proceedings of the 27th European Photovoltaic Solar Energy Conference and Exhibition, Frankfurt, Germany, pp. 224–227 (2012)
12. Siaw, F.L., Chong, K.K., Wong, C.W.: A comprehensive study of dense-array concentrator photovoltaic system using non-imaging planar concentrator. Renewable Energy **62**, 542–555 (2014)

Adaptive Solar Photovoltaic Method to Overcome Partial Shading Effect on Solar Photovoltaic System

Fei Lu Siaw$^{(\boxtimes)}$, Tzer Hwai Gilbert Thio, and Chee Yee Ng

Centre for Advanced Electrical and Electronic Systems (CAEES), Faculty of Engineering,
Built Environment and Information Technology, SEGi University, Kota Damansara,
47810 Petaling Jaya, Malaysia
{siawfeilu,gilbertthioth}@segi.edu.my

Abstract. When there is non-uniform solar irradiation on a solar photovoltaic (PV) system, the output power will decrease due to mismatch effects because shaded cells produce lower current and voltage compared to non-shaded cells. Therefore, an adaptive method is designed to overcome this limitation which makes it suitable to be used as mobile system. Solar modules of the prototype are connected in total-cross-tied and sub solar cells in the solar modules are controlled by relay module which is programmed with an Arduino Uno. When a solar cell is shaded, the output voltage will decrease to set threshold value and sub cells will be activated to recover the voltage loss. Hence, the maximum output power of solar module is increased. MATLAB Simulink is used to simulate the designed solar module under various irradiance levels and partial shading conditions. The prototype measurement values and simulation result values are close with each other.

1 Introduction

Partial shadings commonly occur on solar cells due to the presence of nearby cloud, dust, leaves, and tall objects such as buildings. Partial shadings lead to mismatch in current and voltage and eventually affect the output power of solar cells. The power loss due to partial shading depends on the shading pattern position in an array, and the size of shaded area [1]. Mismatch occurs when the output voltage of solar array varies with the expected results [2, 3]. There are two types of PV mismatch: structural mismatch and functional mismatch [4]. Structural mismatch occurs when there are differences in the characteristics of solar cells even though the manufacturing and operation conditions of the solar cells are the same. Functional mismatch occurs when there are different operating conditions between solar cells in a solar module. Functional mismatch usually occurs when solar cells in the solar module are shaded, hence receiving lesser solar irradiation [5]. Solar cells that receive lesser irradiation will produce lesser output current as compared to the solar cells that receive full solar irradiation. Hence, the maximum current generated by the array will be restricted to match the current of shaded cell.

Bypass diodes are important to protect solar cells from going into reverse-bias breakdown that may lead to permanent damage. If a solar cell is shaded, the bypass diode

© The Author(s), under exclusive license to Springer Nature Singapore Pte Ltd. 2021
M. Awang and S. S. Emamian (Eds.): *Advances in Material Science and Engineering*, LNME, pp. 395–401, 2021.
https://doi.org/10.1007/978-981-16-3641-7_49

that is connected in parallel to its corresponding solar cell in the opposite polarity will become forward bias. The of bypass diode creates an alternate route for the array current to flow so as the underperforming solar cell can be protected [6, 7]. Although the presence of bypass diode across solar cells is able to prevent the cells from being damaged, reduction of power due to shading is inevitable.

Total-cross-tied (TCT) configuration can generate higher output power at partial shading condition because the interconnection of adjacent modules of all parallel connected string [8]. In this configuration, solar cells are connected in parallel rows to form a module. Then, a few modules are connected in series to form a complete TCT circuit. This arrangement reduces power losses when mismatching occurs due to partial shading. This study proposes an adaptive solar photovoltaic method in TCT configuration to overcome shading effect by reducing power losses. It is important to reduce the number of maximum power points and subsequently achieve higher output power under shading conditions. The PV cells in the module dynamically changes based on shading conditions and irradiation variation. Sub cells are connected to "compensate" the solar cells that are shaded, which then prevent the decrease of power caused by the partially shaded solar photovoltaic module.

2   Methods

A prototype is built using two sets of TCT connected solar modules, two double-channel relay module, liquid crystal display (LCD) and Arduino Uno (Fig. 1). There are twelve solar cells in the prototype, which consists of eight main solar cells and four sub solar cells. These solar cells are divided into two solar modules and connected in TCT connection. When a solar cell is shaded, the output voltage will decrease to set threshold value, and thus sub cells will be activated to recover the voltage loss and hence increase the maximum output power of solar module.

Output voltages obtained from both sets of solar modules are measured using analog input A3 and A4 of the Arduino Uno. Relay is used to control the sub solar cells: S11, S12, S21, and S22. The relationship between relay and irradiation is presented in Table 1.

Table 1. Conditions of relay to activate and deactivate

Irradiation	Relay 1 (IN1)	Relay 1 (IN2)	Relay 2 (IN1)	Relay 2 (IN2)
High	Off	Off	Off	Off
Moderate	On	Off	On	Off
Low	On	On	On	On

Sub solar cells S11 and S21 support solar module A (left side) while sub solar cells S12 and S22 support solar module B (right side). Sub solar cells are used to compensate the voltage loss from partial shading of other solar cells. With aid of sub solar cells, the prototype can supply a more stable output voltage during partial shading conditions. LCD

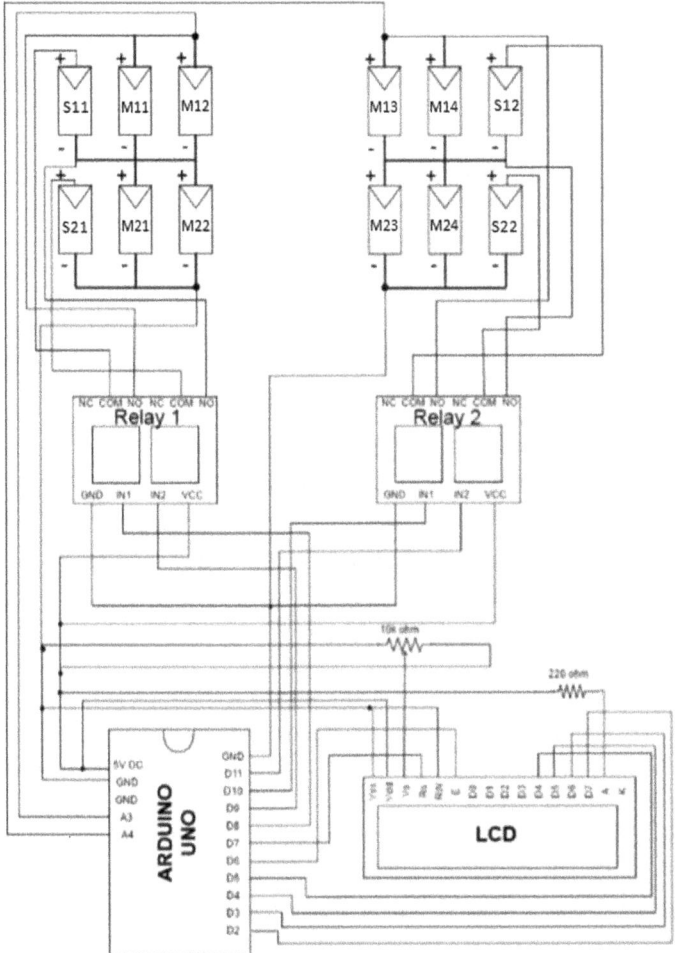

Fig. 1. Schematic diagram of the prototype

is used to display output voltage and irradiation levels received by the solar modules. When the irradiation is below a set reference value after all relays are turned on, "low irradiance" message will pop out from LCD to inform users to move to a place with a better irradiance.

To simulate the performance, MATLAB Simulink is used. The complete circuit consists of components such as solar cell, current sensor, voltage sensor, solver configuration, and ramp. To model the solar cell, open-circuit voltage (V_{oc}) and short-circuit current (I_{sc}) values are obtained from the manufacturer's datasheet. The value of V_{oc} is 2 V and I_{sc} is 60 mA (Table 2). Quality factor, N is set as 1.5. The range of N for polycrystalline silicon solar cell is between 1 and 2 [9]. Therefore, N value of 1.5 is chosen from the average of 1 and 2 as the quality factor.

Table 2. Parameters of solar cell

Characteristics	Details
Type	Polycrystalline
Open-circuit voltage, V_{oc}	2 [V]
Short-circuit current, I_{sc}	60 [mA]
Maximum Power, P_{mp}	0.12 [W]

Table 3. Partial shading conditions on solar module

Scenario	Shaded solar cell	Shaded solar cells
1B	One cell at top row	M12 shaded
2B	One cell at bottom row	M22 shaded
3B	Top row shaded	M11 and M12 shaded
4B	Bottom row shaded	M21 and M22 Shaded
5B	Side column shaded	M12 and M22 shaded

To create partial shading condition in Simulink, a connection point is required in the subsystem of solar module to set a low irradiation input value. All partial shading conditions selected are listed in Table 3. Since both irradiation and partial shading conditions are factors that will affect the output voltage of solar cells negatively, these two factors are studied. A few scenarios are chosen and presented in Table 4.

Table 4. Partial shading and varying irradiation conditions on solar module

Scenario	Shaded solar cell	Irradiation [W/m^2]
1C	1 cell shaded	1000
2C	1 cell shaded	800
3C	1 cell shaded	400
4C	1 row shaded	1000
5C	1 row shaded	800
6C	1 row shaded	400
7C	1 column shaded	1000
8C	1 column shaded	800
9C	1 column shaded	400

3   Results and Discussion

Results obtained from MATLAB simulations are presented in this section. From Table 5, when 1 cell is shaded, the maximum power obtained is 0.222 W. Both scenario 1B and scenario 2B have the same maximum power even though the shaded cells are located at different rows. The percentage of power loss for scenario 1B and 2B is 49% when compared to non-shaded scenario. For scenario 3B and 4B, no power is obtained from the simulations. This is because when the solar cells are connected in series, current across the circuit will follow the lowest current. Hence, when the entire row is shaded, no current is produced. In scenario 5B, the right column, M12 and M22 is shaded completely. The maximum power obtained is 0.215 W. Percentage of power loss for scenario 5B when compared to non-shaded scenario is 51%.

Table 5. Values of simulated scenarios of partial shading conditions

Scenario	Shaded solar cells	P_{mp} [W]	I_{mp} [A]	V_{mp} [V]
1B	M12 shaded	0.222	0.059	3.768
2B	M22 shaded	0.222	0.059	3.768
3B	M11 and M12 shaded	0	0	0
4B	M21 and M22 shaded	0	0	0
5B	M12 and M22 shaded	0.215	0.059	3.645

In addition, the system is also simulated under various partial shading and irradiation conditions (Table 6). The maximum power obtained from scenario 1C, 2C and 3C is 0.222 W, 0.180 W and 0.088 W respectively. For scenario 4C, 5C and 6C, the voltage and current is 0 because the shaded cell is acting like a load. The maximum power obtained from scenario 7C, 8C, and 9C is 0.215 W, 0.171 W, and 0.082 respectively.

Finally, the prototype is tested, and results obtained are recorded. Results obtained from the prototype are compared with results obtained from simulations at irradiance of 800 W/m^2. Table 7 shows the readings obtained from prototype and the data is compared to simulation results of scenario 2C (1 cell shaded in 800 W/m^2) and 8C (1 column shaded in 800 W/m^2).

Based on the data obtained from Table 7, the voltage obtained from the prototype is close with voltage obtained from simulation. The percentage of error obtained from prototype when one cell is shaded is 0.84%. Percentage of error obtained from prototype value when two cells are shaded is 1.89%. Both errors fall between the range of ±2%.

Table 6. Simulated scenarios of partial shading and irradiation conditions

Scenario	Shaded solar cells	P_{mp} [W]	I_{mp} [A]	V_{mp} [V]
1C	1 cell shaded with irradiance of 1000 W/m^2	0.222	0.059	3.768
2C	1 cell shaded with irradiance of 800 W/m^2	0.180	0.048	3.768
3C	1 cell shaded with irradiance of 400 W/m^2	0.088	0.024	3.700
4C	1 row shaded with irradiance of 1000 W/m^2	0	0	0
5C	1 row shaded with irradiance of 800 W/m^2	0	0	0
6C	1 row shaded with irradiance of 400 W/m^2	0	0	0
7C	1 column shaded with irradiance of 1000 W/m^2	0.215	0.059	3.645
8C	1 column shaded with irradiance of 800 W/m^2	0.171	0.047	3.630
9C	1 column shaded with irradiance of 400 W/m^2	0.082	0.023	3.579

Table 7. Comparison of results from the prototype and simulation

Shaded solar cell	Voltage from simulation [V]	Voltage from prototype [V]
Scenario 2C, 1 cell shaded	3.768	3.800
Scenario 8C, 1 column shaded	3.630	3.700

4 Conclusions

Non-uniform sunlight is one of the main factors that affects the performance of solar system. Non-uniform irradiance occurs when shadows from clouds, or other objects such as leaves and dust is on the surface of solar cells. When solar cells in the solar modules are shaded, mismatch will occur, and the output power of the solar system will decrease. An adaptive solar module is developed to reduce power losses when non uniform illumination falls onto it. The designed prototype has twelve solar cells, which consists of 8 main cells and 4 sub cells. These solar cells are divided to two solar modules and connected in TCT connection. When a solar cell is shaded, the output voltage of solar module will decrease to set threshold value. Hence, sub cells will be activated to recover the voltage loss to improve the performance of the solar module. The sub cells are controlled by relay modules connected to an Arduino Uno.

The data obtained from the prototype and MATLAB simulation are compared and analysed. Both results are found to be close with each other. From the results, the proposed adaptive solar module improved the performance of solar module under partial shading conditions. The proposed prototype is compact in size and is able to adapt to partial shading conditions which makes it suitable to be used as mobile PV system that can be carried around.

Acknowledgments. This research was fully funded by SEGi University.

References

1. Meraj, W., Mishra, S.: An approach to increase power in partial shading condition by reconfiguration. In: Annual IEEE India Conference (INDICON), New Delhi, India (2015)
2. Shams El-Dein, M., Kazerani, S.M.: An optimal total cross tied interconnection for reducing mismatch losses in photovoltaic arrays. IEEE Trans. Sustain. Energy **4**, 99–107 (2013)
3. Vijayalekshmy, S., Bindu, G., Rama Iyer, S.: A novel zig-zag scheme for power enhancement of partially shaded solar arrays. Sol. Energy **135**, 92–102 (2016)
4. Boukebbous, S.E., Kerdounm, D.: Study, modelling and simulation of photovoltaic panels under uniform and non-uniform illumination conditions. Revue des Energies Renouvelables **18**, 257–268 (2015)
5. Belhachat, F., Larbes, C.: Modeling, analysis and comparison of solar photovoltaic array configurations under partial shading conditions. Sol. Energy **120**, 399–418 (2015)
6. Siaw, F.L., Chong, K.K., Wong, C.W.: A comprehensive study of dense-array concentrator photovoltaic system using non-imaging planar concentrator. Renewable Energy **62**, 542–555 (2014)
7. Satpathy, P.R., Jena, S., Sharma, R.: Power enhancement from partially shaded modules of solar PV arrays through various interconnections among modules. Energy **144**, 839–850 (2018)
8. Shah, N., Patel, H.: Enhancing output power of PV array operating under non-uniform conditions. In: IEEE 16th International Conference on Environment and Electrical Engineering (EEEIC), Florence, Italy (2016)
9. Raju, P.S., Venkateswarlu, G.: Simscape model of photovoltaic cell . Int. J. Adv. Res. Electr. Electron. Instrum. Eng. **2**, 1766–1772 (2013)

Diffusion Coefficients Determination
for Commercial Grade Perfume

Chan Mieow Kee[1](✉), Chan Yi Shee[1], Ong Chi Siang[2], Lim Lee Fong[1],
Prasilla Kumaran[1], and Tan Woan Giun[1]

[1] Centre for Water Research, Faculty of Engineering and the Built Environment, SEGi
University Jalan Teknologi, Kota Damansara, 47810 Petaling Jaya, Selangor, Malaysia
mkchan@segi.edu.my
[2] Nanyang Technological University, 50 Nanyang Avenue, Singapore 639798, Singapore

Abstract. The demand of perfume is increasing globally due to the high market
demand and the emergence of aromachology, which uses fragrances for stress
reduction. Numerous studies have been done to study the diffusivity performance
of self-formulated perfume. However, the guide to link the self-formulated perfume to marketable perfume remains as a grey area, especially on the diffusivity
performance. This paper reveals the diffusivity of five commercial available perfumes. The main compounds in the perfume were characterised by using gas
chromatography and Fourier Transform Infrared spectroscopy. Results showed
that the diffusion coefficients of commercial grade perfume are within the range
of 0.6 to 1.3×10^{-6} m²/s, which are at least 10 times slower compared to methanol
and ethanol, which are the main compounds in perfume. This is due to the presence
of fragrance, essential oil, antioxidants in the perfume. This also creates long last-
ing fragrance effect on the users. The findings of this research provide a guide to
researchers to formulate perfume, which has the similar diffusivity as commercial
available perfume.

1 Introduction

Perfume is a mixture of fragrance chemicals with different chemical, physical and sen-
sorial properties especially in terms of molecular structure, volatility, water solubility,
polarity, intensity of the fragrance molecule and odour detection threshold (ODT) [1].
These physiochemical properties affect the pyramidal structure of a perfume, which consists of top (head), middle (heart) and base notes. The top note of perfume is the main
body of the perfume and it is most volatile compared to middle and base notes. Top note
is smelled as fast as it is applied and slowly changes to base notes with time depending
on the volatility. In this way, top notes will be strongly perceived in the beginning, after
that is middle notes and finally is base notes after hours and become more intense in
pleasant. Base notes are playing important role as fixatives, which affects the molecular
interaction and evaporation rate of both top and middle notes. Based on the volatility of
fragrance molecule and molecular interaction, all the fragrance particles start to evapo-
rate and diffuse from high concentration area to low concentration with different rate.
When evaporation and diffusion processes take place, fragrance particles are spreading

M. Awang and S. S. Emamian (Eds.): *Advances in Material Science and Engineering*, LNME, pp. 402–409, 2021.
https://doi.org/10.1007/978-981-16-3641-7_50

in the air above the liquid surface of perfume to human nose with different intensity of odour.

Evaporation and diffusion are the key processes in perfume engineering as it determines the fragrance performance and the satisfaction of customers. Teixeira et al. (2013) [2] developed a concentration profile over time to illustrate the diffusivity of mixtures α-pinene, linalool, tonalide and ethanol mixtures. Result showed that only the concentration of α-pinene and ethanol was captured by GC as well as the human nose. The work also concluded that performance of perfume depended on the intrinsic properties of the compounds as well as the molecular interactions in the liquid phase. Pereira et al. (2019) [3] proposed 1D radial diffusion model for pure odorants and the respective mixture. The gas concentration at different time and distance were recorded and the experimental data were fitted closely to the model. The cited works above are mainly focus on the performance of self-fabricated perfume, where the exact composition of mixture is known. Besides, the diffusivity performance was presented by gas concentration profiles of volatile compounds, where the diffusion coefficients of the mixture were not determined. Additionally, Costa et al. (2015) [4] found that research work on the evaporation and diffusivity behaviours of fragrances in mixtures was very limited. Hence, the guideline to link the self-formulated perfume to marketable perfume remains as a grey area. The aim of this study is to determine the diffusion coefficients of commercial grade perfumes. This data served as the guide to researchers to formulate the perfume that meet the desired diffusivity for commercialization purpose.

2 Methods

Diffusivity of Perfume

Appropriate amount of five commercial available perfume, labelled as G1, G2, G3, G4 and G5 (Master Men, Etude House, Eau de Parfum, Avon Black Suede and Giorgio Armani) was injected into the capillary tube in the gaseous diffusion coefficient apparatus (Armfield, CERa) as shown in Fig. 1 below. Capillary tube was connected to silicon tube to supply constant air flow to the system. Microscope with Vernier scale was used to observe the meniscus inside the capillary tube. The change of the meniscus level was recorded every 15 min in the first hour, followed by hourly reading for another 6 h.

The equations to calculate the diffusion coefficient are shown as below [5]:

The rate of mass transfer is given by Eq. (1):

$$N_A = \left(\frac{C_A}{L}\right)\left(\frac{C_T}{C_{BM}}\right)D \tag{1}$$

Where diffusivity (D, m^2/s), saturation concentration at interface (C$_A$, kmol/m^3), effective distance of mass transfer (L, mm), Logarithmic mean molecular concentration of vapour (C$_{BM}$, kmol/m^3) and total molecular concentration (C$_T$, kmol/m^3) are considered. In this case,

$$C_A = \frac{P_{vap}C_T}{P_{atm}} \tag{2}$$

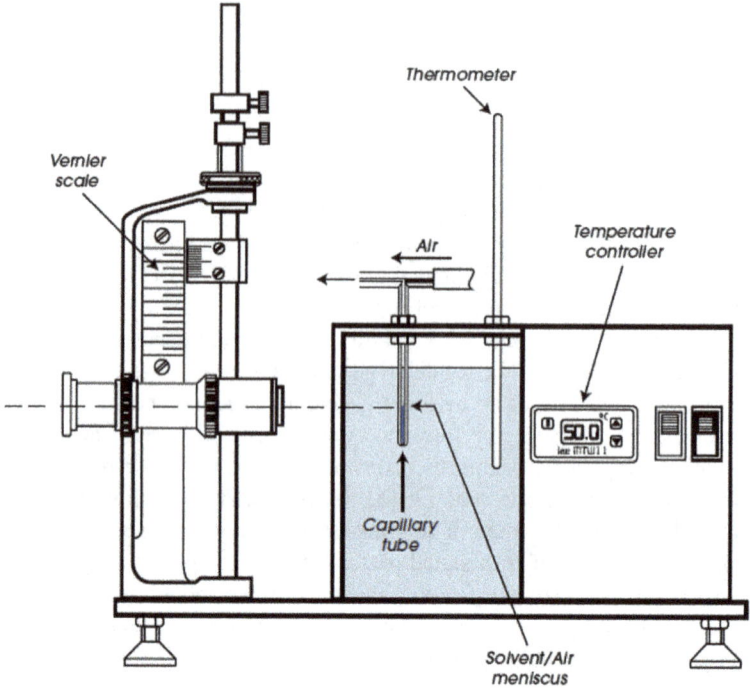

Fig. 1. Gaseous diffusion coefficient apparatus (Armfield, CERa)

$$C_T = \left(\frac{1}{22.414}\right)\left(\frac{T_{ref}}{T_{exp}}\right) \tag{3}$$

$$C_{BM} = \frac{(C_{B1} - C_{B2})}{\ln(C_{B1}/C_{B2})} \tag{4}$$

$$C_{B1} = C_T \tag{5}$$

$$C_{B2} = \frac{C_T(P_{atm} - P_{vap})}{P_{atm}} \tag{6}$$

Where P_{vap} is the vapour pressure, $T_{ref} = 0\,°C$, $T_{exp} = 27\,°C$ and $P_{atm} = 101.3\,kPa$.
The evaporation of liquid is given as Eq. (7):-

$$N_A = \left(\frac{\rho_L}{M}\right)\frac{dL}{dt} \tag{7}$$

Where density of liquid (ρ_L, kg/m³), molecular weight (M, kg/mol) and time (t, s) are considered.

Assume $L = L_o$ at $t = 0$, Eq. (7) becomes:-

$$\frac{t}{(L - L_o)} = \left(\frac{\rho_L}{2MD}\right)\left(\frac{C_{BM}}{C_A C_T}\right)(L - L_o) + \frac{\rho_L C_{BM}}{MDC_A C_T} \tag{8}$$

Equation (8) is in the linear form, $y = mx + c$.

Thus, the diffusion coefficient can be calculated from the slope.

Compounds Identification

The compounds in the perfume were identified by using gas chromatography (GC) (Agilent Technologies, USA) and Fourier transform infrared spectroscopy (FTIR) (SpectrumOne, Perkin Elmer).

1 µL of liquid sample was injected into the GC capillary column in split mode with ratio 10:1 at 240 °C and 4.7282 psi. Helium and air were used as carrier gases with flow rate 30 mL/min and 400 mL/min respectively. The oven temperature was set from 40 °C (held for 5 min) to 240 °C at a rate of 7 °C/min. The injector temperature was 240 °C and FID temperature was 250 °C.

Appropriate amount of sample was placed on the attenuated total reflection (ATR) crystal of FTIR, and the spectrum were recorded within 4000 to 600 cm^{-1} with minimum 16 scans.

3   Results and Discussion

Chromatograms of methanol, ethanol and 5 commercial grade perfumes are shown in Fig. 2, 3, 4, 5, 6, 7 and 8. The retention time of methanol and ethanol are 2.585 min and 2.763 min respectively, as shown in Fig. 2 and 3. This is because the molecular weight of ethanol is higher compared to methanol. Figure 4, 5, 6 and 7 shows the signature peak of ethanol, where the retention time is approximately 2.76 min for sample G1 to G4. This indicates that ethanol is the main carrier in the top note of the respective perfumes to diffuse fragrances. Slight difference in the retention time of ethanol for G1 to G4 may due to the presence of other compounds, such as essential oils and antioxidants in the perfume. The uncountable small peaks indicate the presence of these compounds in the perfume. Meanwhile, methanol was used in formulating G5, as the chromatogram in Fig. 8 shown the signature peak of methanol.

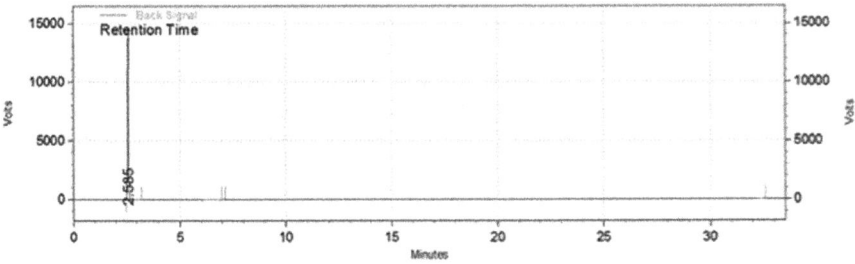

Fig. 2. Gas chromatogram of methanol (retention time: 2.585 min)

FTIR spectra of G1 to G5 are shown in Fig. 9, where the functional groups of the perfumes are identified. Generally, all the samples have the similar functional groups. O-H stretching of methanol and ethanol are represented by the peaks at ~3350 to 3370 cm^{-1}

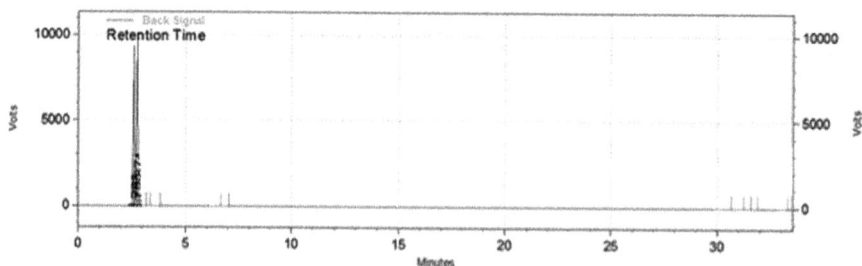

Fig. 3. Gas chromatogram of ethanol (retention time: 2.763 min)

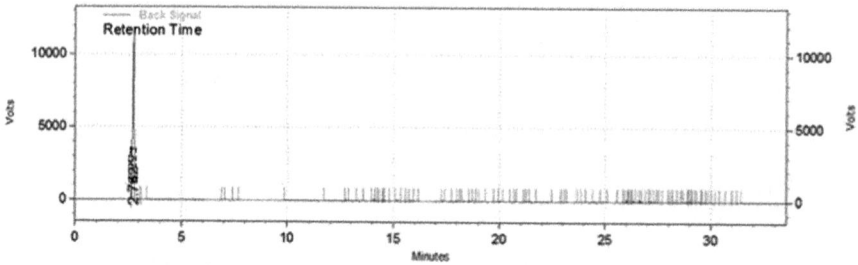

Fig. 4. Gas chromatogram of perfume G1 (retention time: 2.762 min)

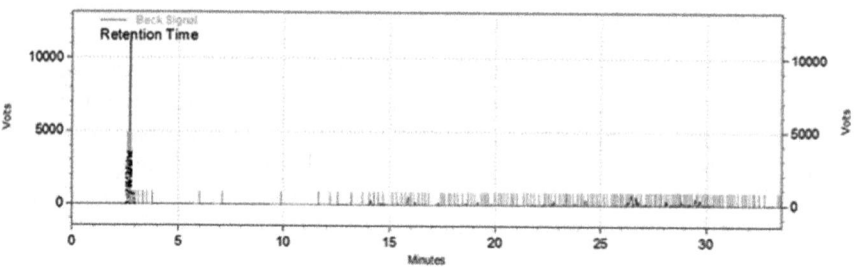

Fig. 5. Gas chromatogram of perfume G2 (retention time: 2.752 min)

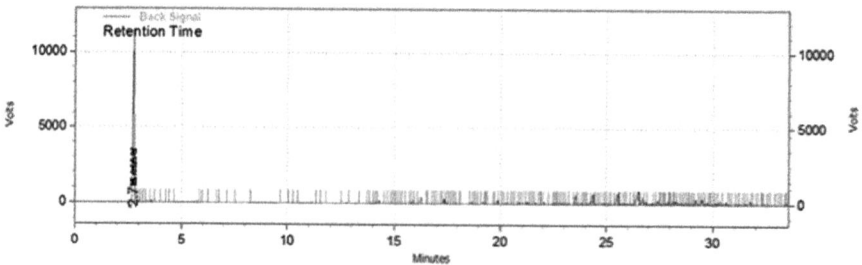

Fig. 6. Gas chromatogram of perfume G3 (retention time: 2.767 min)

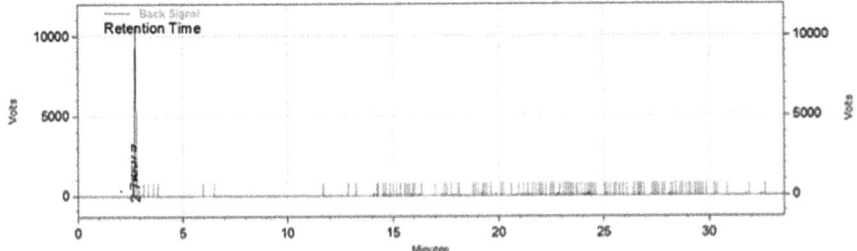

Fig. 7. Gas chromatogram of perfume G4 (retention time: 2.765 min)

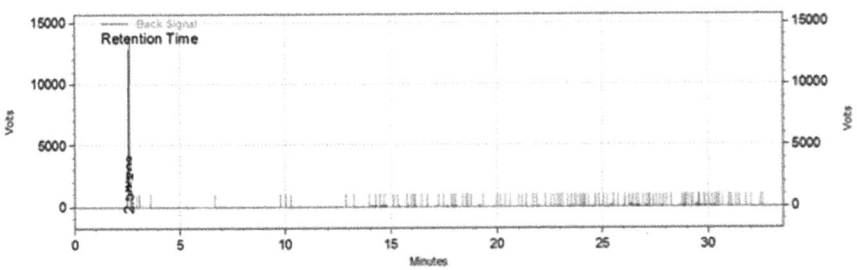

Fig. 8. Gas chromatogram of perfume G5 (retention time: 2.585 min)

for all the perfumes [6]. Functional groups at ~2975 cm^{-1} and 2890 cm^{-1} indicate the presence of C-H stretching of –CH_3 and CH_2 groups. C = O stretching of ester and aldehyde was found in samples G3 and G4 at 1731 and 1728 cm^{-1} [7]. There are a few possible functional groups for the weak peak band at ~1653 cm^{-1} for G5 and G2, include C = O stretching of amine, C = C stretching of alkenes and N-H bending of amide or amine. C-H bending of methyl group was found at ~1378 cm^{-1} in all the samples except G1. C-H bending of CH_2 was found in G1 and G3 at ~1449 cm^{-1}. Lastly, the peak at ~1088 cm^{-1} may due to C-O of alcohol/phenol group. In short, alcohol, aliphatic hydrocarbon, and esters are the main compounds in perfume, and this is in agreement with the finding reported by Aljaff et al. (2013) [6].

Diffusion coefficients of the samples are within 0.6 to 1.3×10^{-6} m^2/s as shown in Table 1. According to Tang et al. (2015) [8], the diffusion coefficient of methanol and ethanol are 1.66×10^{-5} m^2/s and 1.29×10^{-5} m^2/s respectively. Works done by Fuller et al. (1966) [9] showed that the diffusion coefficient of fragrances, such as limomene, α-pinene, linalool and geranyl acetate are within the range of 5.3×10^{-6} to 6.0×10^{-6} m^2/s. Thus, the low diffusion coefficient in both ethanol and methanol based perfume was due to the present of other compounds such as fragrance and essential oil. This is similar to the findings reported by Pereira et al. (2019) [9] where the concentrations (in gas phase) and the respective odour intensities of limonene and α-pinene were lower when these compounds were present in a liquid mixture of limonene, and α-pinene, linalool and geranyl acetate, compared to its pure solution.

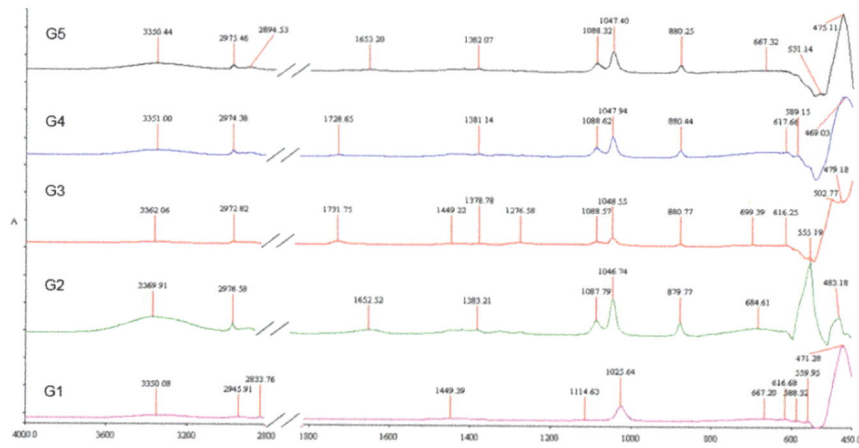

Fig. 9. FTIR spectra of perfume, G1 to G5

Table 1. Diffusion coefficient of commercial available perfume

Samples ID	Diffusion coefficient [m^2/s]	Alcohol (identified from chromatograms)
G1	6.6113×10^{-7}	Ethanol
G2	1.2653×10^{-6}	Ethanol
G3	1.2324×10^{-6}	Ethanol
G4	7.9922×10^{-7}	Ethanol
G5	1.3979×10^{-6}	Methanol

4   Conclusion

Chromatograms showed that ethanol is the main carrier in most of the commercial available perfume. Nevertheless, the diffusivity of ethanol based perfume is very slow compared to pure ethanol. This is due to the present of other compounds such as fragrance and aliphatic compounds, as proved by the FTIR spectrum and chromatograms. It is notable that the commercial available perfume exhibited similar FTIR spectrum. Lastly, the diffusion coefficients of commercial grade perfume are within the range of 0.6 to 1.3×10^{-6} m^2/s. This data served as the reference for formulating commercial grade perfume. Additionally, it also can be used for alcohol recovery for chemical engineering application.

Acknowledgement. Financial support for this research work was provided by SEGi University via research fund number SEGiIRF/2016-03/FOEBE-18/94 was highly acknowledged.

References

1. Vankar, P.S.: Essential oils and fragrances form natural sources. Resonance **9**, 30–41 (2004)

2. Teixeira, M.A., Rodrigues, A.E.: A product engineering approach: diffusion and performance of fragranced products. In: AIChE Meeting (2013)
3. Pereira, J., Costa, P., Loureiro, J.M., Rodregurd, A.E.: Modelling diffusion of fragrances: a radial perspective. Can. J. Chem. Eng. **97**, 351–360 (2019)
4. Costa, P., Teixeira, M.A., Lievre, Y., et al.: Modeling fragrance components release from a simplified matrix used in toiletries and household products. Ind. Eng. Chem. Res **54**, 11720–11731 (2015)
5. Armfield. Gas diffusion coefficient apparatus – Instruction manual CERa (2013)
6. Aliaff, P.M., Manhal, E., Rasheed, B.O.: Identification of synthetic perfume by infrared and optical properties. Pure Appl. Chem. Sci. **1**, 19–30 (2013)
7. Hari, N., Nair, V.P.: FTIR spectroscopic analysis of leaf extract in hexane in Jasminum Azoricum L. Int. J. Sci. Res. Sci. Technol. **4**, 170–172 (2018)
8. Tang, M.J., Shiraiwa, M., Poschl, U., et al.: Compilation and evaluation of gas phase diffusion coefficients of reactive trace gases in the atmosphere: volume 2. Diffusivities of organic compounds, pressure normalised mean fee paths, and average Knudsen numbers for gas uptake calculations. Atmos. Chem. Phys **15**, 5585–5598 (2015)
9. Fuller, E.N., Schettler, P.D., Giddings, J.C.: A new method for prediction of binary gas phase diffusion coefficients. Ind. Eng. Chem. Res **58**, 18–27 (1966)

Author Index

M. Awang and S. S. Emamian (Eds.): *Advances in Material Science and Engineering*, LNME, pp. 411–413, 2021.
https://doi.org/10.1007/978-981-16-3641-7

Lightning Source UK Ltd.
Milton Keynes UK
UKHW020607110722
405674UK00001B/20